STUDENT'S SOLUTIONS MANUAL

CINDY TRIMBLE & ASSOCIATES

PREALGEBRA

SIXTH EDITION

ELAYN MARTIN-GAY

University of New Orleans

Prentice Hall
is an imprint of

The author and publisher of this book have used their best efforts in preparing this book. These efforts include the development, research, and testing of the theories and programs to determine their effectiveness. The author and publisher make no warranty of any kind, expressed or implied, with regard to these programs or the documentation contained in this book. The author and publisher shall not be liable in any event for incidental or consequential damages in connection with, or arising out of, the furnishing, performance, or use of these programs.

Reproduced by Prentice Hall from electronic files supplied by the author.

Copyright © 2011, 2004, 2007 Pearson Education, Inc.
Publishing as Pearson Prentice Hall, 75 Arlington Street, Boston, MA 02116.

All rights reserved. No part of this publication may be reproduced, stored in a retrieval system, or transmitted, in any form or by any means, electronic, mechanical, photocopying, recording, or otherwise, without the prior written permission of the publisher. Printed in the United States of America.

ISBN-13: 978-0-321-63509-9 Standalone
ISBN-10: 0-321-63509-4 Standalone

ISBN-13: 978-0-321-64002-4 Component
ISBN-10: 0-321-64002-0 Component

4 5 6 BRR 14 13 12 11

Prentice Hall
is an imprint of

www.pearsonhighered.com

Contents

Chapter 1

Section 1.2

Practice Problems

1. The place value of the 8 in 38,760,005 is millions.

2. The place value of the 8 in 67,890 is hundreds.

3. The place value of the 8 in 481,922 is ten-thousands.

4. 54 is written as fifty-four.

5. 678 is written as six hundred seventy-eight.

6. 93,205 is written as ninety-three thousand, two hundred five.

7. 679,430,105 is written as six hundred seventy-nine million, four hundred thirty thousand, one hundred five.

8. Thirty-seven in standard form is 37.

9. Two hundred twelve in standard form is 212.

10. Eight thousand, two hundred seventy-four in standard form is 8,274 or 8274.

11. Five million, fifty-seven thousand, twenty-six in standard form is 5,057,026.

12. 4,026,301
$= 4{,}000{,}000 + 20{,}000 + 6000 + 300 + 1$

13. **a.** Find Great Britain in the left-hand column. Read from left to right until the "bronze" column is reached. Great Britain won 15 bronze medals.

b. Find the countries for which the entry in the last column (The United States and China) is greater than 90. The United States and China won more than 90 medals.

Vocabulary and Readiness Check

1. The numbers 0, 1, 2, 3, 4, 5, 6, 7, 8, 9, 10, 11, 12, ... are called whole numbers.

2. The number 1,286 is written in standard form.

3. The number "twenty-one" is written in words.

4. The number 900 + 60 + 5 is written in expanded form.

5. In a whole number, each group of 3 digits is called a period.

6. The place value of the digit 4 in the whole number 264 is ones.

Exercise Set 1.2

1. The place value of the 5 in 657 is tens.

3. The place value of the 5 in 5423 is thousands.

5. The place value of the 5 in 43,526,000 is hundred-thousands.

7. The place value of the 5 in 5,408,092 is millions.

9. 354 is written as three hundred fifty-four.

11. 8279 is written as eight thousand, two hundred seventy-nine.

13. 26,990 is written as twenty-six thousand, nine hundred ninety.

15. 2,388,000 is written as two million, three hundred eighty-eight thousand.

17. 24,350,185 is written as twenty-four million, three hundred fifty thousand, one hundred eighty-five.

19. 304,367 is written as three hundred four thousand, three hundred sixty-seven.

21. 2600 is written as two thousand, six hundred.

23. 15,800,000 is written as fifteen million, eight hundred thousand.

25. 14,433 is written as fourteen thousand, four hundred thirty-three.

27. 13,000,000 is written as thirteen million.

29. Six thousand, five hundred eighty-seven in standard form is 6587.

31. Fifty-nine thousand, eight hundred in standard form is 59,800.

Copyright © 2011 Pearson Education, Inc. Publishing as Prentice Hall.

33. Thirteen million, six hundred one thousand, eleven in standard form is 13,601,011.

35. Seven million, seventeen in standard form is 7,000,017.

37. Two hundred sixty thousand, nine hundred ninety-seven in standard form is 260,997.

39. Three hundred ninety-five in standard form is 395.

41. Thirty thousand, seven hundred fifty in standard form is 30,750.

43. Sixty-six million, four hundred thousand in standard form is 66,400,000.

45. Five hundred sixty-five in standard form is 565.

47. $209 = 200 + 9$

49. $3470 = 3000 + 400 + 70$

51. $80,774 = 80,000 + 700 + 70 + 4$

53. $66,049 = 60,000 + 6000 + 40 + 9$

55. $39,680,000$
$= 30,000,000 + 9,000,000 + 600,000 + 80,000$

57. Mount Shasta erupted in 1786, which is in standard form.

59. Mount Baker has four eruptions listed, which is more eruptions than any other volcano listed in the table.

61. Glacier Peak has an eruption listed in approximately 1750. All other eruptions listed in the table occurred after this one.

63. Boxer has fewer dogs registered than Dachshund.

65. Labrador Retrievers have the most registrations; 123,760 is written as one hundred twenty-three thousand, seven hundred sixty.

67. The maximum weight of an average-size Dachshund is 25 pounds.

69. The largest number is 9861.

71. No; 105.00 should be written as one hundred five.

73. answers may vary

75. 1000 trillion in standard form is 1,000,000,000,000,000.

Section 1.3

Practice Problems

1.
$$\begin{array}{r} 4135 \\ +\ 252 \\ \hline 4387 \end{array}$$

2.
$$\begin{array}{r} {\scriptstyle 1\,1\;11\;\;} \\ 47,364 \\ +\ 135,898 \\ \hline 183,262 \end{array}$$

3. Notice $12 + 8 = 20$ and $4 + 6 = 10$.
$12 + 4 + 8 + 6 + 5 = 20 + 10 + 5 = 35$

4.
$$\begin{array}{r} {\scriptstyle 1\,2\,2\;} \\ 6432 \\ 789 \\ 54 \\ +\quad 28 \\ \hline 7303 \end{array}$$

5. a. $14 - 6 = 8$ because $8 + 6 = 14$.

b. $20 - 8 = 12$ because $12 + 8 = 20$

c. $93 - 93 = 0$ because $0 + 93 = 93$.

d. $42 - 0 = 42$ because $42 + 0 = 42$.

6. a.
$$\begin{array}{r} 9143 \\ -\ 122 \\ \hline 9021 \end{array} \qquad \textit{Check: } \begin{array}{r} 9021 \\ +\ 122 \\ \hline 9143 \end{array}$$

b.
$$\begin{array}{r} 978 \\ -\ 851 \\ \hline 127 \end{array} \qquad \textit{Check: } \begin{array}{r} 127 \\ +\ 851 \\ \hline 978 \end{array}$$

7. a.
$$\begin{array}{r} {\scriptstyle 8\,17} \\ 6\not{9}\not{7} \\ -\ 4\,9 \\ \hline 64\,8 \end{array} \qquad \textit{Check: } \begin{array}{r} 648 \\ +\ 49 \\ \hline 697 \end{array}$$

b.
$$\begin{array}{r} {\scriptstyle 2\,12} \\ \not{3}\not{2}6 \\ -\ 245 \\ \hline 81 \end{array} \qquad \textit{Check: } \begin{array}{r} 81 \\ +\ 245 \\ \hline 326 \end{array}$$

Copyright © 2011 Pearson Education, Inc. Publishing as Prentice Hall.

c. $\begin{array}{r} 1234 \\ -\ 822 \\ \hline 412 \end{array}$ *Check:* $\begin{array}{r} 412 \\ +\ 822 \\ \hline 1234 \end{array}$

8. a. $\begin{array}{r} 9 \\ 3\,\not{1}\!\not{0}\,10 \\ \not{4}\ \not{0}\ \not{0} \\ -\ 1\ 6\ 4 \\ \hline 2\ 3\ 6 \end{array}$ *Check:* $\begin{array}{r} 236 \\ +\ 164 \\ \hline 400 \end{array}$

b. $\begin{array}{r} 9 \\ 9\,\not{1}\!\not{0}\,10 \\ 1\not{0}\ \not{0}\ \not{0} \\ -\ 7\ 6\ 2 \\ \hline 2\ 3\ 8 \end{array}$ *Check:* $\begin{array}{r} 238 \\ +\ 762 \\ \hline 1000 \end{array}$

9. 2 cm + 8 cm + 15 cm + 5 cm = 30 cm
The perimeter is 30 centimeters.

10. 647 + 647 + 647 = 1941
The perimeter is 1941 feet.

11. $\begin{array}{r} 15,759 \\ -\quad 458 \\ \hline 15,301 \end{array}$
The radius of Neptune is 15,301 miles.

12. a. The country with the fewest endangered species corresponds to the shortest bar, which is Australia.

b. To find the total number of endangered species for Brazil, India, and Mexico, we add.
$\begin{array}{r} 73 \\ 89 \\ +\ 72 \\ \hline 234 \end{array}$
The total number of endangered species for Brazil, India, and Mexico is 234.

Calculator Explorations

1. 89 + 45 = 134

2. 76 + 97 = 173

3. 285 + 55 = 340

4. 8773 + 652 = 9425

5. 985 + 1210 + 562 + 77 = 2834

6. 465 + 9888 + 620 + 1550 = 12,523

7. 865 − 95 = 770

8. 76 − 27 = 49

9. 147 − 38 = 109

10. 366 − 87 = 279

11. 9625 − 647 = 8978

12. 10,711 − 8925 = 1786

Vocabulary and Readiness Check

1. The sum of 0 and any number is the same number.

2. In 35 + 20 = 55, the number 55 is called the sum and 35 and 20 are each called an addend.

3. The difference of any number and that same number is 0.

4. The difference of any number and 0 is the same number.

5. In 37 − 19 = 18, the number 37 is the minuend, the 19 is the subtrahend, and the 18 is the difference.

6. The distance around a polygon is called its perimeter.

7. Since 7 + 10 = 10 + 7, we say that changing the order in addition does not change the sum. This property is called the commutative property of addition.

8. Since (3 + 1) + 20 = 3 + (1 + 20), we say that changing the grouping in addition does not change the sum. This property is called the associative property of addition.

Exercise Set 1.3

1. $\begin{array}{r} 14 \\ +\ 22 \\ \hline 36 \end{array}$

3. $\begin{array}{r} 62 \\ +\ 230 \\ \hline 292 \end{array}$

Copyright © 2011 Pearson Education, Inc. Publishing as Prentice Hall.

5. $$\begin{array}{r} 12 \\ 13 \\ +\ 24 \\ \hline 49 \end{array}$$

7. $$\begin{array}{r} 5267 \\ +\ 132 \\ \hline 5399 \end{array}$$

9. $$\begin{array}{r} {\scriptstyle 1\ \ 1\ 1} \\ 22{,}781 \\ +\ 186{,}297 \\ \hline 209{,}078 \end{array}$$

11. $$\begin{array}{r} 8 \\ 9 \\ 2 \\ 5 \\ +\ 1 \\ \hline 25 \end{array}$$

13. $$\begin{array}{r} {\scriptstyle 2\,2} \\ 81 \\ 17 \\ 23 \\ 79 \\ +\ 12 \\ \hline 212 \end{array}$$

15. $$\begin{array}{r} {\scriptstyle 1\ 1\,2} \\ 24 \\ 9006 \\ 489 \\ +\ 2407 \\ \hline 11{,}926 \end{array}$$

17. $$\begin{array}{r} {\scriptstyle 1\,1\ 1} \\ 6\ 820 \\ 4\ 271 \\ +\ 5\ 626 \\ \hline 16{,}717 \end{array}$$

19. $$\begin{array}{r} {\scriptstyle 1\ 1\ 2\ 2} \\ 49 \\ 628 \\ 5\ 762 \\ +\ 29{,}462 \\ \hline 35{,}901 \end{array}$$

21. $$\begin{array}{r} {\scriptstyle 1\ 22\ 2\,1} \\ 121{,}742 \\ 57{,}279 \\ 26{,}586 \\ +\ 426{,}782 \\ \hline 632{,}389 \end{array}$$

23. $$\begin{array}{r} 749 \\ -\ 149 \\ \hline 600 \end{array}$$ *Check:* $$\begin{array}{r} 600 \\ +\ 149 \\ \hline 749 \end{array}$$

25. $$\begin{array}{r} 62 \\ -37 \\ \hline 25 \end{array}$$ *Check:* $$\begin{array}{r} {\scriptstyle 1} \\ 25 \\ +\ 37 \\ \hline 62 \end{array}$$

27. $$\begin{array}{r} 922 \\ -634 \\ \hline 288 \end{array}$$ *Check:* $$\begin{array}{r} {\scriptstyle 1\,1} \\ 288 \\ +\ 634 \\ \hline 922 \end{array}$$

29. $$\begin{array}{r} 600 \\ -\ 432 \\ \hline 168 \end{array}$$ *Check:* $$\begin{array}{r} {\scriptstyle 1\,1} \\ 168 \\ +\ 432 \\ \hline 600 \end{array}$$

31. $$\begin{array}{r} 6283 \\ -\ 560 \\ \hline 5723 \end{array}$$ *Check:* $$\begin{array}{r} {\scriptstyle 1} \\ 5723 \\ +\ 560 \\ \hline 6283 \end{array}$$

33. $$\begin{array}{r} 533 \\ -\ 29 \\ \hline 504 \end{array}$$ *Check:* $$\begin{array}{r} {\scriptstyle 1} \\ 504 \\ +\ 29 \\ \hline 533 \end{array}$$

35. $$\begin{array}{r} 1983 \\ -\ 1904 \\ \hline 79 \end{array}$$ *Check:* $$\begin{array}{r} {\scriptstyle 1} \\ 79 \\ +\ 1904 \\ \hline 1983 \end{array}$$

37. $$\begin{array}{r} 50{,}000 \\ -\ 17{,}289 \\ \hline 32{,}711 \end{array}$$ *Check:* $$\begin{array}{r} {\scriptstyle 1\,1\,1\,1} \\ 32{,}711 \\ +\ 17{,}289 \\ \hline 50{,}000 \end{array}$$

39. $$\begin{array}{r} 7020 \\ -\ 1979 \\ \hline 5041 \end{array}$$ *Check:* $$\begin{array}{r} {\scriptstyle 1\,1\,1} \\ 5041 \\ +\ 1979 \\ \hline 7020 \end{array}$$

Copyright © 2011 Pearson Education, Inc. Publishing as Prentice Hall.

41. $$\begin{array}{r} 51{,}111 \\ -\,19{,}898 \\ \hline 31{,}213 \end{array}$$ *Check*: $$\begin{array}{r} \scriptstyle 1\,1\;1\,1 \\ 31{,}213 \\ +\,19{,}898 \\ \hline 51{,}111 \end{array}$$

43. $$\begin{array}{r} 986 \\ +\,48 \\ \hline 1034 \end{array}$$

45. $$\begin{array}{r} 76 \\ -\,67 \\ \hline 9 \end{array}$$

47. $$\begin{array}{r} 9000 \\ -\,482 \\ \hline 8518 \end{array}$$

49. $$\begin{array}{r} \scriptstyle 1\,1\;1 \\ 10{,}962 \\ 4\,851 \\ +\,7\,063 \\ \hline 22{,}876 \end{array}$$

51. $7 + 8 + 10 = 25$
The perimeter is 25 feet.

53. Opposite sides of a rectangle have the same length.
$4 + 8 + 4 + 8 = 12 + 12 = 24$
The perimeter is 24 inches.

55. $8 + 3 + 5 + 7 + 5 + 1 = 29$
The perimeter is 29 inches.

57. The unknown vertical side has length $12 - 5 = 7$ meters. The unknown horizontal side has length $10 - 5 = 5$ meters.
$10 + 12 + 5 + 7 + 5 + 5 = 44$
The perimeter is 44 meters.

59. "Find the sum" indicates addition.
$$\begin{array}{r} \scriptstyle 1\,1\,1 \\ 297 \\ +\,1796 \\ \hline 2093 \end{array}$$
The sum of 297 and 1796 is 2093.

61. "Find the total" indicates addition.
$$\begin{array}{r} \scriptstyle 1\,3\; \\ 76 \\ 39 \\ 8 \\ 17 \\ +\,126 \\ \hline 266 \end{array}$$
The total of 76, 39, 8, 17, and 126 is 266.

63. "Find the difference" indicates subtraction.
$$\begin{array}{r} 41 \\ -\,21 \\ \hline 20 \end{array}$$
The difference of 41 and 21 is 20.

65. "Increased by" indicates addition.
$$\begin{array}{r} \scriptstyle 1\;\;\; \\ 452 \\ +\,92 \\ \hline 544 \end{array}$$
452 increased by 92 is 544.

67. "Less" indicates subtraction.
$$\begin{array}{r} 108 \\ -\;\;36 \\ \hline 72 \end{array}$$
108 less 36 is 72.

69. "Subtracted from" indicates subtraction.
$$\begin{array}{r} 100 \\ -\,12 \\ \hline 88 \end{array}$$
12 subtracted from 100 is 88.

71. Subtract 19,308 thousand from 22,478 thousand.
$$\begin{array}{r} 22{,}478 \\ -\,19{,}308 \\ \hline 3\,170 \end{array}$$
Florida's projected population increase is 3170 thousand.

73. Subtract the cost of the DVD player from the amount in her savings account.
$$\begin{array}{r} 914 \\ -\,295 \\ \hline 619 \end{array}$$
She will have $619 left.

Copyright © 2011 Pearson Education, Inc. Publishing as Prentice Hall.

75. $$\begin{array}{r} 189{,}000 \\ +\ 75{,}000 \\ \hline 264{,}000 \end{array}$$
The total U.S. land area drained by the Upper Mississippi and Lower Mississippi sub-basins is 264,000 square miles.

77. $$\begin{array}{r} 530{,}000 \\ -\ 247{,}000 \\ \hline 283{,}000 \end{array}$$
The Missouri sub-basin drains 283,000 square miles more than the Arkansas Red-White sub-basin.

79. $70 + 78 + 90 + 102 = 340$
The homeowner needs 340 feet of fencing.

81. $$\begin{array}{r} 503 \\ -\ 239 \\ \hline 264 \end{array}$$
She must read 264 more pages.

83. $$\begin{array}{r} 4{,}280{,}031 \\ +\ 29{,}719{,}969 \\ \hline 34{,}000{,}000 \end{array}$$
The sheep population was 34,000,000.

85. Live rock music has a decibel level of 100 dB.

87. $$\begin{array}{r} 88 \\ -\ 30 \\ \hline 58 \end{array}$$
The sound of snoring is 58 dB louder than normal conversation.

89. $$\begin{array}{r} 2677 \\ +\ 493 \\ \hline 3170 \end{array}$$
There were 3170 stores worldwide.

91. Each side of a square has the same length.
$31 + 31 + 31 + 31 = 124$
The perimeter of the playing board is 124 feet.

93. California has the most Target stores.

95. $$\begin{array}{r} 225 \\ 136 \\ +\ 115 \\ \hline 476 \end{array}$$
The total number of Target stores in California, Texas, and Florida is 476 stores.

97. Florida and Georgia:
$$\begin{array}{r} 115 \\ +\ 51 \\ \hline 166 \end{array}$$
Michigan and Ohio:
$$\begin{array}{r} 57 \\ +\ 63 \\ \hline 120 \end{array}$$
Florida and Georgia have more Target stores.

99. $$\begin{array}{r} {\scriptstyle 1} \\ 2029 \\ +\ 3865 \\ \hline 5894 \end{array}$$
The total highway mileage in Delaware is 5894 miles.

101. The minuend is 48 and the subtrahend is 1.

103. The minuend is 70 and the subtrahend is 7.

105. answers may vary

107. $$\begin{array}{r} {\scriptstyle 1} \\ 566 \\ 932 \\ +\ 871 \\ \hline 2369 \end{array}$$
The given sum is correct.

109. $$\begin{array}{r} {\scriptstyle 2\,2} \\ 14 \\ 173 \\ 86 \\ +\ 257 \\ \hline 530 \end{array}$$
The given sum is incorrect, the correct sum is 530.

111. $$\begin{array}{r} {\scriptstyle 1\,1} \\ 675 \\ +\ 56 \\ \hline 731 \end{array}$$
The given difference is incorrect.
$$\begin{array}{r} 741 \\ -\ 56 \\ \hline 685 \end{array}$$

113. $$\begin{array}{r} 141 \\ +\ 888 \\ \hline 1029 \end{array}$$
The given difference is correct.

Copyright © 2011 Pearson Education, Inc. Publishing as Prentice Hall.

115.
$$\begin{array}{r} 5269 \\ -\ 2385 \\ \hline 2884 \end{array}$$

117. answers may vary

119.
$$\begin{array}{r} \scriptstyle 1\,2\,1\ \,3\,2 \\ 289,462 \\ 369,477 \\ 218,287 \\ +\ 121,685 \\ \hline 998,911 \end{array}$$

Since 998,911 is less than one million, they did not reach their goal.

$$\begin{array}{r} 1,000,000 \\ -\ \ 998,911 \\ \hline 1\,089 \end{array}$$

They need to read 1089 more pages.

Section 1.4

Practice Problems

1. a. To round 57 to the nearest ten, observe that the digit in the ones place is 7. Since the digit is at least 5, we add 1 to the digit in the tens place. The number 57 rounded to the nearest ten is 60.

b. To round 641 to the nearest ten, observe that the digit in the ones place is 1. Since the digit is less than 5, we do not add 1 to the digit in the tens place. The number 641 rounded to the nearest ten is 640.

c. To round 325 to the nearest ten observe that the digit in the ones place is 5. Since the digit is at least 5, we add 1 to the digit in the tens place. The number 325 rounded to the nearest ten is 330.

2. a. To round 72,304 to the nearest thousand, observe that the digit in the hundreds place is 3. Since the digit is less than 5, we do not add 1 to the digit in the thousands place. The number 72,304 rounded to the nearest thousand is 72,000.

b. To round 9222 to the nearest thousand, observe that the digit in the hundreds place is 2. Since the digit is less than 5, we do not add 1 to the digit in the thousands place. The number 9222 rounded to the nearest thousand is 9000.

c. To round 671,800 to the nearest thousand, observe that the digit in the hundreds place is 8. Since this digit is at least 5, we add 1 to the digit in the thousands place. The number 671,800 rounded to the nearest thousand is 672,000.

3. a. To round 3474 to the nearest hundred, observe that the digit in the tens place is 7. Since this digit is at least 5, we add 1 to the digit in the hundreds place. The number 3474 rounded to the nearest hundred is 3500.

b. To round 76,243 to the nearest hundred, observe that the digit in the tens place is 4. Since this digit is less than 5, we do not add 1 to the digit in the hundreds place. The number 76,243 rounded to the nearest hundred is 76,200.

c. To round 978,965 to the nearest hundred, observe that the digit in the tens place is 6. Since this digit is at least 5, we add 1 to the digit in the hundreds place. The number 978,865 rounded to the nearest hundred is 979,000.

4.
$$\begin{array}{rlr} 49 & \text{rounds to} & 50 \\ 25 & \text{rounds to} & 30 \\ 32 & \text{rounds to} & 30 \\ 51 & \text{rounds to} & 50 \\ \underline{98} & \text{rounds to} & +\ 100 \\ \cline{3-3} & & 260 \end{array}$$

5.
$$\begin{array}{rlr} 3785 & \text{rounds to} & 4000 \\ -\ 2479 & \text{rounds to} & -\ 2000 \\ \cline{3-3} & & 2000 \end{array}$$

6.
$$\begin{array}{rlr} 11 & \text{rounds to} & 10 \\ 16 & \text{rounds to} & 20 \\ 19 & \text{rounds to} & 20 \\ +\ 31 & \text{rounds to} & +\ 30 \\ \cline{3-3} & & 80 \end{array}$$

The total distance is approximately 80 miles.

7.
$$\begin{array}{rlr} 48,445 & \text{rounds to} & 48,000 \\ 6,584 & \text{rounds to} & 7,000 \\ +\ 15,632 & \text{rounds to} & +\ 16,000 \\ \cline{3-3} & & 71,000 \end{array}$$

The total number of cases is approximately 71,000.

Copyright © 2011 Pearson Education, Inc. Publishing as Prentice Hall.

Vocabulary and Readiness Check

1. To graph a number on a number line, darken the point representing the location of the number.

2. Another word for approximating a whole number is rounding.

3. The number 65 rounded to the nearest ten is 70 but the number 61 rounded to the nearest ten is 60.

4. An exact number of products is 1265, but an estimate is 1000.

Exercise Set 1.4

1. To round 423 to the nearest ten, observe that the digit in the ones place is 3. Since this digit is less than 5, we do not add 1 to the digit in the tens place. The number 423 rounded to the nearest ten is 420.

3. To round 635 to the nearest ten, observe that the digit in the ones place is 5. Since this digit is at least 5, we add 1 to the digit in the tens place. The number 635 rounded to the nearest ten is 640.

5. To round 2791 to the nearest hundred, observe that the digit in the tens place is 9. Since this digit is at least 5, we add 1 to the digit in the hundreds place. The number 2791 rounded to the nearest hundred is 2800.

7. To round 495 to the nearest ten, observe that the digit in the ones place is 5. Since this digit is at least 5, we add 1 to the digit in the tens place. The number 495 rounded to the nearest ten is 500.

9. To round 21,094 to the nearest thousand, observe that the digit in the hundreds place is 0. Since this digit is less than 5, we do not add 1 to the digit in the thousands place. The number 21,094 rounded to the nearest thousand is 21,000.

11. To round 33,762 to the nearest thousand, observe that the digit in the hundreds place is 7. Since this digit is at least 5, we add 1 to the digit in the thousands place. The number 33,762 rounded to the nearest thousand is 34,000.

13. To round 328,495 to the nearest hundred, observe that the digit in the tens place is 9. Since this digit is at least 5, we add 1 to the digit in the hundreds place. The number 328,495 rounded to the nearest hundred is 328,500.

15. To round 36,499 to the nearest thousand, observe that the digit in the hundreds place is 4. Since this digit is less than 5, we do not add 1 to the digit in the thousands place. The number 36,499 rounded to the nearest thousand is 36,000.

17. To round 39,994 to the nearest ten, observe that the digit in the ones place is 4. Since this digit is less than 5, we do not add 1 to the digit in the tens place. The number 39,994 rounded to the nearest ten is 39,990.

19. To round 29,834,235 to the nearest ten-million, observe that the digit in the millions place is 9. Since this digit is at least 5, we add 1 to the digit in the ten-millions place. The number 29,834,235 rounded to the nearest ten-million is 30,000,000.

21. Estimate 5281 to a given place value by rounding it to that place value. 5281 rounded to the tens place is 5280, to the hundreds place is 5300, and to the thousands place is 5000.

23. Estimate 9444 to a given place value by rounding it to that place value. 9444 rounded to the tens place is 9440, to the hundreds place is 9400, and to the thousands place is 9000.

25. Estimate 14,876 to a given place value by rounding it to that place value. 14,876 rounded to the tens place is 14,880, to the hundreds place is 14,900, and to the thousands place is 15,000.

27. To round 39,786 to the nearest thousand, observe that the digit in the hundreds place is 7. Since this digit is greater than 5, we add 1 to the digit in the thousands place. Therefore, 39,786 rounded to the nearest thousand is 40,000.

29. To round 38,387 to the nearest thousand, observe that the digit in the hundreds place is 3. Since this digit is less than 5, we do not add 1 to the digit in the thousands place. Therefore, 38,387 points rounded to the nearest thousand is 38,000 points.

31. To round 42,570,000,000 to the nearest billion, observe that the digit in the hundred-millions place is 5. Since this digit is at least 5, we add 1 to the digit in the billions place. Therefore, $42,570,000,000 rounded to the nearest billion is $43,000,000,000.

Copyright © 2011 Pearson Education, Inc. Publishing as Prentice Hall.

33. To round 4,934,078 to the nearest hundred-thousand, observe that the digit in the ten-thousands place is 3. Since this digit is less than 5, we do not add 1 to the digit in the hundred-thousands place. Therefore, $4,934,078 rounded to the nearest hundred-thousand is $4,900,000.

35. U.S.: To round 262,700,000 to the nearest million, observe that the digit in the hundred-thousands place is 7. Since this digit is at least 5, we add 1 to the digit in the millions place. The number 262,700,000 rounded to the nearest million is 263,000,000.
India: To round 296,886,000 to the nearest million, observe that the digit in the hundred-thousands place is 8. Since this digit is at least 5, we add 1 to the digit in the millions place. The number 296,886,000 rounded to the nearest million is 297,000,000.

37.

39	rounds to	40
45	rounds to	50
22	rounds to	20
+ 17	rounds to	+ 20
		130

39.

449	rounds to	450
− 373	rounds to	− 370
		80

41.

1913	rounds to	1900
1886	rounds to	1900
+ 1925	rounds to	+ 1900
		5700

43.

1774	rounds to	1800
− 1492	rounds to	− 1500
		300

45.

3995	rounds to	4000
2549	rounds to	2500
+ 4944	rounds to	+ 4900
		11,400

47. 463 + 219 is approximately 460 + 220 = 680.
The answer of 600 is incorrect.

49. 229 + 443 + 606 is approximately
230 + 440 + 610 = 1280.
The answer of 1278 is correct.

51. 7806 + 5150 is approximately
7800 + 5200 = 13,000.
The answer of 12,956 is correct.

53.

899	rounds to	900
1499	rounds to	1500
+ 999	rounds to	+ 1000
		3400

The total cost is approximately $3400.

55.

1429	rounds to	1400
− 530	rounds to	− 500
		900

Boston in approximately 900 miles farther from Kansas City than Chicago is.

57.

20,320	rounds to	20,000
− 14,410	rounds to	− 14,000
		6 000

The difference in elevation is approximately 6000 feet.

59.

142,702	rounds to	140,000
− 75,543	rounds to	− 80,000
		60,000

Joliet was approximately 60,000 larger than Evanston.

61.

908,412	rounds to	908,000
− 905,851	rounds to	− 906,000
		2 000

The increase in enrollment is approximately 2000 children.

63. 761 hundred-thousands is 76,100,000 in standard form. 76,100,000 rounded to the nearest million is 76,000,000. 76,100,000 rounded to the nearest ten-million is 80,000,000.

65. 598 hundred-thousands is 59,800,000 in standard form. 59,800,000 rounded to the nearest million is 60,000,000. 59,800,000 rounded to the nearest ten-million is 60,000,000.

67. 5723, for example, rounded to the nearest hundred is 5700.

69. **a.** The smallest possible number that rounds to 8600 is 8550.

b. The largest possible number that rounds to 8600 is 8649.

Copyright © 2011 Pearson Education, Inc. Publishing as Prentice Hall.

71. answers may vary

73. 54 rounds to 50
17 rounds to 20
50 + 20 + 50 + 20 = 140
The perimeter is approximately 140 meters.

Section 1.5

Practice Problems

1. a. $6 \times 0 = 0$

b. $(1)8 = 8$

c. $(50)(0) = 0$

d. $75 \cdot 1 = 75$

2. a. $6(4 + 5) = 6 \cdot 4 + 6 \cdot 5$

b. $30(2 + 3) = 30 \cdot 2 + 30 \cdot 3$

c. $7(2 + 8) = 7 \cdot 2 + 7 \cdot 8$

3. a.
$$\begin{array}{r} {\scriptstyle 5}\\ 29 \\ \times\ 6 \\ \hline 174 \end{array}$$

b.
$$\begin{array}{r} {\scriptstyle 44}\\ 648 \\ \times\ 5 \\ \hline 3240 \end{array}$$

4.
$$\begin{array}{r} 306 \\ \times\ \ 81 \\ \hline 306 \\ 24\,480 \\ \hline 24,786 \end{array}$$

5.
$$\begin{array}{r} 726 \\ \times\ \ 142 \\ \hline 1\,452 \\ 29\,040 \\ 72\,600 \\ \hline 103,092 \end{array}$$

6. Area = length · width
= (360 miles)(280 miles)
= 100,800 square miles
The area of Wyoming is 100,800 square miles.

7.
$$\begin{array}{r} 16 \\ \times 45 \\ \hline 80 \\ 640 \\ \hline 720 \end{array}$$
The printer can print 720 pages in 45 minutes.

8. $8 \times 11 = 88$
$5 \times 9 = 45$
$$\begin{array}{r} {\scriptstyle 1}\\ 88 \\ +\ 45 \\ \hline 133 \end{array}$$
The total cost is $133.

9. 163 rounds to 200
× 391 rounds to × 400
80,000
There are approximately 80,000 words on 391 pages.

Calculator Explorations

1. $72 \times 48 = 3456$

2. $81 \times 92 = 7452$

3. $163 \cdot 94 = 15,322$

4. $285 \cdot 144 = 41,040$

5. $983(277) = 272,291$

6. $1562(843) = 1,316,766$

Vocabulary and Readiness Check

1. The product of 0 and any number is 0.

2. The product of 1 and any number is the number.

3. In 8 · 12 = 96, the 96 is called the product and 8 and 12 are each called a factor.

4. Since 9 · 10 = 10 · 9, we say that changing the order in multiplication does not change the product. This property is called the commutative property of multiplication.

5. Since (3 · 4) · 6 = 3 · (4 · 6), we say that changing the grouping in multiplication does not change the product. This property is called the associative property of multiplication.

6. Area measures the amount of surface of a region.

7. Area of a rectangle = length · width.

Copyright © 2011 Pearson Education, Inc. Publishing as Prentice Hall.

8. We know $9(10 + 8) = 9 \cdot 10 + 9 \cdot 8$ by the <u>distributive</u> property.

Exercise Set 1.5

1. $1 \cdot 24 = 24$

3. $0 \cdot 19 = 0$

5. $8 \cdot 0 \cdot 9 = 0$

7. $87 \cdot 1 = 87$

9. $6(3 + 8) = 6 \cdot 3 + 6 \cdot 8$

11. $4(3 + 9) = 4 \cdot 3 + 4 \cdot 9$

13. $20(14 + 6) = 20 \cdot 14 + 20 \cdot 6$

15.
$$\begin{array}{r} 64 \\ \times\ 8 \\ \hline 512 \end{array}$$

17.
$$\begin{array}{r} 613 \\ \times\ \ 6 \\ \hline 3678 \end{array}$$

19.
$$\begin{array}{r} 277 \\ \times\ \ 6 \\ \hline 1662 \end{array}$$

21.
$$\begin{array}{r} 1074 \\ \times\ \ 6 \\ \hline 6444 \end{array}$$

23.
$$\begin{array}{r} 89 \\ \times\ 13 \\ \hline 267 \\ 890 \\ \hline 1157 \end{array}$$

25.
$$\begin{array}{r} 421 \\ \times\ \ 58 \\ \hline 3\ 368 \\ 21\ 050 \\ \hline 24{,}418 \end{array}$$

27.
$$\begin{array}{r} 306 \\ \times\ \ 81 \\ \hline 306 \\ 24\ 480 \\ \hline 24{,}786 \end{array}$$

29.
$$\begin{array}{r} 780 \\ \times\ \ 20 \\ \hline 15{,}600 \end{array}$$

31. $(495)(13)(0) = 0$

33. $(640)(1)(10) = (640)(10) = 6400$

35.
$$\begin{array}{r} 1234 \\ \times\ \ 39 \\ \hline 11\ 106 \\ 37\ 020 \\ \hline 48{,}126 \end{array}$$

37.
$$\begin{array}{r} 609 \\ \times\ 234 \\ \hline 2\ 436 \\ 18\ 270 \\ 121\ 800 \\ \hline 142{,}506 \end{array}$$

39.
$$\begin{array}{r} 8649 \\ \times\ \ \ 274 \\ \hline 34\ 596 \\ 605\ 430 \\ 1\ 729\ 800 \\ \hline 2{,}369{,}826 \end{array}$$

41.
$$\begin{array}{r} 589 \\ \times\ 110 \\ \hline 5\ 890 \\ 58\ 900 \\ \hline 64{,}790 \end{array}$$

43.
$$\begin{array}{r} 1941 \\ \times\ \ \ 2035 \\ \hline 9\ 705 \\ 58\ 230 \\ 3\ 882\ 000 \\ \hline 3{,}949{,}935 \end{array}$$

45. Area = (length)(width)
= (9 meters)(7 meters)
= 63 square meters

Perimeter = length + width + length + width
$= 9 + 7 + 9 + 7$
= 32 meters

Copyright © 2011 Pearson Education, Inc. Publishing as Prentice Hall.

47. Area = (length)(width)
= (40 feet)(17 feet)
= 680 square feet

Perimeter = length + width + length + width
= 40 + 17 + 40 + 17
= 114 feet

49. 576 rounds to 600
$\times 354$ rounds to $\times\ 400$
$240{,}000$

51. 604 rounds to 600
$\times 451$ rounds to $\times\ 500$
$300{,}000$

53. 38×42 is approximately 40×40, which is 1600.
The best estimate is c.

55. 612×29 is approximately 600×30, which is 18,000.
The best estimate is c.

57. $80\times 11 = (8\times 10)\times 11$
$= 8\times (10\times 11)$
$= 8\times 110$
$= 880$

59. $6 \times 700 = 4200$

61. $2240 \times 2 = 4480$

63. $125 \times 3 = 375$
There are 375 calories in 3 tablespoons of olive oil.

65. 94×35
470
2820
3290
The total cost is $3290.

67. a. $4 \times 5 = 20$
There are 20 boxes in one layer.

b. $20 \times 5 = 100$
There are 100 boxes on the pallet.

c. $100 \times 20 = 2000$
The weight of the cheese on the pallet is 2000 pounds.

69. Area = (length)(width)
= (110 feet)(80 feet)
= 8800 square feet
The area is 8800 square feet.

71. Area = (length)(width)
= (350 feet)(160 feet)
= 56,000 square feet
The area is 56,000 square feet.

73. 94×62
188
5640
5828
There are 5828 pixels on the screen.

75. 60×35
300
1 800
2 100
There are 2100 characters in 35 lines.

77. $160 \times 8 = 1280$
There are 1280 calories in 8 ounces.

Copyright © 2011 Pearson Education, Inc. Publishing as Prentice Hall.

79.

T-Shirt Size	Number of Shirts Ordered	Cost per Shirt	Cost per Size Ordered
S	4	\$10	\$40
M	6	\$10	\$60
L	20	\$10	\$200
XL	3	\$12	\$36
XXL	3	\$12	\$36
Total Cost			\$372

81. There are 60 minutes in one hour.
$24 \times 60 \times 1000 = 1440 \times 1000 = 1,440,000$
1,440,000 tea bags are produced in one day.

83. $\begin{array}{r} 128 \\ +\ 7 \\ \hline 135 \end{array}$

85. $\begin{array}{r} 134 \\ \times 16 \\ \hline 804 \\ 1340 \\ \hline 2144 \end{array}$

87. $\begin{array}{r} 19 \\ +\ 4 \\ \hline 23 \end{array}$
The sum of 19 and 4 is 23.

89. $\begin{array}{r} 19 \\ -\ 4 \\ \hline 15 \end{array}$
The difference of 19 and 4 is 15.

91. $6 + 6 + 6 + 6 + 6 = 5 \cdot 6$ or $6 \cdot 5$

93. **a.** $3 \cdot 5 = 5 + 5 + 5$ or $3 + 3 + 3 + 3 + 3$

b. answers may vary

95. $\begin{array}{r} 203 \\ \times\ 14 \\ \hline 812 \\ 2030 \\ \hline 2842 \end{array}$

97. $42 \times 3 = 126$
$42 \times 9 = 378$
The problem is $\begin{array}{r} 42 \\ \times 93 \\ \hline \end{array}$

99. answers may vary

101. On a side with 7 windows per row, there are $7 \times 23 = 161$ windows. On a side with 4 windows per row, there are $4 \times 23 = 92$ windows.
$161 + 161 + 92 + 92 = 506$
There are 506 windows on the building.

Section 1.6

Practice Problems

1. a. $9\overline{)72}$ with quotient 8, because $8 \cdot 9 = 72$.

b. $40 \div 5 = 8$ because $8 \cdot 5 = 40$.

c. $\frac{24}{6} = 4$ because $4 \cdot 6 = 24$.

2. a. $\frac{7}{7} = 1$ because $1 \cdot 7 = 7$.

b. $5 \div 1 = 5$ because $5 \cdot 1 = 5$.

c. $1\overline{)11}$ with quotient 11, because $11 \cdot 1 = 11$.

d. $4 \div 1 = 4$ because $4 \cdot 1 = 4$.

e. $\frac{10}{1} = 10$ because $10 \cdot 1 = 10$.

f. $21 \div 21 = 1$ because $1 \cdot 21 = 21$.

3. a. $\frac{0}{7} = 0$ because $0 \cdot 7 = 0$.

b. $8\overline{)0}$ with quotient 0, because $0 \cdot 8 = 0$.

c. $7 \div 0$ is undefined because if $7 \div 0$ is a number, then the number times 0 would be 7.

d. $0 \div 14 = 0$ because $0 \cdot 14 = 0$.

Copyright © 2011 Pearson Education, Inc. Publishing as Prentice Hall.

4. a.

$$\begin{array}{r} 818 \\ 6\overline{)\,4908} \\ \underline{-48} \\ 10 \\ \underline{-6} \\ 48 \\ \underline{-48} \\ 0 \end{array}$$

Check: $\begin{array}{r} 818 \\ \underline{\times\ \ 6} \\ 4908 \end{array}$

b.

$$\begin{array}{r} 553 \\ 4\overline{)\,2212} \\ \underline{-20} \\ 21 \\ \underline{-20} \\ 12 \\ \underline{-12} \\ 0 \end{array}$$

Check: $\begin{array}{r} 553 \\ \underline{\times\ \ 4} \\ 2212 \end{array}$

c.

$$\begin{array}{r} 251 \\ 3\overline{)\,753} \\ \underline{-6} \\ 15 \\ \underline{-15} \\ 03 \\ \underline{-3} \\ 0 \end{array}$$

Check: $\begin{array}{r} 251 \\ \underline{\times\ \ 3} \\ 753 \end{array}$

5. a.

$$\begin{array}{r} 304 \\ 7\overline{)\,2128} \\ \underline{-21} \\ 02 \\ \underline{-0} \\ 28 \\ \underline{-28} \\ 0 \end{array}$$

Check: $304 \times 7 = 2128$

b.

$$\begin{array}{r} 5\,100 \\ 9\overline{)\,45,900} \\ \underline{-45} \\ 0\,9 \\ \underline{-9} \\ 000 \end{array}$$

Check: $5100 \times 9 = 45,900$

6. a.

$$\begin{array}{r} 234 \text{ R } 3 \\ 4\overline{)\,939} \\ \underline{-8} \\ 13 \\ \underline{-12} \\ 19 \\ \underline{-16} \\ 3 \end{array}$$

Check: $234 \cdot 4 + 3 = 939$

b.

$$\begin{array}{r} 657 \text{ R } 2 \\ 5\overline{)\,3287} \\ \underline{-30} \\ 28 \\ \underline{-25} \\ 37 \\ \underline{-35} \\ 2 \end{array}$$

Check: $657 \cdot 5 + 2 = 3287$

7. a.

$$\begin{array}{r} 9067 \text{ R } 2 \\ 9\overline{)\,81,605} \\ \underline{-81} \\ 0\,6 \\ \underline{-0} \\ 60 \\ \underline{-54} \\ 65 \\ \underline{-63} \\ 2 \end{array}$$

Check: $9067 \cdot 9 + 2 = 81,605$

Copyright © 2011 Pearson Education, Inc. Publishing as Prentice Hall.

b.
$$\begin{array}{r} 5827 \text{ R } 2 \\ 4\overline{)\,23,310} \\ \underline{-20} \\ 3\,3 \\ \underline{-3\,2} \\ 11 \\ \underline{-8} \\ 30 \\ \underline{-28} \\ 2 \end{array}$$

Check: $5827 \cdot 4 + 2 = 23,310$

8.
$$\begin{array}{r} 524 \text{ R } 12 \\ 17\overline{)\,8920} \\ \underline{-85} \\ 42 \\ \underline{-34} \\ 80 \\ \underline{-68} \\ 12 \end{array}$$

9.
$$\begin{array}{r} 49 \text{ R } 60 \\ 678\overline{)\,33,282} \\ \underline{-27\,12} \\ 6\,162 \\ \underline{-6\,102} \\ 60 \end{array}$$

10.
$$\begin{array}{r} 57 \\ 3\overline{)\,171} \\ \underline{-15} \\ 21 \\ \underline{-21} \\ 0 \end{array}$$

Each student got 57 CDs.

11.
$$\begin{array}{r} 44 \\ 12\overline{)\,532} \\ \underline{-48} \\ 52 \\ \underline{-48} \\ 4 \end{array}$$

There will be 44 full boxes and 4 printers left over.

12. Find the sum and divide by 7.

$$\begin{array}{r} 4 \\ 7 \\ 35 \\ 16 \\ 9 \\ 3 \\ \underline{+\,52} \\ 126 \end{array} \qquad \begin{array}{r} 18 \\ 7\overline{)\,126} \\ \underline{-7} \\ 56 \\ \underline{-56} \\ 0 \end{array}$$

The average time is 18 minutes.

Calculator Explorations

1. $848 \div 16 = 53$
2. $564 \div 12 = 47$
3. $5890 \div 95 = 62$
4. $1053 \div 27 = 39$
5. $\dfrac{32,886}{126} = 261$
6. $\dfrac{143,088}{264} = 542$
7. $0 \div 315 = 0$
8. $315 \div 0$ is an error.

Vocabulary and Readiness Check

1. In $90 \div 2 = 45$, the answer 45 is called the quotient, 90 is called the dividend, and 2 is called the divisor.
2. The quotient of any number and 1 is the same number.
3. The quotient of any number (except 0) and the same number is 1.
4. The quotient of 0 and any number (except 0) is 0.
5. The quotient of any number and 0 is undefined.
6. The average of a list of numbers is the sum of the numbers divided by the number of numbers.

Exercise Set 1.6

1. $54 \div 9 = 6$

Copyright © 2011 Pearson Education, Inc. Publishing as Prentice Hall.

3. $36 \div 3 = 12$

5. $0 \div 8 = 0$

7. $31 \div 1 = 31$

9. $\frac{18}{18} = 1$

11. $\frac{24}{3} = 8$

13. $26 \div 0$ is undefined

15. $26 \div 26 = 1$

17. $0 \div 14 = 0$

19. $18 \div 2 = 9$

21.
$$\begin{array}{r} 29 \\ 3\overline{)\,87} \\ \underline{-6} \\ 27 \\ \underline{-27} \\ 0 \end{array}$$
Check: $3 \cdot 29 = 87$

23.
$$\begin{array}{r} 74 \\ 3\overline{)\,222} \\ \underline{-21} \\ 12 \\ \underline{-12} \\ 0 \end{array}$$
Check: $74 \cdot 3 = 222$

25.
$$\begin{array}{r} 338 \\ 3\overline{)1014} \\ \underline{-9} \\ 11 \\ \underline{-9} \\ 24 \\ \underline{-24} \\ 0 \end{array}$$
Check: $3 \cdot 338 = 1014$

27. $\frac{30}{0}$ is undefined.

29.
$$\begin{array}{r} 9 \\ 7\overline{)\,63} \\ \underline{-63} \\ 0 \end{array}$$
Check: $7 \cdot 9 = 63$

31.
$$\begin{array}{r} 25 \\ 6\overline{)\,150} \\ \underline{-12} \\ 30 \\ \underline{-30} \\ 0 \end{array}$$
Check: $25 \cdot 6 = 150$

33.
$$\begin{array}{r} 68 \text{ R } 3 \\ 7\overline{)\,479} \\ \underline{-42} \\ 59 \\ \underline{-56} \\ 3 \end{array}$$
Check: $7 \cdot 68 + 3 = 479$

35.
$$\begin{array}{r} 236 \text{ R } 5 \\ 6\overline{)\,1421} \\ \underline{-12} \\ 22 \\ \underline{-18} \\ 41 \\ \underline{-36} \\ 5 \end{array}$$
Check: $236 \cdot 6 + 5 = 1421$

37.
$$\begin{array}{r} 38 \text{ R } 1 \\ 8\overline{)\,305} \\ \underline{-24} \\ 65 \\ \underline{-64} \\ 1 \end{array}$$
Check: $8 \cdot 38 + 1 = 305$

39.
$$\begin{array}{r} 326 \text{ R } 4 \\ 7\overline{)\,2286} \\ \underline{-21} \\ 18 \\ \underline{-14} \\ 46 \\ \underline{-42} \\ 4 \end{array}$$
Check: $326 \cdot 7 + 4 = 2286$

Copyright © 2011 Pearson Education, Inc. Publishing as Prentice Hall.

41.
$$\begin{array}{r} 13 \\ 55\overline{)\,715} \\ \underline{-55} \\ 165 \\ \underline{-165} \\ 0 \end{array}$$
Check: $55 \cdot 13 = 715$

43.
$$\begin{array}{r} 49 \\ 23\overline{)1127} \\ \underline{-92} \\ 207 \\ \underline{-207} \\ 0 \end{array}$$
Check: $49 \cdot 23 = 1127$

45.
$$\begin{array}{r} 97 \text{ R } 8 \\ 97\overline{)\,9417} \\ \underline{-873} \\ 687 \\ \underline{-679} \\ 8 \end{array}$$
Check: $97 \cdot 97 + 8 = 9417$

47.
$$\begin{array}{r} 209 \text{ R } 11 \\ 15\overline{)\,3146} \\ \underline{-30} \\ 14 \\ \underline{-0} \\ 146 \\ \underline{-135} \\ 11 \end{array}$$
Check: $209 \cdot 15 + 11 = 3146$

49.
$$\begin{array}{r} 506 \\ 13\overline{)\,6578} \\ \underline{-65} \\ 07 \\ \underline{-0} \\ 78 \\ \underline{-78} \\ 0 \end{array}$$
Check: $13 \cdot 506 = 6578$

51.
$$\begin{array}{r} 202 \text{ R } 7 \\ 46\overline{)\,9299} \\ \underline{-92} \\ 09 \\ \underline{-0} \\ 99 \\ \underline{-92} \\ 7 \end{array}$$
Check: $202 \cdot 46 + 7 = 9299$

53.
$$\begin{array}{r} 54 \\ 236\overline{)\,12744} \\ \underline{-1180} \\ 944 \\ \underline{-944} \\ 0 \end{array}$$
Check: $236 \cdot 54 = 12,744$

55.
$$\begin{array}{r} 99 \text{ R } 100 \\ 103\overline{)10,297} \\ \underline{-9\ 27} \\ 1\ 027 \\ \underline{-927} \\ 100 \end{array}$$
Check: $99 \cdot 103 + 100 = 10,297$

57.
$$\begin{array}{r} 202 \text{ R } 15 \\ 102\overline{)\,20619} \\ \underline{-204} \\ 21 \\ \underline{-0} \\ 219 \\ \underline{-204} \\ 15 \end{array}$$
Check: $102 \cdot 202 + 15 = 20,619$

59.
$$\begin{array}{r} 579 \text{ R } 72 \\ 423\overline{)\,244,989} \\ \underline{-211\ 5} \\ 33\ 48 \\ \underline{-29\ 61} \\ 3\ 879 \\ \underline{-3\ 807} \\ 72 \end{array}$$
Check: $579 \cdot 423 + 72 = 244,989$

Copyright © 2011 Pearson Education, Inc. Publishing as Prentice Hall.

61. $$\begin{array}{r} 17 \\ 7\overline{)119} \\ \underline{-7} \\ 49 \\ \underline{-49} \\ 0 \end{array}$$

63. $$\begin{array}{r} 511 \text{ R } 3 \\ 7\overline{)3580} \\ \underline{-35} \\ 08 \\ \underline{-7} \\ 10 \\ \underline{-7} \\ 3 \end{array}$$

65. $$\begin{array}{r} 2132 \text{ R } 32 \\ 40\overline{)85312} \\ \underline{-80} \\ 53 \\ \underline{-40} \\ 131 \\ \underline{-120} \\ 112 \\ \underline{-80} \\ 32 \end{array}$$

67. $$\begin{array}{r} 6\,080 \\ 142\overline{)863{,}360} \\ \underline{-852} \\ 11\,3 \\ \underline{-0} \\ 11\,36 \\ \underline{-11\,36} \\ 00 \\ \underline{-0} \\ 0 \end{array}$$

69. $$\begin{array}{r} 23 \text{ R } 2 \\ 5\overline{)117} \\ \underline{-10} \\ 17 \\ \underline{-15} \\ 2 \end{array}$$

The quotient is 23 R 2.

71. $$\begin{array}{r} 5 \text{ R } 25 \\ 35\overline{)200} \\ \underline{-175} \\ 25 \end{array}$$

200 divided by 35 is 5 R 25.

73. $$\begin{array}{r} 20 \text{ R } 2 \\ 3\overline{)62} \\ \underline{-6} \\ 02 \\ \underline{-0} \\ 2 \end{array}$$

The quotient is 20 R 2.

75. $$\begin{array}{r} 33 \\ 65\overline{)2145} \\ \underline{-195} \\ 195 \\ \underline{-195} \\ 0 \end{array}$$

There are 33 students in the group.

77. $$\begin{array}{r} 165 \\ 318\overline{)52470} \\ \underline{-318} \\ 2067 \\ \underline{-1908} \\ 1590 \\ \underline{-1590} \\ 0 \end{array}$$

The person weighs 165 pounds on Earth.

79. $$\begin{array}{r} 310 \\ 18\overline{)5580} \\ \underline{-54} \\ 18 \\ \underline{-18} \\ 0 \end{array}$$

The distance is 310 yards.

81. $$\begin{array}{r} 88 \text{ R } 1 \\ 3\overline{)265} \\ \underline{-24} \\ 25 \\ \underline{-24} \\ 1 \end{array}$$

There are 88 bridges every 3 miles over the 265 miles, plus the first bridge, for a total of 89 bridges.

Copyright © 2011 Pearson Education, Inc. Publishing as Prentice Hall.

83.
$$\begin{array}{r} 10 \\ 492\overline{)\,5280} \\ -492 \\ \hline 360 \end{array}$$

There should be 10 poles, plus the first pole for a total of 11 light poles.

85.
$$\begin{array}{r} 5 \\ 5280\overline{)\,26400} \\ -26400 \\ \hline 0 \end{array}$$

Broad Peak is 5 miles tall.

87.
$$\begin{array}{r} 1760 \\ 3\overline{)\,5280} \\ -3 \\ \hline 22 \\ -21 \\ \hline 18 \\ -18 \\ \hline 0 \end{array}$$

There are 1760 yards in 1 mile.

89.
$$\begin{array}{r} {}^{2} \\ 10 \\ 24 \\ 35 \\ 22 \\ 17 \\ +12 \\ \hline 120 \end{array} \qquad \begin{array}{r} 20 \\ 6\overline{)\,120} \\ -12 \\ \hline 00 \end{array}$$

$$\text{Average} = \frac{120}{6} = 20$$

91.
$$\begin{array}{r} {}^{1} \\ 205 \\ 972 \\ 210 \\ +161 \\ \hline 1548 \end{array} \qquad \begin{array}{r} 387 \\ 4\overline{)\,1548} \\ -12 \\ \hline 34 \\ -32 \\ \hline 28 \\ -28 \\ \hline 0 \end{array}$$

$$\text{Average} = \frac{1548}{4} = 387$$

93.
$$\begin{array}{r} {}^{2} \\ 86 \\ 79 \\ 81 \\ 69 \\ +80 \\ \hline 395 \end{array} \qquad \begin{array}{r} 79 \\ 5\overline{)\,395} \\ -35 \\ \hline 45 \\ -45 \\ \hline 0 \end{array}$$

$$\text{Average} = \frac{395}{5} = 79$$

95.
$$\begin{array}{r} {}^{2} \\ 69 \\ 77 \\ +76 \\ \hline 222 \end{array} \qquad \begin{array}{r} 74 \\ 3\overline{)\,222} \\ -21 \\ \hline 12 \\ -12 \\ \hline 0 \end{array}$$

The average temperature is 74°.

97.
$$\begin{array}{r} {}^{111} \\ 82 \\ 463 \\ 29 \\ +8704 \\ \hline 9278 \end{array}$$

99.
$$\begin{array}{r} 546 \\ \times \quad 28 \\ \hline 4\,368 \\ 10\,920 \\ \hline 15{,}288 \end{array}$$

101.
$$\begin{array}{r} 722 \\ -\;43 \\ \hline 679 \end{array}$$

103. $\frac{45}{0}$ is undefined.

105.
$$\begin{array}{r} 9 \text{ R } 12 \\ 24\overline{)\,228} \\ -216 \\ \hline 12 \end{array}$$

107. The quotient of 40 and 8 is $40 \div 8$, which is choice c.

109. 200 divided by 20 is $200 \div 20$, which is choice b.

Copyright © 2011 Pearson Education, Inc. Publishing as Prentice Hall.

111. $\begin{array}{r} 443{,}135{,}000 \\ +\ 391{,}390{,}600 \\ \hline 834{,}525{,}600 \end{array}$

$\begin{array}{r} 417{,}262{,}800 \\ 2\overline{)834{,}525{,}600} \\ -8 \\ \hline 03 \\ -2 \\ \hline 14 \\ -14 \\ \hline 05 \\ -4 \\ \hline 12 \\ -12 \\ \hline 05 \\ -4 \\ \hline 16 \\ -16 \\ \hline 0 \end{array}$

The top two advertisers spent an average of $417,262,800.

113. The average will increase; answers may vary.

115. No; answers may vary
Possible answer: The average cannot be less than each of the four numbers.

117. $\begin{array}{r} 12 \\ 5\overline{)60} \\ -5 \\ \hline 10 \\ -10 \\ \hline 0 \end{array}$

The length is 12 feet.
Notice that Area = length × width = 12 × 5 = 60.

119. answers may vary

121. $\begin{array}{r} 26 \\ -5 \\ \hline 21 \\ -5 \\ \hline 16 \\ -5 \\ \hline 11 \\ -5 \\ \hline 6 \\ -5 \\ \hline 1 \end{array}$

Therefore, 26 ÷ 5 = 5 R 1

Integrated Review

1. $\begin{array}{r} ^{1}\ \ \\ 42 \\ 63 \\ +\ 89 \\ \hline 194 \end{array}$

2. $\begin{array}{r} 7006 \\ -\ 451 \\ \hline 6555 \end{array}$

3. $\begin{array}{r} 87 \\ \times\ 52 \\ \hline 174 \\ 4350 \\ \hline 4524 \end{array}$

4. $\begin{array}{r} 562 \\ 8\overline{)4496} \\ -40 \\ \hline 49 \\ -48 \\ \hline 16 \\ -16 \\ \hline 0 \end{array}$

5. $1 \cdot 67 = 67$

6. $\frac{36}{0}$ is undefined.

7. $16 \div 16 = 1$

8. $5 \div 1 = 5$

9. $0 \cdot 21 = 0$

10. $7 \cdot 0 \cdot 8 = 0$

11. $0 \div 7 = 0$

12. $12 \div 4 = 3$

13. $9 \cdot 7 = 63$

14. $45 \div 5 = 9$

15. $\begin{array}{r} 207 \\ -\ 69 \\ \hline 138 \end{array}$

Copyright © 2011 Pearson Education, Inc. Publishing as Prentice Hall.

16.
$$\begin{array}{r} {\scriptstyle 1} \\ 207 \\ +\ 69 \\ \hline 276 \end{array}$$

17.
$$\begin{array}{r} 3718 \\ -\ 2549 \\ \hline 1169 \end{array}$$

18.
$$\begin{array}{r} {\scriptstyle 11} \\ 1861 \\ +\ 7965 \\ \hline 9826 \end{array}$$

19.
$$\begin{array}{r} 182 \text{ R } 4 \\ 7\overline{)1278} \\ \underline{-7} \\ 57 \\ \underline{-56} \\ 18 \\ \underline{-14} \\ 4 \end{array}$$

20.
$$\begin{array}{r} 1259 \\ \times\ 63 \\ \hline 3\,777 \\ 75\,540 \\ \hline 79,317 \end{array}$$

21.
$$\begin{array}{r} 1099 \text{ R } 2 \\ 7\overline{)7695} \\ \underline{-7} \\ 06 \\ \underline{-0} \\ 69 \\ \underline{-63} \\ 65 \\ \underline{-63} \\ 2 \end{array}$$

22.
$$\begin{array}{r} 111 \text{ R } 1 \\ 9\overline{)1000} \\ \underline{-9} \\ 10 \\ \underline{-9} \\ 10 \\ \underline{-9} \\ 1 \end{array}$$

23.
$$\begin{array}{r} 663 \text{ R } 24 \\ 32\overline{)21,240} \\ \underline{-19\,2} \\ 2\,04 \\ \underline{-1\,92} \\ 120 \\ \underline{-96} \\ 24 \end{array}$$

24.
$$\begin{array}{r} 1\,076 \text{ R } 60 \\ 65\overline{)70,000} \\ \underline{-65} \\ 5\,0 \\ \underline{-\ 0} \\ 5\,00 \\ \underline{-4\,55} \\ 450 \\ \underline{-390} \\ 60 \end{array}$$

25.
$$\begin{array}{r} 4000 \\ -\ 2963 \\ \hline 1037 \end{array}$$

26.
$$\begin{array}{r} 10,000 \\ -\ \ 101 \\ \hline 9\,899 \end{array}$$

27.
$$\begin{array}{r} 303 \\ \times\ 101 \\ \hline 303 \\ 30\,300 \\ \hline 30,603 \end{array}$$

28. $(475)(100) = 47,500$

29.
$$\begin{array}{r} {\scriptstyle 1} \\ 62 \\ +\ 9 \\ \hline 71 \end{array}$$

The total of 62 and 9 is 71.

30.
$$\begin{array}{r} 62 \\ \times\ 9 \\ \hline 558 \end{array}$$

The product of 62 and 9 is 558.

Copyright © 2011 Pearson Education, Inc. Publishing as Prentice Hall.

31. $\begin{array}{r} 6\text{ R }8 \\ 9\overline{)\ 62} \\ -54 \\ \hline 8 \end{array}$

The quotient of 62 and 9 is 6 R 8.

32. $\begin{array}{r} 62 \\ -\ 9 \\ \hline 53 \end{array}$

The difference of 62 and 9 is 53.

33. $\begin{array}{r} 200 \\ -\ 17 \\ \hline 183 \end{array}$

17 subtracted from 200 is 183.

34. $\begin{array}{r} 432 \\ -\ 201 \\ \hline 231 \end{array}$

The difference of 432 and 201 is 231.

35. 9735 rounded to the nearest ten is 9740.
9735 rounded to the nearest hundred is 9700.
9735 rounded to the nearest thousand is 10,000.

36. 1429 rounded to the nearest ten is 1430.
1429 rounded to the nearest hundred is 1400.
1429 rounded to the nearest thousand is 1000.

37. 20,801 rounded to the nearest ten is 20,800.
20,801 rounded to the nearest hundred is 20,800.
20,801 rounded to the nearest thousand is 21,000.

38. 432,198 rounded to the nearest ten is 432,200.
432,198 rounded to the nearest hundred is 432,200.
432,198 rounded to the nearest thousand is 432,000.

39. $6 + 6 + 6 + 6 = 24$
$6 \times 6 = 36$
The perimeter is 24 feet and the area is 36 square feet.

40. $14 + 7 + 14 + 7 = 42$

$\begin{array}{r} 14 \\ \times\ 7 \\ \hline 98 \end{array}$

The perimeter is 42 inches and the area is 98 square inches.

41. $\begin{array}{r} 13 \\ 9 \\ +\ 6 \\ \hline 28 \end{array}$

The perimeter is 28 miles.

42. The unknown vertical side has length $4 + 3 = 7$ meters. The unknown horizontal side has length $3 + 3 = 6$ meters.

$\begin{array}{r} 3 \\ 4 \\ 3 \\ 7 \\ 6 \\ +\ 3 \\ \hline 26 \end{array}$

The perimeter is 26 meters.

43. $\begin{array}{r} {}^{3} \\ 19 \\ 15 \\ 25 \\ 37 \\ +\ 24 \\ \hline 120 \end{array}$ $\begin{array}{r} 24 \\ 5\overline{)\ 120} \\ -10 \\ \hline 20 \\ -20 \\ \hline 0 \end{array}$

$\text{Average} = \frac{120}{5} = 24$

44. $\begin{array}{r} {}^{1\,2} \\ 108 \\ 131 \\ 98 \\ +\ 159 \\ \hline 496 \end{array}$ $\begin{array}{r} 124 \\ 4\overline{)\ 496} \\ -4 \\ \hline 09 \\ -8 \\ \hline 16 \\ -16 \\ \hline 0 \end{array}$

$\text{Average} = \frac{496}{4} = 124$

45. $\begin{array}{r} 28,547 \\ -\ 26,372 \\ \hline 2\,175 \end{array}$

The Lake Pontchartrain Bridge is longer by 2175 feet.

Copyright © 2011 Pearson Education, Inc. Publishing as Prentice Hall.

46.
$$\begin{array}{r} 485 \\ \times\ 18 \\ \hline 3880 \\ \underline{4850} \\ 8730 \end{array}$$
The amount spent on toys is $8730.

Section 1.7

Practice Problems

1. $8 \cdot 8 \cdot 8 \cdot 8 = 8^4$

2. $3 \cdot 3 \cdot 3 = 3^3$

3. $10 \cdot 10 \cdot 10 \cdot 10 \cdot 10 = 10^5$

4. $5 \cdot 5 \cdot 4 \cdot 4 \cdot 4 \cdot 4 \cdot 4 \cdot 4 = 5^2 \cdot 4^6$

5. $4^2 = 4 \cdot 4 = 16$

6. $7^3 = 7 \cdot 7 \cdot 7 = 343$

7. $11^1 = 11$

8. $2 \cdot 3^2 = 2 \cdot 3 \cdot 3 = 18$

9. $9 \cdot 3 - 8 \div 4 = 27 - 8 \div 4 = 27 - 2 = 25$

10. $48 \div 3 \cdot 2^2 = 48 \div 3 \cdot 4 = 16 \cdot 4 = 64$

11. $$\begin{aligned}(10-7)^4 + 2 \cdot 3^2 &= 3^4 + 2 \cdot 3^2 \\ &= 81 + 2 \cdot 9 \\ &= 81 + 18 \\ &= 99\end{aligned}$$

12. $$\begin{aligned}36 \div [20 - (4 \cdot 2)] + 4^3 - 6 &= 36 \div [20 - 8] + 4^3 - 6 \\ &= 36 \div 12 + 4^3 - 6 \\ &= 36 \div 12 + 64 - 6 \\ &= 3 + 64 - 6 \\ &= 61\end{aligned}$$

13. $$\begin{aligned}\frac{25 + 8 \cdot 2 - 3^3}{2(3-2)} &= \frac{25 + 8 \cdot 2 - 27}{2(1)} \\ &= \frac{25 + 16 - 27}{2} \\ &= \frac{14}{2} \\ &= 7\end{aligned}$$

14. $36 \div 6 \cdot 3 + 5 = 6 \cdot 3 + 5 = 18 + 5 = 23$

15. $$\begin{aligned}\text{Area} &= (\text{side})^2 \\ &= (12 \text{ centimeters})^2 \\ &= 144 \text{ square centimeters}\end{aligned}$$
The area of the square is 144 square centimeters.

Calculator Explorations

1. $4^6 = 4096$

2. $5^6 = 15,625$

3. $5^5 = 3125$

4. $7^6 = 117,649$

5. $2^{11} = 2048$

6. $6^8 = 1,679,616$

7. $7^4 + 5^3 = 2526$

8. $12^4 - 8^4 = 16,640$

9. $63 \cdot 75 - 43 \cdot 10 = 4295$

10. $8 \cdot 22 + 7 \cdot 16 = 288$

11. $4(15 \div 3 + 2) - 10 \cdot 2 = 8$

12. $155 - 2(17 + 3) + 185 = 300$

Vocabulary and Readiness Check

1. In $2^5 = 32$, the 2 is called the base and the 5 is called the exponent.

2. To simplify $8 + 2 \cdot 6$, which operation should be performed first? multiplication

3. To simplify $(8 + 2) \cdot 6$, which operation should be performed first? addition

4. To simplify $9(3 - 2) \div 3 + 6$, which operation should be performed first? subtraction

5. To simplify $8 \div 2 \cdot 6$, which operation should be performed first? division

Copyright © 2011 Pearson Education, Inc. Publishing as Prentice Hall.

Exercise Set 1.7

1. $4 \cdot 4 \cdot 4 = 4^3$

3. $7 \cdot 7 \cdot 7 \cdot 7 \cdot 7 \cdot 7 = 7^6$

5. $12 \cdot 12 \cdot 12 = 12^3$

7. $6 \cdot 6 \cdot 5 \cdot 5 \cdot 5 = 6^2 \cdot 5^3$

9. $9 \cdot 8 \cdot 8 = 9 \cdot 8^2$

11. $3 \cdot 2 \cdot 2 \cdot 2 \cdot 2 = 3 \cdot 2^4$

13. $3 \cdot 2 \cdot 2 \cdot 2 \cdot 2 \cdot 5 \cdot 5 \cdot 5 \cdot 5 \cdot 5 = 3 \cdot 2^4 \cdot 5^5$

15. $8^2 = 8 \cdot 8 = 64$

17. $5^3 = 5 \cdot 5 \cdot 5 = 125$

19. $2^5 = 2 \cdot 2 \cdot 2 \cdot 2 \cdot 2 = 32$

21. $1^{10} = 1 \cdot 1 \cdot 1 \cdot 1 \cdot 1 \cdot 1 \cdot 1 \cdot 1 \cdot 1 \cdot 1 = 1$

23. $7^1 = 7$

25. $2^7 = 2 \cdot 2 \cdot 2 \cdot 2 \cdot 2 \cdot 2 \cdot 2 = 128$

27. $2^8 = 2 \cdot 2 \cdot 2 \cdot 2 \cdot 2 \cdot 2 \cdot 2 \cdot 2 = 256$

29. $4^4 = 4 \cdot 4 \cdot 4 \cdot 4 = 256$

31. $9^3 = 9 \cdot 9 \cdot 9 = 729$

33. $12^2 = 12 \cdot 12 = 144$

35. $10^2 = 10 \cdot 10 = 100$

37. $20^1 = 20$

39. $3^6 = 3 \cdot 3 \cdot 3 \cdot 3 \cdot 3 \cdot 3 = 729$

41. $3 \cdot 2^6 = 3 \cdot 2 \cdot 2 \cdot 2 \cdot 2 \cdot 2 \cdot 2 = 192$

43. $2 \cdot 3^4 = 2 \cdot 3 \cdot 3 \cdot 3 \cdot 3 = 162$

45. $15 + 3 \cdot 2 = 15 + 6 = 21$

47. $14 \div 7 \cdot 2 + 3 = 2 \cdot 2 + 3 = 4 + 3 = 7$

49. $32 \div 4 - 3 = 8 - 3 = 5$

51. $13 + \frac{24}{8} = 13 + 3 = 16$

53. $6 \cdot 5 + 8 \cdot 2 = 30 + 16 = 46$

55. $\frac{5 + 12 \div 4}{1^7} = \frac{5+3}{1} = \frac{8}{1} = 8$

57. $$\begin{aligned}(7+5^2) \div 4 \cdot 2^3 &= (7+25) \div 4 \cdot 2^3 \\ &= 32 \div 4 \cdot 2^3 \\ &= 32 \div 4 \cdot 8 \\ &= 8 \cdot 8 \\ &= 64\end{aligned}$$

59. $$\begin{aligned}5^2 \cdot (10-8) + 2^3 + 5^2 &= 5^2 \cdot 2 + 2^3 + 5^2 \\ &= 25 \cdot 2 + 8 + 25 \\ &= 50 + 8 + 25 \\ &= 83\end{aligned}$$

61. $\frac{18+6}{2^4 - 2^2} = \frac{24}{16-4} = \frac{24}{12} = 2$

63. $(3 + 5) \cdot (9 - 3) = 8 \cdot 6 = 48$

65. $\frac{7(9-6)+3}{3^2 - 3} = \frac{7(3)+3}{9-3} = \frac{21+3}{6} = \frac{24}{6} = 4$

67. $8 \div 0 + 37 =$ undefined

69. $$\begin{aligned}2^4 \cdot 4 - (25 \div 5) &= 2^4 \cdot 4 - 5 \\ &= 16 \cdot 4 - 5 \\ &= 64 - 5 \\ &= 59\end{aligned}$$

71. $$\begin{aligned}3^4 - [35 - (12-6)] &= 3^4 - [35 - 6] \\ &= 3^4 - 29 \\ &= 81 - 29 \\ &= 52\end{aligned}$$

73. $$\begin{aligned}(7 \cdot 5) + [9 \div (3 \div 3)] &= (7 \cdot 5) + [9 \div (1)] \\ &= 35 + 9 \\ &= 44\end{aligned}$$

Copyright © 2011 Pearson Education, Inc. Publishing as Prentice Hall.

75. $8\cdot[2^2+(6-1)\cdot 2]-50\cdot 2 = 8\cdot(2^2+5\cdot 2)-50\cdot 2$
$= 8\cdot(4+5\cdot 2)-50\cdot 2$
$= 8\cdot(4+10)-50\cdot 2$
$= 8\cdot 14-50\cdot 2$
$= 112-50\cdot 2$
$= 112-100$
$= 12$

77. $\dfrac{9^2+2^2-1^2}{8\div 2\cdot 3\cdot 1\div 3} = \dfrac{81+4-1}{4\cdot 3\cdot 1\div 3}$
$= \dfrac{85-1}{12\cdot 1\div 3}$
$= \dfrac{84}{12\div 3}$
$= \dfrac{84}{4}$
$= 21$

79. $\dfrac{2+4^2}{5(20-16)-3^2-5} = \dfrac{2+16}{5(4)-3^2-5}$
$= \dfrac{18}{5(4)-9-5}$
$= \dfrac{18}{20-9-5}$
$= \dfrac{18}{11-5}$
$= \dfrac{18}{6}$
$= 3$

81. $9\div 3+5^2\cdot 2-10 = 9\div 3+25\cdot 2-10$
$= 3+25\cdot 2-10$
$= 3+50-10$
$= 43$

83. $[13\div(20-7)+2^5]-(2+3)^2 = [13\div 13+2^5]-5^2$
$= [13\div 13+32]-5^2$
$= [1+32]-5^2$
$= 33-5^2$
$= 33-25$
$= 8$

85. $7^2-\{18-[40\div(5\cdot 1)+2]+5^2\}$
$= 7^2-\{18-[40\div 5+2]+5^2\}$
$= 7^2-\{18-[8+2]+5^2\}$
$= 7^2-\{18-10+5^2\}$
$= 7^2-\{18-10+25\}$
$= 7^2-33$
$= 49-33$
$= 16$

87. Area of a square = $(\text{side})^2$
$= (7 \text{ meters})^2$
= 49 square meters
Perimeter = 4(side) = 4(7 meters) = 28 meters

89. Area of a square = $(\text{side})^2$
$= (23 \text{ miles})^2$
= 529 square miles
Perimeter = 4(side) = 4(23 miles) = 92 miles

91. The statement is true.

93. $2^5 = 2\cdot 2\cdot 2\cdot 2\cdot 2$
The statement is false.

95. $(2+3)\cdot 6-2 = 5\cdot 6-2 = 30-2 = 28$

97. $24\div(3\cdot 2)+2\cdot 5 = 24\div 6+2\cdot 5 = 4+10 = 14$

99. The unknown vertical length is
30 − 12 = 18 feet. The unknown horizontal
length is 60 − 40 = 20 feet.
Perimeter = 60 + 30 + 40 + 18 + 20 + 12 = 180
The total perimeter of seven homes is
7(180) = 1260 feet.

101. $(7+2^4)^5-(3^5-2^4)^2 = (7+16)^5-(243-16)^2$
$= 23^5-227^2$
$= 6,436,343-51,529$
$= 6,384,814$

103. answers may vary; possible answer:
$(20-10)\cdot 5\div 25+3 = 10\cdot 5\div 25+3$
$= 50\div 25+3$
$= 2+3$
$= 5$

Copyright © 2011 Pearson Education, Inc. Publishing as Prentice Hall.

Section 1.8

Practice Problems

1. $x - 2 = 7 - 2 = 5$

2. $y(x - 3) = 4(8 - 3) = 4(5) = 20$

3. $\frac{y+6}{x} = \frac{18+6}{6} = \frac{24}{6} = 4$

4. $25 - z^3 + x = 25 - 2^3 + 1 = 25 - 8 + 1 = 18$

5. $\frac{5(F-32)}{9} = \frac{5(41-32)}{9} = \frac{5(9)}{9} = \frac{45}{9} = 5$

6. $3(y-6) = 6$
$3(8-6) \stackrel{?}{=} 6$
$3(2) \stackrel{?}{=} 6$
$6 = 6$ True
Yes, 8 is a solution.

7. $5n + 4 = 34$
Let n be 10.
$5(10) + 4 \stackrel{?}{=} 34$
$50 + 4 \stackrel{?}{=} 34$
$54 = 34$ False
No, 10 is not a solution.
Let n be 6.
$5(6) + 4 \stackrel{?}{=} 34$
$30 + 4 \stackrel{?}{=} 34$
$34 = 34$ True
Yes, 6 is a solution.
Let n be 8.
$5(8) + 4 \stackrel{?}{=} 34$
$40 + 4 \stackrel{?}{=} 34$
$44 = 34$ False
No, 8 is not a solution.

8. a. Twice a number is $2x$.

b. 8 increased by a number is $8 + x$ or $x + 8$.

c. 10 minus a number is $10 - x$.

d. 10 subtracted from a number is $x - 10$.

e. The quotient of 6 and a number is $6 \div x$ or $\frac{6}{x}$.

Vocabulary and Readiness Check

1. A combination of operations on letters (variables) and numbers is an expression.

Copyright © 2011 Pearson Education, Inc. Publishing as Prentice Hall.

2. A letter that represents a number is a variable.

3. $3x - 2y$ is called an expression and the letters x and y are variables.

4. Replacing a variable in an expression by a number and then finding the value of the expression is called evaluating the expression.

5. A statement of the form "expression = expression" is called an equation.

6. A value for the variable that makes the equation a true statement is called a solution.

Exercise Set 1.8

1.

a	b	$a+b$	$a-b$	$a \cdot b$	$a \div b$
21	7	$21+7=28$	$21-7=14$	$21 \cdot 7 = 147$	$21 \div 7 = 3$

3.

a	b	$a+b$	$a-b$	$a \cdot b$	$a \div b$
152	0	$152+0=152$	$152-0=152$	$152 \cdot 0 = 0$	$152 \div 0$ is undefined.

5.

a	b	$a+b$	$a-b$	$a \cdot b$	$a \div b$
56	1	$56+1=57$	$56-1=55$	$56 \cdot 1 = 56$	$56 \div 1 = 56$

7. $3 + 2z = 3 + 2(3) = 3 + 6 = 9$

9. $3xz - 5x = 3(2)(3) - 5(2) = 18 - 10 = 8$

11. $z - x + y = 3 - 2 + 5 = 1 + 5 = 6$

13. $4x - z = 4(2) - 3 = 8 - 3 = 5$

15. $y^3 - 4x = 5^3 - 4(2) = 125 - 4(2) = 125 - 8 = 117$

17.
$$\begin{aligned} 2xy^2 - 6 &= 2(2)(5)^2 - 6 \\ &= 2 \cdot 2 \cdot 25 - 6 \\ &= 100 - 6 \\ &= 94 \end{aligned}$$

19. $8 - (y - x) = 8 - (5 - 2) = 8 - 3 = 5$

21.
$$\begin{aligned} x^5 + (y - z) &= 2^5 + (5 - 3) \\ &= 2^5 + 2 \\ &= 32 + 2 \\ &= 34 \end{aligned}$$

23. $\frac{6xy}{z} = \frac{6 \cdot 2 \cdot 5}{3} = \frac{60}{3} = 20$

Copyright © 2011 Pearson Education, Inc. Publishing as Prentice Hall.

25. $\frac{2y-2}{x}=\frac{2(5)-2}{2}=\frac{10-2}{2}=\frac{8}{2}=4$

27. $\frac{x+2y}{z}=\frac{2+2\cdot 5}{3}=\frac{2+10}{3}=\frac{12}{3}=4$

29. $\frac{5x}{y}-\frac{10}{y}=\frac{5(2)}{5}-\frac{10}{5}=\frac{10}{5}-2=2-2=0$

31. $$\begin{aligned}2y^2-4y+3&=2\cdot 5^2-4\cdot 5+3\\&=2\cdot 25-4\cdot 5+3\\&=50-20+3\\&=33\end{aligned}$$

33. $$\begin{aligned}(4y-5z)^3&=(4\cdot 5-5\cdot 3)^3\\&=(20-15)^3\\&=(5)^3\\&=125\end{aligned}$$

35. $(xy+1)^2=(2\cdot 5+1)^2=(10+1)^2=11^2=121$

37. $$\begin{aligned}2y(4z-x)&=2\cdot 5(4\cdot 3-2)\\&=2\cdot 5(12-2)\\&=2\cdot 5(10)\\&=10(10)\\&=100\end{aligned}$$

39. $$\begin{aligned}xy(5+z-x)&=2\cdot 5(5+3-2)\\&=2\cdot 5(6)\\&=10(6)\\&=60\end{aligned}$$

41. $\frac{7x+2y}{3x}=\frac{7(2)+2(5)}{3(2)}=\frac{14+10}{6}=\frac{24}{6}=4$

43.

t	1	2	3	4
$16t^2$	$16\cdot 1^2=16\cdot 1=16$	$16\cdot 2^2=16\cdot 4=64$	$16\cdot 3^2=16\cdot 9=144$	$16\cdot 4^2=16\cdot 16=256$

45. Let n be 10.
$n-8=2$
$10-8\stackrel{?}{=}2$
$2=2$ True
Yes, 10 is a solution.

Copyright © 2011 Pearson Education, Inc. Publishing as Prentice Hall.

47. Let n be 3.
$24 = 80n$
$24 \stackrel{?}{=} 80 \cdot 3$
$24 = 240$ False
No, 3 is not a solution.

49. Let n be 7.
$3n - 5 = 10$
$3(7) - 5 \stackrel{?}{=} 10$
$21 - 5 \stackrel{?}{=} 10$
$16 = 10$ False
No, 7 is not a solution.

51. Let n be 20.
$2(n - 17) = 6$
$2(20 - 17) \stackrel{?}{=} 6$
$2(3) \stackrel{?}{=} 6$
$6 = 6$ True
Yes, 20 is a solution.

53. Let x be 0.
$5x + 3 = 4x + 13$
$5(0) + 3 \stackrel{?}{=} 4(0) + 13$
$0 + 3 \stackrel{?}{=} 0 + 13$
$3 = 13$ False
No, 0 is not a solution.

55. Let f be 8.
$7f = 64 - f$
$7(8) \stackrel{?}{=} 64 - 8$
$56 = 56$ True
Yes, 8 is a solution.

57. $n - 2 = 10$
Let n be 10.
$10 - 2 \stackrel{?}{=} 10$
$8 = 10$ False
Let n be 12.
$12 - 2 \stackrel{?}{=} 10$
$10 = 10$ True
Let n be 14.
$14 - 2 \stackrel{?}{=} 10$
$12 = 10$ False
12 is a solution.

59. $5n = 30$
Let n be 6.
$5 \cdot 6 \stackrel{?}{=} 30$
$30 = 30$ True
Let n be 25.
$5 \cdot 25 \stackrel{?}{=} 30$
$125 = 30$ False
Let n be 30.
$5 \cdot 30 \stackrel{?}{=} 30$
$150 = 30$ False
6 is a solution.

61. $6n + 2 = 26$
Let n be 0.
$6(0) + 2 \stackrel{?}{=} 26$
$0 + 2 \stackrel{?}{=} 26$
$2 = 26$ False
Let n be 2.
$6(2) + 2 \stackrel{?}{=} 26$
$12 + 2 \stackrel{?}{=} 26$
$14 = 26$ False
Let n be 4.
$6(4) + 2 \stackrel{?}{=} 26$
$24 + 2 \stackrel{?}{=} 26$
$26 = 26$ True
4 is a solution.

63. $3(n - 4) = 10$
Let n be 5.
$3(5 - 4) \stackrel{?}{=} 10$
$3(1) \stackrel{?}{=} 10$
$3 = 10$ False
Let n be 7.
$3(7 - 4) \stackrel{?}{=} 10$
$3(3) \stackrel{?}{=} 10$
$9 = 10$ False
Let n be 10.
$3(10 - 4) \stackrel{?}{=} 10$
$3(6) \stackrel{?}{=} 10$
$18 = 10$ False
None are solutions.

65. $7x - 9 = 5x + 13$
Let x be 3.
$7(3) - 9 \stackrel{?}{=} 5(3) + 13$
$21 - 9 \stackrel{?}{=} 15 + 13$
$12 = 28$ False
Let x be 7.
$7(7) - 9 \stackrel{?}{=} 5(7) + 13$
$49 - 9 \stackrel{?}{=} 35 + 13$
$40 = 48$ False
Let x be 11.
$7(11) - 9 \stackrel{?}{=} 5(11) + 13$
$77 - 9 \stackrel{?}{=} 55 + 13$
$68 = 68$ True
11 is a solution.

67. Eight more than a number is $x + 8$.

69. The total of a number and 8 is $x + 8$.

Copyright © 2011 Pearson Education, Inc. Publishing as Prentice Hall.

71. Twenty decreased by a number is $20 - x$.

73. The product of 512 and a number is $512x$.

75. The quotient of eight and a number is $\frac{8}{x}$.

77. The sum of seventeen and a number added to the product of five and the number is $5x + (17 + x)$.

79. The product of five and a number is $5x$.

81. A number subtracted from 11 is $11 - x$.

83. A number less 5 is $x - 5$.

85. 6 divided by a number is $6 \div x$ or $\frac{6}{x}$.

87. Fifty decreased by eight times a number is $50 - 8x$.

89.
$$\begin{aligned} x^4 - y^2 &= 23^4 - 72^2 \\ &= 279,841 - 5184 \\ &= 274,657 \end{aligned}$$

91.
$$\begin{aligned} x^2 + 5y - 112 &= 23^2 + 5(72) - 112 \\ &= 529 + 360 - 112 \\ &= 777 \end{aligned}$$

93. $5x$ is the largest; answers may vary.

95. As t gets larger, $16t^2$ gets larger.

Chapter 1 Vocabulary Check

1. The whole numbers are 0, 1, 2, 3, ...
2. The perimeter of a polygon is its distance around or the sum of the lengths of its sides.
3. The position of each digit in a number determines its place value.
4. An exponent is a shorthand notation for repeated multiplication of the same factor.
5. To find the area of a rectangle, multiply length times width.
6. The digits used to write numbers are 0, 1, 2, 3, 4, 5, 6, 7, 8, and 9.
7. A letter used to represent a number is called a variable.
8. An equation can be written in the form "expression = expression."
9. A combination of operations on variables and numbers is called an expression.
10. A solution of an equation is a value of the variable that makes the equation a true statement.
11. A collection of numbers (or objects) enclosed by braces is called a set.
12. The 21 above is called the sum.
13. The 5 above is called the divisor.
14. The 35 above is called the dividend.
15. The 7 above is called the quotient.
16. The 3 above is called a factor.
17. The 6 above is called the product.
18. The 20 above is called the minuend.
19. The 9 above is called the subtrahend.
20. The 11 above is called the difference.
21. The 4 above is called an addend.

Chapter 1 Review

1. The place value of 4 in 7640 is tens.
2. The place value of 4 in 46,200,120 is ten-millions.
3. 7640 is written as seven thousand, six hundred forty.
4. 46,200,120 is written as forty-six million, two hundred thousand, one hundred twenty.
5. $3158 = 3000 + 100 + 50 + 8$
6. $403,225,000 = 400,000,000 + 3,000,000 + 200,000 + 20,000 + 5000$
7. Eighty-one thousand, nine hundred in standard form is 81,900.

Copyright © 2011 Pearson Education, Inc. Publishing as Prentice Hall.

8. Six billion, three hundred four million in standard form is 6,304,000,000.

9. Locate Europe in the first column and read across to the number in the 2008 column. There were 384,633,765 Internet users in Europe in 2008.

10. Locate Oceania/Australia in the first column and read across to the number in the 2004 column. There were 11,805,500 Internet users in Oceania/Australia in 2004.

11. Locate the smallest number in the 2000 column. Middle East had the fewest Internet users in 2000.

12. Locate the biggest number in the 2008 column. Asia had the greatest number of Internet users in 2008.

13.
$$\begin{array}{r} ^{1} \\ 18 \\ +\,49 \\ \hline 67 \end{array}$$

14.
$$\begin{array}{r} ^{1} \\ 28 \\ +\,39 \\ \hline 67 \end{array}$$

15.
$$\begin{array}{r} 462 \\ -\,397 \\ \hline 65 \end{array}$$

16.
$$\begin{array}{r} 583 \\ -\,279 \\ \hline 304 \end{array}$$

17.
$$\begin{array}{r} 428 \\ +\;\;21 \\ \hline 449 \end{array}$$

18.
$$\begin{array}{r} ^{1} \\ 819 \\ +\;\;21 \\ \hline 840 \end{array}$$

19.
$$\begin{array}{r} 4000 \\ -\;\;86 \\ \hline 3914 \end{array}$$

20.
$$\begin{array}{r} 8000 \\ -\;\;92 \\ \hline 7908 \end{array}$$

21.
$$\begin{array}{r} ^{1\,2\,1} \\ 91 \\ 3623 \\ +\;497 \\ \hline 4211 \end{array}$$

22.
$$\begin{array}{r} ^{1\,1} \\ 82 \\ 1647 \\ +\;238 \\ \hline 1967 \end{array}$$

23.
$$\begin{array}{r} ^{1\,1} \\ 74 \\ 342 \\ +\;918 \\ \hline 1334 \end{array}$$

The sum of 74, 342, and 918 is 1334.

24.
$$\begin{array}{r} ^{2} \\ 49 \\ 529 \\ +\,308 \\ \hline 886 \end{array}$$

The sum of 49, 529, and 308 is 886.

25.
$$\begin{array}{r} 25,862 \\ -\;\;7\,965 \\ \hline 17,897 \end{array}$$

7965 subtracted from 25,862 is 17,897.

26.
$$\begin{array}{r} 39,007 \\ -\;\;4\,349 \\ \hline 34,658 \end{array}$$

4349 subtracted from 39,007 is 34,658.

27.
$$\begin{array}{r} ^{1} \\ 205 \\ +\,7318 \\ \hline 7523 \end{array}$$

The total distance is 7523 miles.

Copyright © 2011 Pearson Education, Inc. Publishing as Prentice Hall.

28. $\begin{array}{r} \scriptstyle 1\,1\,1 \\ 62,589 \\ 65,340 \\ +\ 69,770 \\ \hline 197,699 \end{array}$

Her total earnings were $197,699.

29. 40 + 52 + 52 + 72 = 216
The perimeter is 216 feet.

30. 11 + 20 + 35 = 66
The perimeter is 66 kilometers.

31. $\begin{array}{r} 384,633,765 \\ -\ 241,208,100 \\ \hline 143,425,665 \end{array}$

The number of Internet users in Europe increased by 143,425,665.

32. $\begin{array}{r} 41,939,200 \\ -\ 20,204,331 \\ \hline 21,734,869 \end{array}$

There were 21,734,869 more Internet users in the Middle East than in Oceania/Australia in 2008.

33. Find the shortest bar. The balance was the least in May.

34. Find the tallest bar. The balance was the greatest in August.

35. $\begin{array}{r} 280 \\ -\ 170 \\ \hline 110 \end{array}$

The balance decreased by $110 from February to April.

36. $\begin{array}{r} 490 \\ -\ 250 \\ \hline 240 \end{array}$

The balance increased by $240 from June to August.

37. To round 43 to the nearest ten, observe that the digit in the ones place is 3. Since this digit is less than 5, we do not add 1 to the digit in the tens place. The number 43 rounded to the nearest ten is 40.

38. To round 45 to the nearest ten, observe that the digit in the ones place is 5. Since this digit is at least 5, we add 1 to the digit in the tens place. The number 45 rounded to the nearest ten is 50.

39. To round 876 to the nearest ten, observe that the digit in the ones place is 6. Since this digit is at least 5, we add 1 to the digit in the tens place. The number 876 rounded to the nearest ten is 880.

40. To round 493 to the nearest hundred, observe that the digit in the tens place is 9. Since this digit is at least 5, we add 1 to the digit in the hundreds place. The number 493 rounded to the nearest hundred is 500.

41. To round 3829 to the nearest hundred, observe that the digit in the tens place is 2. Since this digit is less than 5, we do not add 1 to the digit in the hundreds place. The number 3829 rounded to the nearest hundred is 3800.

42. To round 57,534 to the nearest thousand, observe that the digit in the hundreds place is 5. Since this digit is at least 5, we add 1 to the digit in the thousands place. The number 57,534 rounded to the nearest thousand is 58,000.

43. To round 39,583,819 to the nearest million, observe that the digit in the hundred-thousands place is 5. Since this digit is at least 5, we add 1 to the digit in the millions place. The number 39,583,819 rounded to the nearest million is 40,000,000.

44. To round 768,542 to the nearest hundred-thousand, observe that the digit in the ten-thousands place is 6. Since this digit is at least 5, we add 1 to the digit in the hundred-thousands place. The number 768,542 rounded to the nearest hundred-thousand is 800,000.

45. $\begin{array}{rcr} 3785 & \text{rounds to} & \overset{2}{3800} \\ 648 & \text{rounds to} & 600 \\ \underline{+\ 2866} & \text{rounds to} & +\ 2900 \\ & & \overline{7300} \end{array}$

46. $\begin{array}{rcr} 5925 & \text{rounds to} & 5900 \\ \underline{-\ 1787} & \text{rounds to} & -\ 1800 \\ & & \overline{4100} \end{array}$

Copyright © 2011 Pearson Education, Inc. Publishing as Prentice Hall.

47.

630	rounds to	600
192	rounds to	200
271	rounds to	300
56	rounds to	100
703	rounds to	700
454	rounds to	500
+ 329	rounds to	+ 300
		2700

They traveled approximately 2700 miles.

48.

139,009,209	rounds to	139,000,000
− 51,065,630	rounds to	− 51,000,000
		88,000,000

There were approximately 88,000,000 more Internet users in Latin America/Caribbean than in Africa in 2008.

49. $\begin{array}{r} 276 \\ \times\ \ 8 \\ \hline 2208 \end{array}$

50. $\begin{array}{r} 349 \\ \times\ \ 4 \\ \hline 1396 \end{array}$

51. $\begin{array}{r} 57 \\ \times\ 40 \\ \hline 2280 \end{array}$

52. $\begin{array}{r} 69 \\ \times\ 42 \\ \hline 138 \\ 2760 \\ \hline 2898 \end{array}$

53. $20(7)(4) = 140(4) = 560$

54. $25(9)(4) = 225(4) = 900$
or
$25(4)(9) = 100(9) = 900$

55. $26 \cdot 34 \cdot 0 = 0$

56. $62 \cdot 88 \cdot 0 = 0$

57. $\begin{array}{r} 586 \\ \times\ \ 29 \\ \hline 5\,274 \\ 11\,720 \\ \hline 16{,}994 \end{array}$

58. $\begin{array}{r} 242 \\ \times\ 37 \\ \hline 1694 \\ 7260 \\ \hline 8954 \end{array}$

59. $\begin{array}{r} 642 \\ \times\ \ 177 \\ \hline 4\,494 \\ 44\,940 \\ 64\,200 \\ \hline 113{,}634 \end{array}$

60. $\begin{array}{r} 347 \\ \times\ \ 129 \\ \hline 3\,123 \\ 6\,940 \\ 34\,700 \\ \hline 44{,}763 \end{array}$

61. $\begin{array}{r} 1026 \\ \times\ \ 401 \\ \hline 1\,026 \\ 410\,400 \\ \hline 411{,}426 \end{array}$

62. $\begin{array}{r} 2107 \\ \times\ \ 302 \\ \hline 4\,214 \\ 632\,100 \\ \hline 636{,}314 \end{array}$

63. "Product" indicates multiplication.
$\begin{array}{r} 250 \\ \times\ \ 6 \\ \hline 1500 \end{array}$
The product of 6 and 250 is 1500.

64. "Product" indicates multiplication.
$\begin{array}{r} 820 \\ \times\ \ 6 \\ \hline 4920 \end{array}$
The product of 6 and 820 is 4920.

Copyright © 2011 Pearson Education, Inc. Publishing as Prentice Hall.

65. $\begin{array}{r} 32 \\ \times\ 15 \\ \hline 160 \\ 320 \\ \hline 480 \end{array}$ $\begin{array}{r} 38 \\ \times 11 \\ \hline 38 \\ 380 \\ \hline 418 \end{array}$

$\begin{array}{r} 480 \\ +\ 418 \\ \hline 898 \end{array}$

The total cost is $898.

66. $\begin{array}{r} 6112 \\ \times\quad 20 \\ \hline 122,240 \end{array}$

The total cost is $122,240.

67. Area = (length)(width)
= (13 miles)(7 miles)
= 91 square miles

68. Area = (length)(width)
= (25 centimeters)(20 centimeters)
= 500 square centimeters

69. $\frac{49}{7} = 7$ Check: $\begin{array}{r} 7 \\ \times 7 \\ \hline 49 \end{array}$

70. $\frac{36}{9} = 4$ Check: $\begin{array}{r} 9 \\ \times 4 \\ \hline 36 \end{array}$

71. $5\overline{)27}$ = 5 R 2
$\begin{array}{r} 27 \\ \underline{-25} \\ 2 \end{array}$

Check: $5 \times 5 + 2 = 27$

72. $4\overline{)18}$ = 4 R 2
$\begin{array}{r} 18 \\ \underline{-16} \\ 2 \end{array}$

Check: $4 \times 4 + 2 = 18$

73. $918 \div 0$ is undefined.

74. $0 \div 668 = 0$ Check: $0 \cdot 668 = 0$

75. $5\overline{)167}$ = 33 R 2
$\begin{array}{r} 167 \\ \underline{-15} \\ 17 \\ \underline{-15} \\ 2 \end{array}$

Check: $33 \times 5 + 2 = 167$

76. $8\overline{)159}$ = 19 R 7
$\begin{array}{r} 159 \\ \underline{-8} \\ 79 \\ \underline{-72} \\ 7 \end{array}$

Check: $19 \times 8 + 7 = 159$

77. $26\overline{)626}$ = 24 R 2
$\begin{array}{r} 626 \\ \underline{-52} \\ 106 \\ \underline{-104} \\ 2 \end{array}$

Check: $24 \times 26 + 2 = 626$

78. $19\overline{)680}$ = 35 R 15
$\begin{array}{r} 680 \\ \underline{-57} \\ 110 \\ \underline{-95} \\ 15 \end{array}$

Check: $35 \times 19 + 15 = 680$

79. $47\overline{)23,792}$ = 506 R 10
$\begin{array}{r} 23,792 \\ \underline{-23\ 5} \\ 29 \\ \underline{-0} \\ 292 \\ \underline{-282} \\ 10 \end{array}$

Check: $506 \times 47 + 10 = 23,792$

Copyright © 2011 Pearson Education, Inc. Publishing as Prentice Hall.

80. $$\begin{array}{r} 907 \text{ R } 40 \\ 53\overline{)\,48,111} \\ \underline{-47\,7} \\ 41 \\ \underline{-0} \\ 411 \\ \underline{-371} \\ 40 \end{array}$$

Check: $907 \times 53 + 40 = 48,111$

81. $$\begin{array}{r} 2793 \text{ R } 140 \\ 207\overline{)\,578,291} \\ \underline{-414} \\ 164\,2 \\ \underline{-144\,9} \\ 19\,39 \\ \underline{-18\,63} \\ 761 \\ \underline{-621} \\ 140 \end{array}$$

Check: $2793 \times 207 + 140 = 578,291$

82. $$\begin{array}{r} 2012 \text{ R } 60 \\ 306\overline{)\,615,732} \\ \underline{-612} \\ 3\,7 \\ \underline{-0} \\ 3\,73 \\ \underline{-3\,06} \\ 672 \\ \underline{-612} \\ 60 \end{array}$$

Check: $2012 \times 306 + 60 = 615,732$

83. $$\begin{array}{r} 18 \text{ R } 2 \\ 5\overline{)\,92} \\ \underline{-5} \\ 42 \\ \underline{-40} \\ 2 \end{array}$$

The quotient of 92 and 5 is 18 R 2.

84. $$\begin{array}{r} 21 \text{ R } 2 \\ 4\overline{)\,86} \\ \underline{-8} \\ 06 \\ \underline{-4} \\ 2 \end{array}$$

The quotient of 86 and 4 is 21 R 2.

85. $$\begin{array}{r} 27 \\ 24\overline{)\,648} \\ \underline{-48} \\ 168 \\ \underline{-168} \\ 0 \end{array}$$

27 boxes can be filled with cans of corn.

86. $$\begin{array}{r} 13 \\ 1760\overline{)\,22,880} \\ \underline{-17\,60} \\ 5\,280 \\ \underline{-5\,280} \\ 0 \end{array}$$

There are 13 miles in 22,880 yards.

87. Divide the sum by 4.

$$\begin{array}{r} 76 \\ 49 \\ 32 \\ \underline{+\,47} \\ 204 \end{array} \qquad \begin{array}{r} 51 \\ 4\overline{)\,204} \\ \underline{-20} \\ 04 \\ \underline{-4} \\ 0 \end{array}$$

The average is 51.

88. Divide the sum by 4.

$$\begin{array}{r} 23 \\ 85 \\ 62 \\ \underline{+\,66} \\ 236 \end{array} \qquad \begin{array}{r} 59 \\ 4\overline{)\,236} \\ \underline{-20} \\ 36 \\ \underline{-36} \\ 0 \end{array}$$

The average is 59.

89. $8^2 = 8 \cdot 8 = 64$

90. $5^3 = 5 \cdot 5 \cdot 5 = 125$

91. $5 \cdot 9^2 = 5 \cdot 9 \cdot 9 = 405$

92. $4 \cdot 10^2 = 4 \cdot 10 \cdot 10 = 400$

Copyright © 2011 Pearson Education, Inc. Publishing as Prentice Hall.

93. $18 \div 2 + 7 = 9 + 7 = 16$

94. $12 - 8 \div 4 = 12 - 2 = 10$

95. $\frac{5(6^2-3)}{3^2+2} = \frac{5(36-3)}{9+2} = \frac{5(33)}{11} = \frac{165}{11} = 15$

96. $\frac{7(16-8)}{2^3} = \frac{7(8)}{8} = \frac{56}{8} = 7$

97. $48 \div 8 \cdot 2 = 6 \cdot 2 = 12$

98. $27 \div 9 \cdot 3 = 3 \cdot 3 = 9$

99.
$$\begin{aligned} &2+3[1^5+(20-17)\cdot 3]+5\cdot 2 \\ &= 2+3[1^5+3\cdot 3]+5\cdot 2 \\ &= 2+3[1+3\cdot 3]+5\cdot 2 \\ &= 2+3[1+9]+5\cdot 2 \\ &= 2+3\cdot 10+5\cdot 2 \\ &= 2+30+10 \\ &= 42 \end{aligned}$$

100.
$$\begin{aligned} &21-[2^4-(7-5)-10]+8\cdot 2 \\ &= 21-[2^4-2-10]+8\cdot 2 \\ &= 21-[16-2-10]+8\cdot 2 \\ &= 21-4+8\cdot 2 \\ &= 21-4+16 \\ &= 33 \end{aligned}$$

101.
$$\begin{aligned} 19-2(3^2-2^2) &= 19-2(9-4) \\ &= 19-2(5) \\ &= 19-10 \\ &= 9 \end{aligned}$$

102.
$$\begin{aligned} 16-2(4^2-3^2) &= 16-2(16-9) \\ &= 16-2(7) \\ &= 16-14 \\ &= 2 \end{aligned}$$

103. $4 \cdot 5 - 2 \cdot 7 = 20 - 14 = 6$

104. $8 \cdot 7 - 3 \cdot 9 = 56 - 27 = 29$

105.
$$\begin{aligned} (6-4)^3\cdot[10^2\div(3+17)] &= (6-4)^3\cdot[10^2\div 20] \\ &= (6-4)^3\cdot[100\div 20] \\ &= 2^3\cdot 5 \\ &= 8\cdot 5 \\ &= 40 \end{aligned}$$

106.
$$\begin{aligned} (7-5)^3\cdot[9^2\div(2+7)] &= (7-5)^3\cdot[9^2\div 9] \\ &= (7-5)^3\cdot[81\div 9] \\ &= 2^3\cdot 9 \\ &= 8\cdot 9 \\ &= 72 \end{aligned}$$

107. $\frac{5\cdot 7-3\cdot 5}{2(11-3^2)} = \frac{35-15}{2(11-9)} = \frac{20}{2(2)} = \frac{20}{4} = 5$

108. $\frac{4\cdot 8-1\cdot 11}{3(9-2^3)} = \frac{32-11}{3(9-8)} = \frac{21}{3(1)} = \frac{21}{3} = 7$

109. Area = $(\text{side})^2 = (7 \text{ meters})^2 = 49$ square meters

110. Area = $(\text{side})^2 = (3 \text{ inches})^2 = 9$ square inches

111. $\frac{2x}{z} = \frac{2\cdot 5}{2} = \frac{10}{2} = 5$

112. $4x - 3 = 4 \cdot 5 - 3 = 20 - 3 = 17$

113. $\frac{x+7}{y} = \frac{5+7}{0}$ is undefined.

114. $\frac{y}{5x} = \frac{0}{5\cdot 5} = \frac{0}{25} = 0$

115. $x^3 - 2z = 5^3 - 2\cdot 2 = 125 - 2\cdot 2 = 125 - 4 = 121$

116. $\frac{7+x}{3z} = \frac{7+5}{3\cdot 2} = \frac{12}{6} = 2$

117. $(y+z)^2 = (0+2)^2 = 2^2 = 4$

118. $\frac{100}{x} + \frac{y}{3} = \frac{100}{5} + \frac{0}{3} = 20 + 0 = 20$

119. Five subtracted from a number is $x - 5$.

120. Seven more than a number is $x + 7$.

121. Ten divided by a number is $10 \div x$ or $\frac{10}{x}$.

122. The product of 5 and a number is $5x$.

Copyright © 2011 Pearson Education, Inc. Publishing as Prentice Hall.

123. Let n be 5.
$n+12=20-3$
$5+12 \stackrel{?}{=} 20-3$
$17=17$ True
Yes, 5 is a solution.

124. Let n be 23.
$n-8=10+6$
$23-8 \stackrel{?}{=} 10+6$
$15=16$ False
No, 23 is not a solution.

125. Let $n = 14$.
$30=3(n-3)$
$30 \stackrel{?}{=} 3(14-3)$
$30 \stackrel{?}{=} 3(11)$
$30=33$ False
No, 14 is not a solution.

126. Let n be 20.
$5(n-7)=65$
$5(20-7) \stackrel{?}{=} 65$
$5(13) \stackrel{?}{=} 65$
$65=65$ True
Yes, 20 is a solution.

127. $7n=77$
Let n be 6.
$7 \cdot 6 \stackrel{?}{=} 77$
$42=77$ False
Let n be 11.
$7 \cdot 11 \stackrel{?}{=} 77$
$77=77$ True
Let n be 20.
$7 \cdot 20 \stackrel{?}{=} 77$
$140=77$ False
11 is a solution.

128. $n - 25 = 150$
Let n be 125.
$125-25 \stackrel{?}{=} 150$
$100=150$ False
Let n be 145.
$145-25 \stackrel{?}{=} 150$
$120=150$ False
Let n be 175.
$175-25 \stackrel{?}{=} 150$
$150=150$ True
175 is a solution.

129. $5(n + 4) = 90$
Let n be 14.
$5(14+4) \stackrel{?}{=} 90$
$5(18) \stackrel{?}{=} 90$
$90=90$ True
Let n be 16.
$5(16+4) \stackrel{?}{=} 90$
$5(20) \stackrel{?}{=} 90$
$100=90$ False
Let n be 26.
$5(26+4) \stackrel{?}{=} 90$
$5(30) \stackrel{?}{=} 90$
$150=90$ False
14 is a solution.

130. $3n - 8 = 28$
Let n be 3.
$3(3)-8 \stackrel{?}{=} 28$
$9-8 \stackrel{?}{=} 28$
$1=28$ False
Let n be 7.
$3(7)-8 \stackrel{?}{=} 28$
$21-8 \stackrel{?}{=} 28$
$13=28$ False
Let n be 15.
$3(15)-8 \stackrel{?}{=} 28$
$45-8 \stackrel{?}{=} 28$
$37=28$ False
None are solutions.

131.
$$\begin{array}{r} 485 \\ -\ 68 \\ \hline 417 \end{array}$$

132.
$$\begin{array}{r} 729 \\ -\ 47 \\ \hline 682 \end{array}$$

133.
$$\begin{array}{r} 732 \\ \times\ \ 3 \\ \hline 2196 \end{array}$$

134.
$$\begin{array}{r} 629 \\ \times\ \ 4 \\ \hline 2516 \end{array}$$

135.
$$\begin{array}{r} {}^{2\,2} \\ 374 \\ 29 \\ +\ 698 \\ \hline 1101 \end{array}$$

Copyright © 2011 Pearson Education, Inc. Publishing as Prentice Hall.

136.
$$\begin{array}{r} {}^{2\,1} \\ 593 \\ 52 \\ +\ 766 \\ \hline 1411 \end{array}$$

137.
$$\begin{array}{r} 458 \text{ R } 8 \\ 13\overline{)\,5962} \\ -52 \\ \hline 76 \\ -65 \\ \hline 112 \\ -104 \\ \hline 8 \end{array}$$

138.
$$\begin{array}{r} 237 \text{ R } 1 \\ 18\overline{)\,4267} \\ -36 \\ \hline 66 \\ -54 \\ \hline 127 \\ -126 \\ \hline 1 \end{array}$$

139.
$$\begin{array}{r} 1968 \\ \times\ \ 36 \\ \hline 11\,808 \\ 59\,040 \\ \hline 70,848 \end{array}$$

140.
$$\begin{array}{r} 5324 \\ \times\ \ 18 \\ \hline 42\,592 \\ 53\,240 \\ \hline 95,832 \end{array}$$

141.
$$\begin{array}{r} 2000 \\ -\ \ 356 \\ \hline 1644 \end{array}$$

142.
$$\begin{array}{r} 9000 \\ -\ \ 519 \\ \hline 8481 \end{array}$$

143. To round 842 to the nearest ten, observe that the digit in the ones place is 2. Since this digit is less than 5, we do not add 1 to the digit in the tens place. The number 842 rounded to the nearest ten is 840.

144. To round 258,371 to the nearest hundred-thousand, observe that the digit in the ten-thousands place is 5. Since this digit is at least 5, we add 1 to the digit in the hundred-thousands place. The number 258,371 rounded to the nearest hundred-thousand is 300,000.

145. $24 \div 4 \cdot 2 = 6 \cdot 2 = 12$

146. $\dfrac{(15+3)\cdot(8-5)}{2^3+1} = \dfrac{(18)(3)}{8+1} = \dfrac{54}{9} = 6$

147. Let n be 9.

$$\begin{aligned} 5n-6 &= 40 \\ 5\cdot 9-6 &\stackrel{?}{=} 40 \\ 45-6 &\stackrel{?}{=} 40 \\ 39 &= 40 \quad \text{False} \end{aligned}$$

No, 9 is not a solution.

148. Let n be 3.

$$\begin{aligned} 2n-6 &= 5n-15 \\ 2(3)-6 &\stackrel{?}{=} 5(3)-15 \\ 6-6 &\stackrel{?}{=} 15-15 \\ 0 &= 0 \quad \text{True} \end{aligned}$$

Yes, 3 is a solution.

149.
$$\begin{array}{r} 53 \\ 32\overline{)\,1714} \\ -160 \\ \hline 114 \\ -96 \\ \hline 18 \end{array}$$

There are 53 full boxes with 18 left over.

150.
$$\begin{array}{r} 27 \\ \times\ 2 \\ \hline 54 \end{array} \qquad \begin{array}{r} 8 \\ \times 4 \\ \hline 32 \end{array}$$

$$\begin{array}{r} 54 \\ +\ 32 \\ \hline 86 \end{array}$$

The total bill before taxes is $86.

Chapter 1 Test

1. 82,426 in words is eighty-two thousand, four hundred twenty-six.

2. Four hundred two thousand, five hundred fifty in standard form is 402,550.

Copyright © 2011 Pearson Education, Inc. Publishing as Prentice Hall.

3. $\begin{array}{r} {}^{1}\\ 59\\ +\ 82\\ \hline 141 \end{array}$

4. $\begin{array}{r} 600\\ -\ 487\\ \hline 113 \end{array}$

5. $\begin{array}{r} 496\\ \times\ \ 30\\ \hline 14{,}880 \end{array}$

6. $\begin{array}{r} 766 \text{ R } 42\\ 69\overline{)\ 52{,}896}\\ -48\ 3\\ \hline 4\ 59\\ -4\ 14\\ \hline 456\\ -414\\ \hline 42 \end{array}$

7. $2^3 \cdot 5^2 = 2 \cdot 2 \cdot 2 \cdot 5 \cdot 5 = 200$

8. $98 \div 1 = 98$

9. $0 \div 49 = 0$

10. $62 \div 0$ is undefined.

11. $(2^4 - 5) \cdot 3 = (16 - 5) \cdot 3 = 11 \cdot 3 = 33$

12. $\begin{aligned} 16 + 9 \div 3 \cdot 4 - 7 &= 16 + 3 \cdot 4 - 7\\ &= 16 + 12 - 7\\ &= 28 - 7\\ &= 21 \end{aligned}$

13. $6^1 \cdot 2^3 = 6 \cdot 2 \cdot 2 \cdot 2 = 48$

14. $\begin{aligned} 2[(6-4)^2 + (22-19)^2] + 10 &= 2[2^2 + 3^2] + 10\\ &= 2[4+9] + 10\\ &= 2[13] + 10\\ &= 26 + 10\\ &= 36 \end{aligned}$

15. $5698 \cdot 1000 = 5{,}698{,}000$

16. Divide the sum by 5.

$\begin{array}{r} {}^{2}\\ 62\\ 79\\ 84\\ 90\\ +\ 95\\ \hline 410 \end{array}$ $\qquad$ $\begin{array}{r} 82\\ 5\overline{)\ 410}\\ -40\\ \hline 10\\ -10\\ \hline 0 \end{array}$

The average is 82.

17. To round 52,369 to the nearest thousand, observe that the digit in the hundreds place is 3. Since this digit is less than 5, we do not add 1 to the digit in the thousands place. The number 52,369 rounded to the nearest thousand is 52,000.

18. $\begin{array}{rcr} 6289 & \text{rounds to} & 6\ 300\\ 5403 & \text{rounds to} & 5\ 400\\ +\ 1957 & \text{rounds to} & +\ 2\ 000\\ \hline & & 13{,}700 \end{array}$

19. $\begin{array}{rcr} 4267 & \text{rounds to} & 4300\\ -\ 2738 & \text{rounds to} & -\ 2700\\ \hline & & 1600 \end{array}$

20. $\begin{array}{r} 107\\ -\ \ 15\\ \hline 92 \end{array}$

21. $\begin{array}{r} 15\\ +\ 107\\ \hline 122 \end{array}$

22. $\begin{array}{r} 107\\ \times\ 15\\ \hline 535\\ 1070\\ \hline 1605 \end{array}$

23. $\begin{array}{r} 7 \text{ R } 2\\ 15\overline{)\ 107}\\ -105\\ \hline 2 \end{array}$

Copyright © 2011 Pearson Education, Inc. Publishing as Prentice Hall.

24. $\begin{array}{r} 17 \\ 29\overline{)\ 493} \\ \underline{-29} \\ 203 \\ \underline{-203} \\ 0 \end{array}$

Each can cost $17.

25. $\begin{array}{r} 725 \\ \underline{-\ 599} \\ 126 \end{array}$

The higher-priced one is $126 more.

26. $\begin{array}{r} 45 \\ \underline{\times\ 8} \\ 360 \end{array}$

There are 360 calories in 8 tablespoons of white granulated sugar.

27. $\begin{array}{r} 430 \\ \underline{\times\ 16} \\ 2580 \\ \underline{4300} \\ 6880 \end{array}$ $\begin{array}{r} 205 \\ \underline{\times\ \ 5} \\ 1025 \end{array}$

$\begin{array}{r} 6880 \\ \underline{+\ 1025} \\ 7905 \end{array}$

The total cost is $7905.

28. Perimeter = (5 + 5 + 5 + 5) centimeters
= 20 centimeters

Area = $(\text{side})^2$
= $(5 \text{ centimeters})^2$
= 25 square centimeters

29. Perimeter = (20 + 10 + 20 + 10) yards = 60 yards

Area = (length)(width)
= (20 yards)(10 yards)
= 200 square yards

30. Let x be 2.

$5(x^3 - 2) = 5(2^3 - 2) = 5(8 - 2) = 5(6) = 30$

31. Let x be 7 and y be 8.

$\frac{3x-5}{2y} = \frac{3(7)-5}{2 \cdot 8} = \frac{21-5}{16} = \frac{16}{16} = 1$

32. a. The quotient of a number and 17 is $x \div 17$ or $\frac{x}{17}$.

b. Twice a number, decreased by 20 is $2x - 20$.

33. Let n be 6.

$5n - 11 = 19$
$5(6) - 11 \stackrel{?}{=} 19$
$30 - 11 \stackrel{?}{=} 19$
$19 = 19$ True

6 is a solution.

34. $n + 20 = 4n - 10$

Let n be 0.

$0 + 20 \stackrel{?}{=} 4 \cdot 0 - 10$
$20 \stackrel{?}{=} 0 - 10$
$20 = -10$ False

Let n be 10.

$10 + 20 \stackrel{?}{=} 4 \cdot 10 - 10$
$30 \stackrel{?}{=} 40 - 10$
$30 = 30$ True

Let n be 20.

$20 + 20 \stackrel{?}{=} 4 \cdot 20 - 10$
$40 \stackrel{?}{=} 80 - 10$
$40 = 70$ False

10 is a solution.

Copyright © 2011 Pearson Education, Inc. Publishing as Prentice Hall.

Chapter 2

Section 2.1

Practice Problems

1. a. If 0 represents the surface of the earth, then 3805 below the surface of the earth is -3805.

   b. If zero degrees Fahrenheit is represented by 0°F, then 85 degrees below zero, Fahrenheit is represented by -85°F.

2. −5 −4 −3 −2 −1 0 1 2 3 4 5

3. a. $0 > -5$ since 0 is to the right of -5 on a number line.

   b. $-3 < 3$ since -3 is to the left of 3 on a number line.

   c. $-7 > -12$ since -7 is to the right of -12 on a number line.

4. a. $|-6| = 6$ because -6 is 6 units from 0.

   b. $|4| = 4$ because 4 is 4 units from 0.

   c. $|-12| = 12$ because -12 is 12 units from 0.

5. a. The opposite of 14 is -14.

   b. The opposite of -9 is $-(-9)$ or 9.

6. a. $-|-7| = -7$

   b. $-|4| = -4$

   c. $-(-12) = 12$

7. $-|x| = -|-6| = -6$

8. The planet with the highest average temperature is the one that corresponds to the bar that extends the furthest in the positive direction (upward). Venus has the highest average temperature.

Vocabulary and Readiness Check

1. The numbers ...−3, −2, −1, 0, 1, 2, 3, ... are called integers.

2. Positive numbers, negative numbers, and zero, together are called signed numbers.

3. The symbols "<" and ">" are called inequality symbols.

4. Numbers greater than 0 are called positive numbers while numbers less than 0 are called negative numbers.

5. The sign "<" means is less than and ">" means is greater than.

6. On a number line, the greater number is to the right of the lesser number.

7. A number's distance from 0 on the number line is the number's absolute value.

8. The numbers -5 and 5 are called opposites.

Exercise Set 2.1

1. If 0 represents ground level, then 1235 feet underground is -1235.

3. If 0 represents sea level, then 14,433 feet above sea level is +14,433.

5. If 0 represents zero degrees Fahrenheit, then 118 degrees above zero is +118.

7. If 0 represents the surface of the ocean, then 13,000 feet below the surface of the ocean is $-13,000$.

9. If 0 represents a loss of \$0, then a loss of \$2723 million is -2723 million.

11. If 0 represents the surface of the ocean, then 160 feet below the surface is -160 and 147 feet below the surface is -147. Since -160 extends further in the negative direction, Guillermo is deeper.

13. If 0 represents a decrease of 0%, then a 13 percent decrease is -13.

Copyright © 2011 Pearson Education, Inc. Publishing as Prentice Hall.

15. −7 −6 −5 −4 −3 −2 −1 0 1 2 3 4 5 6 7 (points at 0, 3, 4, 6)

17. −7 −6 −5 −4 −3 −2 −1 0 1 2 3 4 5 6 7 (points at −4, −2, −1, 1, 2)

19. 1 9
−14 −12 −10 −8 −6 −4 −2 0 2 4 6 8 10 12 14 (points at 0, 1, 9, 14)

21. −7 −6 −5 −4 −3 −2 −1 0 1 2 3 4 5 6 7 (points at −7, −5, −2, 0)

23. $0 > -7$ since 0 is to the right of −7 on a number line.

25. $-7 < -5$ since −7 is to the left of −5 on a number line.

27. $-30 > -35$ since −30 is to the right of −35 on a number line.

29. $-26 < 26$ since −26 is to the left of 26 on a number line.

31. $|5| = 5$ since 5 is 5 units from 0 on a number line.

33. $|-8| = 8$ since −8 is 8 units from 0 on a number line.

35. $|0| = 0$ since 0 is 0 units from 0 on a number line.

37. $|-55| = 55$ since −55 is 55 units from 0 on a number line.

39. The opposite of 5 is negative 5.
$-(5) = -5$

41. The opposite of negative 4 is 4.
$-(-4) = 4$

43. The opposite of 23 is negative 23.
$-(23) = -23$

45. The opposite of negative 85 is 85.
$-(-85) = 85$

47. $|-7| = 7$

49. $-|20| = -20$

51. $-|-3| = -3$

53. $-(-43) = 43$

55. $|-15| = 15$

57. $-(-33) = 33$

59. $|-x| = |-(-6)| = |6| = 6$

61. $-|-x| = -|-2| = -2$

63. $|x| = |-32| = 32$

Copyright © 2011 Pearson Education, Inc. Publishing as Prentice Hall.

65. $-|x| = -|7| = -7$

67. $-12 < -6$ since -12 is to the left of -6 on a number line.

69. $|-8| = 8$
$|-11| = 11$
Since $8 < 11$, $|-8| < |-11|$.

71. $|-47| = 47$
$-(-47) = 47$
Since $47 = 47$, $|-47| = -(-47)$.

73. $-|-12| = -12$
$-(-12) = 12$
Since $-12 < 12$, $-|-12| < -(-12)$.

75. $0 > -9$ since 0 is to the right of -9 on a number line.

77. $|0| = 0$
$|-9| = 9$
Since $0 < 9$, $|0| < |-9|$.

79. $-|-2| = -2$
$-|-10| = -10$
Since $-2 > -10$, $-|-2| > -|-10|$.

81. $-(-12) = 12$
$-(-18) = 18$
Since $12 < 18$, $-(-12) < -(-18)$.

83. If the number is 31, then the absolute value of 31 is 31 and the opposite of 31 is -31.

85. If the opposite of a number is -28, then the number is 28, and its absolute value is 28.

87. The bar that extends the farthest in the negative direction corresponds to the Caspian Sea, so the Caspian Sea has the lowest elevation.

89. The tallest bar on the graph corresponds to Lake Superior, so Lake Superior has the highest elevation.

91. The positive number on the graph closest to 100°C is 184°C, which corresponds to iodine.

93. The number on the graph closest to -200°C is -186°C, which corresponds to oxygen.

95. $0 + 13 = 13$

97.
$$\begin{array}{r} 15 \\ +\ 20 \\ \hline 35 \end{array}$$

99.
$$\begin{array}{r} {\scriptstyle 1\,2} \\ 47 \\ 236 \\ +\ 77 \\ \hline 360 \end{array}$$

101. $2^2 = 4$, $-|3| = -3$, $-(-5) = 5$, and $-|-8| = -8$, so the numbers in order from least to greatest are $-|-8|, -|3|, 2^2, -(-5)$.

103. $|-1| = 1$, $-|-6| = -6$, $-(-6) = 6$, and $-|1| = -1$, so the numbers in order from least to greatest are $-|-6|, -|1|, |-1|, -(-6)$.

105. $-(-2) = 2$, $5^2 = 25$, $-10 = -10$, $-|-9| = -9$, and $|-12| = 12$, so the numbers in order from least to greatest are $-10, -|-9|, -(-2), |-12|, 5^2$.

107. **a.** $|-9| = 9$; since $9 > 8$, then $|-9| > 8$ is true.

b. $|-5| = 5$; since $5 < 8$, then $|-5| > 8$ is false.

c. $|8| = 8$; since $8 = 8$, then $|8| > 8$ is false.

d. $|-12| = 12$; since $12 > 8$, then $|-12| > 8$ is true.

109. $-(-|-8|) = -(-8) = 8$

111. False; consider $a = -2$ and $b = -3$, then $-2 > -3$.

113. True; a positive number will always be to the right of a negative number on a number line.

115. False; consider $a = -5$, then the opposite of -5 is 5, which is a positive number.

117. answers may vary

119. no; answers may vary

Section 2.2

Practice Problems

1.

Start End
5
−1
−3 −2 −1 0 1 2 3 ④ 5 6 7
$5 + (-1) = 4$

Copyright © 2011 Pearson Education, Inc. Publishing as Prentice Hall.

2. 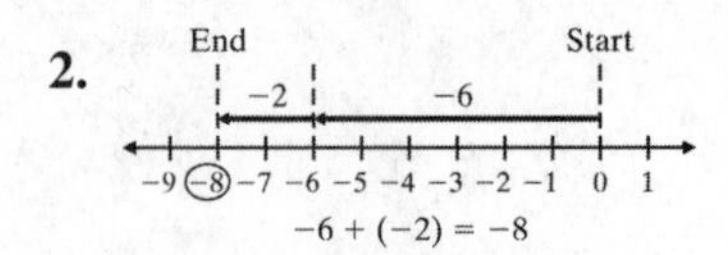

3. End Start

−8

+3

−9 −8 −7 −6 −5 −4 −3 −2 −1 0 1

$-8 + 3 = -5$

4. $|-3| + |-19| = 3 + 19 = 22$
The common sign is negative, so $(-3) + (-19) = -22$.

5. $-12 + (-30) = -42$

6. $9 + 4 = 13$

7. $|-1| = 1$, $|26| = 26$, and $26 - 1 = 25$
$26 > 1$, so the answer is positive.
$-1 + 26 = 25$

8. $|2| = 2$, $|-18| = 18$, and $18 - 2 = 16$
$18 > 2$, so the answer is negative.
$2 + (-18) = -16$

9. $-54 + 20 = -34$

10. $7 + (-2) = 5$

11. $-3 + 0 = -3$

12. $18 + (-18) = 0$

13. $-64 + 64 = 0$

14. $6 + (-2) + (-15) = 4 + (-15) = -11$

15. $$\begin{aligned} 5+(-3)+12+(-14) &= 2+12+(-14) \\ &= 14+(-14) \\ &= 0 \end{aligned}$$

16. $x + 3y = -6 + 3(2) = -6 + 6 = 0$

17. $x + y = -13 + (-9) = -22$

18. $$\begin{aligned} \text{Temperature at 8 a.m.} &= -7+(+4)+(+7) \\ &= -3+(+7) \\ &= 4 \end{aligned}$$
The temperature was 4°F at 8 a.m.

Calculator Explorations

1. $-256 + 97 = -159$

2. $811 + (-1058) = -247$

Copyright © 2011 Pearson Education, Inc. Publishing as Prentice Hall.

3. $6(15) + (-46) = 44$

4. $-129 + 10(48) = 351$

5. $-108{,}650 + (-786{,}205) = -894{,}855$

6. $-196{,}662 + (-129{,}856) = -326{,}518$

Vocabulary and Readiness Check

1. If n is a number, then $-n + n = \underline{0}$.

2. Since $x + n = n + x$, we say that addition is <u>commutative</u>.

3. If a is a number, then $-(-a) = \underline{a}$.

4. Since $n + (x + a) = (n + x) + a$, we say that addition is <u>associative</u>.

Exercise Set 2.2

1.

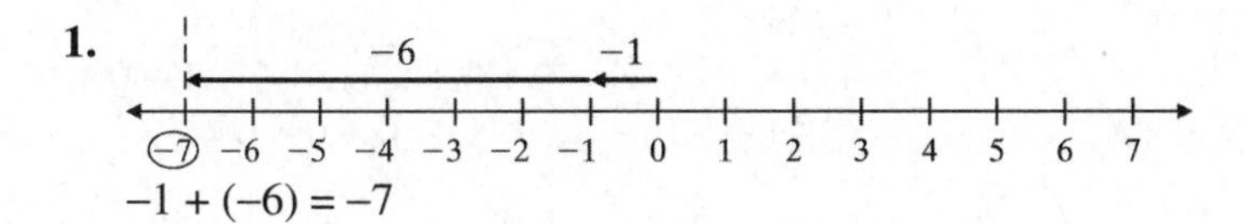

$-1 + (-6) = -7$

3.

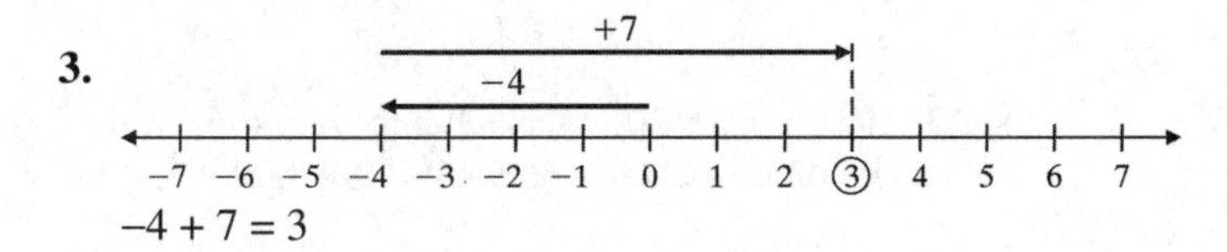

$-4 + 7 = 3$

5.

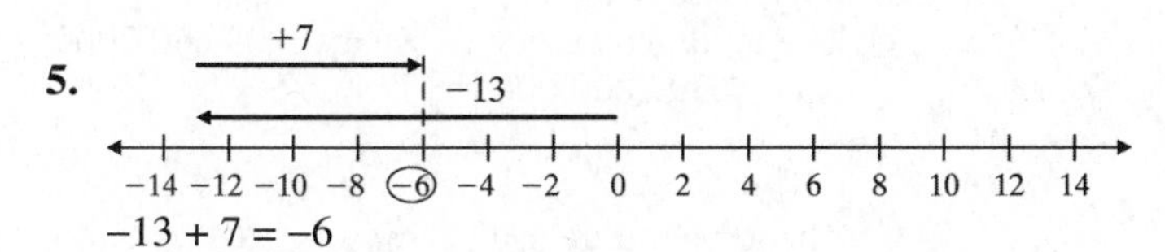

$-13 + 7 = -6$

7. $46 + 21 = 67$

9. $|-8| + |-2| = 8 + 2 = 10$
The common sign is negative, so $(-8) + (-2) = -10$.

11. $-43 + 43 = 0$

13. $|6| - |-2| = 6 - 2 = 4$
$6 > 2$, so the answer is positive.
$6 + (-2) = 4$

15. $-6 + 0 = -6$

17. $|-5| - |3| = 5 - 3 = 2$
$5 > 3$, so the answer is negative.
$3 + (-5) = -2$

19. $|-2| + |-7| = 2 + 7 = 9$
The common sign is negative, so $-2 + (-7) = -9$.

Copyright © 2011 Pearson Education, Inc. Publishing as Prentice Hall.

21. $|-12| + |-12| = 12 + 12 = 24$
The common sign is negative, so
$-12 + (-12) = -24.$

23. $|-640| + |-200| = 640 + 200 = 840$
The common sign is negative, so
$-640 + (-200) = -840.$

25. $|12| - |-5| = 12 - 5 = 7$
$12 > 5$, so the answer is positive.
$12 + (-5) = 7$

27. $|-6| - |3| = 6 - 3 = 3$
$6 > 3$, so the answer is negative.
$-6 + 3 = -3$

29. $|-56| - |26| = 56 - 26 = 30$
$56 > 26$, so the answer is negative.
$-56 + 26 = -30$

31. $|85| - |-45| = 85 - 45 = 40$
$85 > 45$, so the answer is positive.
$-45 + 85 = 40$

33. $|-144| - |124| = 144 - 124 = 20$
$144 > 124$, so the answer is negative.
$124 + (-144) = -20$

35. $|-82| + |-43| = 82 + 43 = 125$
The common sign is negative, so
$-82 + (-43) = -125.$

37. $-4 + 2 + (-5) = -2 + (-5) = -7$

39. $-52 + (-77) + (-117) = -129 + (-117) = -246$

41.
$$\begin{aligned} 12+(-4)+(-4)+12 &= 8+(-4)+12 \\ &= 4+12 \\ &= 16 \end{aligned}$$

43.
$$\begin{aligned} (-10)+14+25+(-16) &= 4+25+(-16) \\ &= 29+(-16) \\ &= 13 \end{aligned}$$

45. $-6 + (-15) + (-7) = -21 + (-7) = -28$

47. $-26 + 15 = -11$

49. $5 + (-2) + 17 = 3 + 17 = 20$

51. $-13 + (-21) = -34$

53. $3 + 14 + (-18) = 17 + (-18) = -1$

55. $-92 + 92 = 0$

57.
$$\begin{aligned} -13+8+(-10)+(-27) &= -5+(-10)+(-27) \\ &= -15+(-27) \\ &= -42 \end{aligned}$$

59. $x + y = -20 + (-50) = -70$

61. $3x + y = 3(2) + (-3) = 6 + (-3) = 3$

63. $3x + y = 3(3) + (-30) = 9 + (-30) = -21$

65. The sum of -6 and 25 is $-6 + 25 = 19$.

67. The sum of -31, -9, and 30 is
$-31 + (-9) + 30 = -40 + 30 = -10.$

69. $0 + (-215) + (-16) = -215 + (-16) = -231$
The diver's final depth is 231 feet below the surface.

71. Stanford:
$$\begin{aligned} &0 + 0 + 0 + (-1) + 0 + 0 + 0 + 1 + 0 + 1 + 0 + 0 \\ &\quad + (-1) + (-1) + (-1) + 0 + 0 + 0 = -2 \end{aligned}$$
Wie:
$$\begin{aligned} &0 + 0 + 0 + (-1) + 0 + 0 + 0 + 0 + (-1) + 0 + 2 \\ &\quad + 0 + 0 + 0 + 0 + 0 + 1 + 0 = +1 \end{aligned}$$

73. The bar for 2007 has a height of 3496, so the net income in 2007 was $3,496,000,000.

75. $69 + 1328 = 1397$
The total net income for years 2003 and 2005 was $1,397,000,000.

77. $-10 + 12 = 2$
The temperature at 11 p.m. was 2°C.

79. $-1786 + 15,395 = 13,609$
The sum of the net incomes for 2006 and 2007 is $13,609.

81. $-55 + 8 = -47$
West Virginia's record low temperature is -47°F.

83. $-10,924 + 3245 = -7679$
The depth of the Aleutian Trench is -7679 meters.

85. $44 - 0 = 44$

87.
$$\begin{array}{r} 200 \\ -\ 59 \\ \hline 141 \end{array}$$

Copyright © 2011 Pearson Education, Inc. Publishing as Prentice Hall.

89. answers may vary

91. $7 + (-10) = -3$

93. $-10 + (-12) = -22$

95. True

97. False; for example, $4 + (-2) = 2 > 0$.

99. answers may vary

Section 2.3

Practice Problems

1. $13 - 4 = 13 + (-4) = 9$

2. $-8 - 2 = -8 + (-2) = -10$

3. $11 - (-15) = 11 + 15 = 26$

4. $-9 - (-1) = -9 + 1 = -8$

5. $6 - 9 = 6 + (-9) = -3$

6. $-14 - 5 = -14 + (-5) = -19$

7. $-3 - (-4) = -3 + 4 = 1$

8. $-15 - 6 = -15 + (-6) = -21$

9. $$\begin{aligned} -6-5-2-(-3) &= -6+(-5)+(-2)+3 \\ &= -11+(-2)+3 \\ &= -13+3 \\ &= -10 \end{aligned}$$

10. $$\begin{aligned} 8+(-2)-9-(-7) &= 8+(-2)+(-9)+7 \\ &= 6+(-9)+7 \\ &= -3+7 \\ &= 4 \end{aligned}$$

11. $x - y = -5 - 13 = -5 + (-13) = -18$

12. $3y - z = 3(9) - (-4) = 27 + 4 = 31$

13. $29,028 - (-1312) = 29,028 + 1312 = 30,340$
Mount Everest is 30,340 feet higher than the Dead Sea.

Vocabulary and Readiness Check

1. It is true that $a - b = \underline{a + (-b)}$. b

2. The opposite of n is $\underline{-n}$. a

3. To evaluate $x - y$ for $x = -10$ and $y = -14$, we replace x with -10 and y with -14 and evaluate $\underline{-10 - (-14)}$. d

4. The expression $-5 - 10$ equals $\underline{-5 + (-10)}$. c

Exercise Set 2.3

1. $-8 - (-8) = -8 + 8 = 0$

3. $19 - 16 = 19 + (-16) = 3$

5. $3 - 8 = 3 + (-8) = -5$

7. $11 - (-11) = 11 + 11 = 22$

9. $-4 - (-7) = -4 + 7 = 3$

11. $-16 - 4 = -16 + (-4) = -20$

13. $3 - 15 = 3 + (-15) = -12$

15. $42 - 55 = 42 + (-55) = -13$

17. $478 - (-30) = 478 + 30 = 508$

19. $-4 - 10 = -4 + (-10) = -14$

21. $-7 - (-3) = -7 + 3 = -4$

23. $17 - 29 = 17 + (-29) = -12$

25. $-25 - 17 = -25 + (-17) = -42$

27. $-22 - (-3) = -22 + 3 = -19$

29. $2 - (-12) = 2 + 12 = 14$

31. $-37 + (-19) = -56$

33. $8 - 13 = 8 + (-13) = -5$

35. $-56 - 89 = -56 + (-89) = -145$

37. $30 - 67 = 30 + (-67) = -37$

39. $8 - 3 - 2 = 8 + (-3) + (-2) = 5 + (-2) = 3$

41. $13-5-7 = 13+(-5)+(-7) = 8+(-7) = 1$

43. $-5 - 8 - (-12) = -5 + (-8) + 12 = -13 + 12 = -1$

45. $$\begin{aligned} -11+(-6)-14 &= -11+(-6)+(-14) \\ &= -17+(-14) \\ &= -31 \end{aligned}$$

Copyright © 2011 Pearson Education, Inc. Publishing as Prentice Hall.

47. $18-(-32)+(-6) = 18+32+(-6)$
$= 50+(-6)$
$= 44$

49. $-(-5)-21+(-16) = 5+(-21)+(-16)$
$= -16+(-16)$
$= -32$

51. $-10-(-12)+(-7)-4 = -10+12+(-7)+(-4)$
$= 2+(-7)+(-4)$
$= -5+(-4)$
$= -9$

53. $-3+4-(-23)-10 = -3+4+23+(-10)$
$= 1+23+(-10)$
$= 24+(-10)$
$= 14$

55. $x - y = -4 - 7 = -4 + (-7) = -11$

57. $x - y = 8 - (-23) = 8 + 23 = 31$

59. $2x - y = 2(4) - (-4) = 8 + 4 = 12$

61. $2x - y = 2(1) - (-18) = 2 + 18 = 20$

63. The temperature in March is 11°F and in February is −4°F.
$11 - (-4) = 11 + 4 = 15$
The difference is 15°F.

65. The two months with the lowest temperatures are January, −10°F, and December, −6°F.
$-6 - (-10) = -6 + 10 = 4$
The difference is 4°F.

67. $136 - (-129) = 136 + 129 = 265$
Therefore, 136°F is 265°F warmer than −129°F.

69. $14 - (-8) = 14 + 8 = 22$
There was a difference of 22 strokes.

71. $-4-3+4-7 = -4+(-3)+4+(-7)$
$= -7+4+(-7)$
$= -3+(-7)$
$= -10$
The temperature at 9 a.m. is −10°C.

73. $-282 - (-436) = -282 + 436 = 154$
The difference in elevation is 154 feet.

75. $-436 - (-505) = -436 + 505 = 69$
The difference in elevation is 69 feet.

77. $600 - (-52) = 600 + 52 = 652$
The difference in elevation is 652 feet.

79. $144 - 0 = 144$
The difference in elevation is 144 feet.

81. $867 - (-330) = 867 + 330 = 1197$
The difference in temperatures is 1197°F.

83. $1646 - 2346 = 1646 + (-2346) = -700$
The trade balance was −$700 billion.

85. The sum of −5 and a number is $-5 + x$.

87. Subtract a number from −20 is $-20 - x$.

89. $\frac{100}{20} = 5$

91.
$$\begin{array}{r} 23 \\ \times\ 46 \\ \hline 138 \\ 920 \\ \hline 1058 \end{array}$$

93. answers may vary

95. $9 - (-7) = 9 + 7 = 16$

97. $10 - 30 = 10 + (-30) = -20$

99. $|-3| - |-7| = 3 - 7 = 3 + (-7) = -4$

101. $|-5| - |5| = 5 - 5 = 0$

103. $|-15| - |-29| = 15 - 29 = 15 + (-29) = -14$

105. $|-8 - 3| = |-8 + (-3)| = |-11| = 11$
$8 - 3 = 8 + (-3) = 5$
Since $11 \neq 5$, the statement is false.

107. answers may vary

Section 2.4

Practice Problems

1. $-3 \cdot 8 = -24$

2. $-5(-2) = 10$

3. $0 \cdot (-20) = 0$

4. $10(-5) = -50$

Copyright © 2011 Pearson Education, Inc. Publishing as Prentice Hall.

5. $8(-6)(-2) = -48(-2) = 96$

6. $(-9)(-2)(-1) = 18(-1) = -18$

7. $(-3)(-4)(-5)(-1) = 12(-5)(-1) = -60(-1) = 60$

8. $$\begin{aligned}(-2)^4 &= (-2)(-2)(-2)(-2)\\ &= 4(-2)(-2)\\ &= -8(-2)\\ &= 16\end{aligned}$$

9. $-8^2 = -(8 \cdot 8) = -64$

10. $\frac{42}{-7} = -6$

11. $-16 \div (-2) = 8$

12. $\frac{-80}{10} = -8$

13. $\frac{-6}{0}$ is undefined.

14. $\frac{0}{-7} = 0$

15. $xy = 5 \cdot (-8) = -40$

16. $\frac{x}{y} = \frac{-12}{-3} = 4$

17. total score $= 4 \cdot (-13) = -52$
The card player's total score was −52.

Vocabulary and Readiness Check

1. The product of a negative number and a positive number is a negative number.

2. The product of two negative numbers is a positive number.

3. The quotient of two negative numbers is a positive number.

4. The quotient of a negative number and a positive number is a negative number.

5. The product of a negative number and zero is 0.

6. The quotient of 0 and a negative number is 0.

7. The quotient of a negative number and 0 is undefined.

Exercise Set 2.4

1. $-6(-2) = 12$

3. $-4(9) = -36$

5. $9(-9) = -81$

7. $0(-11) = 0$

9. $6(-2)(-4) = -12(-4) = 48$

11. $-1(-3)(-4) = 3(-4) = -12$

13. $-4(4)(-5) = -16(-5) = 80$

15. $10(-5)(0)(-7) = 0$

17. $-5(3)(-1)(-1) = -15(-1)(-1) = 15(-1) = -15$

19. $-3^2 = -(3 \cdot 3) = -9$

21. $(-3)^3 = (-3)(-3)(-3) = 9(-3) = -27$

23. $-6^2 = -(6 \cdot 6) = -36$

25. $(-4)^3 = (-4)(-4)(-4) = 16(-4) = -64$

27. $-24 \div 3 = -8$

29. $\frac{-30}{6} = -5$

31. $\frac{-77}{-11} = 7$

33. $\frac{0}{-21} = 0$

35. $\frac{-10}{0}$ is undefined.

37. $\frac{56}{-4} = -14$

39. $-14(0) = 0$

41. $-5(3) = -15$

43. $-9 \cdot 7 = -63$

Copyright © 2011 Pearson Education, Inc. Publishing as Prentice Hall.

45. $-7(-6) = 42$

47. $-3(-4)(-2) = 12(-2) = -24$

49. $(-7)^2 = (-7)(-7) = 49$

51. $-\frac{25}{5} = -5$

53. $-\frac{72}{8} = -9$

55. $-18 \div 3 = -6$

57. $4(-10)(-3) = -40(-3) = 120$

59. $$\begin{aligned} -30(6)(-2)(-3) &= -180(-2)(-3) \\ &= 360(-3) \\ &= -1080 \end{aligned}$$

61. $\frac{-25}{0}$ is undefined.

63. $\frac{120}{-20} = -6$

65. $280 \div (-40) = \frac{280}{-40} = -7$

67. $\frac{-12}{-4} = 3$

69. $-1^4 = -(1 \cdot 1 \cdot 1 \cdot 1) = -1$

71. $$\begin{aligned} (-2)^5 &= (-2)(-2)(-2)(-2)(-2) \\ &= 4(-2)(-2)(-2) \\ &= -8(-2)(-2) \\ &= 16(-2) \\ &= -32 \end{aligned}$$

73. $-2(3)(5)(-6) = -6(5)(-6) = -30(-6) = 180$

75. $(-1)^{32} = 1$, since there are an even number of factors.

77. $-2(-3)(-5) = 6(-5) = -30$

79. $$\begin{array}{r} 48 \\ \times\ 23 \\ \hline 144 \\ 960 \\ \hline 1104 \end{array}$$
$-48 \cdot 23 = -1104$

81. $$\begin{array}{r} 35 \\ \times\ 82 \\ \hline 70 \\ 2800 \\ \hline 2870 \end{array}$$
$35 \cdot (-82) = -2870$

83. $ab = -8 \cdot 7 = -56$

85. $ab = 9(-2) = -18$

87. $ab = (-7)(-5) = 35$

89. $\frac{x}{y} = \frac{5}{-5} = -1$

91. $\frac{x}{y} = \frac{-15}{0}$ is undefined.

93. $\frac{x}{y} = \frac{-36}{-6} = 6$

95. $xy = -8 \cdot (-2) = 16$
$\frac{x}{y} = \frac{-8}{-2} = 4$

97. $xy = 0(-8) = 0$
$\frac{x}{y} = \frac{0}{-8} = 0$

99. $\frac{-54}{9} = -6$
The quotient of -54 and 9 is -6.

101. $$\begin{array}{r} 42 \\ \times\ 6 \\ \hline 252 \end{array}$$
$-42(-6) = 252$
The product of -42 and -6 is 252.

103. The product of -71 and a number is $-71 \cdot x$ or $-71x$.

Copyright © 2011 Pearson Education, Inc. Publishing as Prentice Hall.

105. Subtract a number from −16 is $-16 - x$.

107. −29 increased by a number is $-29 + x$.

109. Divide a number by −33 is $\frac{x}{-33}$ or $x \div (-33)$.

111. A loss of 4 yards is represented by −4.
$3 \cdot (-4) = -12$
The team had a total loss of 12 yards.

113. Each move of 20 feet down is represented by −20.
$5 \cdot (-20) = -100$
The diver is at a depth of 100 feet.

115. $3 \cdot (-70) = -210$
The melting point of nitrogen is −210°C.

117. $-3 \cdot 63 = -189$
The melting point of argon is −189°C.

119. $3 \cdot (-2.1) = -6.3$
The reduction in sales will be −$6.3 million after three months.

121. a. $13{,}922 - 12{,}050 = 1872$
1872 fewer movie titles were released to DVD in 2007 than in 2005. This is a change of −1872 movies.

b. This is a period of 2 years.
$\frac{-1872}{2} = -936$
The average change was −936 movies per year.

123.
$$\begin{aligned} 90 + 12^2 - 5^3 &= 90 + 144 - 125 \\ &= 90 + 144 + (-125) \\ &= 234 + (-125) \\ &= 109 \end{aligned}$$

125. $12 \div 4 - 2 + 7 = 3 - 2 + 7 = 1 + 7 = 8$

127. $-57 \div 3 = -19$

129. $-8 - 20 = -8 + (-20) = -28$

131.
$$\begin{aligned} -4 - 15 - (-11) &= -4 + (-15) + 11 \\ &= -19 + 11 \\ &= -8 \end{aligned}$$

133. The product of an odd number of negative numbers is negative, so the product of seven negative numbers is negative.

135. $(-2)^{12}$ and $(-5)^{12}$ are positive since there are an even number of factors. Note that $(-5)^{12}$ will be the larger of the two since $|-5| > |-2|$ and the exponent on each is the same.
$(-2)^{17}$ and $(-5)^{17}$ are negative since there are an odd number of factors. Note that $\left|(-5)^{17}\right| > \left|(-2)^{17}\right|$ since $|-5| > |-2|$ and the exponent on each is the same. The numbers from least to greatest are $(-5)^{17}$, $(-2)^{17}$, $(-2)^{12}$, $(-5)^{12}$.

137. answers may vary

Integrated Review

1. Let 0 represent 0°F. Then 50 degrees below zero is represented by −50 and 122 degrees above zero is represented by +122 or 122.

2.
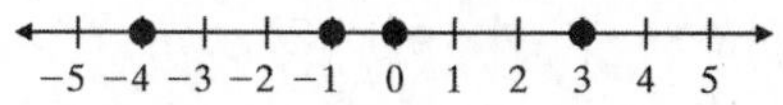

3. $0 > -10$ since 0 is to the right of −10 on a number line.

4. $-4 < 4$ since −4 is to the left of 4 on a number line.

5. $-15 < -5$ since −15 is to the left of −5 on a number line.

6. $-2 > -7$ since −2 is to the right of −7 on a number line.

7. $|-3| = 3$ because −3 is 3 units from 0.

8. $|-9| = 9$ because −9 is 9 units from 0.

9. $-|-4| = -4$

10. $-(-5) = 5$

11. The opposite of 11 is −11.

12. The opposite of −3 is $-(-3) = 3$.

13. The opposite of 64 is −64.

14. The opposite of 0 is $-0 = 0$.

Copyright © 2011 Pearson Education, Inc. Publishing as Prentice Hall.

15. $-3 + 15 = 12$

16. $-9 + (-11) = -20$

17. $-8(-6)(-1) = 48(-1) = -48$

18. $-18 \div 2 = -9$

19. $65 + (-55) = 10$

20. $1000 - 1002 = 1000 + (-1002) = -2$

21. $53 - (-53) = 53 + 53 = 106$

22. $-2 - 1 = -2 + (-1) = -3$

23. $\frac{0}{-47} = 0$

24. $\frac{-36}{-9} = 4$

25. $-17 - (-59) = -17 + 59 = 42$

26. $-8 + (-6) + 20 = -14 + 20 = 6$

27. $\frac{-95}{-5} = 19$

28. $-9(100) = -900$

29. $-12 - 6 - (-6) = -12 + (-6) + 6 = -18 + 6 = -12$

30. $$\begin{aligned} -4+(-8)-16-(-9) &= -4+(-8)+(-16)+9 \\ &= -12+(-16)+9 \\ &= -28+9 \\ &= -19 \end{aligned}$$

31. $\frac{-105}{0}$ is undefined.

32. $7(-16)(0)(-3) = 0$ (since one factor is 0)

33. Subtract -8 from -12 is $-12 - (-8) = -12 + 8 = -4$.

34. The sum of -17 and -27 is $-17 + (-27) = -44$.

35. The product of -5 and -25 is $-5(-25) = 125$.

36. The quotient of -100 and -5 is $\frac{-100}{-5} = 20$.

37. Divide a number by -17 is $\frac{x}{-17}$ or $x \div (-17)$.

38. The sum of -3 and a number is $-3 + x$.

39. A number decreased by -18 is $x - (-18)$.

40. The product of -7 and a number is $-7 \cdot x$ or $-7x$.

41. $x + y = -3 + 12 = 9$

42. $x - y = -3 - 12 = -3 + (-12) = -15$

43. $2y - x = 2(12) - (-3) = 24 - (-3) = 24 + 3 = 27$

44. $3y + x = 3(12) + (-3) = 36 + (-3) = 33$

45. $5x = 5(-3) = -15$

46. $\frac{y}{x} = \frac{12}{-3} = -4$

Section 2.5

Practice Problems

1. $(-2)^4 = (-2)(-2)(-2)(-2) = 16$

2. $-2^4 = -(2)(2)(2)(2) = -16$

3. $3 \cdot 6^2 = 3 \cdot (6 \cdot 6) = 3 \cdot 36 = 108$

4. $\frac{-25}{5(-1)} = \frac{-25}{-5} = 5$

5. $\frac{-18+6}{-3-1} = \frac{-12}{-4} = 3$

6. $$\begin{aligned} 30+50+(-4)^3 &= 30+50+(-64) \\ &= 80+(-64) \\ &= 16 \end{aligned}$$

7. $-2^3 + (-4)^2 + 1^5 = -8 + 16 + 1 = 8 + 1 = 9$

8. $$\begin{aligned} 2(2-9)+(-12)-3 &= 2(-7)+(-12)-3 \\ &= -14+(-12)-3 \\ &= -26-3 \\ &= -29 \end{aligned}$$

Copyright © 2011 Pearson Education, Inc. Publishing as Prentice Hall.

9. $(-5)\cdot|-8|+(-3)+2^3 = (-5)\cdot 8+(-3)+2^3$
$= (-5)\cdot 8+(-3)+8$
$= -40+(-3)+8$
$= -43+8$
$= -35$

10. $-4[-6+5(-3+5)]-7 = -4[-6+5(2)]-7$
$= -4[-6+10]-7$
$= -4(4)-7$
$= -16-7$
$= -23$

11. $x^2 = (-15)^2 = (-15)(-15) = 225$
$-x^2 = -(-15)^2 = -(-15)(-15) = -225$

12. $5y^2 = 5(4)^2 = 5(16) = 80$
$5y^2 = 5(-4)^2 = 5(16) = 80$

13. $x^2 + y = (-6)^2 + (-3) = 36 + (-3) = 33$

14. $4 - x^2 = 4 - (-8)^2 = 4 - 64 = -60$

15. average
$= \dfrac{\text{sum of numbers}}{\text{number of numbers}}$
$= \dfrac{15+(-1)+(-11)+(-14)+(-16)+(-14)+(-1)}{7}$
$= \dfrac{-42}{7}$
$= -6$
The average of the temperatures is −6°F.

Calculator Explorations

1. $\dfrac{-120-360}{-10} = 48$

2. $\dfrac{4750}{-2+(-17)} = -250$

3. $\dfrac{-316+(-458)}{28+(-25)} = -258$

4. $\dfrac{-234+86}{-18+16} = 74$

Vocabulary and Readiness Check

1. To simplify $-2 \div 2 \cdot (3)$ which operation should be performed first? division

2. To simplify $-9 - 3 \cdot 4$, which operation should be performed first? multiplication

3. The average of a list of numbers is $\dfrac{\text{sum of numbers}}{\textit{number} \text{ of numbers}}$.

4. To simplify $5[-9 + (-3)] \div 4$, which operation should be performed first? addition

5. To simplify $-2 + 3(10 - 12) \cdot (-8)$, which operation should be performed first? subtraction

6. To evaluate $x - 3y$ for $x = -7$ and $y = -1$, replace x with −7 and y with −1 and evaluate $-7 - 3(-1)$.

Exercise Set 2.5

1. $(-5)^3 = (-5)(-5)(-5) = -125$

3. $-4^3 = -(4)(4)(4) = -64$

5. $8 \cdot 2^2 = 8 \cdot 4 = 32$

7. $8 - 12 - 4 = -4 - 4 = -8$

9. $7 + 3(-6) = 7 + (-18) = -11$

11. $5(-9) + 2 = -45 + 2 = -43$

13. $-10 + 4 \div 2 = -10 + 2 = -8$

15. $6+7\cdot 3-10 = 6+21-10$
$= 27-10$
$= 17$

17. $\dfrac{16-13}{-3} = \dfrac{3}{-3} = -1$

19. $\dfrac{24}{10+(-4)} = \dfrac{24}{6} = 4$

21. $5(-3) - (-12) = -15 - (-12) = -15 + 12 = -3$

23. $[8+(-4)]^2 = [4]^2 = 16$

Copyright © 2011 Pearson Education, Inc. Publishing as Prentice Hall.

25. $8 \cdot 6 - 3 \cdot 5 + (-20) = 48 - 3 \cdot 5 + (-20)$
$= 48 - 15 + (-20)$
$= 33 + (-20)$
$= 13$

27. $4 - (-3)^4 = 4 - 81 = -77$

29. $|7+3| \cdot 2^3 = |10| \cdot 2^3 = 10 \cdot 2^3 = 10 \cdot 8 = 80$

31. $7 \cdot 6^2 + 4 = 7 \cdot 36 + 4 = 252 + 4 = 256$

33. $7^2 - (4 - 2^3) = 7^2 - (4 - 8)$
$= 7^2 - (-4)$
$= 49 - (-4)$
$= 49 + 4$
$= 53$

35. $|3 - 15| \div 3 = |-12| \div 3 = 12 \div 3 = 4$

37. $-(-2)^6 = -64$

39. $(5-9)^2 \div (4-2)^2 = (-4)^2 \div (2)^2 = 16 \div 4 = 4$

41. $|8 - 24| \cdot (-2) \div (-2) = |-16| \cdot (-2) \div (-2)$
$= 16 \cdot (-2) \div (-2)$
$= -32 \div (-2)$
$= 16$

43. $(-12 - 20) \div 16 - 25 = (-32) \div 16 - 25$
$= -2 - 25$
$= -27$

45. $5(5-2) + (-5)^2 - 6 = 5(3) + (-5)^2 - 6$
$= 5(3) + 25 - 6$
$= 15 + 25 - 6$
$= 40 - 6$
$= 34$

47. $(2-7) \cdot (6-19) = (-5) \cdot (-13)$
$= 65$

49. $(-36 \div 6) - (4 \div 4) = -6 - 1 = -7$

51. $(10 - 4^2)^2 = (10 - 16)^2 = (-6)^2 = 36$

53. $2(8-10)^2 - 5(1-6)^2 = 2(-2)^2 - 5(-5)^2$
$= 2(4) - 5(25)$
$= 8 - 125$
$= -117$

55. $3(-10) \div [5(-3) - 7(-2)] = 3(-10) \div [-15 + 14]$
$= 3(-10) \div (-1)$
$= -30 \div (-1)$
$= 30$

57. $\dfrac{(-7)(-3) - (4)(3)}{3[7 \div (3-10)]} = \dfrac{21 - 12}{3[7 \div (-7)]}$
$= \dfrac{9}{3(-1)}$
$= \dfrac{9}{-3}$
$= -3$

59. $-3[5 + 2(-4 + 9)] + 15 = -3[5 + 2(5)] + 15$
$= -3[5 + 10] + 15$
$= -3(15) + 15$
$= -45 + 15$
$= -30$

61. $x + y + z = -2 + 4 + (-1) = 2 + (-1) = 1$

63. $2x - 3y - 4z = 2(-2) - 3(4) - 4(-1)$
$= -4 - 12 + 4$
$= -16 + 4$
$= -12$

65. $x^2 - y = (-2)^2 - 4 = 4 - 4 = 0$

67. $\dfrac{5y}{z} = \dfrac{5(4)}{-1} = \dfrac{20}{-1} = -20$

69. $x^2 = (-3)^2 = 9$

71. $-z^2 = -(-4)^2 = -16$

73. $2z^3 = 2(-4)^3 = 2(-64) = -128$

75. $10 - x^2 = 10 - (-3)^2 = 10 - 9 = 1$

77. $2x^3 - z = 2(-3)^3 - (-4)$
$= 2(-27) - (-4)$
$= -54 - (-4)$
$= -54 + 4$
$= -50$

Copyright © 2011 Pearson Education, Inc. Publishing as Prentice Hall.

79. $\text{average} = \frac{-10+8+(-4)+2+7+(-5)+(-12)}{7} = \frac{-14}{7} = -2$

81. $\text{average} = \frac{-17+(-26)+(-20)+(-13)}{4} = \frac{-76}{4} = -19$

83. The lowest score is −11 and the highest is 8.
$8 - (-11) = 8 + 11 = 19$
The difference between the lowest score and the highest score is 19.

85. $\text{average} = \frac{-11+(-7)+3}{3} = \frac{-15}{3} = -5$
The average of the scores is −5.

87. no; answers may vary

89. $45 \times 90 = 4050$

91. $90 - 45 = 45$

93. $8 + 8 + 8 + 8 = 32$
The perimeter is 32 inches.

95. $9 + 6 + 9 + 6 = 30$
The perimeter is 30 feet.

97. $2 \cdot (7 - 5) \cdot 3 = 2 \cdot 2 \cdot 3 = 4 \cdot 3 = 12$

99. $-6 \cdot (10 - 4) = -6 \cdot 6 = -36$

101. answers may vary

103. answers may vary

105. $(-12)^4 = (-12)(-12)(-12)(-12) = 20,736$

107. $x^3 - y^2 = (21)^3 - (-19)^2 = 9261 - 361 = 8900$

109. $(xy+z)^x = [2(-5)+7]^2 = [-10+7]^2 = [-3]^2 = 9$

Section 2.6

Practice Problems

1.
$-4x - 3 = 5$
$-4(-2) - 3 \stackrel{?}{=} 5$
$8 - 3 \stackrel{?}{=} 5$
$5 = 5$ True
Since 5 = 5 is true, −2 is a solution of the equation.

2.
$y - 6 = -2$
$y - 6 + 6 = -2 + 6$
$y = 4$
Check: $y - 6 = -2$
$4 - 6 \stackrel{?}{=} -2$
$-2 = -2$ True
The solution is 4.

3.
$-2 = z + 8$
$-2 - 8 = z + 8 - 8$
$-10 = z$
Check: $-2 = z + 8$
$-2 \stackrel{?}{=} -10 + 8$
$-2 = -2$ True
The solution is −10.

4.
$10x = -2 + 9x$
$10x - 9x = -2 + 9x - 9x$
$x = -2$
Check: $10x = -2 + 9x$
$10(-2) \stackrel{?}{=} -2 + 9(-2)$
$-20 \stackrel{?}{=} -2 + (-18)$
$-20 = -20$ True
The solution is −2.

5.
$3y = -18$
$\frac{3y}{3} = \frac{-18}{3}$
$\frac{3}{3} \cdot y = \frac{-18}{3}$
$y = -6$
Check: $3y = -18$
$3(-6) \stackrel{?}{=} -18$
$-18 = -18$ True
The solution is −6.

Copyright © 2011 Pearson Education, Inc. Publishing as Prentice Hall.

6. $-32 = 8x$

$\frac{-32}{8} = \frac{8x}{8}$

$\frac{-32}{8} = \frac{8}{8} \cdot x$

$-4 = x$

Check: $-32 = 8x$

$-32 \stackrel{?}{=} 8(-4)$

$-32 = -32$ True

The solution is −4.

7. $-3y = -27$

$\frac{-3y}{-3} = \frac{-27}{-3}$

$\frac{-3}{-3} \cdot y = \frac{-27}{-3}$

$y = 9$

Check: $-3y = -27$

$-3 \cdot 9 \stackrel{?}{=} -27$

$-27 = -27$ True

The solution is 9.

8. $\frac{x}{-4} = 7$

$-4 \cdot \frac{x}{-4} = -4 \cdot 7$

$\frac{-4}{-4} \cdot x = -4 \cdot 7$

$x = -28$

Check: $\frac{x}{-4} = 7$

$\frac{-28}{-4} \stackrel{?}{=} 7$

$7 = 7$ True

The solution is −28.

Vocabulary and Readiness Check

1. A combination of operations on variables and numbers is called an <u>expression</u>.

2. A statement of the form "expression = expression" is called an <u>equation</u>.

3. An <u>equation</u> contains an equal sign (=) while an <u>expression</u> does not.

4. An <u>expression</u> may be simplified and evaluated while an <u>equation</u> may be solved.

5. A <u>solution</u> of an equation is a number that when substituted for a variable makes the equation a true statement.

6. <u>Equivalent</u> equations have the same solution.

7. By the <u>addition</u> property of equality, the same number may be added to or subtracted from both sides of an equation without changing the solution of the equation.

8. By the <u>multiplication</u> property of equality, the same nonzero number may be multiplied or divided by both sides of an equation without changing the solution of the equation.

Exercise Set 2.6

1. $x - 8 = -2$

$6 - 8 \stackrel{?}{=} -2$

$-2 = -2$ True

Since −2 = −2 is true, 6 is a solution of the equation.

3. $x + 12 = 17$

$-5 + 12 \stackrel{?}{=} 17$

$7 = 17$ False

Since 7 = 17 is false, −5 is not a solution of the equation.

5. $-9f = 64 - f$

$-9(-8) \stackrel{?}{=} 64 - (-8)$

$72 \stackrel{?}{=} 64 + 8$

$72 = 72$ True

Since 72 = 72 is true, −8 is a solution of the equation.

7. $5(c - 5) = -10$

$5(3 - 5) \stackrel{?}{=} -10$

$5(-2) \stackrel{?}{=} -10$

$-10 = -10$ True

Since −10 = −10 is true, 3 is a solution of the equation.

9. $a + 5 = 23$

$a + 5 - 5 = 23 - 5$

$a = 18$

Check: $a + 5 = 23$

$18 + 5 \stackrel{?}{=} 23$

$23 = 23$ True

The solution is 18.

Copyright © 2011 Pearson Education, Inc. Publishing as Prentice Hall.

11.
$$d-9=-21$$
$$d-9+9=-21+9$$
$$d=-12$$
Check: $d-9=-21$
$$-12-9 \stackrel{?}{=} -21$$
$$-21=-21 \quad \text{True}$$
The solution is −12.

13.
$$7=y-2$$
$$7+2=y-2+2$$
$$9=y$$
Check: $7=y-2$
$$7 \stackrel{?}{=} 9-2$$
$$7=7 \quad \text{True}$$
The solution is 9.

15.
$$11x=10x-17$$
$$11x-10x=10x-10x-17$$
$$1x=-17$$
$$x=-17$$
Check: $11x=10x-17$
$$11(-17) \stackrel{?}{=} 10(-17)-17$$
$$-187 \stackrel{?}{=} -170-17$$
$$-187=-187 \quad \text{True}$$
The solution is −17.

17.
$$5x=20$$
$$\frac{5x}{5}=\frac{20}{5}$$
$$\frac{5}{5}\cdot x=\frac{20}{5}$$
$$x=4$$
Check: $5x=20$
$$5(4) \stackrel{?}{=} 20$$
$$20=20 \quad \text{True}$$
The solution is 4.

19.
$$-3z=12$$
$$\frac{-3z}{-3}=\frac{12}{-3}$$
$$\frac{-3}{-3}\cdot z=\frac{12}{-3}$$
$$z=-4$$
Check: $-3z=12$
$$-3(-4) \stackrel{?}{=} 12$$
$$12=12 \quad \text{True}$$
The solution is −4.

21.
$$\frac{n}{7}=-2$$
$$7\cdot\frac{n}{7}=7\cdot(-2)$$
$$\frac{7}{7}\cdot n=7\cdot(-2)$$
$$n=-14$$
Check: $\frac{n}{7}=-2$
$$\frac{-14}{7} \stackrel{?}{=} -2$$
$$-2=-2 \quad \text{True}$$
The solution is −14.

23.
$$2z=-34$$
$$\frac{2z}{2}=\frac{-34}{2}$$
$$\frac{2}{2}\cdot z=\frac{-34}{2}$$
$$z=-17$$
Check: $2z=-34$
$$2\cdot(-17) \stackrel{?}{=} -34$$
$$-34=-34 \quad \text{True}$$
The solution is −17.

25.
$$-4y=0$$
$$\frac{-4y}{-4}=\frac{0}{-4}$$
$$\frac{-4}{-4}\cdot y=\frac{0}{-4}$$
$$y=0$$
Check: $-4y=0$
$$-4(0) \stackrel{?}{=} 0$$
$$0=0 \quad \text{True}$$
The solution is 0.

27.
$$-10x=-10$$
$$\frac{-10x}{-10}=\frac{-10}{-10}$$
$$\frac{-10}{-10}\cdot x=\frac{-10}{-10}$$
$$x=1$$
Check: $-10x=-10$
$$-10\cdot 1 \stackrel{?}{=} -10$$
$$-10=-10 \quad \text{True}$$
The solution is 1.

Copyright © 2011 Pearson Education, Inc. Publishing as Prentice Hall.

29.
$$5x = -35$$
$$\frac{5x}{5} = \frac{-35}{5}$$
$$\frac{5}{5} \cdot x = \frac{-35}{5}$$
$$x = -7$$
The solution is –7.

31.
$$n - 5 = -55$$
$$n - 5 + 5 = -55 + 5$$
$$n = -50$$
The solution is –50.

33.
$$-15 = y + 10$$
$$-15 - 10 = y + 10 - 10$$
$$-25 = y$$
The solution is –25.

35.
$$\frac{x}{-6} = -6$$
$$-6 \cdot \frac{x}{-6} = -6 \cdot (-6)$$
$$\frac{-6}{-6} \cdot x = -6 \cdot (-6)$$
$$x = 36$$
The solution is 36.

37.
$$11n = 10n + 21$$
$$11n - 10n = 10n - 10n + 21$$
$$1n = 21$$
$$n = 21$$
The solution is 21.

39.
$$-12y = -144$$
$$\frac{-12y}{-12} = \frac{-144}{-12}$$
$$\frac{-12}{-12} \cdot y = \frac{-144}{-12}$$
$$y = 12$$
The solution is 12.

41.
$$\frac{n}{4} = -20$$
$$4 \cdot \frac{n}{4} = 4 \cdot (-20)$$
$$\frac{4}{4} \cdot n = 4 \cdot (-20)$$
$$n = -80$$
The solution is –80.

43.
$$-64 = 32y$$
$$\frac{-64}{32} = \frac{32y}{32}$$
$$\frac{-64}{32} = \frac{32}{32} \cdot y$$
$$-2 = y$$
The solution is –2.

45. A number decreased by –2 is $x - (-2)$.

47. The product of –6 and a number is $-6 \cdot x$ or $-6x$.

49. The sum of –15 and a number is $-15 + x$.

51. –8 divided by a number is $-8 \div x$ or $\frac{-8}{x}$.

53.
$$n - 42,860 = -1286$$
$$n - 42,860 + 42,860 = -1286 + 42,860$$
$$n = 41,574$$
The solution is 41,574.

55.
$$-38x = 15,542$$
$$\frac{-38x}{-38} = \frac{15,542}{-38}$$
$$\frac{-38}{-38} \cdot x = \frac{15,542}{-38}$$
$$x = -409$$
The solution is –409.

57. answers may vary

59. answers may vary

Chapter 2 Vocabulary Check

1. Two numbers that are the same distance from 0 on the number line but are on opposite sides of 0 are called opposites.
2. The absolute value of a number is that number's distance from 0 on a number line.
3. The integers are ..., –3, –2, –1, 0, 1, 2, 3, ...
4. The negative numbers are numbers less than zero.
5. The positive numbers are numbers greater than zero.
6. The symbols "<" and ">" are called inequality symbols.

Copyright © 2011 Pearson Education, Inc. Publishing as Prentice Hall.

7. A solution of an equation is a number that when substituted for a variable makes the equation a true statement.

8. The average of a list of numbers is $\dfrac{\text{sum of numbers}}{\textit{number}\text{ of numbers}}$.

9. A combination of operations on variables and numbers is called an expression.

10. A statement of the form "expression = expression" is called an equation.

11. The sign "<" means is less than and ">" means is greater than.

12. By the addition property of equality, the same number may be added to or subtracted from both sides of an equation without changing the solution of the equation.

13. By the multiplication property of equality, the same nonzero number may be multiplied or divided by both sides of an equation without changing the solution of the equation.

Chapter 2 Review

1. If 0 represents ground level, then 1572 feet below the ground is -1572.

2. If 0 represents sea level, then an elevation of 11,239 feet is $+11,239$.

3.

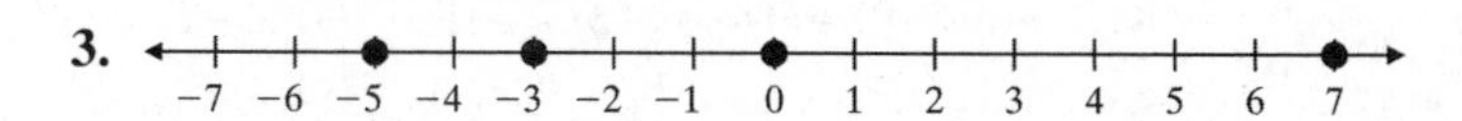

4.

−7 −6 −5 −4 −3 −2 −1 0 1 2 3 4 5 6 7

5. $|-11| = 11$ since -11 is 11 units from 0 on a number line.

6. $|0| = 0$ since 0 is 0 units from 0 on a number line.

7. $-|8| = -8$

8. $-(-9) = 9$

9. $-|-16| = -16$

10. $-(-2) = 2$

11. $-18 > -20$ since -18 is to the right of -20 on a number line.

12. $-5 < 5$ since -5 is to the left of 5 on a number line.

13. $|-123| = 123$
$-|-198| = -198$
Since $123 > -198$, $|-123| > -|-198|$.

14. $|-12| = 12$
$-|-16| = -16$
Since $12 > -16$, $|-12| > -|-16|$.

15. The opposite of -18 is 18.
$-(-18) = 18$

Copyright © 2011 Pearson Education, Inc. Publishing as Prentice Hall.

16. The opposite of 42 is negative 42.
$-(42) = -42$

17. False; consider $a = 1$ and $b = 2$, then $1 < 2$.

18. True

19. True

20. True

21. $|y| = |-2| = 2$

22. $|-x| = |-(-3)| = |3| = 3$

23. $-|-z| = -|-(-5)| = -|5| = -5$

24. $-|-n| = -|-(-10)| = -|10| = -10$

25. The bar that extends the farthest in the negative direction corresponds to Elevator D, so Elevator D extends the farthest below ground.

26. The bar that extends the farthest in the positive direction corresponds to Elevator B, so Elevator B extends the highest above ground.

27. $|5| - |-3| = 5 - 3 = 2$
$5 > 3$, so the answer is positive.
$5 + (-3) = 2$

28. $|18| - |-4| = 18 - 4 = 14$
$18 > 4$, so the answer is positive.
$18 + (-4) = 14$

29. $|16| - |-12| = 16 - 12 = 4$
$16 > 12$, so the answer is positive.
$-12 + 16 = 4$

30. $|40| - |-23| = 40 - 23 = 17$
$40 > 23$, so the answer is positive.
$-23 + 40 = 17$

31. $|-8| + |-15| = 8 + 15 = 23$
The common sign is negative, so
$-8 + (-15) = -23$.

32. $|-5| + |-17| = 5 + 17 = 22$
The common sign is negative, so
$-5 + (-17) = -22$.

33. $|-24| - |3| = 24 - 3 = 21$
$24 > 3$, so the answer is negative.
$-24 + 3 = -21$

34. $|-89| - |19| = 89 - 19 = 70$
$89 > 19$, so the answer is negative.
$-89 + 19 = -70$

35. $15 + (-15) = 0$

36. $-24 + 24 = 0$

37. $|-43| + |-108| = 43 + 108 = 151$
The common sign is negative, so
$-43 + (-108) = -151$.

38. $|-100| + |-506| = 100 + 506 = 606$
The common sign is negative, so
$-100 + (-506) = -606$.

39. $-15 + (-5) = -20$
The temperature at 6 a.m. is -20°C.

40. $-127 + (-23) = -150$
The diver's current depth is -150 feet.

41.
$$\begin{aligned} -6+(-9)+(-4)+(-2) &= -15+(-4)+(-2) \\ &= -19+(-2) \\ &= -21 \end{aligned}$$
His total score was -21.

42. $16 - 4 = 16 + (-4) = 12$
The team's score was 12.

43. $12 - 4 = 12 + (-4) = 8$

44. $-12 - 4 = -12 + (-4) = -16$

45. $-7 - 17 = -7 + (-17) = -24$

46. $7 - 17 = 7 + (-17) = -10$

47. $7 - (-13) = 7 + 13 = 20$

48. $-6 - (-14) = -6 + 14 = 8$

49. $16 - 16 = 16 + (-16) = 0$

50. $-16 - 16 = -16 + (-16) = -32$

51. $-12 - (-12) = -12 + 12 = 0$

52. $-5 - (-12) = -5 + 12 = 7$

53.
$$\begin{aligned} -(-5)-12+(-3) &= 5+(-12)+(-3) \\ &= -7+(-3) \\ &= -10 \end{aligned}$$

Copyright © 2011 Pearson Education, Inc. Publishing as Prentice Hall.

54. $-8+(-12)-10-(-3) = -8+(-12)+(-10)+3$
$= -20+(-10)+3$
$= -30+3$
$= -27$

55. $600 - (-92) = 600 + 92 = 692$
The difference in elevations is 692 feet.

56. $142-125+43-85 = 142+(-125)+43+(-85)$
$= 17+43+(-85)$
$= 60+(-85)$
$= -25$
The balance in his account is –25.

57. $85 - 99 = 85 + (-99) = -14$
You are –14 feet or 14 feet below ground at the end of the drop.

58. $66 - (-16) = 66 + 16 = 82$
The total length of the elevator shaft for Elevator C is 82 feet.

59. $|-5| - |-6| = 5 - 6 = 5 + (-6) = -1$
$5 - 6 = 5 + (-6) = -1$
$|-5| - |-6| = 5 - 6$ is true.

60. $|-5 - (-6)| = |-5 + 6| = |1| = 1$
$5 + 6 = 11$
Since $1 \neq 11$, the statement is false.

61. $-3(-7) = 21$

62. $-6(3) = -18$

63. $-4(16) = -64$

64. $-5(-12) = 60$

65. $(-5)^2 = (-5)(-5) = 25$

66. $(-1)^5 = (-1)(-1)(-1)(-1)(-1) = -1$

67. $12(-3)(0) = 0$

68. $-1(6)(2)(-2) = -6(2)(-2) = -12(-2) = 24$

69. $-15 \div 3 = -5$

70. $\frac{-24}{-8} = 3$

71. $\frac{0}{-3} = 0$

72. $\frac{-46}{0}$ is undefined.

73. $\frac{100}{-5} = -20$

74. $\frac{-72}{8} = -9$

75. $\frac{-38}{-1} = 38$

76. $\frac{45}{-9} = -5$

77. A loss of 5 yards is represented by –5.
$(-5)(2) = -10$
The total loss is 10 yards.

78. A loss of \$50 is represented by –50.
$(-50)(4) = -200$
The total loss is \$200.

79. A debt of \$1024 is represented by –1024.
$-1024 \div 4 = -256$
Each payment is \$256.

80. A drop of 45 degrees is represented by –45.
$\frac{-45}{9} = -5$ or $-45 \div 9 = -5$
The average drop each hour is 5°F.

81. $(-7)^2 = (-7)(-7) = 49$

82. $-7^2 = -(7 \cdot 7) = -49$

83. $5 - 8 + 3 = -3 + 3 = 0$

84. $-3+12+(-7)-10 = 9+(-7)-10 = 2-10 = -8$

85. $-10 + 3 \cdot (-2) = -10 + (-6) = -16$

86. $5 - 10 \cdot (-3) = 5 - (-30) = 5 + 30 = 35$

87. $16 \div (-2) \cdot 4 = -8 \cdot 4 = -32$

88. $-20 \div 5 \cdot 2 = -4 \cdot 2 = -8$

89. $16+(-3)\cdot 12 \div 4 = 16+(-36) \div 4$
$= 16+(-9)$
$= 7$

Copyright © 2011 Pearson Education, Inc. Publishing as Prentice Hall.

90. $-12+10\div(-5)=-12+(-2)=-14$

91. $4^3-(8-3)^2=4^3-(5)^2=64-25=39$

92. $(-3)^3-90=-27-90=-117$

93. $\frac{(-4)(-3)-(-2)(-1)}{-10+5}=\frac{12-2}{-5}=\frac{10}{-5}=-2$

94. $\frac{4(12-18)}{-10\div(-2-3)}=\frac{4(-6)}{-10\div(-5)}=\frac{-24}{2}=-12$

95. $\text{average}=\frac{-18+25+(-30)+7+0+(-2)}{6}$
$=\frac{-18}{6}$
$=-3$

96. $\text{average}=\frac{-45+(-40)+(-30)+(-25)}{4}$
$=\frac{-140}{4}$
$=-35$

97. $2x-y=2(-2)-1=-4-1=-5$

98. $y^2+x^2=1^2+(-2)^2=1+4=5$

99. $\frac{3x}{6}=\frac{3(-2)}{6}=\frac{-6}{6}=-1$

100. $\frac{5y-x}{-y}=\frac{5(1)-(-2)}{-1}=\frac{5+2}{-1}=\frac{7}{-1}=-7$

101. $2n-6=16$
$2(-5)-6\stackrel{?}{=}16$
$-10-6\stackrel{?}{=}16$
$-16=16$ False
Since $-16=16$ is false, -5 is not a solution of the equation.

102. $2(c-8)=-20$
$2(-2-8)\stackrel{?}{=}-20$
$2(-10)\stackrel{?}{=}-20$
$-20=-20$ True
Since $-20=-20$ is true, -2 is a solution of the equation.

103. $n-7=-20$
$n-7+7=-20+7$
$n=-13$
The solution is -13.

104. $-5=n+15$
$-5-15=n+15-15$
$-20=n$
The solution is -20.

105. $10x=-30$
$\frac{10x}{10}=\frac{-30}{10}$
$\frac{10}{10}\cdot x=\frac{-30}{10}$
$x=-3$
The solution is -3.

106. $-8x=72$
$\frac{-8x}{-8}=\frac{72}{-8}$
$\frac{-8}{-8}\cdot x=\frac{72}{-8}$
$x=-9$
The solution is -9.

107. $9y=8y-13$
$9y-8y=8y-8y-13$
$1y=-13$
$y=-13$
The solution is -13.

108. $6x-31=7x$
$6x-6x-31=7x-6x$
$-31=1x$
$-31=x$
The solution is -31.

109. $\frac{n}{-4}=-11$
$-4\cdot\frac{n}{-4}=-4\cdot(-11)$
$\frac{-4}{-4}\cdot n=-4\cdot(-11)$
$n=44$
The solution is 44.

Copyright © 2011 Pearson Education, Inc. Publishing as Prentice Hall.

110. $\frac{x}{-2} = 13$

$-2 \cdot \frac{x}{-2} = -2 \cdot 13$

$\frac{-2}{-2} \cdot x = -2 \cdot 13$

$x = -26$

The solution is −26.

111. $n + 12 = -7$

$n + 12 - 12 = -7 - 12$

$n = -19$

The solution is −19.

112. $n - 40 = -2$

$n - 40 + 40 = -2 + 40$

$n = 38$

The solution is 38.

113. $-36 = -6x$

$\frac{-36}{-6} = \frac{-6x}{-6}$

$\frac{-36}{-6} = \frac{-6}{-6} \cdot x$

$6 = x$

The solution is 6.

114. $-40 = 8y$

$\frac{-40}{8} = \frac{8y}{8}$

$\frac{-40}{8} = \frac{8}{8} \cdot y$

$-5 = y$

The solution is −5.

115. $-6 + (-9) = -15$

116. $-16 - 3 = -16 + (-3) = -19$

117. $-4(-12) = 48$

118. $\frac{84}{-4} = -21$

119. $-76 - (-97) = -76 + 97 = 21$

120. $-9 + 4 = -5$

121. $-32 + 23 = -9$

His financial situation can be represented by −$9.

122. $-11 + 17 = 6$

The temperature at noon on Tuesday was 6°C.

123. $12{,}923 - (-195) = 12{,}923 + 195 = 13{,}118$

The difference in elevations is 13,118 feet.

124. $-18 - 9 = -27$

The temperature on Friday was −27°C.

125. $(3-7)^2 \div (6-4)^3 = (-4)^2 \div (2)^3 = 16 \div 8 = 2$

126.
$$\begin{aligned} 3(4+2)+(-6)-3^2 &= 3(6)+(-6)-3^2 \\ &= 3(6)+(-6)-9 \\ &= 18+(-6)-9 \\ &= 12-9 \\ &= 3 \end{aligned}$$

127. $2 - 4 \cdot 3 + 5 = 2 - 12 + 5 = -10 + 5 = -5$

128. $4 - 6 \cdot 5 + 1 = 4 - 30 + 1 = -26 + 1 = -25$

129. $\frac{-|-14|-6}{7+2(-3)} = \frac{-14-6}{7+(-6)} = \frac{-20}{1} = -20$

130.
$$\begin{aligned} 5(7-6)^3 - 4(2-3)^2 + 2^4 &= 5(1)^3 - 4(-1)^2 + 2^4 \\ &= 5(1) - 4(1) + 16 \\ &= 5 - 4 + 16 \\ &= 1 + 16 \\ &= 17 \end{aligned}$$

131. $n - 9 = -30$

$n - 9 + 9 = -30 + 9$

$n = -21$

The solution is −21.

132. $n + 18 = 1$

$n + 18 - 18 = 1 - 18$

$n = -17$

The solution is −17.

133. $-4x = -48$

$\frac{-4x}{-4} = \frac{-48}{-4}$

$\frac{-4}{-4} \cdot x = \frac{-48}{-4}$

$x = 12$

The solution is 12.

Copyright © 2011 Pearson Education, Inc. Publishing as Prentice Hall.

134. $9x = -81$
$\frac{9x}{9} = \frac{-81}{9}$
$\frac{9}{9} \cdot x = \frac{-81}{9}$
$x = -9$
The solution is −9.

135. $\frac{n}{-2} = 100$
$-2 \cdot \frac{n}{-2} = -2 \cdot 100$
$\frac{-2}{-2} \cdot n = -2 \cdot 100$
$n = -200$
The solution is −200.

136. $\frac{y}{-1} = -3$
$-1 \cdot \frac{y}{-1} = -1(-3)$
$\frac{-1}{-1} \cdot y = -1 \cdot (-3)$
$y = 3$
The solution is 3.

Chapter 2 Test

1. $-5 + 8 = 3$

2. $18 - 24 = 18 + (-24) = -6$

3. $5 \cdot (-20) = -100$

4. $-16 \div (-4) = 4$

5. $-18 + (-12) = -30$

6. $-7 - (-19) = -7 + 19 = 12$

7. $-5 \cdot (-13) = 65$

8. $\frac{-25}{-5} = 5$

9. $|-25| + (-13) = 25 + (-13) = 12$

10. $14 - |-20| = 14 - 20 = 14 + (-20) = -6$

11. $|5| \cdot |-10| = 5 \cdot 10 = 50$

12. $\frac{|-10|}{-|-5|} = \frac{10}{-5} = -2$

13. $-8 + 9 \div (-3) = -8 + (-3) = -11$

14. $-7 + (-32) - 12 + 5 = -7 + (-32) + (-12) + 5$
$= -39 + (-12) + 5$
$= -51 + 5$
$= -46$

15. $(-5)^3 - 24 \div (-3) = -125 - 24 \div (-3)$
$= -125 - (-8)$
$= -125 + 8$
$= -117$

16. $(5-9)^2 \cdot (8-2)^3 = (-4)^2 \cdot (6)^3 = 16 \cdot 216 = 3456$

17. $-(-7)^2 \div 7 \cdot (-4) = -49 \div 7 \cdot (-4) = -7 \cdot (-4) = 28$

18. $3 - (8-2)^3 = 3 - 6^3$
$= 3 - 216$
$= 3 + (-216)$
$= -213$

19. $\frac{4}{2} - \frac{8^2}{16} = \frac{4}{2} - \frac{64}{16} = 2 - 4 = 2 + (-4) = -2$

20. $\frac{-3(-2)+12}{-1(-4-5)} = \frac{6+12}{-1(-9)} = \frac{18}{9} = 2$

21. $\frac{|25-30|^2}{2(-6)+7} = \frac{|-5|^2}{-12+7} = \frac{(5)^2}{-5} = \frac{25}{-5} = -5$

22. $5(-8) - [6 - (2-4)] + (12-16)^2$
$= 5(-8) - [6 - (-2)] + (12-16)^2$
$= 5(-8) - (6+2) + (-4)^2$
$= 5(-8) - 8 + (-4)^2$
$= 5(-8) - 8 + 16$
$= -40 - 8 + 16$
$= -48 + 16$
$= -32$

23. $7x + 3y - 4z = 7(0) + 3(-3) - 4(2)$
$= 0 + (-9) - 8$
$= -9 - 8$
$= -17$

24. $10 - y^2 = 10 - (-3)^2 = 10 - 9 = 1$

25. $\frac{3z}{2y} = \frac{3(2)}{2(-3)} = \frac{6}{-6} = -1$

Copyright © 2011 Pearson Education, Inc. Publishing as Prentice Hall.

26. A descent of 22 feet is represented by –22.
$4(-22) = -88$
Mary is 88 feet below sea level.

27. $129+(-79)+(-40)+35 = 50+(-40)+35$
$= 10+35$
$= 45$
His new balance can be represented by 45.

28. Subtract the elevation of the Romanche Gap from the elevation of Mt. Washington.
$6288 - (-25,354) = 6288 + 25,354 = 31,642$
The difference in elevations is 31,642 feet.

29. Subtract the depth of the lake from the elevation of the surface.
$1495 - 5315 = 1495 + (-5315) = -3820$
The deepest point of the lake is 3820 feet below sea level.

30. $\text{average} = \frac{-12+(-13)+0+9}{4} = \frac{-16}{4} = -4$

31. **a.** The product of a number and 17 is $17x$.

b. Twice a number subtracted from 20 is $20 - 2x$.

32.
$$-9n = -45$$
$$\frac{-9n}{-9} = \frac{-45}{-9}$$
$$\frac{-9}{-9} \cdot n = \frac{-45}{-9}$$
$$n = 5$$
The solution is 5.

33.
$$\frac{n}{-7} = 4$$
$$-7 \cdot \frac{n}{-7} = -7 \cdot 4$$
$$\frac{-7}{-7} \cdot n = -7 \cdot 4$$
$$n = -28$$
The solution is –28.

34.
$$x - 16 = -36$$
$$x - 16 + 16 = -36 + 16$$
$$x = -20$$
The solution is –20.

35.
$$9x = 8x - 4$$
$$9x - 8x = 8x - 8x - 4$$
$$1x = -4$$
$$x = -4$$
The solution is –4.

Cumulative Review Chapters 1–2

1. The place value of 3 in 396,418 is hundred-thousands.

2. The place value of 3 in 4308 is hundreds.

3. The place value of 3 in 93,192 is thousands.

4. The place value of 3 is 693,298 is thousands.

5. The place value of 3 in 534,275,866 is ten-millions.

6. The place value of 3 in 267,301,818 is hundred-thousands.

7. **a.** $-7 < 7$ since –7 is to the left of 7 on a number line.

b. $0 > -4$ since 0 is to the right of –4 on a number line.

c. $-9 > -11$ since –9 is to the right of –11 on a number line.

8. **a.** $12 > -4$ since 12 is to the right of –4 on a number line.

b. $-13 > -31$ since –13 is to the right of –31 on a number line.

c. $-82 < 79$ since –82 is to the left of 79 on a number line.

9. $13+2+7+8+9 = (13+7)+(2+8)+9$
$= 20+10+9$
$= 39$

10. $11+3+9+16 = (11+9)+(3+16) = 20+19 = 39$

11.
$$\begin{array}{r} 7826 \\ -\ \ 505 \\ \hline 7321 \end{array}$$
Check:
$$\begin{array}{r} 7321 \\ +\ \ 505 \\ \hline 7826 \end{array}$$

Copyright © 2011 Pearson Education, Inc. Publishing as Prentice Hall.

12. $\begin{array}{r} 3285 \\ -\ 272 \\ \hline 3013 \end{array}$

Check: $\begin{array}{r} 3013 \\ +\ 272 \\ \hline 3285 \end{array}$

13. Subtract 7257 from the radius of Jupiter.

$\begin{array}{r} 43,441 \\ -\ 7\ 257 \\ \hline 36,184 \end{array}$

The radius of Saturn is 36,184 miles.

14. Subtract the cost of the camera from the amount in her account.

$\begin{array}{r} 762 \\ -\ 237 \\ \hline 525 \end{array}$

She will have $525 left in her account after buying the camera.

15. To round 568 to the nearest ten, observe that the digit in the ones place is 8. Since this digit is at least 5, we add 1 to the digit in the tens place. The number 568 rounded to the nearest ten is 570.

16. To round 568 to the nearest hundred, observe that the digit in the tens place is 6. Since this digit is at least 5, we add 1 to the digit in the hundreds place. The number 568 rounded to the nearest hundred is 600.

17. $\begin{array}{rcr} 4725 & \text{rounds to} & 4700 \\ -\ 2879 & \text{rounds to} & -\ 2900 \\ \hline & & 1800 \end{array}$

18. $\begin{array}{rcr} 8394 & \text{rounds to} & 8000 \\ -\ 2913 & \text{rounds to} & -\ 3000 \\ \hline & & 5000 \end{array}$

19. a. $5(6+5) = 5 \cdot 6 + 5 \cdot 5$

b. $20(4+7) = 20 \cdot 4 + 20 \cdot 7$

c. $2(7+9) = 2 \cdot 7 + 2 \cdot 9$

20. a. $5(2+12) = 5 \cdot 2 + 5 \cdot 12$

b. $9(3+6) = 9 \cdot 3 + 9 \cdot 6$

c. $4(8+1) = 4 \cdot 8 + 4 \cdot 1$

21. $\begin{array}{r} 631 \\ \times\ 125 \\ \hline 3\ 155 \\ 12\ 620 \\ 63\ 100 \\ \hline 78,875 \end{array}$

22. $\begin{array}{r} 299 \\ \times\ 104 \\ \hline 1\ 196 \\ 29\ 900 \\ \hline 31,096 \end{array}$

23. a. $42 \div 7 = 6$ because $6 \cdot 7 = 42$.

b. $\frac{64}{8} = 8$ because $8 \cdot 8 = 64$.

c. $3\overline{)21}$ with quotient 7, because $7 \cdot 3 = 21$.

24. a. $\frac{35}{5} = 7$ because $7 \cdot 5 = 35$.

b. $64 \div 8 = 8$ because $8 \cdot 8 = 64$.

c. $4\overline{)48}$ with quotient 12, because $12 \cdot 4 = 48$.

25. $5\overline{)3705}$ with quotient 741

$\begin{array}{r} 3705 \\ -35 \\ \hline 20 \\ -20 \\ \hline 05 \\ -5 \\ \hline 0 \end{array}$

Check: $\begin{array}{r} 741 \\ \times\ 5 \\ \hline 3705 \end{array}$

Copyright © 2011 Pearson Education, Inc. Publishing as Prentice Hall.

26. $\begin{array}{r} 456 \\ 8\overline{)\,3648} \\ \underline{-32} \\ 44 \\ \underline{-40} \\ 48 \\ \underline{-48} \\ 0 \end{array}$

Check: $\begin{array}{r} 456 \\ \underline{\times\ \ 8} \\ 3648 \end{array}$

27. $\text{number of cards for each person} = \text{number of cards} \div \text{number of friends}$
$= 238 \div 19$

$\begin{array}{r} 12 \text{ R } 10 \\ 19\overline{)\,238} \\ \underline{-19} \\ 48 \\ \underline{-38} \\ 10 \end{array}$

Each friend will receive 12 cards. There will be 10 cards left over.

28. $\text{Cost of each ticket} = \text{total cost} \div \text{number of tickets}$
$= 324 \div 36$

$\begin{array}{r} 9 \\ 36\overline{)\,324} \\ \underline{-324} \\ 0 \end{array}$

Each ticket cost \$9.

29. $9^2 = 9 \cdot 9 = 81$

30. $5^3 = 5 \cdot 5 \cdot 5 = 125$

31. $6^1 = 6$

32. $4^1 = 4$

33. $5 \cdot 6^2 = 5 \cdot 6 \cdot 6 = 180$

34. $2^3 \cdot 7 = 2 \cdot 2 \cdot 2 \cdot 7 = 56$

35. $\dfrac{7-2 \cdot 3+3^2}{5(2-1)} = \dfrac{7-2 \cdot 3+9}{5(1)} = \dfrac{7-6+9}{5} = \dfrac{10}{5} = 2$

36. $\dfrac{6^2+4 \cdot 4+2^3}{37-5^2} = \dfrac{36+4 \cdot 4+8}{37-25}$
$= \dfrac{36+16+8}{12}$
$= \dfrac{60}{12}$
$= 5$

37. $x + 6 = 8 + 6 = 14$

38. $5 + x = 5 + 9 = 14$

39. **a.** $|-9| = 9$ because -9 is 9 units from 0.

b. $|8| = 8$ because 8 is 8 units from 0.

c. $|0| = 0$ because 0 is 0 units from 0.

40. **a.** $|4| = 4$ because 4 is 4 units from 0.

b. $|-7| = 7$ because -7 is 7 units from 0.

41. $-2 + 25 = 23$

42. $8 + (-3) = 5$

43. $2a - b = 2(8) - (-6) = 16 - (-6) = 16 + 6 = 22$

44. $x - y = -2 - (-7) = -2 + 7 = 5$

45. $-7 \cdot 3 = -21$

46. $5(-2) = -10$

47. $0 \cdot (-4) = 0$

48. $-6 \cdot 9 = -54$

49. $3(4-7)+(-2)-5 = 3(-3)+(-2)-5$
$= -9+(-2)-5$
$= -11-5$
$= -16$

50. $4-8(7-3)-(-1) = 4-8(4)-(-1)$
$= 4-32-(-1)$
$= 4-32+1$
$= -28+1$
$= -27$

Copyright © 2011 Pearson Education, Inc. Publishing as Prentice Hall.

Chapter 3

Section 3.1

Practice Problems

1. a. $8m - 14m = (8 - 14)m = -6m$

   b. $6a + a = 6a + 1a = (6 + 1)a = 7a$

   c. $-y^2 + 3y^2 + 7 = -1y^2 + 3y^2 + 7$
   $= (-1+3)y^2 + 7$
   $= 2y^2 + 7$

2. $6z + 5 + z - 4 = 6z + 5 + 1z + (-4)$
   $= 6z + 1z + 5 + (-4)$
   $= (6+1)z + 5 + (-4)$
   $= 7z + 1$

3. $6y + 12y - 6 = 18y - 6$

4. $7y - 5 + y + 8 = 7y + (-5) + 1y + 8$
   $= 7y + 1y + (-5) + 8$
   $= 8y + 3$

5. $-7y + 2 - 2y - 9x + 12 - x$
   $= -7y + 2 + (-2y) + (-9x) + 12 + (-1x)$
   $= -7y + (-2y) + (-9x) + (-1x) + 2 + 12$
   $= -9y - 10x + 14$

6. $6(4a) = (6 \cdot 4)a = 24a$

7. $-8(9x) = (-8 \cdot 9)x = -72x$

8. $8(y + 2) = 8 \cdot y + 8 \cdot 2 = 8y + 16$

9. $3(7a - 5) = 3 \cdot 7a - 3 \cdot 5 = 21a - 15$

10. $6(5 - y) = 6 \cdot 5 - 6 \cdot y = 30 - 6y$

11. $5(2y - 3) - 8 = 5(2y) - 5(3) - 8$
    $= 10y - 15 - 8$
    $= 10y - 23$

12. $-7(x - 1) + 5(2x + 3)$
    $= -7(x) - (-7)(1) + 5(2x) + 5(3)$
    $= -7x + 7 + 10x + 15$
    $= 3x + 22$

13. $-(y+1) + 3y - 12 = -1(y+1) + 3y - 12$
    $= -1 \cdot y + (-1)(1) + 3y - 12$
    $= -y - 1 + 3y - 12$
    $= 2y - 13$

14. $4(2x) = (4 \cdot 2)x = 8x$
    The perimeter is $8x$ centimeters.

15. $3(12y + 9) = 3 \cdot 12y + 3 \cdot 9 = 36y + 27$
    The area of the garden is $(36y + 27)$ square yards.

Vocabulary and Readiness Check

1. $14y^2 + 2x - 23$ is called an expression while $14y^2$, $2x$, and -23 are each called a term.

2. To multiply $3(-7x + 1)$, we use the distributive property.

3. To simplify an expression like $y + 7y$, we combine like terms.

4. By the commutative properties, the *order* of adding or multiplying two numbers can be changed without changing their sum or product.

5. The term $5x$ is called a variable term while the term 7 is called a constant term.

6. The term z has an understood numerical coefficient of 1.

7. By the associative properties, the grouping of adding or multiplying numbers can be changed without changing their sum or product.

8. The terms $-x$ and $5x$ are like terms and the terms $5x$ and $5y$ are unlike terms.

9. For the term $-3x^2y$, -3 is called the numerical coefficient.

Exercise Set 3.1

1. $3x + 5x = (3 + 5)x = 8x$

3. $2n - 3n = (2 - 3)n = -1n = -n$

5. $4c + c - 7c = (4 + 1 - 7)c = -2c$

7. $4x - 6x + x - 5x = (4 - 6 + 1 - 5)x = -6x$

Copyright © 2011 Pearson Education, Inc. Publishing as Prentice Hall.

9. $3a + 2a + 7a - 5 = (3 + 2 + 7)a - 5 = 12a - 5$

11. $6(7x) = (6 \cdot 7)x = 42x$

13. $-3(11y) = (-3 \cdot 11)y = -33y$

15. $12(6a) = (12 \cdot 6)a = 72a$

17. $2(y + 3) = 2 \cdot y + 2 \cdot 3 = 2y + 6$

19. $3(a - 6) = 3 \cdot a - 3 \cdot 6 = 3a - 18$

21. $-4(3x + 7) = -4 \cdot 3x + (-4) \cdot 7 = -12x - 28$

23.
$$\begin{aligned} 2(x+4)-7 &= 2\cdot x+2\cdot 4-7 \\ &= 2x+8-7 \\ &= 2x+1 \end{aligned}$$

25.
$$\begin{aligned} 8+5(3c-1) &= 8+5\cdot 3c-5\cdot 1 \\ &= 8+15c-5 \\ &= 15c+8-5 \\ &= 15c+3 \end{aligned}$$

27.
$$\begin{aligned} -4(6n-5)+3n &= -4\cdot 6n-(-4)\cdot 5+3n \\ &= -24n+20+3n \\ &= -24n+3n+20 \\ &= -21n+20 \end{aligned}$$

29.
$$\begin{aligned} 3+6(w+2)+w &= 3+6\cdot w+6\cdot 2+w \\ &= 3+6w+12+w \\ &= 6w+w+3+12 \\ &= 7w+15 \end{aligned}$$

31.
$$\begin{aligned} 2(3x+1)+5(x-2) &= 2(3x)+2(1)+5(x)-5(2) \\ &= 6x+2+5x-10 \\ &= 6x+5x+2-10 \\ &= 11x-8 \end{aligned}$$

33.
$$\begin{aligned} -(2y-6)+10 &= -1(2y-6)+10 \\ &= -1\cdot 2y-(-1)\cdot(6)+10 \\ &= -2y+6+10 \\ &= -2y+16 \end{aligned}$$

35. $18y - 20y = (18 - 20)y = -2y$

37. $z - 8z = (1 - 8)z = -7z$

39.
$$\begin{aligned} 9d-3c-d &= 9d-d-3c \\ &= (9-1)d-3c \\ &= 8d-3c \end{aligned}$$

41.
$$\begin{aligned} 2y-6+4y-8 &= 2y+4y-6-8 \\ &= (2+4)y-6-8 \\ &= 6y-14 \end{aligned}$$

43.
$$\begin{aligned} 5q+p-6q-p &= 5q-6q+p-p \\ &= (5-6)q+(1-1)p \\ &= -1q+0p \\ &= -q \end{aligned}$$

45.
$$\begin{aligned} 2(x+1)+20 &= 2\cdot x+2\cdot 1+20 \\ &= 2x+2+20 \\ &= 2x+22 \end{aligned}$$

47.
$$\begin{aligned} 5(x-7)-8x &= 5\cdot x-5\cdot 7-8x \\ &= 5x-35-8x \\ &= 5x-8x-35 \\ &= -3x-35 \end{aligned}$$

49.
$$\begin{aligned} -5(z+3)+2z &= -5\cdot z+(-5)\cdot 3+2z \\ &= -5z-15+2z \\ &= -5z+2z-15 \\ &= -3z-15 \end{aligned}$$

51.
$$\begin{aligned} 8-x+4x-2-9x &= -x+4x-9x+8-2 \\ &= (-1+4-9)x+8-2 \\ &= -6x+6 \end{aligned}$$

53. $-7(x + 5) + 5(2x + 1)$
$$\begin{aligned} &= -7 \cdot x + (-7) \cdot 5 + 5 \cdot 2x + 5 \cdot 1 \\ &= -7x - 35 + 10x + 5 \\ &= -7x + 10x - 35 + 5 \\ &= 3x - 30 \end{aligned}$$

55.
$$\begin{aligned} 3r-5r+8+r &= 3r-5r+r+8 \\ &= (3-5+1)r+8 \\ &= -1r+8 \\ &= -r+8 \end{aligned}$$

57.
$$\begin{aligned} -3(n-1)-4n &= -3\cdot n-(-3)\cdot 1-4n \\ &= -3n+3-4n \\ &= -3n-4n+3 \\ &= -7n+3 \end{aligned}$$

59.
$$\begin{aligned} 4(z-3)+5z-2 &= 4\cdot z-4\cdot 3+5z-2 \\ &= 4z-12+5z-2 \\ &= 4z+5z-12-2 \\ &= 9z-14 \end{aligned}$$

61.
$$\begin{aligned} 6(2x-1)-12x &= 6\cdot 2x-6\cdot 1-12x \\ &= 12x-6-12x \\ &= 12x-12x-6 \\ &= -6 \end{aligned}$$

Copyright © 2011 Pearson Education, Inc. Publishing as Prentice Hall.

63. $-(4x-5)+5 = -1(4x-5)+5$
$= -1\cdot 4x-(-1)(5)+5$
$= -4x+5+5$
$= -4x+10$

65. $-(4x-10)+2(3x+5)$
$= -1(4x-10)+2(3x+5)$
$= -1\cdot 4x-(-1)\cdot 10+2\cdot 3x+2\cdot 5$
$= -4x+10+6x+10$
$= -4x+6x+10+10$
$= 2x+20$

67. $3a+4(a+3) = 3a+4\cdot a+4\cdot 3$
$= 3a+4a+12$
$= 7a+12$

69. $5y-2(y-1)+3 = 5y+(-2)(y)-(-2)(1)+3$
$= 5y-2y+2+3$
$= 3y+5$

71. $3y+4y+2y+6+5y+16$
$= 3y+4y+2y+5y+6+16$
$= (3+4+2+5)y+6+16$
$= 14y+22$
The perimeter is $(14y + 22)$ meters.

73. $2a+2a+6+5a+6+2a$
$= 2a+2a+5a+2a+6+6$
$= (2+2+5+2)a+6+6$
$= 11a+12$
The perimeter is $(11a + 12)$ feet.

75. $5(-5x+11) = 5\cdot(-5x)+5\cdot 11 = -25x+55$
The perimeter is $(-25x + 55)$ inches.

77. Area = (length) · (width)
$= (4y)\cdot(9)$
$= (4\cdot 9)y$
$= 36y$
The area is $36y$ square inches.

79. Area = (length) · (width)
$= (x-2)\cdot(32)$
$= x\cdot 32-2\cdot 32$
$= 32x-64$
The area is $(32x - 64)$ square kilometers.

81. Area = (length) · (width)
$= (3y+1)\cdot(20)$
$= 3y\cdot 20+1\cdot 20$
$= (3\cdot 20)y+20$
$= 60y+20$
The area is $(60y + 20)$ square miles.

83. Area = (length) · (width) = 94 · 50 = 4700
The area is 4700 square feet.

85. Perimeter = 2 · (length) + 2 · (width)
$= 2\cdot(18)+2\cdot(14)$
$= 36+28$
$= 64$
The perimeter is 64 feet.

87. $5+x+2x+1 = x+2x+5+1$
$= (1+2)x+5+1$
$= 3x+6$
The perimeter is $(3x + 6)$ feet.

89. $-13 + 10 = -3$

91. $-4 - (-12) = -4 + 12 = 8$

93. $-4 + 4 = 0$

95. $5(3x-2) = 5\cdot 3x - 5\cdot 2 = 15x - 10$
The expressions are not equivalent.

97. $7x-(x+2) = 7x+(-1)\cdot(x+2)$
$= 7x+(-1)\cdot x+(-1)\cdot 2$
$= 7x-x-2$
The expressions are equivalent.

99. Since multiplication is distributed over subtraction in $6(2x-3) = 12x-18$, this is the distributive property.

101. The order of the terms is not changed, but the grouping is. This is the associative property of addition.

103. Add the areas of each rectangle.
$7(2x+1)+3(2x+3) = 7\cdot 2x+7\cdot 1+3\cdot 2x+3\cdot 3$
$= 14x+7+6x+9$
$= 14x+6x+7+9$
$= 20x+16$
The total area is $(20x + 16)$ square miles.

105. $9684q - 686 - 4860q + 12,960$
$= 9684q - 4860q - 686 + 12,960$
$= 4824q + 12,274$

Copyright © 2011 Pearson Education, Inc. Publishing as Prentice Hall.

107. answers may vary

109. answers may vary

Section 3.2

Practice Problems

1.
$$\begin{aligned} x+6 &= 1-3 \\ x+6 &= -2 \\ x+6-6 &= -2-6 \\ x &= -8 \end{aligned}$$

2.
$$\begin{aligned} 10 &= 2m-4m \\ 10 &= -2m \\ \frac{10}{-2} &= \frac{-2m}{-2} \\ -5 &= m \end{aligned}$$

3.
$$\begin{aligned} -8+6 &= \frac{a}{3} \\ -2 &= \frac{a}{3} \\ 3\cdot(-2) &= 3\cdot\frac{a}{3} \\ 3\cdot(-2) &= \frac{3}{3}\cdot a \\ -6 &= a \end{aligned}$$

4.
$$\begin{aligned} -6y-1+7y &= 17+2 \\ -6y+7y-1 &= 17+2 \\ y-1 &= 19 \\ y-1+1 &= 19+1 \\ y &= 20 \end{aligned}$$

5.
$$\begin{aligned} -4-10 &= 4y-5y \\ -14 &= -y \\ \frac{-14}{-1} &= \frac{-1y}{-1} \\ 14 &= y \end{aligned}$$

6.
$$\begin{aligned} 13x &= 4(3x-1) \\ 13x &= 4\cdot 3x - 4\cdot 1 \\ 13x &= 12x-4 \\ 13x-12x &= 12x-4-12x \\ x &= -4 \end{aligned}$$

7.
$$\begin{aligned} 5y+2 &= 17 \\ 5y+2-2 &= 17-2 \\ 5y &= 15 \\ \frac{5y}{5} &= \frac{15}{5} \\ y &= 3 \end{aligned}$$

8.
$$\begin{aligned} -4(x+2)-60 &= 2-10 \\ -4x-8-60 &= 2-10 \\ -4x-68 &= -8 \\ -4x-68+68 &= -8+68 \\ -4x &= 60 \\ \frac{-4x}{-4} &= \frac{60}{-4} \\ x &= -15 \end{aligned}$$

9. **a.** The sum of −3 and a number is $-3 + x$.

b. −5 decreased by a number is $-5 - x$.

c. Three times a number is $3x$.

d. A number subtracted from 83 is $83 - x$.

e. The quotient of a number and −4 is $\frac{x}{-4}$ or $-\frac{x}{4}$.

10. **a.** The product of 5 and a number, decreased by 25, is $5x - 25$.

b. Twice the sum of a number and 3 is $2(x + 3)$.

c. The quotient of 39 and twice a number is $39 \div (2x)$ or $\frac{39}{2x}$.

Vocabulary and Readiness Check

1. The equations $-3x = 51$ and $\frac{-3x}{-3} = \frac{51}{-3}$ are called equivalent equations.

2. The difference between an equation and an expression is that an equation contains an equal sign, while an expression does not.

3. The process of writing $-3x + 10x$ as $7x$ is called simplifying the expression.

Copyright © 2011 Pearson Education, Inc. Publishing as Prentice Hall.

4. For the equation $-5x - 1 = -21$, the process of finding that 4 is the solution is called solving the equation.

5. By the addition property of equality, $x = -2$ and $x + 7 = -2 + 7$ are equivalent equations.

6. By the multiplication property of equality, $y = 8$ and $3 \cdot y = 3 \cdot 8$ are equivalent equations.

Exercise Set 3.2

1.
$$\begin{aligned} x-3&=-1+4\\ x-3&=3\\ x-3+3&=3+3\\ x&=6 \end{aligned}$$

3.
$$\begin{aligned} -7+10&=m-5\\ 3&=m-5\\ 3+5&=m-5+5\\ 8&=m \end{aligned}$$

5.
$$\begin{aligned} 2w-12w&=40\\ -10w&=40\\ \frac{-10w}{-10}&=\frac{40}{-10}\\ w&=-4 \end{aligned}$$

7.
$$\begin{aligned} 24&=t+3t\\ 24&=4t\\ \frac{24}{4}&=\frac{4t}{4}\\ 6&=t \end{aligned}$$

9.
$$\begin{aligned} 2z&=12-14\\ 2z&=-2\\ \frac{2z}{2}&=\frac{-2}{2}\\ z&=-1 \end{aligned}$$

11.
$$\begin{aligned} 4-10&=\frac{z}{-3}\\ -6&=\frac{z}{-3}\\ -3\cdot(-6)&=-3\cdot\frac{z}{-3}\\ 18&=z \end{aligned}$$

13.
$$\begin{aligned} -3x-3x&=50-2\\ -6x&=48\\ \frac{-6x}{-6}&=\frac{48}{-6}\\ x&=-8 \end{aligned}$$

15.
$$\begin{aligned} \frac{x}{5}&=-26+16\\ \frac{x}{5}&=-10\\ 5\cdot\frac{x}{5}&=5\cdot(-10)\\ x&=-50 \end{aligned}$$

17.
$$\begin{aligned} 7x+7-6x&=10\\ 7x-6x+7&=10\\ x+7&=10\\ x+7-7&=10-7\\ x&=3 \end{aligned}$$

19.
$$\begin{aligned} -8-9&=3x+5-2x\\ -17&=x+5\\ -17-5&=x+5-5\\ -22&=x \end{aligned}$$

21.
$$\begin{aligned} 2(5x-3)&=11x\\ 2\cdot5x-2\cdot3&=11x\\ 10x-6&=11x\\ 10x-10x-6&=11x-10x\\ -6&=x \end{aligned}$$

23.
$$\begin{aligned} 3y&=2(y+12)\\ 3y&=2\cdot y+2\cdot12\\ 3y&=2y+24\\ 3y-2y&=2y-2y+24\\ y&=24 \end{aligned}$$

25.
$$\begin{aligned} 21y&=5(4y-6)\\ 21y&=5\cdot4y-5\cdot6\\ 21y&=20y-30\\ 21y-20y&=20y-20y-30\\ y&=-30 \end{aligned}$$

27.
$$\begin{aligned} -3(-4-2z)&=7z\\ -3\cdot(-4)-(-3)(2z)&=7z\\ 12+6z&=7z\\ 12+6z-6z&=7z-6z\\ 12&=z \end{aligned}$$

29.
$$\begin{aligned} 2x-8&=0\\ 2x-8+8&=0+8\\ 2x&=8\\ \frac{2x}{2}&=\frac{8}{2}\\ x&=4 \end{aligned}$$

Copyright © 2011 Pearson Education, Inc. Publishing as Prentice Hall.

31.
$$\begin{aligned} 7y+3&=24 \\ 7y+3-3&=24-3 \\ 7y&=21 \\ \frac{7y}{7}&=\frac{21}{7} \\ y&=3 \end{aligned}$$

33.
$$\begin{aligned} -7&=2x-1 \\ -7+1&=2x-1+1 \\ -6&=2x \\ \frac{-6}{2}&=\frac{2x}{2} \\ -3&=x \end{aligned}$$

35.
$$\begin{aligned} 6(6-4y)&=12y \\ 6\cdot 6-6\cdot 4y&=12y \\ 36-24y&=12y \\ 36-24y+24y&=12y+24y \\ 36&=36y \\ \frac{36}{36}&=\frac{36y}{36} \\ 1&=y \end{aligned}$$

37.
$$\begin{aligned} 11(x-6)&=-4-7 \\ 11\cdot x-11\cdot 6&=-11 \\ 11x-66&=-11 \\ 11x-66+66&=-11+66 \\ 11x&=55 \\ \frac{11x}{11}&=\frac{55}{11} \\ x&=5 \end{aligned}$$

39.
$$\begin{aligned} -3(x+1)-10&=12+8 \\ -3\cdot x-3\cdot 1-10&=20 \\ -3x-3-10&=20 \\ -3x-13&=20 \\ -3x-13+13&=20+13 \\ -3x&=33 \\ \frac{-3x}{-3}&=\frac{33}{-3} \\ x&=-11 \end{aligned}$$

41.
$$\begin{aligned} y-20&=6y \\ y-y-20&=6y-y \\ -20&=5y \\ \frac{-20}{5}&=\frac{5y}{5} \\ -4&=y \end{aligned}$$

43.
$$\begin{aligned} 22-42&=4(x-1)-4 \\ -20&=4\cdot x-4\cdot 1-4 \\ -20&=4x-4-4 \\ -20&=4x-8 \\ -20+8&=4x-8+8 \\ -12&=4x \\ \frac{-12}{4}&=\frac{4x}{4} \\ -3&=x \end{aligned}$$

45.
$$\begin{aligned} -2-3&=-4+x \\ -5&=-4+x \\ -5+4&=-4+4+x \\ -1&=x \end{aligned}$$

47.
$$\begin{aligned} y+1&=-3+4 \\ y+1&=1 \\ y+1-1&=1-1 \\ y&=0 \end{aligned}$$

49.
$$\begin{aligned} 3w-12w&=-27 \\ -9w&=-27 \\ \frac{-9w}{-9}&=\frac{-27}{-9} \\ w&=3 \end{aligned}$$

51.
$$\begin{aligned} -4x&=20-(-4) \\ -4x&=20+4 \\ -4x&=24 \\ \frac{-4x}{-4}&=\frac{24}{-4} \\ x&=-6 \end{aligned}$$

53.
$$\begin{aligned} 18-11&=\frac{x}{-5} \\ 7&=\frac{x}{-5} \\ -5\cdot 7&=-5\cdot\frac{x}{-5} \\ -5\cdot 7&=\frac{-5}{-5}\cdot x \\ -35&=x \end{aligned}$$

55.
$$\begin{aligned} 9x-12&=78 \\ 9x-12+12&=78+12 \\ 9x&=90 \\ \frac{9x}{9}&=\frac{90}{9} \\ x&=10 \end{aligned}$$

Copyright © 2011 Pearson Education, Inc. Publishing as Prentice Hall.

57. $10 = 7t - 12t$
$10 = -5t$
$\frac{10}{-5} = \frac{-5t}{-5}$
$-2 = t$

59. $5 - 5 = 3x + 2x$
$0 = 5x$
$\frac{0}{5} = \frac{5x}{5}$
$0 = x$

61. $50y = 7(7y + 4)$
$50y = 7 \cdot 7y + 7 \cdot 4$
$50y = 49y + 28$
$50y - 49y = 49y - 49y + 28$
$y = 28$

63. $8x = 2(6x + 10)$
$8x = 2 \cdot 6x + 2 \cdot 10$
$8x = 12x + 20$
$8x - 12x = 12x - 12x + 20$
$-4x = 20$
$\frac{-4x}{-4} = \frac{20}{-4}$
$x = -5$

65. $7x + 14 - 6x = -4 - 10$
$7x - 6x + 14 = -14$
$x + 14 = -14$
$x + 14 - 14 = -14 - 14$
$x = -28$

67. $\frac{x}{-4} = -1 - (-8)$
$\frac{x}{-4} = -1 + 8$
$\frac{x}{-4} = 7$
$-4 \cdot \frac{x}{-4} = -4 \cdot 7$
$x = -28$

69. $23x + 8 - 25x = 7 - 9$
$23x - 25x + 8 = -2$
$-2x + 8 = -2$
$-2x + 8 - 8 = -2 - 8$
$-2x = -10$
$\frac{-2x}{-2} = \frac{-10}{-2}$
$x = 5$

71. $-3(x + 9) - 41 = 4 - 60$
$-3 \cdot x - 3 \cdot 9 - 41 = 4 - 60$
$-3x - 27 - 41 = -56$
$-3x - 68 = -56$
$-3x - 68 + 68 = -56 + 68$
$-3x = 12$
$\frac{-3x}{-3} = \frac{12}{-3}$
$x = -4$

73. The sum of −7 and a number is $-7 + x$.

75. Eleven subtracted from a number is $x - 11$.

77. The product of −13 and a number is $-13x$.

79. A number divided by −12 is $\frac{x}{-12}$ or $-\frac{x}{12}$.

81. The product of −11 and a number, increased by 5 is $-11x + 5$.

83. Negative ten decreased by 7 times a number is $-10 - 7x$.

85. Seven added to the product of 4 and a number is $4x + 7$.

87. Twice a number, decreased by 17 is $2x - 17$.

89. The product of −6 and the sum of a number and 15 is $-6(x + 15)$.

91. The quotient of 45 and the product of a number and −5 is $\frac{45}{-5x}$ or $-\frac{45}{5x}$.

93. The quotient of seventeen and a number, increased by −15 is $\frac{17}{x} + (-15)$ or $\frac{17}{x} - 15$.

95. The longest bar corresponds to the year in which the number of trumpeter swans was the greatest. The year is 2005.

97. From the length of the bar, there were approximately 35,000 trumpeter swans in 2005.

99. answers may vary

101. no; answers may vary

103. answers may vary

Copyright © 2011 Pearson Education, Inc. Publishing as Prentice Hall.

105.
$$\frac{y}{72} = -86 - (-1029)$$
$$\frac{y}{72} = -86 + 1029$$
$$\frac{y}{72} = 943$$
$$72 \cdot \frac{y}{72} = 72 \cdot 943$$
$$\frac{72}{72} \cdot y = 72 \cdot 943$$
$$y = 67,896$$

107.
$$\frac{x}{-2} = 5^2 - |-10| - (-9)$$
$$\frac{x}{-2} = 25 - |-10| - (-9)$$
$$\frac{x}{-2} = 25 - 10 - (-9)$$
$$\frac{x}{-2} = 25 - 10 + 9$$
$$\frac{x}{-2} = 24$$
$$-2 \cdot \frac{x}{-2} = -2 \cdot 24$$
$$\frac{-2}{-2} \cdot x = -2 \cdot 24$$
$$x = -48$$

109.
$$|-13| + 3^2 = 100y - |-20| - 99y$$
$$13 + 3^2 = 100y - 20 - 99y$$
$$13 + 9 = 100y - 99y - 20$$
$$22 = y - 20$$
$$22 + 20 = y - 20 + 20$$
$$42 = y$$

Integrated Review

1. $7x - 5y + 14$ is an expression because it does not contain an equal sign.

2. $7x = 35 + 14$ is an equation because it contains an equal sign.

3. $3(x - 2) = 5(x + 1) - 17$ is an equation because it contains an equal sign.

4. $-9(2x + 1) - 4(x - 2) + 14$ is an expression because it does not contain an equal sign.

5. To simplify an expression, we combine any like terms.

6. To solve an equation, we use properties of equality to find any value of the variable that makes the equation a true statement.

7. $7x + x = (7 + 1)x = 8x$

8. $6y - 10y = (6 - 10)y = -4y$

9. $2a + 5a - 9a - 2 = (2 + 5 - 9)a - 2 = -2a - 2$

10.
$$6a - 12 - a - 14 = 6a - a - 12 - 14$$
$$= (6 - 1)a - 12 - 14$$
$$= 5a - 26$$

11. $-2(4x + 7) = -2 \cdot 4x + (-2) \cdot 7 = -8x - 14$

12. $-3(2x - 10) = -3(2x) - (-3)(10) = -6x + 30$

13.
$$5(y + 2) - 20 = 5 \cdot y + 5 \cdot 2 - 20$$
$$= 5y + 10 - 20$$
$$= 5y - 10$$

14.
$$12x + 3(x - 6) - 13 = 12x + 3 \cdot x - 3 \cdot 6 - 13$$
$$= 12x + 3x - 18 - 13$$
$$= (12 + 3)x - 18 - 13$$
$$= 15x - 31$$

15.
$$\text{Area} = (\text{length}) \cdot (\text{width})$$
$$= 3(4x - 2)$$
$$= 3 \cdot 4x - 3 \cdot 2$$
$$= 12x - 6$$
The area is $(12x - 6)$ square meters.

16. Perimeter $= x + x + 2 + 7 = 2x + 9$
The perimeter is $(2x + 9)$ feet.

17.
$$12 = 11x - 14x$$
$$12 = -3x$$
$$\frac{12}{-3} = \frac{-3x}{-3}$$
$$-4 = x$$
Check: $12 = 11x - 14x$
$$12 \stackrel{?}{=} 11(-4) - 14(-4)$$
$$12 \stackrel{?}{=} -44 + 56$$
$$12 = 12 \quad \text{True}$$
The solution is -4.

Copyright © 2011 Pearson Education, Inc. Publishing as Prentice Hall.

18. $8y+7y=-45$
$$15y=-45$$
$$\frac{15y}{15}=\frac{-45}{15}$$
$$y=-3$$
Check: $8y+7y=-45$
$$8(-3)+7(-3) \stackrel{?}{=} -45$$
$$-24+(-21) \stackrel{?}{=} -45$$
$$-45=-45 \quad \text{True}$$
The solution is −3.

19. $x-12=-45+23$
$$x-12=-22$$
$$x-12+12=-22+12$$
$$x=-10$$
Check: $x-12=-45+23$
$$-10-12 \stackrel{?}{=} -45+23$$
$$-22=-22 \quad \text{True}$$
The solution is −10.

20. $6-(-5)=x+5$
$$6+5=x+5$$
$$11=x+5$$
$$11-5=x+5-5$$
$$6=x$$
Check: $6-(-5)=x+5$
$$6-(-5) \stackrel{?}{=} 6+5$$
$$6+5 \stackrel{?}{=} 6+5$$
$$11=11 \quad \text{True}$$
The solution is 6.

21. $\frac{x}{3}=-14+9$
$$\frac{x}{3}=-5$$
$$3\cdot\frac{x}{3}=3(-5)$$
$$x=-15$$
Check: $\frac{x}{3}=-14+9$
$$\frac{-15}{3} \stackrel{?}{=} -14+9$$
$$-5=-5 \quad \text{True}$$
The solution is −15.

22. $\frac{z}{4}=-23-7$
$$\frac{z}{4}=-30$$
$$4\cdot\frac{z}{4}=4\cdot(-30)$$
$$z=-120$$
Check: $\frac{z}{4}=-23-7$
$$\frac{-120}{4} \stackrel{?}{=} -23-7$$
$$-30=-30 \quad \text{True}$$
The solution is −120.

23. $-6+2=4x+1-3x$
$$-4=x+1$$
$$-4-1=x+1-1$$
$$-5=x$$
Check: $-6+2=4x+1-3x$
$$-6+2 \stackrel{?}{=} 4(-5)+1-3(-5)$$
$$-4 \stackrel{?}{=} -20+1+15$$
$$-4=-4 \quad \text{True}$$
The solution is −5.

24. $5-8=5x+10-4x$
$$-3=x+10$$
$$-3-10=x+10-10$$
$$-13=x$$
Check: $5-8=5x+10-4x$
$$5-8 \stackrel{?}{=} 5(-13)+10-4(-13)$$
$$-3 \stackrel{?}{=} -65+10+52$$
$$-3=-3 \quad \text{True}$$
The solution is −13.

25. $6(3x-4)=19x$
$$6\cdot 3x-6\cdot 4=19x$$
$$18x-24=19x$$
$$18x-18x-24=19x-18x$$
$$-24=x$$
Check: $6(3x-4)=19x$
$$6[3(-24)-4] \stackrel{?}{=} 19(-24)$$
$$6(-72-4) \stackrel{?}{=} -456$$
$$6(-76) \stackrel{?}{=} -456$$
$$-456=-456 \quad \text{True}$$
The solution is −24.

Copyright © 2011 Pearson Education, Inc. Publishing as Prentice Hall.

26.
$$25x = 6(4x-9)$$
$$25x = 6 \cdot 4x - 6 \cdot 9$$
$$25x = 24x - 54$$
$$25x - 24x = 24x - 24x - 54$$
$$x = -54$$
Check: $25x = 6(4x-9)$
$$25(-54) \stackrel{?}{=} 6[4(-54)-9]$$
$$-1350 \stackrel{?}{=} 6(-216-9)$$
$$-1350 \stackrel{?}{=} 6(-225)$$
$$-1350 = -1350 \quad \text{True}$$
The solution is −54.

27.
$$-36x - 10 + 37x = -12 - (-14)$$
$$x - 10 = -12 + 14$$
$$x - 10 = 2$$
$$x - 10 + 10 = 2 + 10$$
$$x = 12$$
Check: $-36x - 10 + 37x = -12 - (-14)$
$$-36(12) - 10 + 37(12) \stackrel{?}{=} -12 - (-14)$$
$$-432 - 10 + 444 \stackrel{?}{=} -12 + 14$$
$$2 = 2 \quad \text{True}$$
The solution is 12.

28.
$$-8 + (-14) = -80y + 20 + 81y$$
$$-22 = y + 20$$
$$-22 - 20 = y + 20 - 20$$
$$-42 = y$$
Check: $-8 + (-14) = -80y + 20 + 81y$
$$-8 + (-14) \stackrel{?}{=} -80(-42) + 20 + 81(-42)$$
$$-22 \stackrel{?}{=} 3360 + 20 - 3402$$
$$-22 = -22 \quad \text{True}$$
The solution is −42.

29.
$$3x - 16 = -10$$
$$3x - 16 + 16 = -10 + 16$$
$$3x = 6$$
$$\frac{3x}{3} = \frac{6}{3}$$
$$x = 2$$
Check: $3x - 16 = -10$
$$3(2) - 16 \stackrel{?}{=} -10$$
$$6 - 16 \stackrel{?}{=} -10$$
$$-10 = -10 \quad \text{True}$$
The solution is 2.

30.
$$4x - 21 = -13$$
$$4x - 21 + 21 = -13 + 21$$
$$4x = 8$$
$$\frac{4x}{4} = \frac{8}{4}$$
$$x = 2$$
Check: $4x - 21 = -13$
$$4 \cdot 2 - 21 \stackrel{?}{=} -13$$
$$8 - 21 \stackrel{?}{=} -13$$
$$-13 = -13 \quad \text{True}$$
The solution is 2.

31.
$$-8z - 2z = 26 - (-4)$$
$$-10z = 26 + 4$$
$$-10z = 30$$
$$\frac{-10z}{-10} = \frac{30}{-10}$$
$$z = -3$$
Check: $-8z - 2z = 26 - (-4)$
$$-8(-3) - 2(-3) \stackrel{?}{=} 26 - (-4)$$
$$24 + 6 \stackrel{?}{=} 26 + 4$$
$$30 = 30 \quad \text{True}$$
The solution is −3.

32.
$$-12 + (-13) = 5x - 10x$$
$$-25 = -5x$$
$$\frac{-25}{-5} = \frac{-5x}{-5}$$
$$5 = x$$
Check: $-12 + (-13) = 5x - 10x$
$$-12 + (-13) \stackrel{?}{=} 5(5) - 10(5)$$
$$-25 \stackrel{?}{=} 25 - 50$$
$$-25 = -25 \quad \text{True}$$
The solution is 5.

33.
$$-4(x+8) - 11 = 3 - 26$$
$$-4 \cdot x + (-4) \cdot 8 - 11 = 3 - 26$$
$$-4x - 32 - 11 = -23$$
$$-4x - 43 = -23$$
$$-4x - 43 + 43 = -23 + 43$$
$$-4x = 20$$
$$\frac{-4x}{-4} = \frac{20}{-4}$$
$$x = -5$$
Check: $-4(x+8) - 11 = 3 - 26$
$$-4(-5+8) - 11 \stackrel{?}{=} 3 - 26$$
$$-4(3) - 11 \stackrel{?}{=} -23$$
$$-12 - 11 \stackrel{?}{=} -23$$
$$-23 = -23 \quad \text{True}$$
The solution is −5.

Copyright © 2011 Pearson Education, Inc. Publishing as Prentice Hall.

34.
$$\begin{aligned}-6(x-2)+10 &= -4-10\\ -6\cdot x-(-6)(2)+10 &= -4-10\\ -6x+12+10 &= -14\\ -6x+22 &= -14\\ -6x+22-22 &= -14-22\\ -6x &= -36\\ \frac{-6x}{-6} &= \frac{-36}{-6}\\ x &= 6\end{aligned}$$

Check:
$$\begin{aligned}-6(x-2)+10 &= -4-10\\ -6(6-2)+10 &\stackrel{?}{=} -4-10\\ -6(4)+10 &\stackrel{?}{=} -14\\ -24+10 &\stackrel{?}{=} -14\\ -14 &= -14 \quad \text{True}\end{aligned}$$

The solution is 6.

35. The difference of a number and 10 is $x - 10$.

36. The sum of −20 and a number is $-20 + x$.

37. The product of 10 and a number is $10x$.

38. The quotient of 10 and a number is $\frac{10}{x}$.

39. Five added to the product of −2 and a number is $-2x + 5$.

40. The product of −4 and the difference of a number and 1 is $-4(x - 1)$.

Section 3.3

Practice Problems

1.
$$\begin{aligned}7x+12 &= 3x-4\\ 7x+12-12 &= 3x-4-12\\ 7x &= 3x-16\\ 7x-3x &= 3x-16-3x\\ 4x &= -16\\ \frac{4x}{4} &= \frac{-16}{4}\\ x &= -4\end{aligned}$$

2.
$$\begin{aligned}40-5y+5 &= -2y-10-4y\\ 45-5y &= -6y-10\\ 45-5y-45 &= -6y-10-45\\ -5y &= -6y-55\\ -5y+6y &= -6y-55+6y\\ y &= -55\end{aligned}$$

3.
$$\begin{aligned}6(a-5) &= 4a+4\\ 6a-30 &= 4a+4\\ 6a-30-4a &= 4a+4-4a\\ 2a-30 &= 4\\ 2a-30+30 &= 4+30\\ 2a &= 34\\ \frac{2a}{2} &= \frac{34}{2}\\ a &= 17\end{aligned}$$

4.
$$\begin{aligned}4(x+3)+1 &= 13\\ 4x+12+1 &= 13\\ 4x+13 &= 13\\ 4x+13-13 &= 13-13\\ 4x &= 0\\ \frac{4x}{4} &= \frac{0}{4}\\ x &= 0\end{aligned}$$

5. a. The difference of 110 and 80 is 30 translates to $110 - 80 = 30$.

b. The product of 3 and the sum of −9 and 11 amounts to 6 translates to $3(-9 + 11) = 6$.

c. The quotient of 24 and −6 yields −4 translates to $\frac{24}{-6} = -4$.

Calculator Explorations

1. Replace x with 12.
$76(12 - 25) = -988$
Yes

2. Replace x with 35.
$-47 \cdot 35 + 862 = -783$
Yes

3. Replace x with −170.
$-170 + 562 = 392$
$3 \cdot (-170) + 900 = 390$
No

4. Replace x with −18.
$55(-18 + 10) = -440$
$75 \cdot (-18) + 910 = -440$
Yes

5. Replace x with −21.
$29 \cdot (-21) - 1034 = -1643$
$61 \cdot (-21) - 362 = -1643$
Yes

Copyright © 2011 Pearson Education, Inc. Publishing as Prentice Hall.

6. Replace x with 25.
$-38 \cdot 25 + 205 = -745$
$25 \cdot 25 + 120 = 745$
No

Vocabulary and Readiness Check

1. An example of an expression is $\underline{3x - 9 + x - 16}$ while an example of an equation is $\underline{5(2x + 6) - 1 = 39}$.

2. To solve $\frac{x}{-7} = -10$, we use the <u>multiplication</u> property of equality.

3. To solve $x - 7 = -10$, we use the <u>addition</u> property of equality.

4. To solve $9x - 6x = 10 + 6$, first <u>combine like terms</u>.

5. To solve $5(x - 1) = 25$, first use the <u>distributive</u> property.

6. To solve $4x + 3 = 19$, first use the <u>addition</u> property of equality.

Exercise Set 3.3

1.
$$\begin{aligned} 3x-7&=4x+5\\ 3x-3x-7&=4x-3x+5\\ -7&=x+5\\ -7-5&=x+5-5\\ -12&=x \end{aligned}$$

3.
$$\begin{aligned} 10x+15&=6x+3\\ 10x+15-15&=6x+3-15\\ 10x&=6x-12\\ 10x-6x&=6x-6x-12\\ 4x&=-12\\ \frac{4x}{4}&=\frac{-12}{4}\\ x&=-3 \end{aligned}$$

5.
$$\begin{aligned} 19-3x&=14+2x\\ 19-3x+3x&=14+2x+3x\\ 19&=14+5x\\ 19-14&=14-14+5x\\ 5&=5x\\ \frac{5}{5}&=\frac{5x}{5}\\ 1&=x \end{aligned}$$

7.
$$\begin{aligned} -14x-20&=-12x+70\\ -14x+12x-20&=-12x+12x+70\\ -2x-20&=70\\ -2x-20+20&=70+20\\ -2x&=90\\ \frac{-2x}{-2}&=\frac{90}{-2}\\ x&=-45 \end{aligned}$$

9.
$$\begin{aligned} x+20+2x&=-10-2x-15\\ x+2x+20&=-10-15-2x\\ 3x+20&=-25-2x\\ 3x+2x+20&=-25-2x+2x\\ 5x+20&=-25\\ 5x+20-20&=-25-20\\ 5x&=-45\\ \frac{5x}{5}&=\frac{-45}{5}\\ x&=-9 \end{aligned}$$

11.
$$\begin{aligned} 40+4y-16&=13y-12-3y\\ 40-16+4y&=13y-3y-12\\ 24+4y&=10y-12\\ 24+4y-4y&=10y-4y-12\\ 24&=6y-12\\ 24+12&=6y-12+12\\ 36&=6y\\ \frac{36}{6}&=\frac{6y}{6}\\ 6&=y \end{aligned}$$

13.
$$\begin{aligned} 35-17&=3(x-2)\\ 18&=3x-6\\ 18+6&=3x-6+6\\ 24&=3x\\ \frac{24}{3}&=\frac{3x}{3}\\ 8&=x \end{aligned}$$

15.
$$\begin{aligned} 3(x-1)-12&=0\\ 3x-3-12&=0\\ 3x-15&=0\\ 3x-15+15&=0+15\\ 3x&=15\\ \frac{3x}{3}&=\frac{15}{3}\\ x&=5 \end{aligned}$$

Copyright © 2011 Pearson Education, Inc. Publishing as Prentice Hall.

17.
$$\begin{aligned} 2(y-3) &= y-6 \\ 2y-6 &= y-6 \\ 2y-y-6 &= y-y-6 \\ y-6 &= -6 \\ y-6+6 &= -6+6 \\ y &= 0 \end{aligned}$$

19.
$$\begin{aligned} -2(y+4) &= 2 \\ -2y-8 &= 2 \\ -2y-8+8 &= 2+8 \\ -2y &= 10 \\ \frac{-2y}{-2} &= \frac{10}{-2} \\ y &= -5 \end{aligned}$$

21.
$$\begin{aligned} 2t-1 &= 3(t+7) \\ 2t-1 &= 3t+21 \\ 2t-2t-1 &= 3t-2t+21 \\ -1 &= t+21 \\ -1-21 &= t+21-21 \\ -22 &= t \end{aligned}$$

23.
$$\begin{aligned} 3(5c+1)-12 &= 13c+3 \\ 15c+3-12 &= 13c+3 \\ 15c-9 &= 13c+3 \\ 15c-13c-9 &= 13c-13c+3 \\ 2c-9 &= 3 \\ 2c-9+9 &= 3+9 \\ 2c &= 12 \\ \frac{2c}{2} &= \frac{12}{2} \\ c &= 6 \end{aligned}$$

25.
$$\begin{aligned} -4x &= 44 \\ \frac{-4x}{-4} &= \frac{44}{-4} \\ x &= -11 \end{aligned}$$

27.
$$\begin{aligned} x+9 &= 2 \\ x+9-9 &= 2-9 \\ x &= -7 \end{aligned}$$

29.
$$\begin{aligned} 8-b &= 13 \\ 8-8-b &= 13-8 \\ -b &= 5 \\ \frac{-b}{-1} &= \frac{5}{-1} \\ b &= -5 \end{aligned}$$

31.
$$\begin{aligned} -20-(-50) &= \frac{x}{9} \\ -20+50 &= \frac{x}{9} \\ 30 &= \frac{x}{9} \\ 9\cdot 30 &= 9\cdot\frac{x}{9} \\ 270 &= x \end{aligned}$$

33.
$$\begin{aligned} 3r+4 &= 19 \\ 3r+4-4 &= 19-4 \\ 3r &= 15 \\ \frac{3r}{3} &= \frac{15}{3} \\ r &= 5 \end{aligned}$$

35.
$$\begin{aligned} -7c+1 &= -20 \\ -7c+1-1 &= -20-1 \\ -7c &= -21 \\ \frac{-7c}{-7} &= \frac{-21}{-7} \\ c &= 3 \end{aligned}$$

37.
$$\begin{aligned} 8y-13y &= -20-25 \\ -5y &= -45 \\ \frac{-5y}{-5} &= \frac{-45}{-5} \\ y &= 9 \end{aligned}$$

39.
$$\begin{aligned} 6(7x-1) &= 43x \\ 42x-6 &= 43x \\ 42x-42x-6 &= 43x-42x \\ -6 &= x \end{aligned}$$

41.
$$\begin{aligned} -4+12 &= 16x-3-15x \\ 8 &= x-3 \\ 8+3 &= x-3+3 \\ 11 &= x \end{aligned}$$

43.
$$\begin{aligned} -10(x+3)+28 &= -16-16 \\ -10x-30+28 &= -32 \\ -10x-2 &= -32 \\ -10x-2+2 &= -32+2 \\ -10x &= -30 \\ \frac{-10x}{-10} &= \frac{-30}{-10} \\ x &= 3 \end{aligned}$$

Copyright © 2011 Pearson Education, Inc. Publishing as Prentice Hall.

45.
$$\begin{aligned} 4x+3&=2x+11\\ 4x-2x+3&=2x-2x+11\\ 2x+3&=11\\ 2x+3-3&=11-3\\ 2x&=8\\ \frac{2x}{2}&=\frac{8}{2}\\ x&=4 \end{aligned}$$

47.
$$\begin{aligned} -2y-10&=5y+18\\ -2y-10-5y&=5y+18-5y\\ -7y-10&=18\\ -7y-10+10&=18+10\\ -7y&=28\\ \frac{-7y}{-7}&=\frac{28}{-7}\\ y&=-4 \end{aligned}$$

49.
$$\begin{aligned} -8n+1&=-6n-5\\ -8n+8n+1&=-6n+8n-5\\ 1&=2n-5\\ 1+5&=2n-5+5\\ 6&=2n\\ \frac{6}{2}&=\frac{2n}{2}\\ 3&=n \end{aligned}$$

51.
$$\begin{aligned} 9-3x&=14+2x\\ 9-3x-2x&=14+2x-2x\\ 9-5x&=14\\ 9-5x-9&=14-9\\ -5x&=5\\ \frac{-5x}{-5}&=\frac{5}{-5}\\ x&=-1 \end{aligned}$$

53.
$$\begin{aligned} 9a+29+7&=0\\ 9a+36&=0\\ 9a+36-36&=0-36\\ 9a&=-36\\ \frac{9a}{9}&=\frac{-36}{9}\\ a&=-4 \end{aligned}$$

55.
$$\begin{aligned} 7(y-2)&=4y-29\\ 7y-14&=4y-29\\ 7y-4y-14&=4y-4y-29\\ 3y-14&=-29\\ 3y-14+14&=-29+14\\ 3y&=-15\\ \frac{3y}{3}&=\frac{-15}{3}\\ y&=-5 \end{aligned}$$

57.
$$\begin{aligned} 12+5t&=6(t+2)\\ 12+5t&=6t+12\\ 12+5t-5t&=6t-5t+12\\ 12&=t+12\\ 12-12&=t+12-12\\ 0&=t \end{aligned}$$

59.
$$\begin{aligned} 3(5c-1)-2&=13c+3\\ 15c-3-2&=13c+3\\ 15c-5&=13c+3\\ 15c-13c-5&=13c-13c+3\\ 2c-5&=3\\ 2c-5+5&=3+5\\ 2c&=8\\ \frac{2c}{2}&=\frac{8}{2}\\ c&=4 \end{aligned}$$

61.
$$\begin{aligned} 10+5(z-2)&=4z+1\\ 10+5z-10&=4z+1\\ 5z&=4z+1\\ 5z-4z&=4z-4z+1\\ z&=1 \end{aligned}$$

63.
$$\begin{aligned} 7(6+w)&=6(2+w)\\ 42+7w&=12+6w\\ 42+7w-6w&=12+6w-6w\\ 42+w&=12\\ 42-42+w&=12-42\\ w&=-30 \end{aligned}$$

65. The sum of -42 and 16 is -26 translates to $-42 + 16 = -26$.

67. The product of -5 and -29 gives 145 translates to $-5(-29) = 145$.

69. Three times the difference of -14 and 2 amounts to -48 translates to $3(-14 - 2) = -48$.

Copyright © 2011 Pearson Education, Inc. Publishing as Prentice Hall.

71. The quotient of 100 and twice 50 is equal to 1 translates to $\frac{100}{2(50)} = 1$.

73. From the height of the bar, approximately 97 million returns are expected to be filed electronically in 2010.

75. Subtract the number of returns in 2006 from the number in 2009.
95 million − 81 million = 14 million
The increase from 2006 to 2009 was approximately 14 million returns.

77. $$\begin{aligned} x^3 - 2xy &= 3^3 - 2(3)(-1) \\ &= 27 - 2(3)(-1) \\ &= 27 - (-6) \\ &= 27 + 6 \\ &= 33 \end{aligned}$$

79. $$\begin{aligned} y^5 - 4x^2 &= (-1)^5 - 4(3)^2 \\ &= -1 - 4(9) \\ &= -1 - 36 \\ &= -37 \end{aligned}$$

81. The first step in solving $2x - 5 = -7$ is to add 5 to both sides, which is choice b.

83. The first step in solving $-3x = -12$ is to divide both sides by −3, which is choice a.

85. The error is in the second line.
$$\begin{aligned} 2(3x-5) &= 5x - 7 \\ 6x - 10 &= 5x - 7 \\ 6x - 10 + 10 &= 5x - 7 + 10 \\ 6x &= 5x + 3 \\ 6x - 5x &= 5x + 3 - 5x \\ x &= 3 \end{aligned}$$

87. $$\begin{aligned} (-8)^2 + 3x &= 5x + 4^3 \\ 64 + 3x &= 5x + 64 \\ 64 + 3x - 3x &= 5x - 3x + 64 \\ 64 &= 2x + 64 \\ 64 - 64 &= 2x + 64 - 64 \\ 0 &= 2x \\ \frac{0}{2} &= \frac{2x}{2} \\ 0 &= x \end{aligned}$$

89. $$\begin{aligned} 2^3(x+4) &= 3^2(x+4) \\ 8(x+4) &= 9(x+4) \\ 8x + 32 &= 9x + 36 \\ 8x - 8x + 32 &= 9x - 8x + 36 \\ 32 &= x + 36 \\ 32 - 36 &= x + 36 - 36 \\ -4 &= x \end{aligned}$$

91. no; answers may vary

Section 3.4

Practice Problems

1. a. "Four times a number is 20" is $4x = 20$.

b. "The sum of a number and −5 yields 32" is $x + (-5) = 32$.

c. "Fifteen subtracted from a number amounts to −23" is $x - 15 = -23$.

d. "Five times the difference of a number and 7 is equal to −8" is $5(x - 7) = -8$.

e. "The quotient of triple a number and 5 gives 1" is $\frac{3x}{5} = 1$.

2. "The sum of a number and 2 equals 6 added to three times the number" is
$$\begin{aligned} x + 2 &= 6 + 3x \\ x - x + 2 &= 6 + 3x - x \\ 2 &= 6 + 2x \\ 2 - 6 &= 6 - 6 + 2x \\ -4 &= 2x \\ \frac{-4}{2} &= \frac{2x}{2} \\ -2 &= x \end{aligned}$$

3. Let x be the distance from Denver to San Francisco. Since the distance from Cincinnati to Denver is 71 miles less than the distance from Denver to San Francisco, the distance from Cincinnati to Denver is $x - 71$. Since the total of the two distances is 2399, the sum of x and $x - 71$ is 2399.

Copyright © 2011 Pearson Education, Inc. Publishing as Prentice Hall.

$$x + x - 71 = 2399$$
$$2x - 71 = 2399$$
$$2x - 71 + 71 = 2399 + 71$$
$$2x = 2470$$
$$\frac{2x}{2} = \frac{2470}{2}$$
$$x = 1235$$

The distance from Denver to San Francisco is 1235 miles.

4. Let x be the amount her son receives. Since her husband receives twice as much as her son, her husband receives $2x$. Since the total estate is \$57,000, the sum of x and $2x$ is 57,000.

$$x + 2x = 57,000$$
$$3x = 57,000$$
$$\frac{3x}{3} = \frac{57,000}{3}$$
$$x = 19,000$$

$2x = 2(19,000) = 38,000$

Her husband will receive \$38,000 and her son will receive \$19,000.

Exercise Set 3.4

1. "A number added to –5 is –7" is $-5 + x = -7$.

3. "Three times a number yields 27" is $3x = 27$.

5. "A number subtracted from –20 amounts to 104" is $-20 - x = 104$.

7. "Twice a number gives 108" is $2x = 108$.

9. "The product of 5 and the sum of –3 and a number is –20" is $5(-3 + x) = -20$.

11. "Three times a number, added to 9 is 33" is

$$9 + 3x = 33$$
$$9 - 9 + 3x = 33 - 9$$
$$3x = 24$$
$$\frac{3x}{3} = \frac{24}{3}$$
$$x = 8$$

13. "The sum of 3, 4, and a number amounts to 16" is

$$3 + 4 + x = 16$$
$$7 + x = 16$$
$$7 - 7 + x = 16 - 7$$
$$x = 9$$

15. "The difference of a number and 3 is equal to the quotient of 10 and 5" is

$$x - 3 = \frac{10}{5}$$
$$x - 3 = 2$$
$$x - 3 + 3 = 2 + 3$$
$$x = 5$$

17. "Thirty less a number is equal to the product of 3 and the sum of the number and 6" is

$$30 - x = 3(x + 6)$$
$$30 - x = 3x + 18$$
$$30 - 30 - x = 3x + 18 - 30$$
$$-x = 3x - 12$$
$$-x - 3x = 3x - 3x - 12$$
$$-4x = -12$$
$$\frac{-4x}{-4} = \frac{-12}{-4}$$
$$x = 3$$

19. "40 subtracted from five times a number is 8 more than the number" is

$$5x - 40 = x + 8$$
$$5x - x - 40 = x - x + 8$$
$$4x - 40 = 8$$
$$4x - 40 + 40 = 8 + 40$$
$$4x = 48$$
$$\frac{4x}{4} = \frac{48}{4}$$
$$x = 12$$

21. "Three times the difference of some number and 5 amounts to the quotient of 108 and 12" is

$$3(x - 5) = \frac{108}{12}$$
$$3x - 15 = 9$$
$$3x - 15 + 15 = 9 + 15$$
$$3x = 24$$
$$\frac{3x}{3} = \frac{24}{3}$$
$$x = 8$$

23. "The product of 4 and a number is the same as 30 less twice that same number" is

$$4x = 30 - 2x$$
$$4x + 2x = 30 - 2x + 2x$$
$$6x = 30$$
$$\frac{6x}{6} = \frac{30}{6}$$
$$x = 5$$

Copyright © 2011 Pearson Education, Inc. Publishing as Prentice Hall.

25. The equation is $x + x - 28 = 82$.

$$\begin{aligned} x+x-28 &= 82 \\ 2x-28 &= 82 \\ 2x-28+28 &= 82+28 \\ 2x &= 110 \\ \frac{2x}{2} &= \frac{110}{2} \\ x &= 55 \end{aligned}$$

California has 55 votes and Florida has $55 - 28 = 27$ votes.

27. Let x be the fastest speed of the pheasant. Since a falcon's fastest speed is five times as fast as a pheasant, a falcon's fastest speed is $5x$. Since the total speeds for these two birds is 222 miles per hour, the sum of x and $5x$ is 222.

$$\begin{aligned} x+5x &= 222 \\ 6x &= 222 \\ \frac{6x}{6} &= \frac{222}{6} \\ x &= 37 \end{aligned}$$

The pheasant's fastest speed is 37 miles per hour and the falcon's fastest speed is $5(37) = 185$ miles per hour.

29. Let x be the enrollment (in thousands) at the largest university in India. Since the largest university in Pakistan has 306 thousand more students than the one in India, its enrollment is $x + 306$. Since the combined enrollment is 3306 thousand students, the sum of x and $x + 306$ is 3306.

$$\begin{aligned} x+x+306 &= 3306 \\ 2x+306 &= 3306 \\ 2x+306-306 &= 3306-306 \\ 2x &= 3000 \\ \frac{2x}{2} &= \frac{3000}{2} \\ x &= 1500 \end{aligned}$$

The enrollment at the largest university in India is 1500 thousand students, and the enrollment at the largest university in Pakistan is $1500 + 306 = 1806$ thousand students.

31. Let x be the cost of the games. Since the cost of the Xbox 360 is 3 times as much as the games, the cost of the Xbox 360 is $3x$. Since the total cost is \$560, the sum of x and $3x$ is 560.

$$\begin{aligned} x+3x &= 560 \\ 4x &= 560 \\ \frac{4x}{4} &= \frac{560}{4} \\ x &= 140 \end{aligned}$$

The games cost \$140 and the Xbox 360 costs $3(140) = \$420$.

33. Let x be the distance from Los Angeles to Tokyo. Since the distance from New York to London is 2001 miles less than the distance from Los Angeles to Tokyo, the distance from New York to London is $x - 2001$. Since the total of the two distances is 8939, the sum of x and $x - 2001$ is 8939.

$$\begin{aligned} x+x-2001 &= 8939 \\ 2x-2001 &= 8939 \\ 2x-2001+2001 &= 8939+2001 \\ 2x &= 10{,}940 \\ \frac{2x}{2} &= \frac{10{,}940}{2} \\ x &= 5470 \end{aligned}$$

The distance from Los Angeles to Tokyo is 5470 miles.

35. Let x be the capacity of Michigan Stadium. Since the capacity of Beaver Stadium is 1081 more than that of Michigan Stadium, the capacity of Beaver Stadium is $x + 1081$. Since the combined capacity is 213,483, the sum of x and $x + 1081$ is 213,483.

$$\begin{aligned} x+x+1081 &= 213{,}483 \\ 2x+1081 &= 213{,}483 \\ 2x+1081-1081 &= 213{,}483-1081 \\ 2x &= 212{,}402 \\ \frac{2x}{2} &= \frac{212{,}402}{2} \\ x &= 106{,}201 \end{aligned}$$

The capacity of Michigan Stadium is 106,201 and the capacity of Beaver Stadium is $106{,}201 + 1081 = 107{,}282$.

37. Let x be the number of tourists projected to visit Spain in 2020. Since the number of tourists projected to visit China in 2020 is twice the number projected for Spain, the number projected for China is $2x$. Since the total number of tourists projected for the two countries is 210 million, the sum of x and $2x$ is 210 million.

$$\begin{aligned} x+2x &= 210 \\ 3x &= 210 \\ \frac{3x}{3} &= \frac{210}{3} \\ x &= 70 \end{aligned}$$

$2x = 2(70) = 140$

70 million tourists are projected to visit Spain in 2020; 140 million are projected to visit China in 2020.

Copyright © 2011 Pearson Education, Inc. Publishing as Prentice Hall.

39. Let x be the number of cars manufactured in Spain. Since the number of cars manufactured in Germany is twice as many as those manufactured in Spain, the number of cars manufactured in Germany is $2x$. Since the total number of these cars is 19,827, the sum of x and $2x$ is 19,827.

$$x + 2x = 19,827$$
$$3x = 19,827$$
$$\frac{3x}{3} = \frac{19,827}{3}$$
$$x = 6609$$

The number of cars manufactured in Spain is 6609 per day and the number in Germany is 2(6609) = 13,218 per day.

41. Let x be the amount the biker received for the accessories. Since he received five times as much for the bike, he received $5x$ for the bike. Since he received a total of \$270 for the bike and accessories, the sum of x and $5x$ is 270.

$$x + 5x = 270$$
$$6x = 270$$
$$\frac{6x}{6} = \frac{270}{6}$$
$$x = 45$$

Thus, the biker received 5 · \$45 = \$225 for the bike.

43. Let x be the points scored by the Stanford Cardinal. Since the Tennessee Lady Volunteers scored 16 points more than the Stanford Cardinal, the Tennessee Lady Volunteers scored x + 16 points. Since both teams scored a total of 132 points, the sum of x and x + 16 is 132.

$$x + x + 16 = 132$$
$$2x + 16 = 132$$
$$2x + 16 - 16 = 132 - 16$$
$$2x = 116$$
$$\frac{2x}{2} = \frac{116}{2}$$
$$x = 58$$

The Tennessee Lady Volunteers scored 58 + 16 = 74 points.

45. Let x be the number of computers in Japan. Since the USA has 162,550 million more computers than Japan, the number of computers in the USA is x + 162,550. Since the total number of computers in the two countries is 318,450 million, the sum of x and x + 162,550 is 318,450.

$$x + x + 162,550 = 318,450$$
$$2x + 162,550 = 318,450$$
$$2x + 162,550 - 162,550 = 318,450 - 162,550$$
$$2x = 155,900$$
$$\frac{2x}{2} = \frac{155,900}{2}$$
$$x = 77,950$$

Japan has 77,950 million computers, and the USA has 77,950 + 162,550 = 240,500 million computers.

47. To round 586 to the nearest ten, observe that the digit in the ones place is 6. Since this digit is at least 5, we add 1 to the digit in the tens place. 586 rounded to the nearest ten is 590.

49. To round 1026 to the nearest hundred, observe that the digit in the tens place is 2. Since this digit is less than 5, we do not add 1 to the digit in the hundreds place. 1026 rounded to the nearest hundred is 1000.

51. To round 2986 to the nearest thousand, observe that the digit in the hundreds place is 9. Since this digit is at least 5, we add 1 to the digit in the thousands place. 2986 rounded to the nearest thousand is 3000.

53. Yes; answers may vary

55. Use $P = A + C$, where P = 230,000 and C = 13,800.

$$P = A + C$$
$$230,000 = A + 13,800$$
$$230,000 - 13,800 = A + 13,800 - 13,800$$
$$216,200 = A$$

The seller received \$216,200.

57. Use $P = C + M$ where P = 999 and M = 450.

$$P = C + M$$
$$999 = C + 450$$
$$999 - 450 = C + 450 - 450$$
$$549 = C$$

The wholesale cost is \$549.

Chapter 3 Vocabulary Check

1. An algebraic expression is <u>simplified</u> when all like terms have been <u>combined</u>.

2. Terms that are exactly the same, except that they may have different numerical coefficients, are called <u>like</u> terms.

Copyright © 2011 Pearson Education, Inc. Publishing as Prentice Hall.

3. A letter used to represent a number is called a variable.

4. A combination of operations on variables and numbers is called an algebraic expression.

5. The addends of an algebraic expression are called the terms of the expression.

6. The number factor of a variable term is called the numerical coefficient.

7. Replacing a variable in an expression by a number and then finding the value of the expression is called evaluating the expression for the variable.

8. A term that is a number only is called a constant.

9. An equation is of the form expression = expression.

10. A solution of an equation is a value for the variable that makes an equation a true statement.

11. To multiply $-3(2x + 1)$, we use the distributive property.

12. By the multiplication property of equality, we may multiply or divide both sides of an equation by any nonzero number without changing the solution of the equation.

13. By the addition property of equality, the same number may be added to or subtracted from both sides of an equation without changing the solution of the equation.

Chapter 3 Review

1. $3y + 7y - 15 = (3 + 7)y - 15 = 10y - 15$

2. $$\begin{aligned} 2y-10-8y &= 2y-8y-10 \\ &= (2-8)y-10 \\ &= -6y-10 \end{aligned}$$

3. $$\begin{aligned} 8a+a-7-15a &= 8a+a-15a-7 \\ &= (8+1-15)a-7 \\ &= -6a-7 \end{aligned}$$

4. $$\begin{aligned} y+3-9y-1 &= y-9y+3-1 \\ &= (1-9)y+3-1 \\ &= -8y+2 \end{aligned}$$

5. $2(x + 5) = 2 \cdot x + 2 \cdot 5 = 2x + 10$

6. $-3(y + 8) = -3 \cdot y + (-3) \cdot 8 = -3y - 24$

7. $$\begin{aligned} 7x+3(x-4)+x &= 7x+3\cdot x-3\cdot 4+x \\ &= 7x+3x-12+x \\ &= 7x+3x+x-12 \\ &= (7+3+1)x-12 \\ &= 11x-12 \end{aligned}$$

8. $$\begin{aligned} -(3m+2)-m-10 &= -1(3m+2)-m-10 \\ &= -1\cdot 3m+(-1)\cdot 2-m-10 \\ &= -3m-2-m-10 \\ &= -3m-m-2-10 \\ &= (-3-1)m-2-10 \\ &= -4m-12 \end{aligned}$$

9. $$\begin{aligned} 3(5a-2)-20a+10 &= 3\cdot 5a-3\cdot 2-20a+10 \\ &= 15a-6-20a+10 \\ &= 15a-20a-6+10 \\ &= (15-20)a-6+10 \\ &= -5a+4 \end{aligned}$$

10. $$\begin{aligned} 6y+3+2(3y-6) &= 6y+3+2\cdot 3y-2\cdot 6 \\ &= 6y+3+6y-12 \\ &= 6y+6y+3-12 \\ &= (6+6)y+3-12 \\ &= 12y-9 \end{aligned}$$

11. $$\begin{aligned} 6y-7+11y-y+2 &= 6y+11y-y-7+2 \\ &= (6+11-1)y-7+2 \\ &= 16y-5 \end{aligned}$$

12. $$\begin{aligned} 10-x+5x-12-3x &= -x+5x-3x+10-12 \\ &= (-1+5-3)x+10-12 \\ &= 1x-2 \\ &= x-2 \end{aligned}$$

13. Perimeter $= 2(2x + 3) = 2 \cdot 2x + 2 \cdot 3 = 4x + 6$
The perimeter is $(4x + 6)$ yards.

14. Perimeter $= 4 \cdot 5y = 20y$
The perimeter is $20y$ meters.

15. $$\begin{aligned} \text{Area} &= (\text{length})\cdot(\text{width}) \\ &= 3\cdot(2x-1) \\ &= 3\cdot 2x-3\cdot 1 \\ &= 6x-3 \end{aligned}$$
The area is $(6x - 3)$ square yards.

Copyright © 2011 Pearson Education, Inc. Publishing as Prentice Hall.

16. Add the areas of the two rectangles.

$$\begin{aligned}10(x-2)+7(5x+4) &= 10\cdot x-10\cdot 2+7\cdot 5x+7\cdot 4\\ &= 10x-20+35x+28\\ &= 10x+35x-20+28\\ &= (10+35)x-20+28\\ &= 45x+8\end{aligned}$$

The area is $(45x + 8)$ square centimeters.

17.
$$\begin{aligned}z-5&=-7\\ z-5+5&=-7+5\\ z&=-2\end{aligned}$$

18.
$$\begin{aligned}3x+10&=4x\\ 3x-3x+10&=4x-3x\\ 10&=x\end{aligned}$$

19.
$$\begin{aligned}3y&=-21\\ \frac{3y}{3}&=\frac{-21}{3}\\ y&=-7\end{aligned}$$

20.
$$\begin{aligned}-3a&=-15\\ \frac{-3a}{-3}&=\frac{-15}{-3}\\ a&=5\end{aligned}$$

21.
$$\begin{aligned}\frac{x}{-6}&=2\\ -6\cdot\frac{x}{-6}&=-6\cdot 2\\ x&=-12\end{aligned}$$

22.
$$\begin{aligned}\frac{y}{-15}&=-3\\ -15\cdot\frac{y}{-15}&=-15\cdot(-3)\\ y&=45\end{aligned}$$

23.
$$\begin{aligned}n+18&=10-(-2)\\ n+18&=10+2\\ n+18&=12\\ n+18-18&=12-18\\ n&=-6\end{aligned}$$

24.
$$\begin{aligned}c-5&=-13+7\\ c-5&=-6\\ c-5+5&=-6+5\\ c&=-1\end{aligned}$$

25.
$$\begin{aligned}7x+5-6x&=-20\\ x+5&=-20\\ x+5-5&=-20-5\\ x&=-25\end{aligned}$$

26.
$$\begin{aligned}17x&=2(8x-4)\\ 17x&=2\cdot 8x-2\cdot 4\\ 17x&=16x-8\\ 17x-16x&=16x-16x-8\\ x&=-8\end{aligned}$$

27.
$$\begin{aligned}5x+7&=-3\\ 5x+7-7&=-3-7\\ 5x&=-10\\ \frac{5x}{5}&=\frac{-10}{5}\\ x&=-2\end{aligned}$$

28.
$$\begin{aligned}-14&=9y+4\\ -14-4&=9y+4-4\\ -18&=9y\\ \frac{-18}{9}&=\frac{9y}{9}\\ -2&=y\end{aligned}$$

29.
$$\begin{aligned}\frac{z}{4}&=-8-(-6)\\ \frac{z}{4}&=-8+6\\ \frac{z}{4}&=-2\\ 4\cdot\frac{z}{4}&=4\cdot(-2)\\ z&=-8\end{aligned}$$

30.
$$\begin{aligned}-1+(-8)&=\frac{x}{5}\\ -9&=\frac{x}{5}\\ 5\cdot(-9)&=5\cdot\frac{x}{5}\\ -45&=x\end{aligned}$$

31.
$$\begin{aligned}6y-7y&=100-105\\ -y&=-5\\ \frac{-y}{-1}&=\frac{-5}{-1}\\ y&=5\end{aligned}$$

Copyright © 2011 Pearson Education, Inc. Publishing as Prentice Hall.

32. $$\begin{aligned} 19x-16x &= 45-60 \\ 3x &= -15 \\ \frac{3x}{3} &= \frac{-15}{3} \\ x &= -5 \end{aligned}$$

33. $$\begin{aligned} 9(2x-7) &= 19x \\ 9\cdot 2x - 9\cdot 7 &= 19x \\ 18x-63 &= 19x \\ 18x-18x-63 &= 19x-18x \\ -63 &= x \end{aligned}$$

34. $$\begin{aligned} -5(3x+3) &= -14x \\ -5\cdot 3x - 5\cdot 3 &= -14x \\ -15x-15 &= -14x \\ -15x+15x-15 &= -14x+15x \\ -15 &= x \end{aligned}$$

35. $$\begin{aligned} 3x-4 &= 11 \\ 3x-4+4 &= 11+4 \\ 3x &= 15 \\ \frac{3x}{3} &= \frac{15}{3} \\ x &= 5 \end{aligned}$$

36. $$\begin{aligned} 6y+1 &= 73 \\ 6y+1-1 &= 73-1 \\ 6y &= 72 \\ \frac{6y}{6} &= \frac{72}{6} \\ y &= 12 \end{aligned}$$

37. $$\begin{aligned} 2(x+4)-10 &= -2(7) \\ 2\cdot x + 2\cdot 4 - 10 &= -14 \\ 2x+8-10 &= -14 \\ 2x-2 &= -14 \\ 2x-2+2 &= -14+2 \\ 2x &= -12 \\ \frac{2x}{2} &= \frac{-12}{2} \\ x &= -6 \end{aligned}$$

38. $$\begin{aligned} -3(x-6)+13 &= 20-1 \\ -3\cdot x - (-3)\cdot 6 + 13 &= 19 \\ -3x+18+13 &= 19 \\ -3x+31 &= 19 \\ -3x+31-31 &= 19-31 \\ -3x &= -12 \\ \frac{-3x}{-3} &= \frac{-12}{-3} \\ x &= 4 \end{aligned}$$

39. The product of –5 and a number is $-5x$.

40. Three subtracted from a number is $x - 3$.

41. The sum of –5 and a number is $-5 + x$.

42. The quotient of –2 and a number is $\frac{-2}{x}$ or $-\frac{2}{x}$.

43. The product of –5 and a number, decreased by 50 is $-5x - 50$.

44. Eleven added to twice a number is $2x + 11$.

45. The quotient of 70 and the sum of a number and 6 is $\frac{70}{x+6}$.

46. Twice the difference of a number and 13 is $2(x - 13)$.

47. $$\begin{aligned} 2x+5 &= 7x-100 \\ 2x-2x+5 &= 7x-2x-100 \\ 5 &= 5x-100 \\ 5+100 &= 5x-100+100 \\ 105 &= 5x \\ \frac{105}{5} &= \frac{5x}{5} \\ 21 &= x \end{aligned}$$

48. $$\begin{aligned} -6x-4 &= x+66 \\ -6x+6x-4 &= x+6x+66 \\ -4 &= 7x+66 \\ -4-66 &= 7x+66-66 \\ -70 &= 7x \\ \frac{-70}{7} &= \frac{7x}{7} \\ -10 &= x \end{aligned}$$

49. $$\begin{aligned} 2x+7 &= 6x-1 \\ 2x-2x+7 &= 6x-2x-1 \\ 7 &= 4x-1 \\ 7+1 &= 4x-1+1 \\ 8 &= 4x \\ \frac{8}{4} &= \frac{4x}{4} \\ 2 &= x \end{aligned}$$

Copyright © 2011 Pearson Education, Inc. Publishing as Prentice Hall.

50.
$$\begin{aligned} 5x-18&=-4x\\ 5x-5x-18&=-4x-5x\\ -18&=-9x\\ \frac{-18}{-9}&=\frac{-9x}{-9}\\ 2&=x \end{aligned}$$

51.
$$\begin{aligned} 5(n-3)&=7+3n\\ 5n-15&=7+3n\\ 5n-3n-15&=7+3n-3n\\ 2n-15&=7\\ 2n-15+15&=7+15\\ 2n&=22\\ \frac{2n}{2}&=\frac{22}{2}\\ n&=11 \end{aligned}$$

52.
$$\begin{aligned} 7(2+x)&=4x-1\\ 14+7x&=4x-1\\ 14+7x-4x&=4x-4x-1\\ 14+3x&=-1\\ 14-14+3x&=-1-14\\ 3x&=-15\\ \frac{3x}{3}&=\frac{-15}{3}\\ x&=-5 \end{aligned}$$

53.
$$\begin{aligned} 6x+3-(-x)&=-20+5x-7\\ 6x+3+x&=-20+5x-7\\ 7x+3&=-27+5x\\ 7x-5x+3&=-27+5x-5x\\ 2x+3&=-27\\ 2x+3-3&=-27-3\\ 2x&=-30\\ \frac{2x}{2}&=\frac{-30}{2}\\ x&=-15 \end{aligned}$$

54.
$$\begin{aligned} x-25+2x&=-5+2x-10\\ -25+3x&=-15+2x\\ -25+3x-2x&=-15+2x-2x\\ -25+x&=-15\\ -25+25+x&=-15+25\\ x&=10 \end{aligned}$$

55.
$$\begin{aligned} 3(x-4)&=5x-8\\ 3x-12&=5x-8\\ 3x-3x-12&=5x-3x-8\\ -12&=2x-8\\ -12+8&=2x-8+8\\ -4&=2x\\ \frac{-4}{2}&=\frac{2x}{2}\\ -2&=x \end{aligned}$$

56.
$$\begin{aligned} 4(x-3)&=-2x-48\\ 4x-12&=-2x-48\\ 4x+2x-12&=-2x+2x-48\\ 6x-12&=-48\\ 6x-12+12&=-48+12\\ 6x&=-36\\ \frac{6x}{6}&=\frac{-36}{6}\\ x&=-6 \end{aligned}$$

57.
$$\begin{aligned} 6(2n-1)+18&=0\\ 12n-6+18&=0\\ 12n+12&=0\\ 12n+12-12&=0-12\\ 12n&=-12\\ \frac{12n}{12}&=\frac{-12}{12}\\ n&=-1 \end{aligned}$$

58.
$$\begin{aligned} 7(3y-2)-7&=0\\ 21y-14-7&=0\\ 21y-21&=0\\ 21y-21+21&=0+21\\ 21y&=21\\ \frac{21y}{21}&=\frac{21}{21}\\ y&=1 \end{aligned}$$

59.
$$\begin{aligned} 95x-14&=20x-10+10x-4\\ 95x-14&=30x-14\\ 95x-14+14&=30x-14+14\\ 95x&=30x\\ 95x-30x&=30x-30x\\ 65x&=0\\ \frac{65x}{65}&=\frac{0}{65}\\ x&=0 \end{aligned}$$

Copyright © 2011 Pearson Education, Inc. Publishing as Prentice Hall.

60. $$\begin{aligned} 32z+11-28z &= 50+2z-(-1) \\ 4z+11 &= 50+2z+1 \\ 4z+11 &= 51+2z \\ 4z-2z+11 &= 51+2z-2z \\ 2z+11 &= 51 \\ 2z+11-11 &= 51-11 \\ 2z &= 40 \\ \frac{2z}{2} &= \frac{40}{2} \\ z &= 20 \end{aligned}$$

61. The difference of 20 and –8 is 28 translates to $20-(-8)=28$.

62. Nineteen subtracted from –2 amounts to –21 translates to $-2-19=-21$.

63. The quotient of –75 and the sum of 5 and 20 is equal to –3 translates to $\frac{-75}{5+20}=-3$.

64. Five times the sum of 2 and –6 yields –20 translates to $5[2+(-6)]=-20$.

65. "Twice a number minus 8 is 40" is $2x-8=40$.

66. "The product of a number and 6 is equal to the sum of the number and 2*a*" is $6x=x+2a$.

67. "Twelve subtracted from the quotient of a number and 2 is 10" is $\frac{x}{2}-12=10$.

68. "The difference of a number and 3 is the quotient of 8 and 4" is $x-3=\frac{8}{4}$.

69. "Five times a number subtracted from 40 is the same as three times the number" is

$$\begin{aligned} 40-5x &= 3x \\ 40-5x+5x &= 3x+5x \\ 40 &= 8x \\ \frac{40}{8} &= \frac{8x}{8} \\ 5 &= x \end{aligned}$$

The number is 5.

70. "The product of a number and 3 is twice the difference of that number and 8" is

$$\begin{aligned} 3x &= 2(x-8) \\ 3x &= 2x-16 \\ 3x-2x &= 2x-2x-16 \\ x &= -16 \end{aligned}$$

The number is –16.

71. Let x be the number of votes for the Independent candidate. Since the Democratic candidate received 272 more votes than the Independent candidate, the Democratic candidate received $x+272$ votes. The total number of votes is 18,500.

$$\begin{aligned} x+x+272+14,000 &= 18,500 \\ 2x+14,272 &= 18,500 \\ 2x+14,272-14,272 &= 18,500-14,272 \\ 2x &= 4228 \\ \frac{2x}{2} &= \frac{4228}{2} \\ x &= 2114 \end{aligned}$$

The Democratic candidate received $2114+272=2386$ votes.

72. Let x be the number of movies on videotapes. Since the number of movies on DVDs is twice the number on videotapes, the number of movies on DVDs is $2x$. Since the total number of movies is 126, the sum of x and $2x$ is 126.

$$\begin{aligned} x+2x &= 126 \\ 3x &= 126 \\ \frac{3x}{3} &= \frac{126}{3} \\ x &= 42 \end{aligned}$$

He has $2(42)=84$ movies on DVDs.

73. $9x-20x=(9-20)x=-11x$

74. $-5(7x)=(-5\cdot 7)x=-35x$

75. $$\begin{aligned} 12x+5(2x-3)-4 &= 12x+10x-15-4 \\ &= 22x-19 \end{aligned}$$

76. $$\begin{aligned} -7(x+6)-2(x-5) &= -7x-42-2x+10 \\ &= -7x-2x-42+10 \\ &= -9x-32 \end{aligned}$$

77. $$\begin{aligned} c-5 &= -13+7 \\ c-5 &= -6 \\ c-5+5 &= -6+5 \\ c &= -1 \end{aligned}$$

Copyright © 2011 Pearson Education, Inc. Publishing as Prentice Hall.

78.
$$\begin{aligned} 7x+5-6x &= -20 \\ x+5 &= -20 \\ x+5-5 &= -20-5 \\ x &= -25 \end{aligned}$$

79.
$$\begin{aligned} -7x+3x &= -50-2 \\ -4x &= -52 \\ \frac{-4x}{-4} &= \frac{-52}{-4} \\ x &= 13 \end{aligned}$$

80.
$$\begin{aligned} -x+8x &= -38-4 \\ 7x &= -42 \\ \frac{7x}{7} &= \frac{-42}{7} \\ x &= -6 \end{aligned}$$

81.
$$\begin{aligned} 9x+12-8x &= -6+(-4) \\ x+12 &= -10 \\ x+12-12 &= -10-12 \\ x &= -22 \end{aligned}$$

82.
$$\begin{aligned} -17x+14+20x-2x &= 5-(-3) \\ -17x+20x-2x+14 &= 5+3 \\ x+14 &= 8 \\ x+14-14 &= 8-14 \\ x &= -6 \end{aligned}$$

83.
$$\begin{aligned} 5(2x-3) &= 11x \\ 10x-15 &= 11x \\ 10x-10x-15 &= 11x-10x \\ -15 &= x \end{aligned}$$

84.
$$\begin{aligned} \frac{y}{-3} &= -1-5 \\ \frac{y}{-3} &= -6 \\ -3\cdot\frac{y}{-3} &= -3\cdot(-6) \\ y &= 18 \end{aligned}$$

85.
$$\begin{aligned} 12y-10 &= -70 \\ 12y-10+10 &= -70+10 \\ 12y &= -60 \\ \frac{12y}{12} &= \frac{-60}{12} \\ y &= -5 \end{aligned}$$

86.
$$\begin{aligned} 4n-8 &= 2n+14 \\ 4n-2n-8 &= 2n-2n+14 \\ 2n-8 &= 14 \\ 2n-8+8 &= 14+8 \\ 2n &= 22 \\ \frac{2n}{2} &= \frac{22}{2} \\ n &= 11 \end{aligned}$$

87.
$$\begin{aligned} -6(x-3) &= x+4 \\ -6x+18 &= x+4 \\ -6x+6x+18 &= x+6x+4 \\ 18 &= 7x+4 \\ 18-4 &= 7x+4-4 \\ 14 &= 7x \\ \frac{14}{7} &= \frac{7x}{7} \\ 2 &= x \end{aligned}$$

88.
$$\begin{aligned} 9(3x-4)+63 &= 0 \\ 27x-36+63 &= 0 \\ 27x+27 &= 0 \\ 27x+27-27 &= 0-27 \\ 27x &= -27 \\ \frac{27x}{27} &= \frac{-27}{27} \\ x &= -1 \end{aligned}$$

89.
$$\begin{aligned} -5z+3z-7 &= 8z-1-6 \\ -2z-7 &= 8z-7 \\ -2z-8z-7 &= 8z-8z-7 \\ -10z-7 &= -7 \\ -10z-7+7 &= -7+7 \\ -10z &= 0 \\ \frac{-10z}{-10} &= \frac{0}{-10} \\ z &= 0 \end{aligned}$$

90.
$$\begin{aligned} 4x-3+6x &= 5x-3-30 \\ 10x-3 &= 5x-33 \\ 10x-5x-3 &= 5x-5x-33 \\ 5x-3 &= -33 \\ 5x-3+3 &= -33+3 \\ 5x &= -30 \\ \frac{5x}{5} &= \frac{-30}{5} \\ x &= -6 \end{aligned}$$

Copyright © 2011 Pearson Education, Inc. Publishing as Prentice Hall.

91. "Three times a number added to twelve is 27" is

$$12+3x=27$$
$$12-12+3x=27-12$$
$$3x=15$$
$$\frac{3x}{3}=\frac{15}{3}$$
$$x=5$$

The number is 5.

92. "Twice the sum of a number and four is ten" is

$$2(x+4)=10$$
$$2x+8=10$$
$$2x+8-8=10-8$$
$$2x=2$$
$$\frac{2x}{2}=\frac{2}{2}$$
$$x=1$$

The number is 1.

93. Let x be the number of roadway miles in Hawaii. Since Delaware has 1585 more roadway miles than Hawaii, Delaware has $x + 1585$ roadway miles. Since the total number of roadway miles is 10,203, the sum of x and $x + 1585$ is 10,203.

$$x+x+1585=10,203$$
$$2x+1585=10,203$$
$$2x+1585-1585=10,203-1585$$
$$2x=8618$$
$$\frac{2x}{2}=\frac{8618}{2}$$
$$x=4309$$

Hawaii has 4309 roadway miles and Delaware has 4309 + 1585 = 5894 roadway miles.

94. Let x be the number of roadway miles in South Dakota. Since North Dakota has 3094 more roadway miles than South Dakota, North Dakota has $x + 3094$ roadway miles. Since the total number of roadway miles is 170,470, the sum of x and $x + 3094$ is 170,470.

$$x+x+3094=170,470$$
$$2x+3094=170,470$$
$$2x+3094-3094=170,470-3094$$
$$2x=167,376$$
$$\frac{2x}{2}=\frac{167,376}{2}$$
$$x=83,688$$

South Dakota has 83,688 roadway miles and North Dakota has 83,688 + 3094 = 86,782 roadway miles.

Chapter 3 Test

1. $7x-5-12x+10=7x-12x-5+10$
$=(7-12)x-5+10$
$=-5x+5$

2. $-2(3y + 7) = -2 \cdot 3y + (-2) \cdot 7 = -6y - 14$

3. $-(3z+2)-5z-18=-1(3z+2)-5z-18$
$=-1\cdot 3z+(-1)\cdot 2-5z-18$
$=-3z-2-5z-18$
$=-3z-5z-2-18$
$=-8z-20$

4. perimeter $= 3(5x + 5) = 3 \cdot 5x + 3 \cdot 5 = 15x + 15$
The perimeter is $(15x + 15)$ inches.

5. Area = (length) $\cdot$ (width)
$=4\cdot(3x-1)$
$=4\cdot 3x-4\cdot 1$
$=12x-4$
The area is $(12x - 4)$ square meters.

6.
$$12=y-3y$$
$$12=-2y$$
$$\frac{12}{-2}=\frac{-2y}{-2}$$
$$-6=y$$

7.
$$\frac{x}{2}=-5-(-2)$$
$$\frac{x}{2}=-5+2$$
$$\frac{x}{2}=-3$$
$$2\cdot\frac{x}{2}=2\cdot(-3)$$
$$x=-6$$

8.
$$5x+12-4x-14=22$$
$$x-2=22$$
$$x-2+2=22+2$$
$$x=24$$

9.
$$-4x+7=15$$
$$-4x+7-7=15-7$$
$$-4x=8$$
$$\frac{-4x}{-4}=\frac{8}{-4}$$
$$x=-2$$

Copyright © 2011 Pearson Education, Inc. Publishing as Prentice Hall.

10.
$$\begin{aligned} 2(x-6) &= 0 \\ 2x-12 &= 0 \\ 2x-12+12 &= 0+12 \\ 2x &= 12 \\ \frac{2x}{2} &= \frac{12}{2} \\ x &= 6 \end{aligned}$$

11.
$$\begin{aligned} -4(x-11)-34 &= 10-12 \\ -4x+44-34 &= 10-12 \\ -4x+10 &= -2 \\ -4x+10-10 &= -2-10 \\ -4x &= -12 \\ \frac{-4x}{-4} &= \frac{-12}{-4} \\ x &= 3 \end{aligned}$$

12.
$$\begin{aligned} 5x-2 &= x-10 \\ 5x-x-2 &= x-x-10 \\ 4x-2 &= -10 \\ 4x-2+2 &= -10+2 \\ 4x &= -8 \\ \frac{4x}{4} &= \frac{-8}{4} \\ x &= -2 \end{aligned}$$

13.
$$\begin{aligned} 4(5x+3) &= 2(7x+6) \\ 20x+12 &= 14x+12 \\ 20x+12-14x &= 14x+12-14x \\ 6x+12 &= 12 \\ 6x+12-12 &= 12-12 \\ 6x &= 0 \\ \frac{6x}{6} &= \frac{0}{6} \\ x &= 0 \end{aligned}$$

14.
$$\begin{aligned} 6+2(3n-1) &= 28 \\ 6+6n-2 &= 28 \\ 6n+4 &= 28 \\ 6n+4-4 &= 28-4 \\ 6n &= 24 \\ \frac{6n}{6} &= \frac{24}{6} \\ n &= 4 \end{aligned}$$

15. The sum of −23 and a number translates to $-23 + x$.

16. Three times a number, subtracted from −2 translates to $-2 - 3x$.

17. The sum of twice 5 and −15 is −5 translates to $2 \cdot 5 + (-15) = -5$.

18. Six added to three times a number equals −30 translates to $3x + 6 = -30$.

19. The difference of three times a number and five times the same number is 4 translates to
$$\begin{aligned} 3x-5x &= 4 \\ -2x &= 4 \\ \frac{-2x}{-2} &= \frac{4}{-2} \\ x &= -2 \end{aligned}$$
The number is −2.

20. Let x be the number of free throws Maria made. Since Paula made twice as many free throws as Maria, Paula made $2x$ free throws. Since the total number of free throws was 12, the sum of x and $2x$ is 12.
$$\begin{aligned} x+2x &= 12 \\ 3x &= 12 \\ \frac{3x}{3} &= \frac{12}{3} \\ x &= 4 \end{aligned}$$
Paula made $2(4) = 8$ free throws.

21. Let x be the number of women runners entered in the race. Since the number of men entered in the race is 112 more than the number of women, the number of men is $x + 112$. Since the total number of runners in the race is 600, the sum of x and $x + 112$ is 600.
$$\begin{aligned} x+x+112 &= 600 \\ 2x+112 &= 600 \\ 2x+112-112 &= 600-112 \\ 2x &= 488 \\ \frac{2x}{2} &= \frac{488}{2} \\ x &= 244 \end{aligned}$$
244 women entered the race.

Cumulative Review Chapters 1–3

1. 308,063,557 in words is three hundred eight million, sixty-three thousand, five hundred fifty-seven.

2. 276,004 in words is two hundred seventy-six thousand, four.

3. $2 + 3 + 1 + 3 + 4 = 13$
The perimeter is 13 inches.

Copyright © 2011 Pearson Education, Inc. Publishing as Prentice Hall.

4. $6 + 3 + 6 + 3 = 18$
The perimeter is 18 inches.

5. $$\begin{array}{r} 900 \\ -\ 174 \\ \hline 726 \end{array} \qquad \textit{Check:}\ \begin{array}{r} 726 \\ +\ 174 \\ \hline 900 \end{array}$$

6. $$\begin{array}{r} 17,801 \\ -\ 8216 \\ \hline 9585 \end{array} \qquad \textit{Check:}\ \begin{array}{r} 9585 \\ +\ 8216 \\ \hline 17,801 \end{array}$$

7. To round 248,982 to the nearest hundred, observe that the digit in the tens place is 8. Since this digit is at least 5, we add 1 to the digit in the hundreds place. The number 248,982 rounded to the nearest hundred is 249,000.

8. To round 844,497 to the nearest thousand, observe that the digit in the hundreds place is 4. Since this digit is less than 5, we do not add 1 to the digit in the thousands place. The number 844,497 rounded to the nearest thousand is 844,000.

9. $$\begin{array}{r} 25 \\ \times\ 8 \\ \hline 200 \end{array}$$

10. $$\begin{array}{r} 395 \\ \times\ 74 \\ \hline 1\,580 \\ 27\,650 \\ \hline 29,230 \end{array}$$

11. $$\begin{array}{r} 208 \\ 9\overline{)\ 1872} \\ -18 \\ \hline 07 \\ -0 \\ \hline 72 \\ -72 \\ \hline 0 \end{array}$$

Check: $$\begin{array}{r} 208 \\ \times\ 9 \\ \hline 1872 \end{array}$$

12. $$\begin{array}{r} 86 \\ 46\overline{)\ 3956} \\ -368 \\ \hline 276 \\ -276 \\ \hline 0 \end{array}$$

Check: $$\begin{array}{r} 86 \\ \times\ 46 \\ \hline 516 \\ 3440 \\ \hline 3956 \end{array}$$

13. $2 \cdot 4 - 3 \div 3 = 8 - 3 \div 3 = 8 - 1 = 7$

14. $8 \cdot 4 + 9 \div 3 = 32 + 9 \div 3 = 32 + 3 = 35$

15. $$\begin{aligned} x^2 + z - 3 &= 5^2 + 4 - 3 \\ &= 25 + 4 - 3 \\ &= 29 - 3 \\ &= 26 \end{aligned}$$

16. $$\begin{aligned} 2a^2 + 5 - c &= 2 \cdot 2^2 + 5 - 3 \\ &= 2 \cdot 4 + 5 - 3 \\ &= 8 + 5 - 3 \\ &= 13 - 3 \\ &= 10 \end{aligned}$$

17. $2n - 30 = 10$
Let $n = 26$.
$$\begin{aligned} 2 \cdot 26 - 30 &\stackrel{?}{=} 10 \\ 52 - 30 &\stackrel{?}{=} 10 \\ 22 &= 10 \quad \text{False} \end{aligned}$$
26 is not a solution.
Let $n = 40$.
$$\begin{aligned} 2 \cdot 40 - 30 &\stackrel{?}{=} 10 \\ 80 - 30 &\stackrel{?}{=} 10 \\ 50 &= 10 \quad \text{False} \end{aligned}$$
40 is not a solution.
Let $n = 20$.
$$\begin{aligned} 2 \cdot 20 - 30 &\stackrel{?}{=} 10 \\ 40 - 30 &\stackrel{?}{=} 10 \\ 10 &= 10 \quad \text{True} \end{aligned}$$
20 is a solution.

18. a. $-14 < 0$ because -14 is to the left of 0 on the number line.

b. $-(-7) = 7$, so $-(-7) > -8$ because 7 is to the right of -8 on the number line.

19. $5 + (-2) = 3$

Copyright © 2011 Pearson Education, Inc. Publishing as Prentice Hall.

20. $-3 + (-4) = -7$

21. $-15 + (-10) = -25$

22. $3 + (-7) = -4$

23. $-2 + 25 = 23$

24. $21 + 15 + (-19) = 36 + (-19) = 17$

25. $-4 - 10 = -4 + (-10) = -14$

26. $-2 - 3 = -2 + (-3) = -5$

27. $6 - (-5) = 6 + 5 = 11$

28. $19 - (-10) = 19 + 10 = 29$

29. $-11 - (-7) = -11 + 7 = -4$

30. $-16 - (-13) = -16 + 13 = -3$

31. $\frac{-12}{6} = -2$

32. $\frac{-30}{-5} = 6$

33. $-20 \div (-4) = 5$

34. $26 \div (-2) = -13$

35. $\frac{48}{-3} = -16$

36. $\frac{-120}{12} = -10$

37. $(-3)^2 = (-3)(-3) = 9$

38. $-2^5 = -(2 \cdot 2 \cdot 2 \cdot 2 \cdot 2) = -32$

39. $-3^2 = -(3 \cdot 3) = -9$

40. $(-5)^2 = (-5)(-5) = 25$

41.
$$\begin{aligned} 2y-6+4y+8 &= 2y+4y-6+8 \\ &= (2+4)y-6+8 \\ &= 6y+2 \end{aligned}$$

42.
$$\begin{aligned} 6x+2-3x+7 &= 6x-3x+2+7 \\ &= (6-3)x+2+7 \\ &= 3x+9 \end{aligned}$$

43.
$$\begin{aligned} 3y+1 &= 3 \\ 3(-1)+1 &\stackrel{?}{=} 3 \\ -3+1 &\stackrel{?}{=} 3 \\ -2 &= 3 \quad \text{False} \end{aligned}$$
Since $-2 = 3$ is false, -1 is not a solution of the equation.

44.
$$\begin{aligned} 5x-3 &= 7 \\ 5(2)-3 &\stackrel{?}{=} 7 \\ 10-3 &\stackrel{?}{=} 7 \\ 7 &= 7 \quad \text{True} \end{aligned}$$
Since $7 = 7$ is true, 2 is a solution of the equation.

45.
$$\begin{aligned} -12x &= -36 \\ \frac{-12x}{-12} &= \frac{-36}{-12} \\ x &= 3 \end{aligned}$$

46.
$$\begin{aligned} -3y &= 15 \\ \frac{-3y}{-3} &= \frac{15}{-3} \\ y &= -5 \end{aligned}$$

47.
$$\begin{aligned} 2x-6 &= 18 \\ 2x-6+6 &= 18+6 \\ 2x &= 24 \\ \frac{2x}{2} &= \frac{24}{2} \\ x &= 12 \end{aligned}$$

48.
$$\begin{aligned} 3a+5 &= -1 \\ 3a+5-5 &= -1-5 \\ 3a &= -6 \\ \frac{3a}{3} &= \frac{-6}{3} \\ a &= -2 \end{aligned}$$

49. Let x be the price of the software. Since the price of the computer system is four times the price of the software, the price of the computer system is $4x$. Since the combined price is \$2100, the sum of x and $4x$ is 2100.
$$\begin{aligned} x+4x &= 2100 \\ 5x &= 2100 \\ \frac{5x}{5} &= \frac{2100}{5} \\ x &= 420 \end{aligned}$$
The price of the software is \$420 and the price of the computer system is 4(\$420) = \$1680.

Copyright © 2011 Pearson Education, Inc. Publishing as Prentice Hall.

50. Let x be the number. "Two times the number plus four is the same amount as three times the number minus seven" translates to

$$2x+4=3x-7$$
$$2x-2x+4=3x-2x-7$$
$$4=x-7$$
$$4+7=x-7+7$$
$$11=x$$

The number is 11.

Copyright © 2011 Pearson Education, Inc. Publishing as Prentice Hall.

Chapter 4

Section 4.1

Practice Problems

1. In the fraction $\frac{11}{2}$, the numerator is 11 and the denominator is 2.

2. In the fraction $\frac{10y}{17}$, the numerator is $10y$ and the denominator is 17.

3. 3 out of 8 equal parts are shaded: $\frac{3}{8}$

4. 1 out of 6 equal parts is shaded: $\frac{1}{6}$

5. 7 out of 10 equal parts are shaded: $\frac{7}{10}$

6. 9 out of 16 equal parts are shaded: $\frac{9}{16}$

7. answers may vary; for example,

8. answers may vary; for example,

9. number of planets farther $\rightarrow 5$
number of planets in our solar system $\rightarrow 8$

$\frac{5}{8}$ of the planets in our solar system are farther from the Sun than Earth is.

10. Each part is $\frac{1}{3}$ of a whole and there are 8 parts shaded, or 2 wholes and 2 more parts.

$\frac{8}{3}$; $2\frac{2}{3}$

11. Each part is $\frac{1}{4}$ of a whole and there are 5 parts shaded, or 1 whole and 1 more part.

$\frac{5}{4}$; $1\frac{1}{4}$

12. a. 0, $\frac{5}{7}$, 1

b. 0, $\frac{2}{3}$, 1

c. 0, $\frac{4}{6}$, 1

13. a. 0, 1, 2, $\frac{8}{3}$, 3

b. 0, 1, $\frac{5}{4}$, 2

c. 0, 1 or $\frac{7}{7}$

14. $\frac{9}{9}=1$

15. $\frac{-6}{-6}=1$

16. $\frac{0}{-1}=0$

17. $\frac{4}{1}=4$

18. $\frac{-13}{0}$ is undefined.

19. $\frac{-13}{1}=-13$

20. a. $5\frac{2}{7}=\frac{7\cdot 5+2}{7}=\frac{35+2}{7}=\frac{37}{7}$

b. $6\frac{2}{3}=\frac{3\cdot 6+2}{3}=\frac{18+2}{3}=\frac{20}{3}$

c. $10\frac{9}{10}=\frac{10\cdot 10+9}{10}=\frac{100+9}{10}=\frac{109}{10}$

d. $4\frac{1}{5}=\frac{5\cdot 4+1}{5}=\frac{20+1}{5}=\frac{21}{5}$

Copyright © 2011 Pearson Education, Inc. Publishing as Prentice Hall.

21. a. $\begin{array}{r} 1 \\ 5\overline{)9} \\ \underline{5} \\ 4 \end{array}$

$\frac{9}{5} = 1\frac{4}{5}$

b. $\begin{array}{r} 2 \\ 9\overline{)23} \\ \underline{18} \\ 5 \end{array}$

$\frac{23}{9} = 2\frac{5}{9}$

c. $\begin{array}{r} 12 \\ 4\overline{)48} \\ \underline{4} \\ 8 \\ \underline{8} \\ 0 \end{array}$

$\frac{48}{4} = 12$

d. $\begin{array}{r} 4 \\ 13\overline{)62} \\ \underline{52} \\ 10 \end{array}$

$\frac{62}{13} = 4\frac{10}{13}$

e. $\begin{array}{r} 7 \\ 7\overline{)51} \\ \underline{49} \\ 2 \end{array}$

$\frac{51}{7} = 7\frac{2}{7}$

f. $\begin{array}{r} 1 \\ 20\overline{)21} \\ \underline{20} \\ 1 \end{array}$

$\frac{21}{20} = 1\frac{1}{20}$

Vocabulary and Readiness Check

1. The number $\frac{17}{31}$ is called a <u>fraction</u>. The number 31 is called its <u>denominator</u> and 17 is called its <u>numerator</u>.

2. If we simplify each fraction, $\frac{-9}{-9} = \underline{1}$, $\frac{0}{-4} = \underline{0}$, and we say $\frac{-4}{0}$ <u>is undefined</u>.

3. The fraction $\frac{8}{3}$ is called an <u>improper</u> fraction, the fraction $\frac{3}{8}$ is called a <u>proper</u> fraction, and $10\frac{3}{8}$ is called a <u>mixed number</u>.

4. The value of an improper fraction is always $\underline{\geq 1}$ and the value of a proper fraction is always $\underline{< 1}$.

Exercise Set 4.1

1. In the fraction $\frac{1}{2}$, the numerator is 1 and the denominator is 2. Since $1 < 2$, the fraction is proper.

3. In the fraction $\frac{10}{3}$, the numerator is 10 and the denominator is 3. Since $10 > 3$, the fraction is improper.

5. In the fraction $\frac{15}{15}$, the numerator is 15 and the denominator is 15. Since $15 \geq 15$, the fraction is improper.

7. 1 out of 3 equal parts is shaded: $\frac{1}{3}$

9. Each part is $\frac{1}{4}$ of a whole and there are 11 parts shaded, or 2 wholes and 3 more parts.

a. $\frac{11}{4}$

b. $2\frac{3}{4}$

Copyright © 2011 Pearson Education, Inc. Publishing as Prentice Hall.

11. Each part is $\frac{1}{6}$ of a whole and there are 23 parts shaded, or 3 wholes and 5 more parts.

a. $\frac{23}{6}$

b. $3\frac{5}{6}$

13. 7 out of 12 equal parts are shaded: $\frac{7}{12}$

15. 3 out of 7 equal parts are shaded: $\frac{3}{7}$

17. 4 out of 9 equal parts are shaded: $\frac{4}{9}$

19. Each part is $\frac{1}{3}$ of a whole and there are 4 parts shaded, or 1 whole and 1 more part.

a. $\frac{4}{3}$

b. $1\frac{1}{3}$

21. Each part is $\frac{1}{2}$ of a whole and there are 11 parts shaded, or 5 wholes and 1 more part.

a. $\frac{11}{2}$

b. $5\frac{1}{2}$

23. 1 out of 6 equal parts is shaded: $\frac{1}{6}$

25. 5 of 8 equal parts are shaded: $\frac{5}{8}$

27. answers may vary; for example,

29. answers may vary; for example,

31. answers may vary; for example,

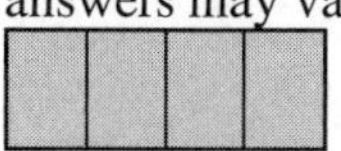

33. freshmen $\rightarrow$ 42
students $\rightarrow$ 131

$\frac{42}{131}$ of the students are freshmen.

35. a. number of students not freshmen $= 131 - 42$
$= 89$

b. not freshmen $\rightarrow$ 89
students $\rightarrow$ 131

$\frac{89}{131}$ of the students are not freshmen.

37. born in Ohio $\rightarrow$ 7
U.S. presidents $\rightarrow$ 44

$\frac{7}{44}$ of U.S. presidents were born in Ohio.

39. number turned into hurricanes $\rightarrow$ 15
number of tropical storms $\rightarrow$ 28

$\frac{15}{28}$ of the tropical storms turned into hurricanes.

41. 11 of 31 days of March is $\frac{11}{31}$ of the month.

43. number of sophomores $\rightarrow$ 10
number of students in class $\rightarrow$ 31

$\frac{10}{31}$ of the class is sophomores.

45. There are 50 states total. 33 states contain federal Indian reservations.

a. $\frac{33}{50}$ of the states contain federal Indian reservations.

b. $50 - 33 = 17$
17 states do not contain federal Indian reservations.

c. $\frac{17}{50}$ of the states do not contain federal Indian reservations.

Copyright © 2011 Pearson Education, Inc. Publishing as Prentice Hall.

47. a. $\begin{array}{l}\text{blue} \to 21 \\ \text{total} \to \overline{50}\end{array}$

$\frac{21}{50}$ of the marbles are blue.

b. $50 - 21 = 29$
29 of the marbles are red.

c. $\begin{array}{l}\text{red} \to 29 \\ \text{total} \to \overline{50}\end{array}$

$\frac{29}{50}$ of the marbles are red.

49. Number line from 0 to 3 with a point at $\frac{1}{4}$.

51. Number line from 0 to 2 with a point at $\frac{4}{7}$.

53. Number line from 0 to 2 with a point at $\frac{8}{5}$.

55. Number line from 0 to 4 with a point at $\frac{7}{3}$.

57. $\frac{12}{12} = 1$

59. $\frac{-5}{1} = -5$

61. $\frac{0}{-2} = 0$

63. $\frac{-8}{-8} = 1$

65. $\frac{-9}{0}$ is undefined

67. $\frac{3}{1} = 3$

Copyright © 2011 Pearson Education, Inc. Publishing as Prentice Hall.

69. $2\frac{1}{3} = \frac{3 \cdot 2 + 1}{3} = \frac{6+1}{3} = \frac{7}{3}$

71. $3\frac{3}{5} = \frac{5 \cdot 3 + 3}{5} = \frac{15+3}{5} = \frac{18}{5}$

73. $6\frac{5}{8} = \frac{8 \cdot 6 + 5}{8} = \frac{48+5}{8} = \frac{53}{8}$

75. $11\frac{6}{7} = \frac{7 \cdot 11 + 6}{7} = \frac{77+6}{7} = \frac{83}{7}$

77. $9\frac{7}{20} = \frac{20 \cdot 9 + 7}{20} = \frac{180+7}{20} = \frac{187}{20}$

79. $166\frac{2}{3} = \frac{3 \cdot 166 + 2}{3} = \frac{498+2}{3} = \frac{500}{3}$

81. $\begin{array}{r} 3 \\ 5\overline{)17} \\ \underline{15} \\ 2 \end{array}$

$\frac{17}{5} = 3\frac{2}{5}$

83. $\begin{array}{r} 4 \\ 8\overline{)37} \\ \underline{32} \\ 5 \end{array}$

$\frac{37}{8} = 4\frac{5}{8}$

85. $\begin{array}{r} 3 \\ 15\overline{)47} \\ \underline{45} \\ 2 \end{array}$

$\frac{47}{15} = 3\frac{2}{15}$

87. $\begin{array}{r} 15 \\ 15\overline{)225} \\ \underline{15} \\ 75 \\ \underline{75} \\ 0 \end{array}$

$\frac{225}{15} = 15$

89. $\begin{array}{r} 1 \\ 175\overline{)182} \\ \underline{175} \\ 7 \end{array}$

$\frac{182}{175} = 1\frac{7}{175}$

91. $\begin{array}{r} 6 \\ 112\overline{)737} \\ \underline{672} \\ 65 \end{array}$

$\frac{737}{112} = 6\frac{65}{112}$

93. $3^2 = 3 \cdot 3 = 9$

95. $5^3 = 5 \cdot 5 \cdot 5 = 125$

97. $-\frac{11}{2} = \frac{-11}{2} = \frac{11}{-2}$

99. $\frac{-13}{15} = \frac{13}{-15} = -\frac{13}{15}$

101. answers may vary

103. ●●●○○○○○○

105. 1 is not close to 8, so $\frac{1}{8}$ is closer to 0 than to 1.

$7\frac{1}{8}$ rounded to the nearest whole number is 7.

107. 1532 + 576 + 1059 = 3167
576 of the 3167 stores are named Banana
Republic: $\frac{576}{3167}$

109. 1700 + 550 = 2250
1700 of the 2250 affiliates are located in the
United States: $\frac{1700}{2250}$

Section 4.2

Practice Problems

1. a. $30 = 2 \cdot 15$
$\downarrow \; \downarrow \searrow$
$2 \cdot 3 \cdot 5$

Copyright © 2011 Pearson Education, Inc. Publishing as Prentice Hall.

b. $56 = 2 \cdot 28$
$\downarrow \ \downarrow\searrow$
$2 \cdot 2 \cdot 14$
$\downarrow \ \downarrow \ \downarrow\searrow$
$2 \cdot 2 \cdot 2 \cdot 7 = 2^3 \cdot 7$

c. $72 = 2 \cdot 36$
$\downarrow \ \downarrow\searrow$
$2 \cdot 2 \cdot 18$
$\downarrow \ \downarrow \ \downarrow\searrow$
$2 \cdot 2 \cdot 2 \cdot 9$
$\downarrow \ \downarrow \ \downarrow \ \downarrow\searrow$
$2 \cdot 2 \cdot 2 \cdot 3 \cdot 3 = 2^3 \cdot 3^2$

2. $60 = 2 \cdot 30$
$\downarrow \ \downarrow\searrow$
$2 \cdot 2 \cdot 15$
$\downarrow \ \downarrow \ \downarrow\searrow$
$2 \cdot 2 \cdot 3 \cdot 5 = 2^2 \cdot 3 \cdot 5$

3. $3\overline{)33}$ → 11
$3\overline{)99}$
$3\overline{)297}$

The prime factorization of 297 is $3^3 \cdot 11$.

4. $\frac{30}{45} = \frac{15 \cdot 2}{15 \cdot 3} = \frac{15}{15} \cdot \frac{2}{3} = 1 \cdot \frac{2}{3} = \frac{2}{3}$

5. $\frac{39x}{51} = \frac{3 \cdot 13 \cdot x}{3 \cdot 17} = \frac{3}{3} \cdot \frac{13x}{17} = 1 \cdot \frac{13x}{17} = \frac{13x}{17}$

6. $-\frac{9}{50} = -\frac{3 \cdot 3}{2 \cdot 5 \cdot 5}$

Since 9 and 50 have no common factors, $-\frac{9}{50}$ is already in simplest form.

7. $\frac{49}{112} = \frac{7 \cdot 7}{7 \cdot 16} = \frac{7}{7} \cdot \frac{7}{16} = 1 \cdot \frac{7}{16} = \frac{7}{16}$

8. $-\frac{64}{20} = -\frac{4 \cdot 16}{4 \cdot 5} = -\frac{\overset{1}{\cancel{4}} \cdot 16}{\underset{1}{\cancel{4}} \cdot 5} = -\frac{16}{5}$

9. $\frac{7a^3}{56a^2} = \frac{7 \cdot a \cdot a \cdot a}{7 \cdot 8 \cdot a \cdot a} = \frac{\overset{1}{\cancel{7}} \cdot a \cdot \overset{1}{\cancel{a}} \cdot \overset{1}{\cancel{a}}}{\underset{1}{\cancel{7}} \cdot 8 \cdot \underset{1}{\cancel{a}} \cdot \underset{1}{\cancel{a}}} = \frac{a}{8}$

10. $\frac{7}{9}$ is in simplest form.

$\frac{21}{27} = \frac{3 \cdot 7}{3 \cdot 9} = \frac{\overset{1}{\cancel{3}} \cdot 7}{\underset{1}{\cancel{3}} \cdot 9} = \frac{7}{9}$

Since $\frac{7}{9}$ and $\frac{21}{27}$ both simplify to $\frac{7}{9}$, they are equivalent.

11. Not equivalent, since the cross products are not equal: $13 \cdot 5 = 65$ and $4 \cdot 18 = 72$

12. $\frac{\text{6 parks in Washington}}{\text{58 national parks}} = \frac{2 \cdot 3}{2 \cdot 29} = \frac{\overset{1}{\cancel{2}} \cdot 3}{\underset{1}{\cancel{2}} \cdot 29} = \frac{3}{29}$

$\frac{3}{29}$ of the national parks are in Washington state.

Calculator Explorations

1. $\frac{128}{224} = \frac{4}{7}$

2. $\frac{231}{396} = \frac{7}{12}$

3. $\frac{340}{459} = \frac{20}{27}$

4. $\frac{999}{1350} = \frac{37}{50}$

5. $\frac{432}{810} = \frac{8}{15}$

6. $\frac{225}{315} = \frac{5}{7}$

7. $\frac{54}{243} = \frac{2}{9}$

Copyright © 2011 Pearson Education, Inc. Publishing as Prentice Hall.

8. $\frac{455}{689} = \frac{35}{53}$

Vocabulary and Readiness Check

1. The number 40 equals $2 \cdot 2 \cdot 2 \cdot 5$. Since each factor is prime, we call $2 \cdot 2 \cdot 2 \cdot 5$ the prime factorization of 40.

2. A natural number, other than 1, that is not prime is called a composite number.

3. A natural number that has exactly two different factors, 1 and itself, is called a prime number.

4. In $\frac{11}{48}$, since 11 and 48 have no common factors other than 1, $\frac{11}{48}$ is in simplest form.

5. Fractions that represent the same portion of a whole are called equivalent fractions.

6. In the statement $\frac{5}{12} = \frac{15}{36}$, $5 \cdot 36$ and $12 \cdot 15$ are called cross products.

Exercise Set 4.2

1. $20 = 2 \cdot 10$
↓ ↙↘
$2 \cdot 2 \cdot 5 = 2^2 \cdot 5$

3. $48 = 2 \cdot 24$
↓ ↓↘
$2 \cdot 2 \cdot 12$
↓ ↓ ↓↘
$2 \cdot 2 \cdot 2 \cdot 6$
↓ ↓ ↓ ↓↘
$2 \cdot 2 \cdot 2 \cdot 2 \cdot 3 = 2^4 \cdot 3$

5. $81 = 9 \cdot 9$
↙↘ ↙↘
$3 \cdot 3 \cdot 3 \cdot 3 = 3^4$

7. $162 = 2 \cdot 81$
↓ ↙ ↘
$2 \cdot 9 \cdot 9$
↓ ↙↘ ↙↘
$2 \cdot 3 \cdot 3 \cdot 3 \cdot 3 = 2 \cdot 3^4$

9. $110 = 2 \cdot 55$
↓ ↓↘
$2 \cdot 5 \cdot 11 = 2 \cdot 5 \cdot 11$

11. $85 = 5 \cdot 17$

13. $240 = 2 \cdot 120$
↓ ↓↘
$2 \cdot 2 \cdot 60$
↓ ↓ ↓↘
$2 \cdot 2 \cdot 2 \cdot 30$
↓ ↓ ↓ ↓↘
$2 \cdot 2 \cdot 2 \cdot 2 \cdot 15$
↓ ↓ ↓ ↓ ↓↘
$2 \cdot 2 \cdot 2 \cdot 2 \cdot 3 \cdot 5 = 2^4 \cdot 3 \cdot 5$

15. $828 = 2 \cdot 414$
↓ ↓↘
$2 \cdot 2 \cdot 207$
↓ ↓ ↓↘
$2 \cdot 2 \cdot 3 \cdot 69$
↓ ↓ ↓ ↓↘
$2 \cdot 2 \cdot 3 \cdot 3 \cdot 23 = 2^2 \cdot 3^2 \cdot 23$

17. $\frac{3}{12} = \frac{3 \cdot 1}{3 \cdot 4} = \frac{1}{4}$

19. $\frac{4x}{42} = \frac{2 \cdot 2 \cdot x}{2 \cdot 3 \cdot 7} = \frac{2 \cdot x}{3 \cdot 7} = \frac{2x}{21}$

21. $\frac{14}{16} = \frac{2 \cdot 7}{2 \cdot 8} = \frac{7}{8}$

23. $\frac{20}{30} = \frac{2 \cdot 10}{3 \cdot 10} = \frac{2}{3}$

25. $\frac{35a}{50a} = \frac{5 \cdot 7 \cdot a}{5 \cdot 10 \cdot a} = \frac{7}{10}$

27. $-\frac{63}{81} = -\frac{9 \cdot 7}{9 \cdot 9} = -\frac{7}{9}$

29. $\frac{30x^2}{36x} = \frac{5 \cdot 6 \cdot x \cdot x}{6 \cdot 6 \cdot x} = \frac{5 \cdot x}{6} = \frac{5x}{6}$

Copyright © 2011 Pearson Education, Inc. Publishing as Prentice Hall.

31. $\frac{27}{64} = \frac{3 \cdot 3 \cdot 3}{4 \cdot 4 \cdot 4}$

Since 27 and 64 have no common factors, $\frac{27}{64}$ is already in simplest form.

33. $\frac{25xy}{40y} = \frac{5 \cdot 5 \cdot x \cdot y}{5 \cdot 8 \cdot y} = \frac{5 \cdot x}{8} = \frac{5x}{8}$

35. $-\frac{40}{64} = -\frac{8 \cdot 5}{8 \cdot 8} = -\frac{5}{8}$

37. $\frac{36x^3y^2}{24xy} = \frac{3 \cdot 12 \cdot x \cdot x \cdot x \cdot y \cdot y}{2 \cdot 12 \cdot x \cdot y} = \frac{3 \cdot x \cdot x \cdot y}{2} = \frac{3x^2y}{2}$

39. $\frac{90}{120} = \frac{30 \cdot 3}{30 \cdot 4} = \frac{3}{4}$

41. $\frac{40xy}{64xyz} = \frac{5 \cdot 8 \cdot x \cdot y}{8 \cdot 8 \cdot x \cdot y \cdot z} = \frac{5}{8 \cdot z} = \frac{5}{8z}$

43. $\frac{66}{308} = \frac{22 \cdot 3}{22 \cdot 14} = \frac{3}{14}$

45. $-\frac{55}{85y} = -\frac{5 \cdot 11}{5 \cdot 17 \cdot y} = -\frac{11}{17 \cdot y} = -\frac{11}{17y}$

47. $\frac{189z}{216z} = \frac{7 \cdot 27 \cdot z}{8 \cdot 27 \cdot z} = \frac{7}{8}$

49. $$\begin{aligned}\frac{224a^3b^4c^2}{16ab^4c^2} &= \frac{14 \cdot 16 \cdot a \cdot a \cdot a \cdot b \cdot b \cdot b \cdot b \cdot c \cdot c}{1 \cdot 16 \cdot a \cdot b \cdot b \cdot b \cdot b \cdot c \cdot c} \\ &= \frac{14 \cdot a \cdot a}{1} \\ &= 14a^2\end{aligned}$$

51. Equivalent, since the cross products are equal: $2 \cdot 12 = 24$ and $6 \cdot 4 = 24$

53. Not equivalent, since the cross products are not equal: $7 \cdot 8 = 56$ and $5 \cdot 11 = 55$

55. Equivalent, since the cross products are equal: $10 \cdot 9 = 90$ and $15 \cdot 6 = 90$

57. Equivalent, since the cross products are equal: $3 \cdot 18 = 54$ and $9 \cdot 6 = 54$

59. Not equivalent, since the cross products are not equal: $10 \cdot 15 = 150$ and $13 \cdot 13 = 169$

61. Not equivalent, since the cross products are not equal: $8 \cdot 24 = 192$ and $12 \cdot 18 = 216$

63. $\frac{2 \text{ hours}}{8 \text{ hours}} = \frac{1 \cdot 2}{4 \cdot 2} = \frac{1}{4}$

2 hours represents $\frac{1}{4}$ of a work shift.

65. $\frac{2640 \text{ feet}}{5280 \text{ feet}} = \frac{2640 \cdot 1}{2640 \cdot 2} = \frac{1}{2}$

2640 feet represents $\frac{1}{2}$ of a mile.

67. a. $\frac{16}{50} = \frac{8 \cdot 2}{25 \cdot 2} = \frac{8}{25}$

$\frac{8}{25}$ of the states can claim at least one Ritz-Carlton hotel.

b. $50 - 16 = 34$
34 states do not have a Ritz-Carlton hotel.

c. $\frac{34}{50} = \frac{17 \cdot 2}{25 \cdot 2} = \frac{17}{25}$

$\frac{17}{25}$ of the states do not have a Ritz-Carlton hotel.

69. $\frac{10 \text{ inches}}{24 \text{ inches}} = \frac{2 \cdot 5}{2 \cdot 12} = \frac{5}{12}$

$\frac{5}{12}$ of the wall is concrete.

71. a. $50 - 22 = 28$
28 states do not have this type of Web site.

b. $\frac{28}{50} = \frac{2 \cdot 14}{2 \cdot 25} = \frac{14}{25}$

$\frac{14}{25}$ of the states do not have this type of Web site.

73. $\frac{22 \text{ individuals}}{320 \text{ individuals}} = \frac{22}{320} = \frac{2 \cdot 11}{2 \cdot 160} = \frac{11}{160}$

$\frac{11}{160}$ of the U.S. astronauts who had flown in space were born in Texas.

Copyright © 2011 Pearson Education, Inc. Publishing as Prentice Hall.

75. $\frac{x^3}{9} = \frac{(-3)^3}{9} = \frac{-27}{9} = -3$

77. $2y = 2(-7) = -14$

79. answers may vary

81. $\frac{3975}{6625} = \frac{3 \cdot 1325}{5 \cdot 1325} = \frac{3}{5}$

83. 36 blood donors have blood type A Rh-positive.

$\frac{36 \text{ donors}}{100 \text{ donors}} = \frac{4 \cdot 9}{4 \cdot 25} = \frac{9}{25}$

$\frac{9}{25}$ of blood donors have type A Rh-positive blood type.

85. $3 + 1 = 4$ blood donors have an AB blood type.

$\frac{4 \text{ donors}}{100 \text{ donors}} = \frac{4 \cdot 1}{4 \cdot 25} = \frac{1}{25}$

$\frac{1}{25}$ of blood donors have an AB blood type.

87.
$$
\begin{array}{l}
34{,}020 = 2 \cdot 17{,}010 \\
\qquad 2 \cdot 2 \cdot 8505 \\
\qquad 2 \cdot 2 \cdot 3 \cdot 2835 \\
\qquad 2 \cdot 2 \cdot 3 \cdot 3 \cdot 945 \\
\qquad 2 \cdot 2 \cdot 3 \cdot 3 \cdot 3 \cdot 315 \\
\qquad 2 \cdot 2 \cdot 3 \cdot 3 \cdot 3 \cdot 3 \cdot 105 \\
\qquad 2 \cdot 2 \cdot 3 \cdot 3 \cdot 3 \cdot 3 \cdot 3 \cdot 35 \\
\qquad 2 \cdot 2 \cdot 3 \cdot 3 \cdot 3 \cdot 3 \cdot 3 \cdot 5 \cdot 7
\end{array}
$$

$34{,}020 = 2^2 \cdot 3^5 \cdot 5 \cdot 7$

89. answers may vary

91. no; answers may vary

93. The piece representing education is labeled $\frac{1}{10}$, so $\frac{1}{10}$ of entering college freshmen plan to major in education.

95. answers may vary

97. The piece representing National Memorials is labeled $\frac{2}{25}$, so $\frac{2}{25}$ of National Park Service areas are National Memorials.

99. answers may vary

101. 8691, 786, 2235, 105, 222, 900, and 1470 are divisible by 3 because the sum of each number's digits is divisible by 3. 786, 22, 222, 900, and 1470 are divisible by 2 because they are even numbers. 786, 222, 900, and 1470 are divisible by both 2 and 3.

103. 6; answers may vary

Section 4.3

Practice Problems

1. $\frac{3}{7} \cdot \frac{5}{11} = \frac{3 \cdot 5}{7 \cdot 11} = \frac{15}{77}$

2. $\frac{1}{3} \cdot \frac{1}{9} = \frac{1 \cdot 1}{3 \cdot 9} = \frac{1}{27}$

3. $\frac{6}{77} \cdot \frac{7}{8} = \frac{6 \cdot 7}{77 \cdot 8} = \frac{2 \cdot 3 \cdot 7}{7 \cdot 11 \cdot 2 \cdot 4} = \frac{3}{11 \cdot 4} = \frac{3}{44}$

4. $\frac{4}{27} \cdot \frac{3}{8} = \frac{4 \cdot 3}{27 \cdot 8} = \frac{1 \cdot 4 \cdot 3}{3 \cdot 9 \cdot 4 \cdot 2} = \frac{1}{9 \cdot 2} = \frac{1}{18}$

5. $\frac{1}{2} \cdot \left(-\frac{11}{28}\right) = -\frac{1 \cdot 11}{2 \cdot 28} = -\frac{11}{56}$

6. $\left(-\frac{4}{11}\right)\left(-\frac{33}{16}\right) = \frac{4 \cdot 33}{11 \cdot 16} = \frac{4 \cdot 3 \cdot 11}{11 \cdot 4 \cdot 4} = \frac{3}{4}$

7.
$$
\begin{aligned}
\frac{1}{6} \cdot \frac{3}{10} \cdot \frac{25}{16} &= \frac{1 \cdot 3 \cdot 25}{6 \cdot 10 \cdot 16} \\
&= \frac{1 \cdot 3 \cdot 5 \cdot 5}{2 \cdot 3 \cdot 2 \cdot 5 \cdot 16} \\
&= \frac{1 \cdot 5}{2 \cdot 2 \cdot 16} \\
&= \frac{5}{64}
\end{aligned}
$$

8. $\frac{2}{3} \cdot \frac{3y}{2} = \frac{2 \cdot 3 \cdot y}{3 \cdot 2 \cdot 1} = \frac{y}{1} = y$

Copyright © 2011 Pearson Education, Inc. Publishing as Prentice Hall.

9. $\frac{a^3}{b^2}\cdot\frac{b}{a^2}=\frac{a^3\cdot b}{b^2\cdot a^2}=\frac{a\cdot a\cdot a\cdot b}{b\cdot b\cdot a\cdot a}=\frac{a}{b}$

10. a. $\left(\frac{3}{4}\right)^3=\frac{3}{4}\cdot\frac{3}{4}\cdot\frac{3}{4}=\frac{3\cdot 3\cdot 3}{4\cdot 4\cdot 4}=\frac{27}{64}$

b. $\left(-\frac{4}{5}\right)^2=\left(-\frac{4}{5}\right)\cdot\left(-\frac{4}{5}\right)=\frac{4\cdot 4}{5\cdot 5}=\frac{16}{25}$

11. $\frac{8}{7}\div\frac{2}{9}=\frac{8}{7}\cdot\frac{9}{2}=\frac{8\cdot 9}{7\cdot 2}=\frac{2\cdot 4\cdot 9}{7\cdot 2}=\frac{4\cdot 9}{7}=\frac{36}{7}$

12. $\frac{4}{9}\div\frac{1}{2}=\frac{4}{9}\cdot\frac{2}{1}=\frac{4\cdot 2}{9\cdot 1}=\frac{8}{9}$

13.
$$\begin{aligned}-\frac{10}{4}\div\frac{2}{9}&=-\frac{10}{4}\cdot\frac{9}{2}\\&=-\frac{10\cdot 9}{4\cdot 2}\\&=-\frac{2\cdot 5\cdot 9}{4\cdot 2}\\&=-\frac{5\cdot 9}{4}\\&=-\frac{45}{4}\end{aligned}$$

14.
$$\begin{aligned}\frac{3y}{4}\div 5y^3&=\frac{3y}{4}\div\frac{5y^3}{1}\\&=\frac{3y}{4}\cdot\frac{1}{5y^3}\\&=\frac{3y\cdot 1}{4\cdot 5y^3}\\&=\frac{3\cdot y\cdot 1}{4\cdot 5\cdot y\cdot y\cdot y}\\&=\frac{3\cdot 1}{4\cdot 5\cdot y\cdot y}\\&=\frac{3}{20y^2}\end{aligned}$$

15.
$$\begin{aligned}\left(-\frac{2}{3}\cdot\frac{9}{14}\right)\div\frac{7}{15}&=\left(-\frac{2\cdot 9}{3\cdot 14}\right)\div\frac{7}{15}\\&=\left(-\frac{2\cdot 3\cdot 3}{3\cdot 7\cdot 2}\right)\div\frac{7}{15}\\&=\left(-\frac{3}{7}\right)\div\frac{7}{15}\\&=\left(-\frac{3}{7}\right)\cdot\frac{15}{7}\\&=-\frac{3\cdot 15}{7\cdot 7}\\&=-\frac{45}{49}\end{aligned}$$

16. a. $xy=-\frac{3}{4}\cdot\frac{9}{2}=-\frac{3\cdot 9}{4\cdot 2}=-\frac{27}{8}$

b.
$$\begin{aligned}x\div y&=-\frac{3}{4}\div\frac{9}{2}\\&=-\frac{3}{4}\cdot\frac{2}{9}\\&=-\frac{3\cdot 2}{4\cdot 9}\\&=-\frac{1\cdot 3\cdot 2}{2\cdot 2\cdot 3\cdot 3}\\&=-\frac{1}{2\cdot 3}\\&=-\frac{1}{6}\end{aligned}$$

17.
$$\begin{aligned}2x&=-\frac{9}{4}\\2\left(-\frac{9}{8}\right)&\stackrel{?}{=}-\frac{9}{4}\\\frac{2}{1}\cdot\left(-\frac{9}{8}\right)&\stackrel{?}{=}-\frac{9}{4}\\-\frac{2\cdot 9}{1\cdot 8}&\stackrel{?}{=}-\frac{9}{4}\\-\frac{2\cdot 9}{1\cdot 2\cdot 4}&\stackrel{?}{=}-\frac{9}{4}\\-\frac{9}{4}&=-\frac{9}{4}\quad\text{True}\end{aligned}$$

Yes, $-\frac{9}{8}$ is a solution of the equation.

18. $\frac{1}{6}\cdot 60=\frac{1}{6}\cdot\frac{60}{1}=\frac{1\cdot 60}{6\cdot 1}=\frac{1\cdot 6\cdot 10}{6\cdot 1}=10$

Thus, there are 10 roller coasters in Hershey Park.

Copyright © 2011 Pearson Education, Inc. Publishing as Prentice Hall.

Vocabulary and Readiness Check

1. To multiply two fractions, we write $\frac{a}{b}\cdot\frac{c}{d}=\underline{\frac{a\cdot c}{b\cdot d}}$.

2. Two numbers are <u>reciprocals</u> of each other if their product is 1.

3. The expression $\frac{2^3}{7}=\underline{\frac{2\cdot 2\cdot 2}{7}}$ while $\left(\frac{2}{7}\right)^3=\underline{\frac{2}{7}\cdot\frac{2}{7}\cdot\frac{2}{7}}$.

4. Every number has a reciprocal expect <u>0</u>.

5. To divide two fractions, we write $\frac{a}{b}\div\frac{c}{d}=\underline{\frac{a\cdot d}{b\cdot c}}$.

6. The word "of" indicates <u>multiplication</u>.

Exercise Set 4.3

1. $\frac{6}{11}\cdot\frac{3}{7}=\frac{6\cdot 3}{11\cdot 7}=\frac{18}{77}$

3. $-\frac{2}{7}\cdot\frac{5}{8}=-\frac{2\cdot 5}{7\cdot 8}=-\frac{2\cdot 5}{7\cdot 2\cdot 4}=-\frac{5}{7\cdot 4}=-\frac{5}{28}$

5. $-\frac{1}{2}\cdot-\frac{2}{15}=\frac{1\cdot 2}{2\cdot 15}=\frac{1}{15}$

7. $$\begin{aligned}\frac{18x}{20}\cdot\frac{36}{99}&=\frac{18x\cdot 36}{20\cdot 99}\\&=\frac{2\cdot 9\cdot x\cdot 2\cdot 18}{2\cdot 2\cdot 5\cdot 9\cdot 11}\\&=\frac{18\cdot x}{5\cdot 11}\\&=\frac{18x}{55}\end{aligned}$$

9. $3a^2\cdot\frac{1}{4}=\frac{3a^2}{1}\cdot\frac{1}{4}=\frac{3a^2\cdot 1}{1\cdot 4}=\frac{3a^2}{4}$

11. $\frac{x^3}{y^3}\cdot\frac{y^2}{x}=\frac{x^3\cdot y^2}{y^3\cdot x}=\frac{x\cdot x\cdot x\cdot y\cdot y}{y\cdot y\cdot y\cdot x}=\frac{x\cdot x}{y}=\frac{x^2}{y}$

13. $0\cdot\frac{8}{9}=0$

15. $$\begin{aligned}-\frac{17y}{20}\cdot\frac{4}{5y}&=-\frac{17y\cdot 4}{20\cdot 5y}\\&=-\frac{17\cdot y\cdot 4}{5\cdot 4\cdot 5\cdot y}\\&=-\frac{17}{5\cdot 5}\\&=-\frac{17}{25}\end{aligned}$$

17. $$\begin{aligned}\frac{11}{20}\cdot\frac{1}{7}\cdot\frac{5}{22}&=\frac{11\cdot 1\cdot 5}{20\cdot 7\cdot 22}\\&=\frac{11\cdot 1\cdot 5}{5\cdot 2\cdot 2\cdot 7\cdot 11\cdot 2}\\&=\frac{1}{2\cdot 2\cdot 7\cdot 2}\\&=\frac{1}{56}\end{aligned}$$

19. $\left(\frac{1}{5}\right)^3=\left(\frac{1}{5}\right)\left(\frac{1}{5}\right)\left(\frac{1}{5}\right)=\frac{1\cdot 1\cdot 1}{5\cdot 5\cdot 5}=\frac{1}{125}$

21. $\left(-\frac{2}{3}\right)^2=-\frac{2}{3}\cdot-\frac{2}{3}=\frac{2\cdot 2}{3\cdot 3}=\frac{4}{9}$

23. $$\begin{aligned}\left(-\frac{2}{3}\right)^3\cdot\frac{1}{2}&=\left(-\frac{2}{3}\right)\left(-\frac{2}{3}\right)\left(-\frac{2}{3}\right)\cdot\frac{1}{2}\\&=-\frac{2\cdot 2\cdot 2\cdot 1}{3\cdot 3\cdot 3\cdot 2}\\&=-\frac{2\cdot 2\cdot 1}{3\cdot 3\cdot 3}\\&=-\frac{4}{27}\end{aligned}$$

25. $\frac{2}{3}\div\frac{5}{6}=\frac{2}{3}\cdot\frac{6}{5}=\frac{2\cdot 6}{3\cdot 5}=\frac{2\cdot 2\cdot 3}{3\cdot 5}=\frac{2\cdot 2}{5}=\frac{4}{5}$

27. $$\begin{aligned}-\frac{6}{15}\div\frac{12}{5}&=-\frac{6}{15}\cdot\frac{5}{12}\\&=-\frac{6\cdot 5}{15\cdot 12}\\&=-\frac{6\cdot 5\cdot 1}{5\cdot 3\cdot 6\cdot 2}\\&=-\frac{1}{3\cdot 2}\\&=-\frac{1}{6}\end{aligned}$$

29. $-\frac{8}{9}\div\frac{x}{2}=-\frac{8}{9}\cdot\frac{2}{x}=-\frac{8\cdot 2}{9\cdot x}=-\frac{16}{9x}$

Copyright © 2011 Pearson Education, Inc. Publishing as Prentice Hall.

31. $\frac{11y}{20} \div \frac{3}{11} = \frac{11y}{20} \cdot \frac{11}{3} = \frac{11y \cdot 11}{20 \cdot 3} = \frac{121y}{60}$

33. $-\frac{2}{3} \div 4 = -\frac{2}{3} \div \frac{4}{1}$
$= -\frac{2}{3} \cdot \frac{1}{4}$
$= -\frac{2 \cdot 1}{3 \cdot 4}$
$= -\frac{2 \cdot 1}{3 \cdot 2 \cdot 2}$
$= -\frac{1}{3 \cdot 2}$
$= -\frac{1}{6}$

35. $\frac{1}{5x} \div \frac{5}{x^2} = \frac{1}{5x} \cdot \frac{x^2}{5}$
$= \frac{1 \cdot x^2}{5x \cdot 5}$
$= \frac{1 \cdot x \cdot x}{5 \cdot x \cdot 5}$
$= \frac{1 \cdot x}{5 \cdot 5}$
$= \frac{x}{25}$

37. $\frac{2}{3} \cdot \frac{5}{9} = \frac{2 \cdot 5}{3 \cdot 9} = \frac{10}{27}$

39. $\frac{3x}{7} \div \frac{5}{6x} = \frac{3x}{7} \cdot \frac{6x}{5} = \frac{3x \cdot 6x}{7 \cdot 5} = \frac{18x^2}{35}$

41. $\frac{16}{27y} \div \frac{8}{15y} = \frac{16}{27y} \cdot \frac{15y}{8}$
$= \frac{16 \cdot 15y}{27y \cdot 8}$
$= \frac{8 \cdot 2 \cdot 3 \cdot 5 \cdot y}{3 \cdot 9 \cdot y \cdot 8}$
$= \frac{2 \cdot 5}{9}$
$= \frac{10}{9}$

43. $-\frac{5}{28} \cdot \frac{35}{25} = -\frac{5 \cdot 35}{28 \cdot 25} = -\frac{5 \cdot 7 \cdot 5 \cdot 1}{7 \cdot 4 \cdot 5 \cdot 5} = -\frac{1}{4}$

45. $\left(-\frac{3}{4}\right)^2 = -\frac{3}{4} \cdot -\frac{3}{4} = \frac{3 \cdot 3}{4 \cdot 4} = \frac{9}{16}$

47. $\frac{x^2}{y} \cdot \frac{y^3}{x} = \frac{x^2 \cdot y^3}{y \cdot x} = \frac{x \cdot x \cdot y \cdot y \cdot y}{y \cdot x \cdot 1} = \frac{x \cdot y \cdot y}{1} = xy^2$

49. $7 \div \frac{2}{11} = \frac{7}{1} \div \frac{2}{11} = \frac{7}{1} \cdot \frac{11}{2} = \frac{7 \cdot 11}{1 \cdot 2} = \frac{77}{2}$

51. $-3x \div \frac{x^2}{12} = -\frac{3x}{1} \div \frac{x^2}{12}$
$= -\frac{3x}{1} \cdot \frac{12}{x^2}$
$= -\frac{3x \cdot 12}{1 \cdot x^2}$
$= -\frac{3 \cdot x \cdot 12}{1 \cdot x \cdot x}$
$= -\frac{3 \cdot 12}{1 \cdot x}$
$= -\frac{36}{x}$

53. $\left(\frac{2}{7} \div \frac{7}{2}\right) \cdot \frac{3}{4} = \left(\frac{2}{7} \cdot \frac{2}{7}\right) \cdot \frac{3}{4}$
$= \frac{4}{49} \cdot \frac{3}{4}$
$= \frac{4 \cdot 3}{49 \cdot 4}$
$= \frac{3}{49}$

55. $-\frac{19}{63y} \cdot 9y^2 = -\frac{19}{63y} \cdot \frac{9y^2}{1}$
$= -\frac{19 \cdot 9y^2}{63y \cdot 1}$
$= -\frac{19 \cdot 9 \cdot y \cdot y}{9 \cdot 7 \cdot y \cdot 1}$
$= -\frac{19 \cdot y}{7 \cdot 1}$
$= -\frac{19y}{7}$

57. $-\frac{2}{3} \cdot -\frac{6}{11} = \frac{2 \cdot 6}{3 \cdot 11} = \frac{2 \cdot 2 \cdot 3}{3 \cdot 11} = \frac{2 \cdot 2}{11} = \frac{4}{11}$

59. $\frac{4}{8} \div \frac{3}{16} = \frac{4}{8} \cdot \frac{16}{3} = \frac{4 \cdot 16}{8 \cdot 3} = \frac{4 \cdot 8 \cdot 2}{8 \cdot 3} = \frac{4 \cdot 2}{3} = \frac{8}{3}$

Copyright © 2011 Pearson Education, Inc. Publishing as Prentice Hall.

61. $$\frac{21x^2}{10y} \div \frac{14x}{25y} = \frac{21x^2}{10y} \cdot \frac{25y}{14x} = \frac{21x^2 \cdot 25y}{10y \cdot 14x} = \frac{3 \cdot 7 \cdot x \cdot x \cdot 5 \cdot 5 \cdot y}{2 \cdot 5 \cdot y \cdot 2 \cdot 7 \cdot x} = \frac{3 \cdot x \cdot 5}{2 \cdot 2} = \frac{15x}{4}$$

63. $$\left(1 \div \frac{3}{4}\right) \cdot \frac{2}{3} = \left(\frac{1}{1} \div \frac{3}{4}\right) \cdot \frac{2}{3} = \left(\frac{1}{1} \cdot \frac{4}{3}\right) \cdot \frac{2}{3} = \frac{1 \cdot 4}{1 \cdot 3} \cdot \frac{2}{3} = \frac{1 \cdot 4 \cdot 2}{1 \cdot 3 \cdot 3} = \frac{8}{9}$$

65. $$\frac{a^3}{2} \div 30a^3 = \frac{a^3}{2} \div \frac{30a^3}{1} = \frac{a^3}{2} \cdot \frac{1}{30a^3} = \frac{a^3 \cdot 1}{2 \cdot 30a^3} = \frac{a \cdot a \cdot a \cdot 1}{2 \cdot 30 \cdot a \cdot a \cdot a} = \frac{1}{2 \cdot 30} = \frac{1}{60}$$

67. $$\frac{ab^2}{c} \cdot \frac{c}{ab} = \frac{ab^2 \cdot c}{c \cdot ab} = \frac{a \cdot b \cdot b \cdot c}{c \cdot a \cdot b} = b$$

69. $$\left(\frac{1}{2} \cdot \frac{2}{3}\right) \div \frac{5}{6} = \left(\frac{1 \cdot 2}{2 \cdot 3}\right) \div \frac{5}{6} = \frac{1}{3} \div \frac{5}{6} = \frac{1}{3} \cdot \frac{6}{5} = \frac{1 \cdot 6}{3 \cdot 5} = \frac{1 \cdot 2 \cdot 3}{3 \cdot 5} = \frac{1 \cdot 2}{5} = \frac{2}{5}$$

71. $$-\frac{4}{7} \div \left(\frac{4}{5} \cdot \frac{3}{7}\right) = -\frac{4}{7} \div \left(\frac{4 \cdot 3}{5 \cdot 7}\right) = -\frac{4}{7} \div \frac{12}{35} = -\frac{4}{7} \cdot \frac{35}{12} = -\frac{4 \cdot 35}{7 \cdot 12} = -\frac{4 \cdot 7 \cdot 5}{7 \cdot 4 \cdot 3} = -\frac{5}{3}$$

73. a. $$xy = \frac{2}{5} \cdot \frac{5}{6} = \frac{2 \cdot 5}{5 \cdot 6} = \frac{2 \cdot 5}{5 \cdot 2 \cdot 3} = \frac{1}{3}$$

b. $$x \div y = \frac{2}{5} \div \frac{5}{6} = \frac{2}{5} \cdot \frac{6}{5} = \frac{2 \cdot 6}{5 \cdot 5} = \frac{12}{25}$$

75. a. $$xy = -\frac{4}{5} \cdot \frac{9}{11} = -\frac{4 \cdot 9}{5 \cdot 11} = -\frac{36}{55}$$

b. $$x \div y = -\frac{4}{5} \div \frac{9}{11} = -\frac{4}{5} \cdot \frac{11}{9} = -\frac{4 \cdot 11}{5 \cdot 9} = -\frac{44}{45}$$

Copyright © 2011 Pearson Education, Inc. Publishing as Prentice Hall.

77.
$$3x = -\frac{5}{6}$$
$$3\left(-\frac{5}{18}\right) \stackrel{?}{=} -\frac{5}{6}$$
$$\frac{3}{1}\cdot -\frac{5}{18} \stackrel{?}{=} -\frac{5}{6}$$
$$-\frac{3\cdot 5}{1\cdot 18} \stackrel{?}{=} -\frac{5}{6}$$
$$-\frac{3\cdot 5}{3\cdot 6} \stackrel{?}{=} -\frac{5}{6}$$
$$-\frac{5}{6} = -\frac{5}{6} \quad \text{True}$$

Yes, $-\frac{5}{18}$ is a solution of the equation.

79.
$$-\frac{1}{2}z = \frac{1}{10}$$
$$-\frac{1}{2}\cdot\frac{2}{5} \stackrel{?}{=} \frac{1}{10}$$
$$-\frac{1\cdot 2}{2\cdot 5} \stackrel{?}{=} \frac{1}{10}$$
$$-\frac{1}{5} = \frac{1}{10} \quad \text{False}$$

No, $\frac{2}{5}$ is not a solution of the equation.

81.
$$\frac{1}{4} \text{ of } 200 = \frac{1}{4}\cdot 200 = \frac{1}{4}\cdot\frac{200}{1} = \frac{1\cdot 200}{4\cdot 1} = \frac{1\cdot 4\cdot 50}{4\cdot 1} = \frac{50}{1} = 50$$

83.
$$\frac{5}{6} \text{ of } 24 = \frac{5}{6}\cdot 24 = \frac{5}{6}\cdot\frac{24}{1} = \frac{5\cdot 24}{6\cdot 1} = \frac{5\cdot 6\cdot 4}{6\cdot 1} = \frac{5\cdot 4}{1} = \frac{20}{1} = 20$$

85.
$$\frac{4}{25} \text{ of } 800 = \frac{4}{25}\cdot 800 = \frac{4}{25}\cdot\frac{800}{1} = \frac{4\cdot 800}{25\cdot 1} = \frac{4\cdot 25\cdot 32}{25\cdot 1} = \frac{4\cdot 32}{1} = 128$$

128 of the students would be expected to major in business.

87.
$$\frac{7}{25} \text{ of 175 million} = \frac{7}{25}\cdot 175{,}000{,}000 = \frac{7}{25}\cdot\frac{175{,}000{,}000}{1} = \frac{7\cdot 175{,}000{,}000}{25\cdot 1} = \frac{7\cdot 25\cdot 7{,}000{,}000}{25\cdot 1} = 49{,}000{,}000$$

Approximately 49 million people ages 16–24 attended the movies.

Copyright © 2011 Pearson Education, Inc. Publishing as Prentice Hall.

89. $\frac{2}{5}$ of $2170 = \frac{2}{5}\cdot 2170$
$$= \frac{2}{5}\cdot\frac{2170}{1} = \frac{2\cdot 2170}{5\cdot 1} = \frac{2\cdot 5\cdot 434}{5\cdot 1} = \frac{2\cdot 434}{1} = 868$$
He hiked 868 miles.

91. $\frac{1}{2}$ of $\frac{3}{8} = \frac{1}{2}\cdot\frac{3}{8} = \frac{1\cdot 3}{2\cdot 8} = \frac{3}{16}$

The radius of the circle is $\frac{3}{16}$ inch.

93. $\frac{2}{3}$ of $2757 = \frac{2}{3}\cdot 2757$
$$= \frac{2}{3}\cdot\frac{2757}{1} = \frac{2\cdot 2757}{3\cdot 1} = \frac{2\cdot 3\cdot 919}{3\cdot 1} = \frac{2\cdot 919}{1} = 1838$$
The sale price is \$1838.

95. $\frac{1}{184}$ of $9200 = \frac{1}{184}\cdot 9200$
$$= \frac{1}{184}\cdot\frac{9200}{1} = \frac{1\cdot 9200}{184\cdot 1} = \frac{1\cdot 184\cdot 50}{184\cdot 1} = 50$$
There are 50 libraries in Mississippi.

97. area = length · width $= \frac{5}{14}\cdot\frac{1}{5} = \frac{5\cdot 1}{14\cdot 5} = \frac{1}{14}$

The area is $\frac{1}{14}$ square foot.

99. $\frac{8}{25}\cdot 12{,}000 = \frac{8}{25}\cdot\frac{12{,}000}{1}$
$$= \frac{8\cdot 25\cdot 480}{25\cdot 1} = \frac{8\cdot 480}{1} = 3840$$
The family drove 3840 miles for work.

101. $\frac{1}{5}$ of $12{,}000 = \frac{1}{5}\cdot 12{,}000$
$$= \frac{1}{5}\cdot\frac{12{,}000}{1} = \frac{1\cdot 5\cdot 2400}{5\cdot 1} = 2400$$
The family drove 2400 miles for family business.

103.
$$\begin{array}{r} 27 \\ 76 \\ +\,98 \\ \hline 201 \end{array}$$

105.
$$\begin{array}{r} 968 \\ -\,772 \\ \hline 196 \end{array}$$

107. answers may vary

109. $\frac{42}{25}\cdot\frac{125}{36}\div\frac{7}{6} = \frac{42}{25}\cdot\frac{125}{36}\cdot\frac{6}{7}$
$$= \frac{42\cdot 125\cdot 6}{25\cdot 36\cdot 7} = \frac{6\cdot 7\cdot 5\cdot 25\cdot 6}{25\cdot 6\cdot 6\cdot 7} = 5$$

111. $\frac{1}{8}$ of $305{,}400{,}000 = \frac{1}{8}\cdot 305{,}400{,}000$
$$= \frac{1}{8}\cdot\frac{305{,}400{,}000}{1} = \frac{1\cdot 305{,}400{,}000}{8\cdot 1} = \frac{1\cdot 8\cdot 38{,}175{,}000}{8\cdot 1} = \frac{38{,}175{,}000}{1} = 38{,}175{,}000$$
The population of California is approximately 38,175,000.

Copyright © 2011 Pearson Education, Inc. Publishing as Prentice Hall.

113. $\frac{63}{200}$ of $27{,}000 = \frac{63}{200} \cdot 27{,}000$
$= \frac{63}{200} \cdot \frac{27{,}000}{1}$
$= \frac{63 \cdot 27{,}000}{200 \cdot 1}$
$= \frac{63 \cdot 200 \cdot 135}{200 \cdot 1}$
$= \frac{63 \cdot 135}{1}$
$= 8505$

The National Park Service is charged with maintaining 8505 monuments and statues.

Section 4.4

Practice Problems

1. $\frac{6}{13} + \frac{2}{13} = \frac{6+2}{13} = \frac{8}{13}$

2. $\frac{5}{8x} + \frac{1}{8x} = \frac{5+1}{8x} = \frac{6}{8x} = \frac{2 \cdot 3}{2 \cdot 4x} = \frac{3}{4x}$

3. $\frac{20}{11} + \frac{6}{11} + \frac{7}{11} = \frac{20+6+7}{11} = \frac{33}{11}$ or 3

4. $\frac{11}{12} - \frac{6}{12} = \frac{11-6}{12} = \frac{5}{12}$

5. $\frac{7}{15} - \frac{2}{15} = \frac{7-2}{15} = \frac{5}{15} = \frac{1 \cdot 5}{3 \cdot 5} = \frac{1}{3}$

6. $-\frac{8}{17} + \frac{4}{17} = \frac{-8+4}{17} = \frac{-4}{17}$ or $-\frac{4}{17}$

7. $\frac{2}{5} - \frac{7y}{5} = \frac{2-7y}{5}$

8. $\frac{4}{11} - \frac{6}{11} - \frac{3}{11} = \frac{4-6-3}{11} = \frac{-5}{11}$ or $-\frac{5}{11}$

9. $x + y = -\frac{10}{12} + \frac{5}{12} = \frac{-10+5}{12} = \frac{-5}{12}$ or $-\frac{5}{12}$

10. perimeter $= \frac{3}{20} + \frac{3}{20} + \frac{3}{20} + \frac{3}{20}$
$= \frac{3+3+3+3}{20}$
$= \frac{12}{20}$
$= \frac{3 \cdot 4}{5 \cdot 4}$
$= \frac{3}{5}$

The perimeter is $\frac{3}{5}$ mile.

11. $\frac{13}{4} - \frac{11}{4} = \frac{13-11}{4} = \frac{2}{4} = \frac{1 \cdot 2}{2 \cdot 2} = \frac{1}{2}$

He ran $\frac{1}{2}$ mile farther on Monday.

12. 16 is a multiple of 8, so the LCD of $\frac{7}{8}$ and $\frac{11}{16}$ is 16.

13. $30 \cdot 1 = 30$ Not a multiple of 25.
$30 \cdot 2 = 60$ Not a multiple of 25.
$30 \cdot 3 = 90$ Not a multiple of 25.
$30 \cdot 4 = 120$ Not a multiple of 25.
$30 \cdot 5 = 150$ A multiple of 25.

The LCD of $\frac{23}{25}$ and $\frac{1}{30}$ is 150.

14. $40 = \boxed{2 \cdot 2 \cdot 2 \cdot 5}$
$108 = 2 \cdot 2 \cdot \boxed{3 \cdot 3 \cdot 3}$
$\text{LCD} = 2 \cdot 2 \cdot 2 \cdot 3 \cdot 3 \cdot 3 \cdot 5 = 1080$

The LCD of $-\frac{3}{40}$ and $\frac{11}{108}$ is 1080.

15. $20 = 2 \cdot 2 \cdot \boxed{5}$
$24 = \boxed{2 \cdot 2 \cdot 2} \cdot 3$
$45 = \boxed{3 \cdot 3} \cdot 5$
$\text{LCD} = 2 \cdot 2 \cdot 2 \cdot 3 \cdot 3 \cdot 5 = 360$

The LCD of $\frac{7}{20}$, $\frac{1}{24}$, and $\frac{13}{45}$ is 360.

16. $y = \boxed{y}$
$11 = \boxed{11}$

The LCD of $\frac{7}{y}$ and $\frac{6}{11}$ is $11y$.

Copyright © 2011 Pearson Education, Inc. Publishing as Prentice Hall.

17. $\frac{7}{8}=\frac{7}{8}\cdot\frac{7}{7}=\frac{7\cdot 7}{8\cdot 7}=\frac{49}{56}$

18. $\frac{1}{4}=\frac{1}{4}\cdot\frac{5}{5}=\frac{1\cdot 5}{4\cdot 5}=\frac{5}{20}$

19. $\frac{3x}{7}=\frac{3x}{7}\cdot\frac{6}{6}=\frac{3x\cdot 6}{7\cdot 6}=\frac{18x}{42}$

20. $4=\frac{4}{1}\cdot\frac{6}{6}=\frac{4\cdot 6}{1\cdot 6}=\frac{24}{6}$

21. $\frac{9}{4x}=\frac{9}{4x}\cdot\frac{9}{9}=\frac{9\cdot 9}{4x\cdot 9}=\frac{81}{36x}$

Vocabulary and Readiness Check

1. The fractions $\frac{9}{11}$ and $\frac{13}{11}$ are called like fractions while $\frac{3}{4}$ and $\frac{1}{3}$ are called unlike fractions.

2. $\frac{a}{b}+\frac{c}{b}=\frac{a+c}{\underline{b}}$ and $\frac{a}{b}-\frac{c}{b}=\frac{a-c}{\underline{b}}$.

3. As long as b is not 0, $\frac{-a}{b}=\frac{a}{-b}=-\frac{a}{\underline{b}}$.

4. The distance around a figure is called its perimeter.

5. The smallest positive number divisible by all the denominators of a list of fractions is called the least common denominator (LCD).

6. Fractions that represent the same portion of a whole are called equivalent fractions.

Exercise Set 4.4

1. $\frac{5}{11}+\frac{2}{11}=\frac{5+2}{11}=\frac{7}{11}$

3. $\frac{2}{9}+\frac{4}{9}=\frac{2+4}{9}=\frac{6}{9}=\frac{3\cdot 2}{3\cdot 3}=\frac{2}{3}$

5. $-\frac{6}{20}+\frac{1}{20}=\frac{-6+1}{20}=\frac{-5}{20}=-\frac{1\cdot 5}{4\cdot 5}=-\frac{1}{4}$

7. $-\frac{3}{14}+\left(-\frac{4}{14}\right)=\frac{-3+(-4)}{14}=\frac{-7}{14}=-\frac{1\cdot 7}{2\cdot 7}=-\frac{1}{2}$

9. $\frac{2}{9x}+\frac{4}{9x}=\frac{2+4}{9x}=\frac{6}{9x}=\frac{2\cdot 3}{3\cdot 3\cdot x}=\frac{2}{3x}$

11.
$$-\frac{7x}{18}+\frac{3x}{18}+\frac{2x}{18}=\frac{-7x+3x+2x}{18}=\frac{-2x}{18}=-\frac{2\cdot x}{2\cdot 9}=-\frac{x}{9}$$

13. $\frac{10}{11}-\frac{4}{11}=\frac{10-4}{11}=\frac{6}{11}$

15. $\frac{7}{8}-\frac{1}{8}=\frac{7-1}{8}=\frac{6}{8}=\frac{3\cdot 2}{4\cdot 2}=\frac{3}{4}$

17. $\frac{1}{y}-\frac{4}{y}=\frac{1-4}{y}=\frac{-3}{y}=-\frac{3}{y}$

19.
$$-\frac{27}{33}-\left(-\frac{8}{33}\right)=-\frac{27}{33}+\frac{8}{33}=\frac{-27+8}{33}=\frac{-19}{33}=-\frac{19}{33}$$

21.
$$\frac{20}{21}-\frac{10}{21}-\frac{17}{21}=\frac{20-10-17}{21}=\frac{-7}{21}=-\frac{1\cdot 7}{3\cdot 7}=-\frac{1}{3}$$

23. $\frac{7a}{4}-\frac{3}{4}=\frac{7a-3}{4}$

25. $-\frac{9}{100}+\frac{99}{100}=\frac{-9+99}{100}=\frac{90}{100}=\frac{9\cdot 10}{10\cdot 10}=\frac{9}{10}$

Copyright © 2011 Pearson Education, Inc. Publishing as Prentice Hall.

27. $-\frac{13x}{28}-\frac{13x}{28}=\frac{-13x-13x}{28}$
$=\frac{-26x}{28}$
$=-\frac{2\cdot 13\cdot x}{2\cdot 14}$
$=-\frac{13x}{14}$

29. $\frac{9x}{15}+\frac{1}{15}=\frac{9x+1}{15}$

31. $\frac{7x}{16}-\frac{15x}{16}=\frac{7x-15x}{16}=\frac{-8x}{16}=-\frac{8\cdot x}{8\cdot 2}=-\frac{x}{2}$

33. $\frac{9}{12}-\frac{7}{12}-\frac{10}{12}=\frac{9-7-10}{12}=\frac{-8}{12}=-\frac{2\cdot 4}{3\cdot 4}=-\frac{2}{3}$

35. $\frac{x}{4}+\frac{3x}{4}-\frac{2x}{4}+\frac{x}{4}=\frac{x+3x-2x+x}{4}=\frac{3x}{4}$

37. $x+y=\frac{3}{4}+\frac{2}{4}=\frac{3+2}{4}=\frac{5}{4}$

39. $x-y=-\frac{1}{5}-\left(-\frac{3}{5}\right)=-\frac{1}{5}+\frac{3}{5}=\frac{-1+3}{5}=\frac{2}{5}$

41. $\frac{4}{20}+\frac{7}{20}+\frac{9}{20}=\frac{4+7+9}{20}=\frac{20}{20}=1$
The perimeter is 1 inch.

43. $\frac{5}{12}+\frac{7}{12}+\frac{5}{12}+\frac{7}{12}=\frac{5+7+5+7}{12}=\frac{24}{12}=2$
The perimeter is 2 meters.

45. To find the remaining amount of track to be inspected, subtract the $\frac{5}{20}$ mile that has already been inspected from the $\frac{19}{20}$ mile total that must be inspected.
$\frac{19}{20}-\frac{5}{20}=\frac{19-5}{20}=\frac{14}{20}=\frac{2\cdot 7}{2\cdot 10}=\frac{7}{10}$
$\frac{7}{10}$ of a mile of track remains to be inspected.

47. To find the fraction that had speed limits less than 70 mph, subtract the $\frac{17}{50}$ that have 70 mph speed limits from the $\frac{37}{50}$ that have speed limits up to and including 70.
$\frac{37}{50}-\frac{17}{50}=\frac{37-17}{50}=\frac{20}{50}=\frac{2\cdot 10}{5\cdot 10}=\frac{2}{5}$
$\frac{2}{5}$ of the states have speed limits that were less than 70 mph.

49. North America makes up $\frac{16}{100}$ of the world's land area, while South America makes up $\frac{12}{100}$ of the land area.
$\frac{16}{100}+\frac{12}{100}=\frac{16+12}{100}=\frac{28}{100}=\frac{4\cdot 7}{4\cdot 25}=\frac{7}{25}$
$\frac{7}{25}$ of the world's land area is within North America and South America.

51. Antarctica makes up $\frac{9}{100}$ of the world's land area, while Europe makes up $\frac{7}{100}$ of the world's land area.
$\frac{9}{100}-\frac{7}{100}=\frac{9-7}{100}=\frac{2}{100}=\frac{1\cdot 2}{50\cdot 2}=\frac{1}{50}$
Antarctica's land area is $\frac{1}{50}$ greater than that of Europe.

53. Multiples of 15:
$15\cdot 1=15$, not a multiple of 9
$15\cdot 2=30$, not a multiple of 9
$15\cdot 3=45$, a multiple of 9
LCD: 45

55. Multiples of 36:
$36\cdot 1=36$, not a multiple of 24
$36\cdot 2=72$, a multiple of 24
LCD: 72

57. $6=\boxed{2}\cdot 3$
$15=\boxed{3}\cdot 5$
$25=\boxed{5\cdot 5}$
$\text{LCD}=2\cdot 3\cdot 5\cdot 5=150$

Copyright © 2011 Pearson Education, Inc. Publishing as Prentice Hall.

59. $24 = \boxed{2 \cdot 2 \cdot 2 \cdot 3}$
$x = \boxed{x}$
$\text{LCD} = 2 \cdot 2 \cdot 2 \cdot 3 \cdot x = 24x$

61. $18 = \boxed{2 \cdot 3 \cdot 3}$
$21 = 3 \cdot \boxed{7}$
$\text{LCD} = 2 \cdot 3 \cdot 3 \cdot 7 = 126$

63. $3 = \boxed{3}$
$21 = 3 \cdot \boxed{7}$
$56 = \boxed{2 \cdot 2 \cdot 2} \cdot 7$
$\text{LCD} = 2 \cdot 2 \cdot 2 \cdot 3 \cdot 7 = 168$

65. $\frac{2}{3} = \frac{2}{3} \cdot \frac{7}{7} = \frac{2 \cdot 7}{3 \cdot 7} = \frac{14}{21}$

67. $\frac{4}{7} = \frac{4}{7} \cdot \frac{5}{5} = \frac{4 \cdot 5}{7 \cdot 5} = \frac{20}{35}$

69. $\frac{1}{2} = \frac{1}{2} \cdot \frac{25}{25} = \frac{1 \cdot 25}{2 \cdot 25} = \frac{25}{50}$

71. $\frac{14x}{17} = \frac{14x}{17} \cdot \frac{4}{4} = \frac{14x \cdot 4}{17 \cdot 4} = \frac{56x}{68}$

73. $\frac{2y}{3} = \frac{2y}{3} \cdot \frac{4}{4} = \frac{2y \cdot 4}{3 \cdot 4} = \frac{8y}{12}$

75. $\frac{5}{9} = \frac{5}{9} \cdot \frac{4a}{4a} = \frac{5 \cdot 4a}{9 \cdot 4a} = \frac{20a}{36a}$

77. books & magazines: $\frac{27}{50} = \frac{27 \cdot 2}{50 \cdot 2} = \frac{54}{100}$

clothing & accessories: $\frac{1}{2} = \frac{1 \cdot 50}{2 \cdot 50} = \frac{50}{100}$

computer hardware: $\frac{23}{50} = \frac{23 \cdot 2}{50 \cdot 2} = \frac{46}{100}$

computer software: $\frac{1}{2} = \frac{1 \cdot 50}{2 \cdot 50} = \frac{50}{100}$

drugs, health & beauty aids: $\frac{3}{20} = \frac{3 \cdot 5}{20 \cdot 5} = \frac{15}{100}$

electronics and appliances: $\frac{13}{20} = \frac{13 \cdot 5}{20 \cdot 5} = \frac{65}{100}$

food, beer, wine: $\frac{9}{20} = \frac{9 \cdot 5}{20 \cdot 5} = \frac{45}{100}$

home furnishings: $\frac{13}{25} = \frac{13 \cdot 4}{25 \cdot 4} = \frac{52}{100}$

music and videos: $\frac{3}{5} = \frac{3 \cdot 20}{5 \cdot 20} = \frac{60}{100}$

office equipment & supplies: $\frac{61}{100}$

sporting goods: $\frac{12}{25} = \frac{12 \cdot 4}{25 \cdot 4} = \frac{48}{100}$

toys and hobbies and games: $\frac{1}{2} = \frac{1 \cdot 50}{2 \cdot 50} = \frac{50}{100}$

79. $\frac{15}{100}$ is the smallest fraction, so drugs, health and beauty aids has the smallest fraction sold online.

81. $3^2 = 3 \cdot 3 = 9$

83. $5^3 = 5 \cdot 5 \cdot 5 = 125$

85. $7^2 = 7 \cdot 7 = 49$

87. $2^3 \cdot 3 = 2 \cdot 2 \cdot 2 \cdot 3 = 24$

89. $\frac{2}{7} + \frac{9}{7} = \frac{2+9}{7} = \frac{11}{7}$

91. answers may vary

93. $\frac{16}{100} + \frac{12}{100} + \frac{7}{100} + \frac{20}{100} + \frac{30}{100} + \frac{6}{100} + \frac{9}{100}$
$= \frac{16+12+7+20+30+6+9}{100}$
$= \frac{100}{100}$
$= 1$
answers may vary

95. $\frac{37x}{165} = \frac{37x}{165} \cdot \frac{22}{22} = \frac{37x \cdot 22}{165 \cdot 22} = \frac{814x}{3630}$

97. answers may vary

99. $\frac{2}{3} = \frac{2 \cdot 5}{3 \cdot 5} = \frac{10}{15}$
$\frac{2}{3} = \frac{2 \cdot 20}{3 \cdot 20} = \frac{40}{60}$
$\frac{2}{3} = \frac{2 \cdot 100}{3 \cdot 100} = \frac{200}{300}$

a, b, and d are equivalent to $\frac{2}{3}$.

Copyright © 2011 Pearson Education, Inc. Publishing as Prentice Hall.

Section 4.5

Practice Problems

1. The LCD is 21.
$\frac{2}{7}+\frac{8}{21}=\frac{2\cdot 3}{7\cdot 3}+\frac{8}{21}=\frac{6}{21}+\frac{8}{21}=\frac{14}{21}=\frac{2\cdot 7}{3\cdot 7}=\frac{2}{3}$

2. The LCD is 18.
$\frac{5y}{6}+\frac{2y}{9}=\frac{5y\cdot 3}{6\cdot 3}+\frac{2y\cdot 2}{9\cdot 2}=\frac{15y}{18}+\frac{4y}{18}=\frac{19y}{18}$

3. The LCD is 20.
$$\begin{aligned}-\frac{1}{5}+\frac{9}{20}&=-\frac{1\cdot 4}{5\cdot 4}+\frac{9}{20}\\&=-\frac{4}{20}+\frac{9}{20}\\&=\frac{5}{20}\\&=\frac{1\cdot 5}{4\cdot 5}\\&=\frac{1}{4}\end{aligned}$$

4. The LCD is 70.
$\frac{5}{7}-\frac{9}{10}=\frac{5\cdot 10}{7\cdot 10}-\frac{9\cdot 7}{10\cdot 7}=\frac{50}{70}-\frac{63}{70}=-\frac{13}{70}$

5. The LCD is 24.
$$\begin{aligned}\frac{5}{8}-\frac{1}{3}-\frac{1}{12}&=\frac{5\cdot 3}{8\cdot 3}-\frac{1\cdot 8}{3\cdot 8}-\frac{1\cdot 2}{12\cdot 2}\\&=\frac{15}{24}-\frac{8}{24}-\frac{2}{24}\\&=\frac{5}{24}\end{aligned}$$

6. Recall that $5=\frac{5}{1}$. The LCD is 4.
$\frac{5}{1}-\frac{y}{4}=\frac{5\cdot 4}{1\cdot 4}-\frac{y}{4}=\frac{20}{4}-\frac{y}{4}=\frac{20-y}{4}$

7. The LCD is 40. Write each fraction as an equivalent fraction with a denominator of 40.
$\frac{5}{8}=\frac{5\cdot 5}{8\cdot 5}=\frac{25}{40}$
$\frac{11}{20}=\frac{11\cdot 2}{20\cdot 2}=\frac{22}{40}$
Since $25>22$, $\frac{25}{40}>\frac{22}{40}$, so $\frac{5}{8}>\frac{11}{20}$.

8. The LCD is 20. Write each fraction as an equivalent fraction with a denominator of 20.
$-\frac{17}{20}$
$-\frac{4}{5}=-\frac{4\cdot 4}{5\cdot 4}=-\frac{16}{20}$
Since $-17<-16$, $-\frac{17}{20}<-\frac{16}{20}$, so $-\frac{17}{20}<-\frac{4}{5}$.

9. The LCD is 99.
$x-y=\frac{5}{11}-\frac{4}{9}=\frac{5\cdot 9}{11\cdot 9}-\frac{4\cdot 11}{9\cdot 11}=\frac{45}{99}-\frac{44}{99}=\frac{1}{99}$

10. The LCD is 30.
$$\begin{aligned}\frac{3}{5}+\frac{3}{10}+\frac{1}{15}&=\frac{3\cdot 6}{5\cdot 6}+\frac{3\cdot 3}{10\cdot 3}+\frac{1\cdot 2}{15\cdot 2}\\&=\frac{18}{30}+\frac{9}{30}+\frac{2}{30}\\&=\frac{29}{30}\end{aligned}$$
The homeowner needs $\frac{29}{30}$ cubic yard of cement.

11. The LCD is 12.
$\frac{3}{4}-\frac{2}{3}=\frac{3\cdot 3}{4\cdot 3}-\frac{2\cdot 4}{3\cdot 4}=\frac{9}{12}-\frac{8}{12}=\frac{1}{12}$
The difference in length is $\frac{1}{12}$ foot.

Calculator Explorations

1. $\frac{1}{16}+\frac{2}{5}=\frac{37}{80}$

2. $\frac{3}{20}+\frac{2}{25}=\frac{23}{100}$

3. $\frac{4}{9}+\frac{7}{8}=\frac{95}{72}$

4. $\frac{9}{11}+\frac{5}{12}=\frac{163}{132}$

5. $\frac{10}{17}+\frac{12}{19}=\frac{394}{323}$

6. $\frac{14}{31}+\frac{15}{21}=\frac{253}{217}$

Copyright © 2011 Pearson Education, Inc. Publishing as Prentice Hall.

Vocabulary and Readiness Check

1. To add or subtract unlike fractions, we first write the fractions as equivalent fractions with a common denominator. The common denominator we use is called the least common denominator.

2. The LCD for $\frac{1}{6}$ and $\frac{5}{8}$ is 24.

3. $\frac{1}{6}+\frac{5}{8}=\frac{1}{6}\cdot\frac{4}{4}+\frac{5}{8}\cdot\frac{3}{3}=\frac{4}{24}+\frac{15}{24}=\frac{19}{24}$.

4. $\frac{1}{6}-\frac{5}{8}=\frac{1}{6}\cdot\frac{4}{4}-\frac{5}{8}\cdot\frac{3}{3}=\frac{4}{24}-\frac{15}{24}=-\frac{11}{24}$.

5. $x-y$ is an expression while $3x=\frac{1}{5}$ is an equation.

6. Since $-10<-1$, we know that $-\frac{10}{13}<-\frac{1}{13}$.

Exercise Set 4.5

1. The LCD is 6.
$$\frac{2}{3}+\frac{1}{6}=\frac{2\cdot 2}{3\cdot 2}+\frac{1}{6}=\frac{4}{6}+\frac{1}{6}=\frac{5}{6}$$

3. The LCD is 6.
$$\frac{1}{2}-\frac{1}{3}=\frac{1\cdot 3}{2\cdot 3}-\frac{1\cdot 2}{3\cdot 2}=\frac{3}{6}-\frac{2}{6}=\frac{1}{6}$$

5. The LCD is 33.
$$-\frac{2}{11}+\frac{2}{33}=-\frac{2\cdot 3}{11\cdot 3}+\frac{2}{33}=-\frac{6}{33}+\frac{2}{33}=-\frac{4}{33}$$

7. The LCD is 14.
$$\frac{3}{14}-\frac{3}{7}=\frac{3}{14}-\frac{3\cdot 2}{7\cdot 2}=\frac{3}{14}-\frac{6}{14}=-\frac{3}{14}$$

9. The LCD is 35.
$$\begin{aligned}\frac{11x}{35}+\frac{2x}{7}&=\frac{11x}{35}+\frac{2x\cdot 5}{7\cdot 5}\\&=\frac{11x}{35}+\frac{10x}{35}\\&=\frac{21x}{35}\\&=\frac{3\cdot 7\cdot x}{5\cdot 7}\\&=\frac{3x}{5}\end{aligned}$$

11. The LCD is 12.
$$2-\frac{y}{12}=\frac{2}{1}-\frac{y}{12}=\frac{2\cdot 12}{1\cdot 12}-\frac{y}{12}=\frac{24}{12}-\frac{y}{12}=\frac{24-y}{12}$$

13. The LCD is 36.
$$\frac{5}{12}-\frac{1}{9}=\frac{5\cdot 3}{12\cdot 3}-\frac{1\cdot 4}{9\cdot 4}=\frac{15}{36}-\frac{4}{36}=\frac{11}{36}$$

15. The LCD is 7.
$$-7+\frac{5}{7}=-\frac{7}{1}+\frac{5}{7}=-\frac{7\cdot 7}{1\cdot 7}+\frac{5}{7}=-\frac{49}{7}+\frac{5}{7}=-\frac{44}{7}$$

17. The LCD is 99.
$$\begin{aligned}\frac{5a}{11}+\frac{4a}{9}&=\frac{5a\cdot 9}{11\cdot 9}+\frac{4a\cdot 11}{9\cdot 11}\\&=\frac{45a}{99}+\frac{44a}{99}\\&=\frac{89a}{99}\end{aligned}$$

19. The LCD is 6.
$$\frac{2y}{3}-\frac{1}{6}=\frac{2y\cdot 2}{3\cdot 2}-\frac{1}{6}=\frac{4y}{6}-\frac{1}{6}=\frac{4y-1}{6}$$

21. The LCD is $2x$.
$$\frac{1}{2}+\frac{3}{x}=\frac{1\cdot x}{2\cdot x}+\frac{3\cdot 2}{x\cdot 2}=\frac{x}{2x}+\frac{6}{2x}=\frac{x+6}{2x}$$

23. The LCD is 33.
$$-\frac{2}{11}-\frac{2}{33}=-\frac{2\cdot 3}{11\cdot 3}-\frac{2}{33}=-\frac{6}{33}-\frac{2}{33}=-\frac{8}{33}$$

25. The LCD is 14.
$$\frac{9}{14}-\frac{3}{7}=\frac{9}{14}-\frac{3\cdot 2}{7\cdot 2}=\frac{9}{14}-\frac{6}{14}=\frac{3}{14}$$

Copyright © 2011 Pearson Education, Inc. Publishing as Prentice Hall.

27. The LCD is 35.

$$\frac{11y}{35}-\frac{2}{7}=\frac{11y}{35}-\frac{2\cdot 5}{7\cdot 5}=\frac{11y}{35}-\frac{10}{35}=\frac{11y-10}{35}$$

29. The LCD is 36.

$$\frac{1}{9}-\frac{5}{12}=\frac{1\cdot 4}{9\cdot 4}-\frac{5\cdot 3}{12\cdot 3}=\frac{4}{36}-\frac{15}{36}=-\frac{11}{36}$$

31. The LCD is 60.

$$\begin{aligned}\frac{7}{15}-\frac{5}{12}&=\frac{7\cdot 4}{15\cdot 4}-\frac{5\cdot 5}{12\cdot 5}\\&=\frac{28}{60}-\frac{25}{60}\\&=\frac{3}{60}\\&=\frac{1\cdot 3}{20\cdot 3}\\&=\frac{1}{20}\end{aligned}$$

33. The LCD is 56.

$$\frac{5}{7}-\frac{1}{8}=\frac{5\cdot 8}{7\cdot 8}-\frac{1\cdot 7}{8\cdot 7}=\frac{40}{56}-\frac{7}{56}=\frac{33}{56}$$

35. The LCD is 16.

$$\frac{7}{8}+\frac{3}{16}=\frac{7\cdot 2}{8\cdot 2}+\frac{3}{16}=\frac{14}{16}+\frac{3}{16}=\frac{17}{16}$$

37. $\frac{3}{9}-\frac{5}{9}=\frac{3-5}{9}=-\frac{2}{9}$

39. The LCD is 30.

$$\begin{aligned}-\frac{2}{5}+\frac{1}{3}-\frac{3}{10}&=-\frac{2\cdot 6}{5\cdot 6}+\frac{1\cdot 10}{3\cdot 10}-\frac{3\cdot 3}{10\cdot 3}\\&=-\frac{12}{30}+\frac{10}{30}-\frac{9}{30}\\&=-\frac{11}{30}\end{aligned}$$

41. The LCD is 33.

$$\begin{aligned}\frac{5}{11}+\frac{y}{3}&=\frac{5\cdot 3}{11\cdot 3}+\frac{y\cdot 11}{3\cdot 11}\\&=\frac{15}{33}+\frac{11y}{33}\\&=\frac{15+11y}{33}\end{aligned}$$

43. The LCD is 42.

$$-\frac{5}{6}-\frac{3}{7}=-\frac{5\cdot 7}{6\cdot 7}-\frac{3\cdot 6}{7\cdot 6}=-\frac{35}{42}-\frac{18}{42}=-\frac{53}{42}$$

45. The LCD is 16.

$$\begin{aligned}\frac{x}{2}+\frac{x}{4}+\frac{2x}{16}&=\frac{x\cdot 8}{2\cdot 8}+\frac{x\cdot 4}{4\cdot 4}+\frac{2x}{16}\\&=\frac{8x}{16}+\frac{4x}{16}+\frac{2x}{16}\\&=\frac{14x}{16}\\&=\frac{2\cdot 7x}{2\cdot 8}\\&=\frac{7x}{8}\end{aligned}$$

47. The LCD is 18.

$$\frac{7}{9}-\frac{1}{6}=\frac{7\cdot 2}{9\cdot 2}-\frac{1\cdot 3}{6\cdot 3}=\frac{14}{18}-\frac{3}{18}=\frac{11}{18}$$

49. The LCD is 39.

$$\begin{aligned}\frac{2a}{3}+\frac{6a}{13}&=\frac{2a\cdot 13}{3\cdot 13}+\frac{6a\cdot 3}{13\cdot 3}\\&=\frac{26a}{39}+\frac{18a}{39}\\&=\frac{44a}{39}\end{aligned}$$

51. The LCD is 60.

$$\frac{7}{30}-\frac{5}{12}=\frac{7\cdot 2}{30\cdot 2}-\frac{5\cdot 5}{12\cdot 5}=\frac{14}{60}-\frac{25}{60}=-\frac{11}{60}$$

53. The LCD is $9y$.

$$\frac{5}{9}+\frac{1}{y}=\frac{5\cdot y}{9\cdot y}+\frac{1\cdot 9}{y\cdot 9}=\frac{5y}{9y}+\frac{9}{9y}=\frac{5y+9}{9y}$$

55. The LCD is 20.

$$\begin{aligned}\frac{6}{5}-\frac{3}{4}+\frac{1}{2}&=\frac{6\cdot 4}{5\cdot 4}-\frac{3\cdot 5}{4\cdot 5}+\frac{1\cdot 10}{2\cdot 10}\\&=\frac{24}{20}-\frac{15}{20}+\frac{10}{20}\\&=\frac{19}{20}\end{aligned}$$

57. The LCD is 45.

$$\frac{4}{5}+\frac{4}{9}=\frac{4\cdot 9}{5\cdot 9}+\frac{4\cdot 5}{9\cdot 5}=\frac{36}{45}+\frac{20}{45}=\frac{56}{45}$$

Copyright © 2011 Pearson Education, Inc. Publishing as Prentice Hall.

59. The LCD is $72x$.

$$\frac{5}{9x}+\frac{1}{8}=\frac{5\cdot 8}{9x\cdot 8}+\frac{1\cdot 9x}{8\cdot 9x}=\frac{40}{72x}+\frac{9x}{72x}=\frac{40+9x}{72x}$$

61. The LCD is 24.

$$-\frac{9}{12}+\frac{17}{24}-\frac{1}{6}=-\frac{9\cdot 2}{12\cdot 2}+\frac{17}{24}-\frac{1\cdot 4}{6\cdot 4}=-\frac{18}{24}+\frac{17}{24}-\frac{4}{24}=-\frac{5}{24}$$

63. The LCD is 56.

$$\frac{3x}{8}+\frac{2x}{7}-\frac{5}{14}=\frac{3x\cdot 7}{8\cdot 7}+\frac{2x\cdot 8}{7\cdot 8}-\frac{5\cdot 4}{14\cdot 4}=\frac{21x}{56}+\frac{16x}{56}-\frac{20}{56}=\frac{37x-20}{56}$$

65. The LCD is 70. Write each fraction as an equivalent fraction with a denominator of 70.

$$\frac{2}{7}=\frac{2\cdot 10}{7\cdot 10}=\frac{20}{70}$$

$$\frac{3}{10}=\frac{3\cdot 7}{10\cdot 7}=\frac{21}{70}$$

Since $20 < 21$, $\frac{20}{70}<\frac{21}{70}$, so $\frac{2}{7}<\frac{3}{10}$.

67. A positive fraction is greater than a negative fraction.

$$\frac{5}{6}>-\frac{13}{15}$$

69. The LCD is 28. Write each fraction as an equivalent fraction with a denominator of 28.

$$-\frac{3}{4}=-\frac{3\cdot 7}{4\cdot 7}=-\frac{21}{28}$$

$$-\frac{11}{14}=-\frac{11\cdot 2}{14\cdot 2}=-\frac{22}{28}$$

Since $-21>-22$, $-\frac{21}{28}>-\frac{22}{28}$, so $-\frac{3}{4}>-\frac{11}{14}$.

71. The LCD is 12.

$$x+y=\frac{1}{3}+\frac{3}{4}=\frac{1\cdot 4}{3\cdot 4}+\frac{3\cdot 3}{4\cdot 3}=\frac{4}{12}+\frac{9}{12}=\frac{13}{12}$$

73. $xy=\frac{1}{3}\cdot\frac{3}{4}=\frac{1\cdot 3}{3\cdot 4}=\frac{1}{4}$

75.

$$2y+x=2\left(\frac{3}{4}\right)+\frac{1}{3}=\frac{6}{4}+\frac{1}{3}=\frac{2\cdot 3}{2\cdot 2}+\frac{1}{3}=\frac{3}{2}+\frac{1}{3}=\frac{3\cdot 3}{2\cdot 3}+\frac{1\cdot 2}{3\cdot 2}=\frac{9}{6}+\frac{2}{6}=\frac{11}{6}$$

77. The LCD is 15.

$$\frac{4}{5}+\frac{1}{3}+\frac{4}{5}+\frac{1}{3}=\frac{4}{5}\cdot\frac{3}{3}+\frac{1}{3}\cdot\frac{5}{5}+\frac{4}{5}\cdot\frac{3}{3}+\frac{1}{3}\cdot\frac{5}{5}=\frac{12}{15}+\frac{5}{15}+\frac{12}{15}+\frac{5}{15}=\frac{34}{15}$$

The perimeter is $\frac{34}{15}$ or $2\frac{4}{15}$ centimeters.

79. The LCD is 20.

$$\frac{1}{4}+\frac{1}{5}+\frac{1}{2}+\frac{3}{4}=\frac{1\cdot 5}{4\cdot 5}+\frac{1\cdot 4}{5\cdot 4}+\frac{1\cdot 10}{2\cdot 10}+\frac{3\cdot 5}{4\cdot 5}=\frac{5}{20}+\frac{4}{20}+\frac{10}{20}+\frac{15}{20}=\frac{34}{20}=\frac{17\cdot 2}{10\cdot 2}=\frac{17}{10}\text{ or }1\frac{7}{10}$$

The perimeter is $\frac{17}{10}$ meters or $1\frac{7}{10}$ meters.

81. The sum of a number and $\frac{1}{2}$ translates as $x+\frac{1}{2}$.

Copyright © 2011 Pearson Education, Inc. Publishing as Prentice Hall.

83. A number subtracted from $-\frac{3}{8}$ translates as $-\frac{3}{8}-x$.

85. The LCD is 100.
$\frac{17}{100}-\frac{1}{10}=\frac{17}{100}-\frac{1\cdot 10}{10\cdot 10}=\frac{17}{100}-\frac{10}{100}=\frac{7}{100}$
The sloth can travel $\frac{7}{100}$ mph faster in trees.

87. $1-\frac{3}{16}-\frac{3}{16}=\frac{16}{16}-\frac{3}{16}-\frac{3}{16}=\frac{10}{16}=\frac{5}{8}$
The inner diameter is $\frac{5}{8}$ inch.

89. The LCD is 100.
$\frac{13}{20}-\frac{4}{25}=\frac{13\cdot 5}{20\cdot 5}-\frac{4\cdot 4}{25\cdot 4}=\frac{65}{100}-\frac{16}{100}=\frac{49}{100}$
$\frac{49}{100}$ of the American students ages 10 to 17 name math or science as their favorite subject in school.

91. The LCD is 32.
$$\begin{aligned}\frac{1}{2}+\frac{11}{16}+\frac{9}{32}&=\frac{1}{2}\cdot\frac{16}{16}+\frac{11}{16}\cdot\frac{2}{2}+\frac{9}{32}\\&=\frac{16}{32}+\frac{22}{32}+\frac{9}{32}\\&=\frac{47}{32}\end{aligned}$$
The total length is $\frac{47}{32}$ inches.

93. The LCD is 50.
$$\begin{aligned}\frac{13}{50}+\frac{1}{2}&=\frac{13}{50}+\frac{1}{2}\cdot\frac{25}{25}\\&=\frac{13}{50}+\frac{25}{50}\\&=\frac{38}{50}\\&=\frac{2\cdot 19}{2\cdot 25}\\&=\frac{19}{25}\end{aligned}$$
The Pacific and Atlantic Oceans account for $\frac{19}{25}$ of the world's water surfaces.

95. The piece representing Lakes/Seashores is labeled $\frac{1}{25}$, so $\frac{1}{25}$ of the areas maintained by the National Park Service are National Lakes or National Seashores.

97. $1-\frac{21}{100}=\frac{100}{100}-\frac{21}{100}=\frac{79}{100}$
$\frac{79}{100}$ of areas maintained by the National Park Service are NOT National Monuments.

99. $-50\div 5\cdot 2=-10\cdot 2=-20$

101. $(8-6)\cdot(4-7)=2\cdot(-3)=-6$

103. a.

b. There seems to be an error.

c. $\frac{3}{5}+\frac{4}{5}=\frac{3+4}{5}=\frac{7}{5}\left(\text{or }\frac{14}{10}\text{ not }\frac{7}{10}\right)$

105. The LCD is 540.
$$\begin{aligned}\frac{2}{3}-\frac{1}{4}-\frac{2}{540}&=\frac{2}{3}\cdot\frac{180}{180}-\frac{1}{4}\cdot\frac{135}{135}-\frac{2}{540}\\&=\frac{360}{540}-\frac{135}{540}-\frac{2}{540}\\&=\frac{225}{540}-\frac{2}{540}\\&=\frac{223}{540}\end{aligned}$$

107. The LCD is 1760.
$$\begin{aligned}\frac{30}{55}+\frac{1000}{1760}&=\frac{30\cdot 32}{55\cdot 32}+\frac{1000}{1760}\\&=\frac{960}{1760}+\frac{1000}{1760}\\&=\frac{1960}{1760}\\&=\frac{49\cdot 40}{44\cdot 40}\\&=\frac{49}{44}\end{aligned}$$

109. answers may vary

Copyright © 2011 Pearson Education, Inc. Publishing as Prentice Hall.

111. The LCD is 106. Write each fraction as an equivalent fraction with a denominator of 106.

$\frac{24}{53} = \frac{24 \cdot 2}{53 \cdot 2} = \frac{48}{106}$

$\frac{51}{106}$

Since $51 > 48$, $\frac{51}{106} > \frac{48}{106}$, so $\frac{51}{106} > \frac{24}{53}$.

Standard mail accounted for the greater portion of the mail handled by volume.

Integrated Review

1. 3 out of 7 equal parts are shaded: $\frac{3}{7}$

2. Each part is $\frac{1}{4}$ of a whole and there are 5 parts shaded, or 1 whole and 1 more part: $\frac{5}{4}$ or $1\frac{1}{4}$

3. number that get fewer than 8 $\rightarrow$ 73
total number of people $\rightarrow$ 85

$\frac{73}{85}$ of people get fewer than 8 hours of sleep each night.

4.

5. $\frac{11}{-11} = -1$

6. $\frac{17}{1} = 17$

7. $\frac{0}{-3} = 0$

8. $\frac{7}{0}$ is undefined

9. $65 = 5 \cdot 13$

10. $70 = 2 \cdot 35$

$\downarrow \ \downarrow\searrow$

$2 \cdot 5 \cdot 7$

$70 = 2 \cdot 5 \cdot 7$

11. $315 = 3 \cdot 105$

$\downarrow \ \downarrow\searrow$

$3 \cdot 3 \cdot 35$

$\downarrow \ \downarrow \ \downarrow\searrow$

$3 \cdot 3 \cdot 5 \cdot 7$

$315 = 3^2 \cdot 5 \cdot 7$

12. $441 = 3 \cdot 147$

$\downarrow \ \downarrow\searrow$

$3 \cdot 3 \cdot 49$

$\downarrow \ \downarrow \ \downarrow\searrow$

$3 \cdot 3 \cdot 7 \cdot 7$

$441 = 3^2 \cdot 7^2$

13. $\frac{2}{14} = \frac{2 \cdot 1}{2 \cdot 7} = \frac{1}{7}$

14. $\frac{24}{20} = \frac{6 \cdot 4}{5 \cdot 4} = \frac{6}{5}$

15. $-\frac{56}{60} = -\frac{14 \cdot 4}{15 \cdot 4} = -\frac{14}{15}$

16. $-\frac{72}{80} = -\frac{8 \cdot 9}{8 \cdot 10} = -\frac{9}{10}$

17. $\frac{54x}{135} = \frac{27 \cdot 2 \cdot x}{27 \cdot 5} = \frac{2x}{5}$

18. $\frac{90}{240y} = \frac{30 \cdot 3}{30 \cdot 8 \cdot y} = \frac{3}{8y}$

19. $\frac{165z^3}{210z} = \frac{15 \cdot 11 \cdot z \cdot z \cdot z}{15 \cdot 14 \cdot z} = \frac{11 \cdot z \cdot z}{14} = \frac{11z^2}{14}$

20. $\frac{245ab}{385a^2b^3} = \frac{35 \cdot 7 \cdot a \cdot b}{35 \cdot 11 \cdot a \cdot a \cdot b \cdot b \cdot b}$

$= \frac{7}{11 \cdot a \cdot b \cdot b}$

$= \frac{7}{11ab^2}$

21. Not equivalent, since the cross products are not equal: $7 \cdot 10 = 70$, $8 \cdot 9 = 72$

22. Equivalent, since the cross products are equal: $10 \cdot 18 = 180$, $12 \cdot 15 = 180$

Copyright © 2011 Pearson Education, Inc. Publishing as Prentice Hall.

23. a. number not adjacent $\rightarrow 2$
total number $\rightarrow \overline{50}$

$\frac{2}{50} = \frac{1 \cdot 2}{25 \cdot 2} = \frac{1}{25}$ of the states are not adjacent to any other states.

b. $50 - 2 = 48$; 48 states are adjacent to other states.

c. $\frac{48}{50} = \frac{2 \cdot 24}{2 \cdot 25} = \frac{24}{25}$ of the states are adjacent to other states.

24. a. number not rated $\rightarrow 255$
total number $\rightarrow \overline{725}$

$\frac{255}{725} = \frac{5 \cdot 51}{5 \cdot 145} = \frac{51}{145}$ of the new films were not rated.

b. $725 - 255 = 470$; 470 of the new films were rated.

c. $\frac{470}{725} = \frac{5 \cdot 94}{5 \cdot 145} = \frac{94}{145}$ of the new films were rated.

25. $5 = \boxed{5}$
$6 = \boxed{2 \cdot 3}$
$\text{LCM} = 2 \cdot 3 \cdot 5 = 30$

26. $2 = \boxed{2}$
$14 = 2 \cdot \boxed{7}$
$\text{LCM} = 2 \cdot 7 = 14$

27. $6 = \boxed{2} \cdot 3$
$18 = 2 \cdot \boxed{3 \cdot 3}$
$30 = 2 \cdot 3 \cdot \boxed{5}$
$\text{LCM} = 2 \cdot 3 \cdot 3 \cdot 5 = 90$

28. $\frac{7}{9} = \frac{7}{9} \cdot \frac{4}{4} = \frac{7 \cdot 4}{9 \cdot 4} = \frac{28}{36}$

29. $\frac{11}{15} = \frac{11}{15} \cdot \frac{5}{5} = \frac{11 \cdot 5}{15 \cdot 5} = \frac{55}{75}$

30. $\frac{5}{6} = \frac{5}{6} \cdot \frac{8}{8} = \frac{5 \cdot 8}{6 \cdot 8} = \frac{40}{48}$

31. $\frac{1}{5} + \frac{3}{5} = \frac{1+3}{5} = \frac{4}{5}$

32. $\frac{1}{5} - \frac{3}{5} = \frac{1-3}{5} = \frac{-2}{5} = -\frac{2}{5}$

33. $\frac{1}{5} \cdot \frac{3}{5} = \frac{1 \cdot 3}{5 \cdot 5} = \frac{3}{25}$

34. $\frac{1}{5} \div \frac{3}{5} = \frac{1}{5} \cdot \frac{5}{3} = \frac{1 \cdot 5}{5 \cdot 3} = \frac{1}{3}$

35. $\frac{2}{3} \div \frac{5}{6} = \frac{2}{3} \cdot \frac{6}{5} = \frac{2 \cdot 6}{3 \cdot 5} = \frac{2 \cdot 2 \cdot 3}{3 \cdot 5} = \frac{2 \cdot 2}{5} = \frac{4}{5}$

36. $\frac{2a}{3} \cdot \frac{5}{6a} = \frac{2a \cdot 5}{3 \cdot 6a} = \frac{2 \cdot a \cdot 5}{3 \cdot 2 \cdot 3 \cdot a} = \frac{5}{3 \cdot 3} = \frac{5}{9}$

37. The LCD is $6y$.

$$\begin{aligned} \frac{2}{3y} - \frac{5}{6y} &= \frac{2}{3y} \cdot \frac{2}{2} - \frac{5}{6y} \\ &= \frac{2 \cdot 2}{3y \cdot 2} - \frac{5}{6y} \\ &= \frac{4}{6y} - \frac{5}{6y} \\ &= -\frac{1}{6y} \end{aligned}$$

38. The LCD is 6.

$$\begin{aligned} \frac{2x}{3} + \frac{5x}{6} &= \frac{2x}{3} \cdot \frac{2}{2} + \frac{5x}{6} \\ &= \frac{2x \cdot 2}{3 \cdot 2} + \frac{5x}{6} \\ &= \frac{4x}{6} + \frac{5x}{6} \\ &= \frac{9x}{6} \\ &= \frac{3 \cdot 3 \cdot x}{3 \cdot 2} \\ &= \frac{3x}{2} \end{aligned}$$

39. $-\frac{1}{7} \cdot -\frac{7}{18} = \frac{1 \cdot 7}{7 \cdot 18} = \frac{1}{18}$

40. $-\frac{4}{9} \cdot -\frac{3}{7} = \frac{4 \cdot 3}{9 \cdot 7} = \frac{4 \cdot 3}{3 \cdot 3 \cdot 7} = \frac{4}{3 \cdot 7} = \frac{4}{21}$

Copyright © 2011 Pearson Education, Inc. Publishing as Prentice Hall.

41. $-\frac{7z}{8} \div 6z^2 = -\frac{7z}{8} \div \frac{6z^2}{1}$
$= -\frac{7z}{8} \cdot \frac{1}{6z^2}$
$= -\frac{7z \cdot 1}{8 \cdot 6z^2}$
$= -\frac{7 \cdot z \cdot 1}{8 \cdot 6 \cdot z \cdot z}$
$= -\frac{7}{48z}$

42. $-\frac{9}{10} \div 5 = -\frac{9}{10} \div \frac{5}{1} = -\frac{9}{10} \cdot \frac{1}{5} = -\frac{9 \cdot 1}{10 \cdot 5} = -\frac{9}{50}$

43. The LCD is 40.
$\frac{7}{8} + \frac{1}{20} = \frac{7 \cdot 5}{8 \cdot 5} + \frac{1 \cdot 2}{20 \cdot 2} = \frac{35}{40} + \frac{2}{40} = \frac{37}{40}$

44. The LCD is 36.
$\frac{5}{12} - \frac{1}{9} = \frac{5 \cdot 3}{12 \cdot 3} - \frac{1 \cdot 4}{9 \cdot 4} = \frac{15}{36} - \frac{4}{36} = \frac{11}{36}$

45. The LCD is 18.
$\frac{2}{9} + \frac{1}{18} + \frac{1}{3} = \frac{2 \cdot 2}{9 \cdot 2} + \frac{1}{18} + \frac{1 \cdot 6}{3 \cdot 6}$
$= \frac{4}{18} + \frac{1}{18} + \frac{6}{18}$
$= \frac{11}{18}$

46. The LCD is 50.
$\frac{3y}{10} + \frac{y}{5} + \frac{6}{25} = \frac{3y \cdot 5}{10 \cdot 5} + \frac{y \cdot 10}{5 \cdot 10} + \frac{6 \cdot 2}{25 \cdot 2}$
$= \frac{15y}{50} + \frac{10y}{50} + \frac{12}{50}$
$= \frac{25y}{50} + \frac{12}{50}$
$= \frac{25y + 12}{50}$

47. $\frac{2}{3}$ of a number translates as $\frac{2}{3} \cdot x$ or $\frac{2}{3}x$.

48. The quotient of a number and $-\frac{1}{5}$ translates as $x \div \left(-\frac{1}{5}\right)$.

49. A number subtracted from $-\frac{8}{9}$ translates as $-\frac{8}{9} - x$.

50. $\frac{6}{11}$ increased by a number translates as $\frac{6}{11} + x$.

51. $\frac{2}{3} \cdot 1530 = \frac{2}{3} \cdot \frac{1530}{1}$
$= \frac{2 \cdot 1530}{3 \cdot 1}$
$= \frac{2 \cdot 3 \cdot 510}{3 \cdot 1}$
$= \frac{2 \cdot 510}{1}$
$= 1020$
$\frac{2}{3}$ of 1530 is 1020.

52. $18 \div \frac{3}{4} = \frac{18}{1} \div \frac{3}{4}$
$= \frac{18}{1} \cdot \frac{4}{3}$
$= \frac{18 \cdot 4}{1 \cdot 3}$
$= \frac{3 \cdot 6 \cdot 4}{1 \cdot 3}$
$= \frac{6 \cdot 4}{1}$
$= 24$
He can sell 24 lots.

53. The LCD is 16.
$\frac{7}{8} - \frac{1}{16} - \frac{1}{16} = \frac{7 \cdot 2}{8 \cdot 2} - \frac{1}{16} - \frac{1}{16}$
$= \frac{14}{16} - \frac{1}{16} - \frac{1}{16}$
$= \frac{12}{16}$
$= \frac{4 \cdot 3}{4 \cdot 4}$
$= \frac{3}{4}$
The inner diameter is $\frac{3}{4}$ foot.

Copyright © 2011 Pearson Education, Inc. Publishing as Prentice Hall.

Section 4.6

Practice Problems

1. $\frac{\frac{7y}{10}}{\frac{1}{5}}=\frac{7y}{10}\div\frac{1}{5}=\frac{7y}{10}\cdot\frac{5}{1}=\frac{7y\cdot5}{10\cdot1}=\frac{7\cdot y\cdot5}{2\cdot5\cdot1}=\frac{7y}{2}$

2. $$\frac{\frac{1}{2}+\frac{1}{6}}{\frac{3}{4}-\frac{2}{3}}=\frac{\frac{1\cdot3}{2\cdot3}+\frac{1}{6}}{\frac{3\cdot3}{4\cdot3}-\frac{2\cdot4}{3\cdot4}}=\frac{\frac{3}{6}+\frac{1}{6}}{\frac{9}{12}-\frac{8}{12}}=\frac{\frac{4}{6}}{\frac{1}{12}}=\frac{4}{6}\div\frac{1}{12}=\frac{4}{6}\cdot\frac{12}{1}=\frac{4\cdot2\cdot6}{6\cdot1}=\frac{8}{1}\text{ or }8$$

3. The LCD is 12.
$$\frac{\frac{1}{2}+\frac{1}{6}}{\frac{3}{4}-\frac{2}{3}}=\frac{12\left(\frac{1}{2}+\frac{1}{6}\right)}{12\left(\frac{3}{4}-\frac{2}{3}\right)}=\frac{12\cdot\frac{1}{2}+12\cdot\frac{1}{6}}{12\cdot\frac{3}{4}-12\cdot\frac{2}{3}}=\frac{6+2}{9-8}=\frac{8}{1}\text{ or }8$$

4. The LCD is 20.
$$\frac{\frac{3}{4}}{\frac{x}{5}-1}=\frac{20\left(\frac{3}{4}\right)}{20\left(\frac{x}{5}-1\right)}=\frac{20\cdot\frac{3}{4}}{20\cdot\frac{x}{5}-20\cdot1}=\frac{15}{4x-20}$$

5. $\left(\frac{2}{3}\right)^3-2=\frac{8}{27}-2=\frac{8}{27}-\frac{54}{27}=-\frac{46}{27}$

6. $$\left(-\frac{1}{2}+\frac{1}{5}\right)\left(\frac{7}{8}+\frac{1}{8}\right)=\left(-\frac{1\cdot5}{2\cdot5}+\frac{1\cdot2}{5\cdot2}\right)\left(\frac{7}{8}+\frac{1}{8}\right)=\left(-\frac{5}{10}+\frac{2}{10}\right)\left(\frac{7}{8}+\frac{1}{8}\right)=\left(-\frac{3}{10}\right)\left(\frac{8}{8}\right)=\left(-\frac{3}{10}\right)(1)=-\frac{3}{10}$$

7. $$-\frac{3}{5}-xy=-\frac{3}{5}-\frac{3}{10}\cdot\frac{2}{3}=-\frac{3}{5}-\frac{3\cdot2}{5\cdot2\cdot3}=-\frac{3}{5}-\frac{1}{5}=-\frac{4}{5}$$

Vocabulary and Readiness Check

1. A fraction whose numerator or denominator or both numerator and denominator contain fractions is called a complex fraction.

2. To simplify $-\frac{1}{2}+\frac{2}{3}\cdot\frac{7}{8}$, which operation do we perform first? multiplication

3. To simplify $-\frac{1}{2}\div\frac{2}{3}\cdot\frac{7}{8}$, which operation do we perform first? division

4. To simplify $\frac{7}{8}\cdot\left(\frac{1}{2}-\frac{2}{3}\right)$, which operation do we perform first? subtraction

5. To simplify $\frac{1}{3}\div\frac{1}{4}\cdot\left(\frac{9}{11}+\frac{3}{8}\right)^3$, which operation do we perform first? addition

6. To simplify $9-\left(-\frac{3}{4}\right)^2$, which operation do we perform first? evaluate the exponential expression

Copyright © 2011 Pearson Education, Inc. Publishing as Prentice Hall.

Exercise Set 4.6

1. $\frac{\frac{1}{8}}{\frac{3}{4}} = \frac{1}{8} \div \frac{3}{4} = \frac{1}{8} \cdot \frac{4}{3} = \frac{1 \cdot 4}{8 \cdot 3} = \frac{1 \cdot 4}{2 \cdot 4 \cdot 3} = \frac{1}{6}$

3. $\frac{\frac{2}{3}}{\frac{2}{7}} = \frac{2}{3} \div \frac{2}{7} = \frac{2}{3} \cdot \frac{7}{2} = \frac{2 \cdot 7}{3 \cdot 2} = \frac{7}{3}$

5. $\frac{\frac{2x}{27}}{\frac{4}{9}} = \frac{2x}{27} \div \frac{4}{9} = \frac{2x}{27} \cdot \frac{9}{4} = \frac{2x \cdot 9}{27 \cdot 4} = \frac{2 \cdot x \cdot 9}{3 \cdot 9 \cdot 2 \cdot 2} = \frac{x}{6}$

7. The LCD of 4, 5, and 2 is 20.

$$\begin{aligned}\frac{\frac{3}{4}+\frac{2}{5}}{\frac{1}{2}+\frac{3}{5}} &= \frac{20 \cdot \left(\frac{3}{4}+\frac{2}{5}\right)}{20 \cdot \left(\frac{1}{2}+\frac{3}{5}\right)} \\ &= \frac{20 \cdot \frac{3}{4} + 20 \cdot \frac{2}{5}}{20 \cdot \frac{1}{2} + 20 \cdot \frac{3}{5}} \\ &= \frac{15+8}{10+12} \\ &= \frac{23}{22}\end{aligned}$$

9. The LCD is 8.

$$\begin{aligned}\frac{\frac{3x}{4}}{5-\frac{1}{8}} &= \frac{8 \cdot \left(\frac{3x}{4}\right)}{8 \cdot \left(5-\frac{1}{8}\right)} \\ &= \frac{\frac{8}{1} \cdot \frac{3x}{4}}{8 \cdot 5 - 8 \cdot \frac{1}{8}} \\ &= \frac{\frac{2 \cdot 4 \cdot 3x}{1 \cdot 4}}{40-1} \\ &= \frac{6x}{39} \\ &= \frac{2 \cdot 3 \cdot x}{3 \cdot 13} \\ &= \frac{2x}{13}\end{aligned}$$

11. $$\begin{aligned}\frac{1}{5} + \frac{1}{3} \cdot \frac{1}{4} &= \frac{1}{5} + \frac{1 \cdot 1}{3 \cdot 4} \\ &= \frac{1}{5} + \frac{1}{12} \\ &= \frac{1 \cdot 12}{5 \cdot 12} + \frac{1 \cdot 5}{12 \cdot 5} \\ &= \frac{12}{60} + \frac{5}{60} \\ &= \frac{17}{60}\end{aligned}$$

13. $\frac{5}{6} \div \frac{1}{3} \cdot \frac{1}{4} = \frac{5}{6} \cdot \frac{3}{1} \cdot \frac{1}{4} = \frac{5 \cdot 3 \cdot 1}{2 \cdot 3 \cdot 1 \cdot 2 \cdot 2} = \frac{5}{2 \cdot 2 \cdot 2} = \frac{5}{8}$

15. $$\begin{aligned}2^2 - \left(\frac{1}{3}\right)^2 &= 4 - \frac{1}{9} \\ &= \frac{4}{1} - \frac{1}{9} \\ &= \frac{4 \cdot 9}{1 \cdot 9} - \frac{1}{9} \\ &= \frac{36}{9} - \frac{1}{9} \\ &= \frac{35}{9}\end{aligned}$$

17. $$\begin{aligned}\left(\frac{2}{9}+\frac{4}{9}\right)\left(\frac{1}{3}-\frac{9}{10}\right) &= \left(\frac{6}{9}\right)\left(\frac{1}{3} \cdot \frac{10}{10} - \frac{9}{10} \cdot \frac{3}{3}\right) \\ &= \left(\frac{6}{9}\right)\left(\frac{10}{30} - \frac{27}{30}\right) \\ &= \left(\frac{6}{9}\right)\left(\frac{-17}{30}\right) \\ &= -\frac{6 \cdot 17}{9 \cdot 30} \\ &= -\frac{2 \cdot 3 \cdot 17}{3 \cdot 3 \cdot 2 \cdot 15} \\ &= -\frac{17}{3 \cdot 15} \\ &= -\frac{17}{45}\end{aligned}$$

Copyright © 2011 Pearson Education, Inc. Publishing as Prentice Hall.

19. $\left(\frac{7}{8}-\frac{1}{2}\right)\div\frac{3}{11}=\left(\frac{7}{8}-\frac{1}{2}\cdot\frac{4}{4}\right)\div\frac{3}{11}$
$=\left(\frac{7}{8}-\frac{4}{8}\right)\div\frac{3}{11}$
$=\frac{3}{8}\div\frac{3}{11}$
$=\frac{3}{8}\cdot\frac{11}{3}$
$=\frac{3\cdot 11}{8\cdot 3}$
$=\frac{11}{8}$

21. $2\cdot\left(\frac{1}{4}+\frac{1}{5}\right)+2=2\cdot\left(\frac{1\cdot 5}{4\cdot 5}+\frac{1\cdot 4}{5\cdot 4}\right)+2$
$=2\left(\frac{5}{20}+\frac{4}{20}\right)+2$
$=2\left(\frac{9}{20}\right)+2$
$=\frac{2}{1}\cdot\frac{9}{20}+2$
$=\frac{2\cdot 9}{1\cdot 2\cdot 10}+2$
$=\frac{9}{10}+2$
$=\frac{9}{10}+\frac{2\cdot 10}{1\cdot 10}$
$=\frac{9}{10}+\frac{20}{10}$
$=\frac{29}{10}$

23. $\left(\frac{3}{4}\right)^2\div\left(\frac{3}{4}-\frac{1}{12}\right)=\left(\frac{3}{4}\right)^2\div\left(\frac{3\cdot 3}{4\cdot 3}-\frac{1}{12}\right)$
$=\left(\frac{3}{4}\right)^2\div\left(\frac{9}{12}-\frac{1}{12}\right)$
$=\left(\frac{3}{4}\right)^2\div\frac{8}{12}$
$=\frac{9}{16}\div\frac{8}{12}$
$=\frac{9}{16}\cdot\frac{12}{8}$
$=\frac{9\cdot 3\cdot 4}{4\cdot 4\cdot 8}$
$=\frac{27}{32}$

25. $\left(\frac{2}{5}-\frac{3}{10}\right)^2=\left(\frac{2}{5}\cdot\frac{2}{2}-\frac{3}{10}\right)^2$
$=\left(\frac{4}{10}-\frac{3}{10}\right)^2$
$=\left(\frac{1}{10}\right)^2$
$=\frac{1}{10}\cdot\frac{1}{10}$
$=\frac{1}{100}$

27. $\left(\frac{3}{4}+\frac{1}{8}\right)^2-\left(\frac{1}{2}+\frac{1}{8}\right)=\left(\frac{3\cdot 2}{4\cdot 2}+\frac{1}{8}\right)^2-\left(\frac{1\cdot 4}{2\cdot 4}+\frac{1}{8}\right)$
$=\left(\frac{6}{8}+\frac{1}{8}\right)^2-\left(\frac{4}{8}+\frac{1}{8}\right)$
$=\left(\frac{7}{8}\right)^2-\frac{5}{8}$
$=\frac{49}{64}-\frac{5}{8}$
$=\frac{49}{64}-\frac{5\cdot 8}{8\cdot 8}$
$=\frac{49}{64}-\frac{40}{64}$
$=\frac{9}{64}$

29. $5y-z=5\left(\frac{2}{5}\right)-\frac{5}{6}$
$=2-\frac{5}{6}$
$=\frac{2}{1}\cdot\frac{6}{6}-\frac{5}{6}$
$=\frac{12}{6}-\frac{5}{6}$
$=\frac{7}{6}$

31. $\frac{x}{z}=\frac{-\frac{1}{3}}{\frac{5}{6}}=-\frac{1}{3}\div\frac{5}{6}=-\frac{1}{3}\cdot\frac{6}{5}=-\frac{1\cdot 3\cdot 2}{3\cdot 5}=-\frac{2}{5}$

Copyright © 2011 Pearson Education, Inc. Publishing as Prentice Hall.

33. $x^2 - yz = \left(-\frac{1}{3}\right)^2 - \left(\frac{2}{5}\right)\left(\frac{5}{6}\right)$
$= \left(-\frac{1}{3}\right)\left(-\frac{1}{3}\right) - \left(\frac{2}{5}\right)\left(\frac{5}{6}\right)$
$= \frac{1\cdot 1}{3\cdot 3} - \frac{2\cdot 5}{5\cdot 2\cdot 3}$
$= \frac{1}{9} - \frac{1}{3}$
$= \frac{1}{9} - \frac{1}{3}\cdot\frac{3}{3}$
$= \frac{1}{9} - \frac{3}{9}$
$= -\frac{2}{9}$

35. $(1+x)(1+z) = \left[1+\left(-\frac{1}{3}\right)\right]\left[1+\frac{5}{6}\right]$
$= \left(\frac{3}{3}-\frac{1}{3}\right)\left(\frac{6}{6}+\frac{5}{6}\right)$
$= \left(\frac{2}{3}\right)\left(\frac{11}{6}\right)$
$= \frac{2\cdot 11}{3\cdot 2\cdot 3}$
$= \frac{11}{9}$

37. $\frac{\frac{5a}{24}}{\frac{1}{12}} = \frac{5a}{24} \div \frac{1}{12} = \frac{5a}{24}\cdot\frac{12}{1} = \frac{5\cdot a\cdot 12}{12\cdot 2\cdot 1} = \frac{5a}{2}$

39. $\left(\frac{3}{2}\right)^3 + \left(\frac{1}{2}\right)^3 = \frac{27}{8} + \frac{1}{8} = \frac{28}{8} = \frac{4\cdot 7}{4\cdot 2} = \frac{7}{2}$

41. $\left(-\frac{1}{2}\right)^2 + \frac{1}{5} = \frac{1}{4} + \frac{1}{5} = \frac{1\cdot 5}{4\cdot 5} + \frac{1\cdot 4}{5\cdot 4} = \frac{5}{20} + \frac{4}{20} = \frac{9}{20}$

43. $\frac{2+\frac{1}{6}}{1-\frac{4}{3}} = \frac{6\left(2+\frac{1}{6}\right)}{6\left(1-\frac{4}{3}\right)}$
$= \frac{6\cdot 2+6\cdot\frac{1}{6}}{6\cdot 1-6\cdot\frac{4}{3}}$
$= \frac{12+1}{6-8}$
$= \frac{13}{-2}$
$= -\frac{13}{2}$

45. $\left(1-\frac{2}{5}\right)^2 = \left(\frac{5}{5}-\frac{2}{5}\right)^2 = \left(\frac{3}{5}\right)^2 = \frac{3}{5}\cdot\frac{3}{5} = \frac{9}{25}$

47. $\left(\frac{3}{4}-1\right)\left(\frac{1}{8}+\frac{1}{2}\right) = \left(\frac{3}{4}-\frac{4}{4}\right)\left(\frac{1}{8}+\frac{4}{8}\right)$
$= \left(-\frac{1}{4}\right)\left(\frac{5}{8}\right)$
$= -\frac{1\cdot 5}{4\cdot 8}$
$= -\frac{5}{32}$

49. $\left(-\frac{2}{9}-\frac{7}{9}\right)^4 = \left(-\frac{9}{9}\right)^4$
$= (-1)^4$
$= (-1)(-1)(-1)(-1)$
$= 1$

51. $\frac{\frac{1}{3}-\frac{5}{6}}{\frac{3}{4}+\frac{1}{2}} = \frac{12\left(\frac{1}{3}-\frac{5}{6}\right)}{12\left(\frac{3}{4}+\frac{1}{2}\right)}$
$= \frac{12\cdot\frac{1}{3}-12\cdot\frac{5}{6}}{12\cdot\frac{3}{4}+12\cdot\frac{1}{2}}$
$= \frac{4-10}{9+6}$
$= \frac{-6}{15}$
$= -\frac{2\cdot 3}{3\cdot 5}$
$= -\frac{2}{5}$

53. $\left(\frac{3}{4}\div\frac{6}{5}\right) - \left(\frac{3}{4}\cdot\frac{6}{5}\right) = \left(\frac{3}{4}\cdot\frac{5}{6}\right) - \left(\frac{3}{4}\cdot\frac{6}{5}\right)$
$= \frac{3\cdot 5}{4\cdot 2\cdot 3} - \frac{3\cdot 2\cdot 3}{2\cdot 2\cdot 5}$
$= \frac{5}{4\cdot 2} - \frac{3\cdot 3}{2\cdot 5}$
$= \frac{5}{8} - \frac{9}{10}$
$= \frac{5\cdot 5}{8\cdot 5} - \frac{9\cdot 4}{10\cdot 4}$
$= \frac{25}{40} - \frac{36}{40}$
$= -\frac{11}{40}$

Copyright © 2011 Pearson Education, Inc. Publishing as Prentice Hall.

55. $\dfrac{\frac{x}{3}+2}{5+\frac{1}{3}} = \dfrac{3\left(\frac{x}{3}+2\right)}{3\left(5+\frac{1}{3}\right)}$
$= \dfrac{3\cdot\frac{x}{3}+3\cdot 2}{3\cdot 5+3\cdot\frac{1}{3}}$
$= \dfrac{x+6}{15+1}$
$= \dfrac{x+6}{16}$

57. $3+\dfrac{1}{2} = \dfrac{3}{1}\cdot\dfrac{2}{2}+\dfrac{1}{2} = \dfrac{6}{2}+\dfrac{1}{2} = \dfrac{7}{2}$ or $3\dfrac{1}{2}$

59. $9-\dfrac{5}{6} = \dfrac{9}{1}\cdot\dfrac{6}{6}-\dfrac{5}{6} = \dfrac{54}{6}-\dfrac{5}{6} = \dfrac{49}{6}$ or $8\dfrac{1}{6}$

61. no; answers may vary

63. $\dfrac{\frac{1}{2}+\frac{3}{4}}{2} = \dfrac{4\left(\frac{1}{2}+\frac{3}{4}\right)}{4(2)} = \dfrac{4\cdot\frac{1}{2}+4\cdot\frac{3}{4}}{4\cdot 2} = \dfrac{2+3}{8} = \dfrac{5}{8}$

65. $\dfrac{\frac{1}{4}+\frac{2}{14}}{2} = \dfrac{28\left(\frac{1}{4}+\frac{2}{14}\right)}{28(2)}$
$= \dfrac{28\cdot\frac{1}{4}+28\cdot\frac{2}{14}}{28\cdot 2}$
$= \dfrac{7+4}{56}$
$= \dfrac{11}{56}$

67. The average of *a* and *b* should be halfway between *a* and *b*.

69. False; the average cannot be greater than the greatest number.

71. True

73. True

75. addition; answers may vary

77. Subtraction, multiplication, addition, division

79. Division, multiplication, subtraction, addition

81. $\dfrac{2+x}{y} = \dfrac{2+\frac{3}{4}}{-\frac{4}{7}}$
$= \dfrac{28\left(2+\frac{3}{4}\right)}{28\left(-\frac{4}{7}\right)}$
$= \dfrac{28\cdot 2+28\cdot\frac{3}{4}}{28\cdot\left(-\frac{4}{7}\right)}$
$= \dfrac{56+21}{-16}$
$= -\dfrac{77}{16}$

83. $x^2+7y = \left(\dfrac{3}{4}\right)^2+7\left(-\dfrac{4}{7}\right)$
$= \dfrac{9}{16}-\dfrac{4}{1}$
$= \dfrac{9}{16}-\dfrac{4\cdot 16}{1\cdot 16}$
$= \dfrac{9}{16}-\dfrac{64}{16}$
$= -\dfrac{55}{16}$

Section 4.7

Practice Problems

1.

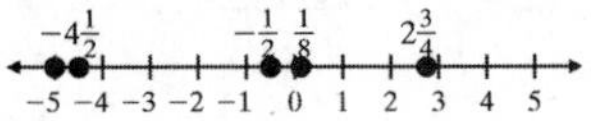

2. $1\dfrac{2}{3}\cdot\dfrac{11}{15} = \dfrac{5}{3}\cdot\dfrac{11}{15} = \dfrac{5\cdot 11}{3\cdot 5\cdot 3} = \dfrac{11}{9}$ or $1\dfrac{2}{9}$

3. $\dfrac{5}{6}\cdot 18 = \dfrac{5}{6}\cdot\dfrac{18}{1} = \dfrac{5\cdot 18}{6\cdot 1} = \dfrac{5\cdot 6\cdot 3}{6\cdot 1} = \dfrac{15}{1}$ or 15

4. $3\dfrac{1}{5}\cdot 2\dfrac{3}{4} = \dfrac{16}{5}\cdot\dfrac{11}{4} = \dfrac{16\cdot 11}{5\cdot 4} = \dfrac{4\cdot 4\cdot 11}{5\cdot 4} = \dfrac{44}{5}$ or $8\dfrac{4}{5}$

Estimate: $3\dfrac{1}{5}$ rounds to 3, $2\dfrac{3}{4}$ rounds to 3.

$3\cdot 3 = 9$, so the answer is reasonable.

5. $3\cdot 6\dfrac{7}{15} = \dfrac{3}{1}\cdot\dfrac{97}{15} = \dfrac{3\cdot 97}{1\cdot 5\cdot 3} = \dfrac{97}{5}$ or $19\dfrac{2}{5}$

Estimate: $6\dfrac{7}{15}$ rounds to 6 and $3\cdot 6 = 18$, so the answer is reasonable.

Copyright © 2011 Pearson Education, Inc. Publishing as Prentice Hall.

6. $\frac{4}{9} \div 7 = \frac{4}{9} \div \frac{7}{1} = \frac{4}{9} \cdot \frac{1}{7} = \frac{4 \cdot 1}{9 \cdot 7} = \frac{4}{63}$

7. $\frac{8}{15} \div 3\frac{4}{5} = \frac{8}{15} \div \frac{19}{5}$
$= \frac{8}{15} \cdot \frac{5}{19}$
$= \frac{8 \cdot 5}{15 \cdot 19}$
$= \frac{8 \cdot 5}{5 \cdot 3 \cdot 19}$
$= \frac{8}{57}$

8. $3\frac{2}{7} \div 2\frac{3}{14} = \frac{23}{7} \div \frac{31}{14}$
$= \frac{23}{7} \cdot \frac{14}{31}$
$= \frac{23 \cdot 7 \cdot 2}{7 \cdot 31}$
$= \frac{46}{31}$ or $1\frac{15}{31}$

9. $\begin{array}{r} 2\frac{1}{6} \\ +\ 4\frac{2}{5} \\ \hline \end{array}$ $\begin{array}{r} 2\frac{5}{30} \\ +\ 4\frac{12}{30} \\ \hline 6\frac{17}{30} \end{array}$

Estimate: $2\frac{1}{6}$ rounds to 2, $4\frac{2}{5}$ rounds to 4, and $2 + 4 = 6$, so the answer is reasonable.

10. $\begin{array}{r} 3\frac{5}{14} \\ +\ 2\frac{6}{7} \\ \hline \end{array}$ $\begin{array}{r} 3\frac{5}{14} \\ +\ 2\frac{12}{14} \\ \hline 5\frac{17}{14} \end{array} = 5 + 1\frac{3}{14} = 6\frac{3}{14}$

11. $\begin{array}{r} 12 \\ 3\frac{6}{7} \\ +\ 2\frac{1}{5} \\ \hline \end{array}$ $\begin{array}{r} 12 \\ 3\frac{30}{35} \\ +\ 2\frac{7}{35} \\ \hline 17\frac{37}{35} \end{array} = 17 + 1\frac{2}{35} = 18\frac{2}{35}$

12. $\begin{array}{r} 32\frac{7}{9} \\ -\ 16\frac{5}{18} \\ \hline \end{array}$ $\begin{array}{r} 32\frac{14}{18} \\ -\ 16\frac{5}{18} \\ \hline 16\frac{9}{18} \end{array} = 16\frac{1}{2}$

13. $\begin{array}{r} 9\frac{7}{15} \\ -\ 4\frac{3}{5} \\ \hline \end{array}$ $\begin{array}{r} 9\frac{7}{15} \\ -\ 4\frac{9}{15} \\ \hline \end{array}$ $\begin{array}{r} 8\frac{22}{15} \\ -\ 4\frac{9}{15} \\ \hline 4\frac{13}{15} \end{array}$

14. $\begin{array}{r} 25 \\ -\ 10\frac{2}{9} \\ \hline \end{array}$ $\begin{array}{r} 24\frac{9}{9} \\ -\ 10\frac{2}{9} \\ \hline 14\frac{7}{9} \end{array}$

15. $\begin{array}{r} 23\frac{1}{4} \\ -\ 19\frac{5}{12} \\ \hline \end{array}$ $\begin{array}{r} 23\frac{3}{12} \\ -\ 19\frac{5}{12} \\ \hline \end{array}$ $\begin{array}{r} 22\frac{15}{12} \\ -\ 19\frac{5}{12} \\ \hline 3\frac{10}{12} \end{array} = 3\frac{5}{6}$

The girth of the largest known American beech tree is $3\frac{5}{6}$ feet larger than the girth of the largest known sugar maple tree.

16. $44 \div 3\frac{1}{7} = 44 \div \frac{22}{7}$
$= \frac{44}{1} \cdot \frac{7}{22}$
$= \frac{2 \cdot 22 \cdot 7}{1 \cdot 22}$
$= \frac{14}{1}$ or 14

Therefore, 14 dresses can be made from 44 yards.

17. $-9\frac{3}{7} = -\frac{7 \cdot 9 + 3}{7} = -\frac{66}{7}$

18. $-5\frac{10}{11} = -\frac{11 \cdot 5 + 10}{11} = -\frac{65}{11}$

Copyright © 2011 Pearson Education, Inc. Publishing as Prentice Hall.

19. $-\frac{37}{8} = -4\frac{5}{8}$

$$\begin{array}{r} 4 \\ 8\overline{)37} \\ \underline{32} \\ 5 \end{array}$$

20. $-\frac{46}{5} = -9\frac{1}{5}$

$$\begin{array}{r} 9 \\ 5\overline{)46} \\ \underline{45} \\ 1 \end{array}$$

21.
$$\begin{aligned} 2\frac{3}{4}\cdot\left(-3\frac{3}{5}\right) &= \frac{11}{4}\cdot\left(-\frac{18}{5}\right) \\ &= -\frac{11\cdot 18}{4\cdot 5} \\ &= -\frac{11\cdot 2\cdot 9}{2\cdot 2\cdot 5} \\ &= -\frac{99}{10} \text{ or } -9\frac{9}{10} \end{aligned}$$

22.
$$\begin{aligned} -4\frac{2}{7}\div 1\frac{1}{4} &= -\frac{30}{7}\div\frac{5}{4} \\ &= -\frac{30}{7}\cdot\frac{4}{5} \\ &= -\frac{30\cdot 4}{7\cdot 5} \\ &= -\frac{5\cdot 6\cdot 4}{7\cdot 5} \\ &= -\frac{24}{7} \text{ or } -3\frac{3}{7} \end{aligned}$$

23.
$$\begin{array}{r} 12\frac{3}{4} \\ \underline{-\ 6\frac{2}{3}} \end{array} \qquad \begin{array}{r} 12\frac{9}{12} \\ \underline{-\ 6\frac{8}{12}} \\ 6\frac{1}{12} \end{array}$$

$6\frac{2}{3} - 12\frac{3}{4} = -6\frac{1}{12}$

24.
$$\begin{array}{r} 9\frac{2}{7} \\ \underline{+\ 30\frac{11}{14}} \end{array} \qquad \begin{array}{r} 9\frac{4}{14} \\ \underline{+\ 30\frac{11}{14}} \\ 39\frac{15}{14} \end{array} = 40\frac{1}{14}$$

$-9\frac{2}{7} - 30\frac{11}{14} = -40\frac{1}{14}$

Calculator Explorations

1. $25\frac{5}{11} = \frac{280}{11}$

2. $67\frac{14}{15} = \frac{1019}{15}$

3. $107\frac{31}{35} = \frac{3776}{35}$

4. $186\frac{17}{21} = \frac{3923}{21}$

5. $\frac{365}{14} = 26\frac{1}{14}$

6. $\frac{290}{13} = 22\frac{4}{13}$

7. $\frac{2769}{30} = 92\frac{3}{10}$

8. $\frac{3941}{17} = 231\frac{14}{17}$

Vocabulary and Readiness Check

1. The number $5\frac{3}{4}$ is called a <u>mixed number</u>.

2. For $5\frac{3}{4}$, the 5 is called the <u>whole number</u> part and $\frac{3}{4}$ is called the <u>fraction</u> part.

3. To estimate operations on mixed numbers, we <u>round</u> mixed numbers to the nearest whole number.

4. The mixed number $2\frac{5}{8}$ written as an <u>improper</u> fraction is $\frac{21}{8}$.

Exercise Set 4.7

1.

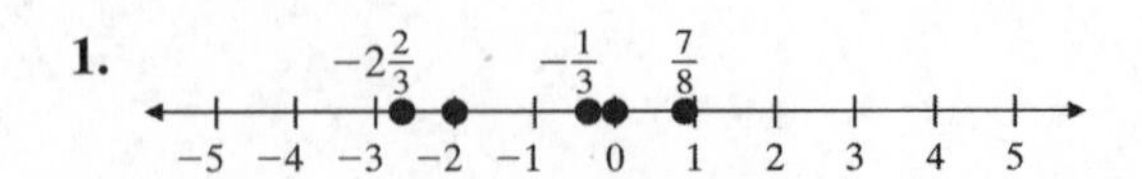

Copyright © 2011 Pearson Education, Inc. Publishing as Prentice Hall.

3. $-3\frac{4}{5}$, $\frac{1}{3}$, $1\frac{1}{3}$

−5 −4 −3 −2 −1 0 1 2 3 4 5

5. $2\frac{11}{12}$ rounds to 3.

$1\frac{1}{4}$ rounds to 1.

$3 \cdot 1 = 3$
The best estimate is b.

7. $12\frac{2}{11}$ rounds to 12.

$3\frac{9}{10}$ rounds to 4.

$12 \div 4 = 3$
The best estimate is a.

9. $2\frac{2}{3} \cdot \frac{1}{7} = \frac{8}{3} \cdot \frac{1}{7} = \frac{8}{21}$

11. $7 \div 1\frac{3}{5} = \frac{7}{1} \div \frac{8}{5} = \frac{7}{1} \cdot \frac{5}{8} = \frac{7 \cdot 5}{1 \cdot 8} = \frac{35}{8}$ or $4\frac{3}{8}$

13. Exact: $2\frac{1}{5} \cdot 3\frac{1}{2} = \frac{11}{5} \cdot \frac{7}{2} = \frac{77}{10}$ or $7\frac{7}{10}$

Estimate: $2\frac{1}{5}$ rounds to 2, $3\frac{1}{2}$ rounds to 4.

$2 \cdot 4 = 8$ so the answer is reasonable.

15. Exact: $3\frac{4}{5} \cdot 6\frac{2}{7} = \frac{19}{5} \cdot \frac{44}{7} = \frac{836}{35}$ or $23\frac{31}{35}$

Estimate: $3\frac{4}{5}$ rounds to 4, $6\frac{2}{7}$ rounds to 6.

$4 \cdot 6 = 24$

17. $5 \cdot 2\frac{1}{2} = \frac{5}{1} \cdot \frac{5}{2} = \frac{25}{2}$ or $12\frac{1}{2}$

19. $3\frac{2}{3} \cdot 1\frac{1}{2} = \frac{11}{3} \cdot \frac{3}{2} = \frac{11 \cdot 3}{3 \cdot 2} = \frac{11}{2}$ or $5\frac{1}{2}$

21. $2\frac{2}{3} \div \frac{1}{7} = \frac{8}{3} \cdot \frac{7}{1} = \frac{56}{3}$ or $18\frac{2}{3}$

23. $3\frac{7}{8}$ rounds to 4.

$2\frac{1}{5}$ rounds to 2.

$4 + 2 = 6$
The best estimate is a.

25. $8\frac{1}{3}$ rounds to 8.

$1\frac{1}{2}$ rounds to 2.

$8 + 2 = 10$
The best estimate is b.

27. Exact:
$$\begin{array}{r} 4\frac{7}{12} \\ +\,2\frac{1}{12} \\ \hline 6\frac{8}{12} = 6\frac{2}{3} \end{array}$$

Estimate: $4\frac{7}{12}$ rounds to 5, $2\frac{1}{12}$ rounds to 2.

$5 + 2 = 7$ so the answer is reasonable.

29. Exact:
$$\begin{array}{r} 10\frac{3}{14} \\ +\,3\frac{4}{7} \\ \hline \end{array} \qquad \begin{array}{r} 10\frac{3}{14} \\ +\,3\frac{8}{14} \\ \hline 13\frac{11}{14} \end{array}$$

Estimate: $10\frac{3}{14}$ rounds to 10, $3\frac{4}{7}$ rounds to 4.

$10 + 4 = 14$ so the answer is reasonable.

31.
$$\begin{array}{r} 9\frac{1}{5} \\ +\,8\frac{2}{25} \\ \hline \end{array} \qquad \begin{array}{r} 9\frac{5}{25} \\ +\,8\frac{2}{25} \\ \hline 17\frac{7}{25} \end{array}$$

33.
$$\begin{array}{r} 12\frac{3}{14} \\ 10 \\ +\,25\frac{5}{12} \\ \hline \end{array} \qquad \begin{array}{r} 12\frac{18}{84} \\ 10 \\ +\,25\frac{35}{84} \\ \hline 47\frac{53}{84} \end{array}$$

Copyright © 2011 Pearson Education, Inc. Publishing as Prentice Hall.

35. $\begin{array}{r} 15\frac{4}{7} \\ +\ 9\frac{11}{14} \\ \hline \end{array}$ $\begin{array}{r} 15\frac{8}{14} \\ +\ 9\frac{11}{14} \\ \hline 24\frac{19}{14} \end{array}$ $= 24+1\frac{5}{14} = 25\frac{5}{14}$

37. $\begin{array}{r} 3\frac{5}{8} \\ 2\frac{1}{6} \\ +\ 7\frac{3}{4} \\ \hline \end{array}$ $\begin{array}{r} 3\frac{15}{24} \\ 2\frac{4}{24} \\ +\ 7\frac{18}{24} \\ \hline 12\frac{37}{24} \end{array}$ $= 12+1\frac{13}{24} = 13\frac{13}{24}$

39. Exact: $\begin{array}{r} 4\frac{7}{10} \\ -\ 2\frac{1}{10} \\ \hline 2\frac{6}{10} \end{array}$ $= 2\frac{3}{5}$

Estimate: $4\frac{7}{10}$ rounds to 5, $2\frac{1}{10}$ rounds to 2.
$5 - 2 = 3$ so the answer is reasonable.

41. Exact: $\begin{array}{r} 10\frac{13}{14} \\ -\ 3\frac{4}{7} \\ \hline \end{array}$ $\begin{array}{r} 10\frac{13}{14} \\ -\ 3\frac{8}{14} \\ \hline 7\frac{5}{14} \end{array}$

Estimate: $10\frac{13}{14}$ rounds to 11, $3\frac{4}{7}$ rounds to 4.
$11 - 4 = 7$ so the answer is reasonable.

43. $\begin{array}{r} 9\frac{1}{5} \\ -\ 8\frac{6}{25} \\ \hline \end{array}$ $\begin{array}{r} 9\frac{5}{25} \\ -\ 8\frac{6}{25} \\ \hline \end{array}$ $\begin{array}{r} 8\frac{30}{25} \\ -\ 8\frac{6}{25} \\ \hline \frac{24}{25} \end{array}$

45. $\begin{array}{r} 6 \\ -\ 2\frac{4}{9} \\ \hline \end{array}$ $\begin{array}{r} 5\frac{9}{9} \\ -\ 2\frac{4}{9} \\ \hline 3\frac{5}{9} \end{array}$

47. $\begin{array}{r} 63\frac{1}{6} \\ -\ 47\frac{5}{12} \\ \hline \end{array}$ $\begin{array}{r} 63\frac{2}{12} \\ -\ 47\frac{5}{12} \\ \hline \end{array}$ $\begin{array}{r} 62\frac{14}{12} \\ -\ 47\frac{5}{12} \\ \hline 15\frac{9}{12} \end{array}$ $= 15\frac{3}{4}$

49. $\begin{array}{r} 2\frac{3}{4} \\ +\ 1\frac{1}{4} \\ \hline 3\frac{4}{4} \end{array}$ $= 3+1 = 4$

51. $\begin{array}{r} 15\frac{4}{7} \\ -\ 9\frac{11}{14} \\ \hline \end{array}$ $\begin{array}{r} 15\frac{8}{14} \\ -\ 9\frac{11}{14} \\ \hline \end{array}$ $\begin{array}{r} 14\frac{22}{14} \\ -\ 9\frac{11}{14} \\ \hline 5\frac{11}{14} \end{array}$

53. $3\frac{1}{9} \cdot 2 = \frac{28}{9} \cdot \frac{2}{1} = \frac{56}{9}$ or $6\frac{2}{9}$

55. $1\frac{2}{3} \div 2\frac{1}{5} = \frac{5}{3} \div \frac{11}{5} = \frac{5}{3} \cdot \frac{5}{11} = \frac{25}{33}$

57. $22\frac{4}{9} + 13\frac{5}{18} = 22\frac{8}{18} + 13\frac{5}{18} = 35\frac{13}{18}$

59. $5\frac{2}{3} - 3\frac{1}{6} = 5\frac{4}{6} - 3\frac{1}{6} = 2\frac{3}{6} = 2\frac{1}{2}$

61. $\begin{array}{r} 15\frac{1}{5} \\ 20\frac{3}{10} \\ +\ 37\frac{2}{15} \\ \hline \end{array}$ $\begin{array}{r} 15\frac{6}{30} \\ 20\frac{9}{30} \\ +\ 37\frac{4}{30} \\ \hline 72\frac{19}{30} \end{array}$

Copyright © 2011 Pearson Education, Inc. Publishing as Prentice Hall.

63. $$\begin{array}{r} 6\frac{4}{7} \\ -5\frac{11}{14} \\ \hline \end{array} \qquad \begin{array}{r} 6\frac{8}{14} \\ -5\frac{11}{14} \\ \hline \end{array} \qquad \begin{array}{r} 5\frac{22}{14} \\ -5\frac{11}{14} \\ \hline \frac{11}{14} \end{array}$$

65. $$\begin{aligned} 4\frac{2}{7}\cdot 1\frac{3}{10} &= \frac{30}{7}\cdot\frac{13}{10} \\ &= \frac{30\cdot 13}{7\cdot 10} \\ &= \frac{3\cdot 10\cdot 13}{7\cdot 10} \\ &= \frac{39}{7} \text{ or } 5\frac{4}{7} \end{aligned}$$

67. $$\begin{array}{r} 6\frac{2}{11} \\ 3 \\ +4\frac{10}{33} \\ \hline \end{array} \qquad \begin{array}{r} 6\frac{6}{33} \\ 3 \\ +4\frac{10}{33} \\ \hline 13\frac{16}{33} \end{array}$$

69. $-5\frac{2}{7}$ decreased by a number translates as $-5\frac{2}{7}-x$.

71. Multiply $1\frac{9}{10}$ by a number translates as $1\frac{9}{10}\cdot x$.

73. $12\frac{3}{4}\div 4 = \frac{51}{4}\div\frac{4}{1} = \frac{51}{4}\cdot\frac{1}{4} = \frac{51}{16}$ or $3\frac{3}{16}$

The patient walked $3\frac{3}{16}$ miles per day.

75. Subtract the standard gauge in the U.S. from the standard gauge in Spain.

$$\begin{array}{r} 65\frac{9}{10} \\ -56\frac{1}{2} \\ \hline \end{array} \qquad \begin{array}{r} 65\frac{9}{10} \\ -56\frac{5}{10} \\ \hline 9\frac{4}{10} = 9\frac{2}{5} \end{array}$$

Spain's standard gauge is $9\frac{2}{5}$ inches wider than the U.S. standard gauge.

77. $$\begin{array}{r} 11\frac{1}{4} \\ -\ 3\frac{3}{5} \\ \hline \end{array} \qquad \begin{array}{r} 11\frac{5}{20} \\ -\ 3\frac{12}{20} \\ \hline \end{array} \qquad \begin{array}{r} 10\frac{25}{20} \\ -\ 3\frac{12}{20} \\ \hline 7\frac{13}{20} \end{array}$$

Tucson gets an average of $7\frac{13}{20}$ inches more rain than Yuma.

79. $2\cdot 1\frac{3}{4} = \frac{2}{1}\cdot\frac{7}{4} = \frac{2\cdot 7}{1\cdot 4} = \frac{2\cdot 7}{1\cdot 2\cdot 2} = \frac{7}{2}$ or $3\frac{1}{2}$

The area is $\frac{7}{2}$ or $3\frac{1}{2}$ square yards.

81. $\frac{3}{4}\cdot 1\frac{1}{4} = \frac{3}{4}\cdot\frac{5}{4} = \frac{3\cdot 5}{4\cdot 4} = \frac{15}{16}$

The area is $\frac{15}{16}$ square inch.

83. $$\begin{array}{r} 5\frac{1}{3} \\ 5 \\ 7\frac{7}{8} \\ +\ 3 \\ \hline \end{array} \qquad \begin{array}{r} 5\frac{8}{24} \\ 5 \\ 7\frac{21}{24} \\ +\ 3 \\ \hline 20\frac{29}{24} = 20+1\frac{5}{24} = 21\frac{5}{24} \end{array}$$

The perimeter is $21\frac{5}{24}$ meters.

85. $$\begin{aligned} 15\frac{2}{3}-\left(3\frac{1}{4}+2\frac{1}{2}\right) &= 15\frac{2}{3}-\left(3\frac{1}{4}+2\frac{2}{4}\right) \\ &= 15\frac{2}{3}-5\frac{3}{4} \end{aligned}$$

$$\begin{array}{r} 15\frac{2}{3} \\ -\ 5\frac{3}{4} \\ \hline \end{array} \qquad \begin{array}{r} 15\frac{8}{12} \\ -\ 5\frac{9}{12} \\ \hline \end{array} \qquad \begin{array}{r} 14\frac{20}{12} \\ -\ 5\frac{9}{12} \\ \hline 9\frac{11}{12} \end{array}$$

No; the remaining board is $9\frac{11}{12}$ feet which is $\frac{1}{12}$ foot short.

Copyright © 2011 Pearson Education, Inc. Publishing as Prentice Hall.

87. $12 \div 2\frac{4}{7} = \frac{12}{1} \div \frac{18}{7} = \frac{12}{1} \cdot \frac{7}{18} = \frac{6 \cdot 2 \cdot 7}{1 \cdot 6 \cdot 3} = \frac{14}{3} = 4\frac{2}{3}$

The length is $4\frac{2}{3}$ meters.

89.
$$\begin{array}{r} 2\frac{2}{3} \\ 4\frac{7}{15} \\ +\ 2\frac{37}{60} \\ \hline \end{array} \qquad \begin{array}{r} 2\frac{40}{60} \\ 4\frac{28}{60} \\ +\ 2\frac{37}{60} \\ \hline 8\frac{105}{60} \end{array}$$
$8\frac{105}{60} = 8\frac{7}{4} = 8 + 1\frac{3}{4} = 9\frac{3}{4}$

The total duration of the eclipses is $9\frac{3}{4}$ minutes.

91.
$$\begin{array}{r} 4\frac{7}{15} \\ -\ 2\frac{2}{3} \\ \hline \end{array} \qquad \begin{array}{r} 4\frac{7}{15} \\ -\ 2\frac{10}{15} \\ \hline \end{array} \qquad \begin{array}{r} 3\frac{22}{15} \\ -\ 2\frac{10}{15} \\ \hline 1\frac{12}{15} \end{array}$$
$1\frac{12}{15} = 1\frac{4}{5}$

It will be $1\frac{4}{5}$ minutes longer.

93.
$$\begin{aligned} -4\frac{2}{5} \cdot 2\frac{3}{10} &= -\frac{22}{5} \cdot \frac{23}{10} \\ &= -\frac{2 \cdot 11 \cdot 23}{5 \cdot 2 \cdot 5} \\ &= -\frac{253}{25} \text{ or } -10\frac{3}{25} \end{aligned}$$

95.
$$\begin{aligned} -5\frac{1}{8} - 19\frac{3}{4} &= -\left(5\frac{1}{8} + 19\frac{3}{4}\right) \\ &= -\left(5\frac{1}{8} + 19\frac{6}{8}\right) \\ &= -\left(24\frac{7}{8}\right) \\ &= -24\frac{7}{8} \end{aligned}$$

97.
$$\begin{aligned} -31\frac{2}{15} + 17\frac{3}{20} &= -31\frac{8}{60} + 17\frac{9}{60} \\ &= -30\frac{68}{60} + 17\frac{9}{60} \\ &= -13\frac{59}{60} \end{aligned}$$

99. $-1\frac{5}{7} \cdot \left(-2\frac{1}{2}\right) = \frac{12}{7} \cdot \frac{5}{2} = \frac{6 \cdot 2 \cdot 5}{7 \cdot 2} = \frac{30}{7} = 4\frac{2}{7}$

101.
$$\begin{aligned} 11\frac{7}{8} - 13\frac{5}{6} &= 11\frac{21}{24} - 13\frac{20}{24} \\ &= -\left(13\frac{20}{24} - 11\frac{21}{24}\right) \\ &= -\left(12\frac{44}{24} - 11\frac{21}{24}\right) \\ &= -1\frac{23}{24} \end{aligned}$$

103.
$$\begin{aligned} -7\frac{3}{10} \div (-100) &= -\frac{73}{10} \div \left(-\frac{100}{1}\right) \\ &= -\frac{73}{10} \cdot \left(-\frac{1}{100}\right) \\ &= \frac{73 \cdot 1}{10 \cdot 100} \\ &= \frac{73}{1000} \end{aligned}$$

105. $\frac{1}{3}(3x) = \left(\frac{1}{3} \cdot 3\right)x = 1 \cdot x = x$

107. $\frac{2}{3}\left(\frac{3}{2}a\right) = \left(\frac{2}{3} \cdot \frac{3}{2}\right)a = 1 \cdot a = a$

109. **a.** $9\frac{5}{5} = 9 + 1 = 10$

b. $9\frac{100}{100} = 9 + 1 = 10$

c. $6\frac{44}{11} = 6 + 4 = 10$

d. $8\frac{13}{13} = 8 + 1 = 9$

a, b, and c are equivalent to 10.

111. Incorrect, to divide mixed numbers, first write each mixed number as an improper fraction.

113. answers may vary

115. answers may vary

117. answers may vary

Copyright © 2011 Pearson Education, Inc. Publishing as Prentice Hall.

Section 4.8

Practice Problems

1.
$$y-\frac{2}{3}=\frac{5}{12}$$
$$y-\frac{2}{3}+\frac{2}{3}=\frac{5}{12}+\frac{2}{3}$$
$$y=\frac{5}{12}+\frac{2\cdot 4}{3\cdot 4}$$
$$y=\frac{5}{12}+\frac{8}{12}$$
$$y=\frac{13}{12}$$

Check: $y-\frac{2}{3}=\frac{5}{12}$
$$\frac{13}{12}-\frac{2}{3}\stackrel{?}{=}\frac{5}{12}$$
$$\frac{13}{12}-\frac{2\cdot 4}{3\cdot 4}\stackrel{?}{=}\frac{5}{12}$$
$$\frac{13}{12}-\frac{8}{12}\stackrel{?}{=}\frac{5}{12}$$
$$\frac{5}{12}=\frac{5}{12} \quad \text{True}$$

The solution is $\frac{13}{12}$.

2.
$$\frac{1}{5}y=2$$
$$5\cdot\frac{1}{5}y=5\cdot 2$$
$$y=10$$

Check: $\frac{1}{5}y=2$
$$\frac{1}{5}\cdot 10\stackrel{?}{=}2$$
$$2=2 \quad \text{True}$$

The solution is 10.

3.
$$\frac{5}{7}b=25$$
$$\frac{7}{5}\cdot\frac{5}{7}b=\frac{7}{5}\cdot 25$$
$$1b=\frac{7\cdot 25}{5}$$
$$b=35$$

Check: $\frac{5}{7}b=25$
$$\frac{5}{7}\cdot 35\stackrel{?}{=}25$$
$$25=25 \quad \text{True}$$

The solution is 35.

4.
$$-\frac{7}{10}x=\frac{2}{5}$$
$$-\frac{10}{7}\cdot-\frac{7}{10}x=-\frac{10}{7}\cdot\frac{2}{5}$$
$$x=-\frac{10\cdot 2}{7\cdot 5}$$
$$x=-\frac{4}{7}$$

Check: $-\frac{7}{10}x=\frac{2}{5}$
$$-\frac{7}{10}\cdot-\frac{4}{7}\stackrel{?}{=}\frac{2}{5}$$
$$\frac{28}{70}\stackrel{?}{=}\frac{2}{5}$$
$$\frac{2}{5}=\frac{2}{5} \quad \text{True}$$

The solution is $-\frac{4}{7}$.

5.
$$5x=-\frac{3}{4}$$
$$\frac{1}{5}\cdot 5x=\frac{1}{5}\cdot-\frac{3}{4}$$
$$x=-\frac{1\cdot 3}{5\cdot 4}$$
$$x=-\frac{3}{20}$$

Check: $5x=-\frac{3}{4}$
$$5\cdot-\frac{3}{20}\stackrel{?}{=}-\frac{3}{4}$$
$$-\frac{15}{20}\stackrel{?}{=}-\frac{3}{4}$$
$$-\frac{3}{4}=-\frac{3}{4} \quad \text{True}$$

The solution is $-\frac{3}{20}$.

Copyright © 2011 Pearson Education, Inc. Publishing as Prentice Hall.

6. The LCD is 15.

$$\frac{11}{15}x = -\frac{3}{5}$$
$$15 \cdot \frac{11}{15}x = 15 \cdot -\frac{3}{5}$$
$$11x = -9$$
$$\frac{11x}{11} = \frac{-9}{11}$$
$$x = -\frac{9}{11}$$

The solution is $-\frac{9}{11}$.

7. The LCD is 8.

$$\frac{y}{8} + \frac{3}{4} = 2$$
$$8\left(\frac{y}{8} + \frac{3}{4}\right) = 8(2)$$
$$8 \cdot \frac{y}{8} + 8 \cdot \frac{3}{4} = 8(2)$$
$$y + 6 = 16$$
$$y + 6 - 6 = 16 - 6$$
$$y = 10$$

Check: $\frac{y}{8} + \frac{3}{4} = 2$

$$\frac{10}{8} + \frac{3}{4} \stackrel{?}{=} 2$$
$$\frac{10}{8} + \frac{6}{8} \stackrel{?}{=} 2$$
$$\frac{16}{8} \stackrel{?}{=} 2$$
$$2 = 2 \quad \text{True}$$

The solution is 10.

8.

$$\frac{x}{5} - x = \frac{1}{5}$$
$$5\left(\frac{x}{5} - x\right) = 5\left(\frac{1}{5}\right)$$
$$5\left(\frac{x}{5}\right) - 5(x) = 5\left(\frac{1}{5}\right)$$
$$x - 5x = 1$$
$$-4x = 1$$
$$\frac{-4x}{-4} = \frac{1}{-4}$$
$$x = -\frac{1}{4}$$

To check, replace x with $-\frac{1}{4}$ in the original equation to see that a true statement results. The solution is $-\frac{1}{4}$.

9. The LCD is 10.

$$\frac{y}{2} = \frac{y}{5} + \frac{3}{2}$$
$$10\left(\frac{y}{2}\right) = 10\left(\frac{y}{5} + \frac{3}{2}\right)$$
$$10\left(\frac{y}{2}\right) = 10\left(\frac{y}{5}\right) + 10\left(\frac{3}{2}\right)$$
$$5y = 2y + 15$$
$$5y - 2y = 2y - 2y + 15$$
$$3y = 15$$
$$\frac{3y}{3} = \frac{15}{3}$$
$$y = 5$$

Check: $\frac{y}{2} = \frac{y}{5} + \frac{3}{2}$

$$\frac{5}{2} \stackrel{?}{=} \frac{5}{5} + \frac{3}{2}$$
$$2\frac{1}{2} \stackrel{?}{=} 1 + 1\frac{1}{2}$$
$$2\frac{1}{2} = 2\frac{1}{2} \quad \text{True}$$

The solution is 5.

10.

$$\frac{9}{10} - \frac{y}{3} = \frac{9}{10} \cdot \frac{3}{3} - \frac{y}{3} \cdot \frac{10}{10}$$
$$= \frac{9 \cdot 3}{10 \cdot 3} + \frac{y \cdot 10}{3 \cdot 10}$$
$$= \frac{27}{30} - \frac{10y}{30}$$
$$= \frac{27 - 10y}{30}$$

Copyright © 2011 Pearson Education, Inc. Publishing as Prentice Hall.

Exercise Set 4.8

1.
$$x+\frac{1}{3}=-\frac{1}{3}$$
$$x+\frac{1}{3}-\frac{1}{3}=-\frac{1}{3}-\frac{1}{3}$$
$$x=-\frac{2}{3}$$

Check: $x+\frac{1}{3}=-\frac{1}{3}$
$$-\frac{2}{3}+\frac{1}{3}\stackrel{?}{=}-\frac{1}{3}$$
$$-\frac{1}{3}=-\frac{1}{3} \quad \text{True}$$

The solution is $-\frac{2}{3}$.

3.
$$y-\frac{3}{13}=-\frac{2}{13}$$
$$y-\frac{3}{13}+\frac{3}{13}=-\frac{2}{13}+\frac{3}{13}$$
$$y=\frac{-2+3}{13}$$
$$y=\frac{1}{13}$$

Check: $y-\frac{3}{13}=-\frac{2}{13}$
$$\frac{1}{13}-\frac{3}{13}\stackrel{?}{=}-\frac{2}{13}$$
$$\frac{1-3}{13}\stackrel{?}{=}-\frac{2}{13}$$
$$-\frac{2}{13}=-\frac{2}{13} \quad \text{True}$$

The solution is $\frac{1}{13}$.

5.
$$3x-\frac{1}{5}-2x=\frac{1}{5}+\frac{2}{5}$$
$$x-\frac{1}{5}=\frac{3}{5}$$
$$x-\frac{1}{5}+\frac{1}{5}=\frac{3}{5}+\frac{1}{5}$$
$$x=\frac{4}{5}$$

Check: $3x-\frac{1}{5}-2x=\frac{1}{5}+\frac{2}{5}$
$$3\cdot\frac{4}{5}-\frac{1}{5}-2\cdot\frac{4}{5}\stackrel{?}{=}\frac{1}{5}+\frac{2}{5}$$
$$\frac{12}{5}-\frac{1}{5}-\frac{8}{5}\stackrel{?}{=}\frac{1}{5}+\frac{2}{5}$$
$$\frac{3}{5}=\frac{3}{5} \quad \text{True}$$

The solution is $\frac{4}{5}$.

7.
$$x-\frac{1}{12}=\frac{5}{6}$$
$$x-\frac{1}{12}+\frac{1}{12}=\frac{5}{6}+\frac{1}{12}$$
$$x=\frac{5\cdot 2}{6\cdot 2}+\frac{1}{12}$$
$$x=\frac{10}{12}+\frac{1}{12}$$
$$x=\frac{11}{12}$$

Check: $x-\frac{1}{12}=\frac{5}{6}$
$$\frac{11}{12}-\frac{1}{12}\stackrel{?}{=}\frac{5}{6}$$
$$\frac{10}{12}\stackrel{?}{=}\frac{5}{6}$$
$$\frac{5}{6}=\frac{5}{6} \quad \text{True}$$

The solution is $\frac{11}{12}$.

9.
$$\frac{2}{5}+y=-\frac{3}{10}$$
$$\frac{2}{5}+y-\frac{2}{5}=-\frac{3}{10}-\frac{2}{5}$$
$$y=-\frac{3}{10}-\frac{2\cdot 2}{5\cdot 2}$$
$$y=-\frac{3}{10}-\frac{4}{10}$$
$$y=-\frac{7}{10}$$

Copyright © 2011 Pearson Education, Inc. Publishing as Prentice Hall.

Check:
$$\frac{2}{5}+y=-\frac{3}{10}$$
$$\frac{2}{5}+\left(-\frac{7}{10}\right)\stackrel{?}{=}-\frac{3}{10}$$
$$\frac{2}{5}\cdot\frac{2}{2}+\left(-\frac{7}{10}\right)\stackrel{?}{=}-\frac{3}{10}$$
$$\frac{4}{10}+\left(-\frac{7}{10}\right)\stackrel{?}{=}-\frac{3}{10}$$
$$-\frac{3}{10}=-\frac{3}{10} \quad \text{True}$$

The solution is $-\frac{7}{10}$.

11.
$$7z+\frac{1}{16}-6z=\frac{3}{4}$$
$$z+\frac{1}{16}=\frac{3}{4}$$
$$z+\frac{1}{16}-\frac{1}{16}=\frac{3}{4}-\frac{1}{16}$$
$$z=\frac{3\cdot 4}{4\cdot 4}-\frac{1}{16}$$
$$z=\frac{12}{16}-\frac{1}{16}$$
$$z=\frac{11}{16}$$

Check:
$$7z+\frac{1}{16}-6z=\frac{3}{4}$$
$$7\left(\frac{11}{16}\right)+\frac{1}{16}-6\left(\frac{11}{16}\right)\stackrel{?}{=}\frac{3}{4}$$
$$\frac{77}{16}+\frac{1}{16}-\frac{66}{16}\stackrel{?}{=}\frac{3}{4}$$
$$\frac{12}{16}\stackrel{?}{=}\frac{3}{4}$$
$$\frac{3}{4}=\frac{3}{4} \quad \text{True}$$

The solution is $\frac{11}{16}$.

13.
$$-\frac{2}{9}=x-\frac{5}{6}$$
$$-\frac{2}{9}+\frac{5}{6}=x-\frac{5}{6}+\frac{5}{6}$$
$$-\frac{2}{9}\cdot\frac{2}{2}+\frac{5}{6}\cdot\frac{3}{3}=x$$
$$-\frac{4}{18}+\frac{15}{18}=x$$
$$\frac{11}{18}=x$$

Check:
$$-\frac{2}{9}=x-\frac{5}{6}$$
$$-\frac{2}{9}\stackrel{?}{=}\frac{11}{18}-\frac{5}{6}$$
$$-\frac{2}{9}\cdot\frac{2}{2}\stackrel{?}{=}\frac{11}{18}-\frac{5}{6}\cdot\frac{3}{3}$$
$$-\frac{4}{18}\stackrel{?}{=}\frac{11}{18}-\frac{15}{18}$$
$$-\frac{4}{18}=-\frac{4}{18} \quad \text{True}$$

The solution is $\frac{11}{18}$.

15.
$$7x=2$$
$$\frac{1}{7}\cdot 7x=\frac{1}{7}\cdot 2$$
$$x=\frac{2}{7}$$

17.
$$\frac{1}{4}x=3$$
$$4\cdot\frac{1}{4}x=4\cdot 3$$
$$x=12$$

19.
$$\frac{2}{9}y=-6$$
$$\frac{9}{2}\cdot\frac{2}{9}y=\frac{9}{2}\cdot -6$$
$$y=-\frac{9\cdot 6}{2}$$
$$y=-\frac{9\cdot 2\cdot 3}{2}$$
$$y=-27$$

21.
$$-\frac{4}{9}z=-\frac{3}{2}$$
$$-\frac{9}{4}\cdot -\frac{4}{9}z=-\frac{9}{4}\cdot -\frac{3}{2}$$
$$z=\frac{27}{8}$$

23.
$$7a=\frac{1}{3}$$
$$\frac{1}{7}\cdot 7a=\frac{1}{7}\cdot\frac{1}{3}$$
$$a=\frac{1}{21}$$

Copyright © 2011 Pearson Education, Inc. Publishing as Prentice Hall.

25.
$$-3x = -\frac{6}{11}$$
$$-\frac{1}{3}\cdot -3x = -\frac{1}{3}\cdot -\frac{6}{11}$$
$$x = \frac{6}{3\cdot 11}$$
$$x = \frac{2\cdot 3}{3\cdot 11}$$
$$x = \frac{2}{11}$$

27.
$$\frac{5}{9}x = -\frac{3}{18}$$
$$18\left(\frac{5}{9}x\right) = 18\left(-\frac{3}{18}\right)$$
$$10x = -3$$
$$\frac{10x}{10} = \frac{-3}{10}$$
$$x = -\frac{3}{10}$$

29.
$$\frac{x}{3} + 2 = \frac{7}{3}$$
$$3\left(\frac{x}{3} + 2\right) = 3\cdot\frac{7}{3}$$
$$3\cdot\frac{x}{3} + 3\cdot 2 = 7$$
$$x + 6 = 7$$
$$x + 6 - 6 = 7 - 6$$
$$x = 1$$

31.
$$\frac{x}{5} - x = -8$$
$$5\left(\frac{x}{5} - x\right) = 5(-8)$$
$$5\left(\frac{x}{5}\right) - 5(x) = -40$$
$$x - 5x = -40$$
$$-4x = -40$$
$$\frac{-4x}{-4} = \frac{-40}{-4}$$
$$x = 10$$

33.
$$\frac{1}{2} - \frac{3}{5} = \frac{x}{10}$$
$$10\left(\frac{1}{2} - \frac{3}{5}\right) = 10\cdot\frac{x}{10}$$
$$10\cdot\frac{1}{2} - 10\cdot\frac{3}{5} = x$$
$$5 - 6 = x$$
$$-1 = x$$

35.
$$\frac{x}{3} = \frac{x}{5} - 2$$
$$15\left(\frac{x}{3}\right) = 15\left(\frac{x}{5} - 2\right)$$
$$5x = 15\cdot\frac{x}{5} - 15\cdot 2$$
$$5x = 3x - 30$$
$$5x - 3x = 3x - 3x - 30$$
$$2x = -30$$
$$\frac{2x}{2} = \frac{-30}{2}$$
$$x = -15$$

37.
$$\frac{x}{7} - \frac{4}{3} = \frac{x}{7}\cdot\frac{3}{3} - \frac{4}{3}\cdot\frac{7}{7} = \frac{3x}{21} - \frac{28}{21} = \frac{3x - 28}{21}$$

39.
$$\frac{y}{2} + 5 = \frac{y}{2} + \frac{5}{1}\cdot\frac{2}{2} = \frac{y}{2} + \frac{10}{2} = \frac{y + 10}{2}$$

41.
$$\frac{3x}{10} + \frac{x}{6} = \frac{3x}{10}\cdot\frac{3}{3} + \frac{x}{6}\cdot\frac{5}{5} = \frac{9x}{30} + \frac{5x}{30} = \frac{14x}{30} = \frac{2\cdot 7\cdot x}{2\cdot 15} = \frac{7x}{15}$$

43.
$$\frac{3}{8}x = \frac{1}{2}$$
$$\frac{8}{3}\cdot\frac{3}{8}x = \frac{8}{3}\cdot\frac{1}{2}$$
$$x = \frac{8}{6}$$
$$x = \frac{4}{3}$$

Copyright © 2011 Pearson Education, Inc. Publishing as Prentice Hall.

45.
$$\frac{2}{3}-\frac{x}{5}=\frac{4}{15}$$
$$15\left(\frac{2}{3}-\frac{x}{5}\right)=15\cdot\frac{4}{15}$$
$$15\cdot\frac{2}{3}-15\cdot\frac{x}{5}=4$$
$$10-3x=4$$
$$10-3x-10=4-10$$
$$-3x=-6$$
$$\frac{-3x}{-3}=\frac{-6}{-3}$$
$$x=2$$

47.
$$\frac{9}{14}z=\frac{27}{20}$$
$$\frac{14}{9}\cdot\frac{9}{14}z=\frac{14}{9}\cdot\frac{27}{20}$$
$$z=\frac{2\cdot 7\cdot 9\cdot 3}{9\cdot 2\cdot 10}$$
$$z=\frac{21}{10}$$

49.
$$-3m-5m=\frac{4}{7}$$
$$-8m=\frac{4}{7}$$
$$-\frac{1}{8}\cdot -8m=-\frac{1}{8}\cdot\frac{4}{7}$$
$$m=-\frac{1\cdot 4}{2\cdot 4\cdot 7}$$
$$m=-\frac{1}{14}$$

51.
$$\frac{x}{4}+1=\frac{1}{4}$$
$$4\left(\frac{x}{4}+1\right)=4\left(\frac{1}{4}\right)$$
$$4\cdot\frac{x}{4}+4\cdot 1=1$$
$$x+4=1$$
$$x+4-4=1-4$$
$$x=-3$$

53. $\frac{5}{9}-\frac{2}{3}=\frac{5}{9}-\frac{2}{3}\cdot\frac{3}{3}=\frac{5}{9}-\frac{6}{9}=-\frac{1}{9}$

55.
$$\frac{1}{5}y=10$$
$$5\cdot\frac{1}{5}y=5\cdot 10$$
$$y=50$$

57.
$$\frac{5}{7}y=-\frac{15}{49}$$
$$\frac{7}{5}\cdot\frac{5}{7}y=\frac{7}{5}\cdot -\frac{15}{49}$$
$$y=-\frac{7\cdot 15}{5\cdot 49}$$
$$y=-\frac{7\cdot 3\cdot 5}{5\cdot 7\cdot 7}$$
$$y=-\frac{3}{7}$$

59.
$$\frac{x}{2}-x=-2$$
$$2\left(\frac{x}{2}-x\right)=2(-2)$$
$$2\cdot\frac{x}{2}-2\cdot x=-4$$
$$x-2x=-4$$
$$-x=-4$$
$$\frac{-x}{-1}=\frac{-4}{-1}$$
$$x=4$$

61.
$$-\frac{5}{8}y=\frac{3}{16}-\frac{9}{16}$$
$$-\frac{5}{8}y=-\frac{6}{16}$$
$$-\frac{5}{8}y=-\frac{3}{8}$$
$$-\frac{8}{5}\cdot -\frac{5}{8}y=-\frac{8}{5}\cdot -\frac{3}{8}$$
$$y=\frac{8\cdot 3}{5\cdot 8}$$
$$y=\frac{3}{5}$$

63.
$$17x-25x=\frac{1}{3}$$
$$-8x=\frac{1}{3}$$
$$-\frac{1}{8}\cdot -8x=-\frac{1}{8}\cdot\frac{1}{3}$$
$$x=-\frac{1}{24}$$

Copyright © 2011 Pearson Education, Inc. Publishing as Prentice Hall.

65.
$$\frac{7}{6}x = \frac{1}{4} - \frac{2}{3}$$
$$12 \cdot \frac{7}{6}x = 12\left(\frac{1}{4} - \frac{2}{3}\right)$$
$$14x = 12 \cdot \frac{1}{4} - 12 \cdot \frac{2}{3}$$
$$14x = 3 - 8$$
$$14x = -5$$
$$\frac{14x}{14} = \frac{-5}{14}$$
$$x = -\frac{5}{14}$$

67.
$$\frac{b}{4} = \frac{b}{12} + \frac{2}{3}$$
$$12 \cdot \frac{b}{4} = 12\left(\frac{b}{12} + \frac{2}{3}\right)$$
$$3b = 12 \cdot \frac{b}{12} + 12 \cdot \frac{2}{3}$$
$$3b = b + 8$$
$$3b - b = b + 8 - b$$
$$2b = 8$$
$$\frac{2b}{2} = \frac{8}{2}$$
$$b = 4$$

69.
$$\frac{x}{3} + 2 = \frac{x}{2} + 8$$
$$6\left(\frac{x}{3} + 2\right) = 6\left(\frac{x}{2} + 8\right)$$
$$6 \cdot \frac{x}{3} + 6 \cdot 2 = 6 \cdot \frac{x}{2} + 6 \cdot 8$$
$$2x + 12 = 3x + 48$$
$$2x + 12 - 2x = 3x + 48 - 2x$$
$$12 = x + 48$$
$$12 - 48 = x + 48 - 48$$
$$-36 = x$$

71. To round 57,236 to the nearest hundred, observe that the digit in the tens place is 3. Since this digit is less than 5, we do not add 1 to the digit in the hundreds place. 57,236 rounded to the nearest hundred is 57,200.

73. To round 327 to the nearest ten, observe that the digit in the ones place is 7. Since this digit is at least 5, we add 1 to the digit in the tens place. 327 rounded to the nearest ten is 330.

75. answers may vary

77.
$$\frac{14}{11} + \frac{3x}{8} = \frac{x}{2}$$
$$88\left(\frac{14}{11} + \frac{3x}{8}\right) = 88 \cdot \frac{x}{2}$$
$$88 \cdot \frac{14}{11} + 88 \cdot \frac{3x}{8} = 44x$$
$$112 + 33x = 44x$$
$$112 + 33x - 33x = 44x - 33x$$
$$112 = 11x$$
$$\frac{112}{11} = \frac{11x}{11}$$
$$\frac{112}{11} = x$$

79. $A = l \cdot w = \frac{3}{4} \cdot \frac{1}{4} = \frac{3}{16}$

The area is $\frac{3}{16}$ square inch.

$P = 2l + 2w = 2 \cdot \frac{3}{4} + 2 \cdot \frac{1}{4} = \frac{3}{2} + \frac{1}{2} = \frac{4}{2} = 2$

The perimeter is 2 inches.

Chapter 4 Vocabulary Check

1. Two numbers are reciprocals of each other if their product is 1.
2. A composite number is a natural number greater than 1 that is not prime.
3. Fractions that represent the same portion of a whole are called equivalent fractions.
4. An improper fraction is a fraction whose numerator is greater than or equal to its denominator.
5. A prime number is a natural number greater than 1 whose only factors are 1 and itself.
6. A fraction is in simplest form when the numerator and the denominator have no factors in common other than 1.
7. A proper fraction is one whose numerator is less than its denominator.
8. A mixed number contains a whole number part and a fraction part.
9. In the fraction $\frac{7}{9}$, the 7 is called the numerator and the 9 is called the denominator.

Copyright © 2011 Pearson Education, Inc. Publishing as Prentice Hall.

10. The prime factorization of a number is the factorization in which all the factors are prime numbers.

11. The fraction $\frac{3}{0}$ is undefined.

12. The fraction $\frac{0}{5} = 0$.

13. Fractions that have the same denominator are called like fractions.

14. The LCM of the denominators in a list of fractions is called the least common denominator.

15. A fraction whose numerator or denominator or both numerator and denominator contain fractions is called a complex fraction.

16. In $\frac{a}{b} = \frac{c}{d}$, $a \cdot d$ and $b \cdot c$ are called cross products.

Chapter 4 Review

1. 2 out of 6 equal parts are shaded: $\frac{2}{6}$

2. 4 out of 7 equal parts are shaded: $\frac{4}{7}$

3. Each part is $\frac{1}{3}$ of a whole and there are 7 parts shaded, or 2 wholes and 1 more part: $\frac{7}{3}$ or $2\frac{1}{3}$

4. Each part is $\frac{1}{4}$ of a whole and there are 13 parts shaded, or 3 wholes and 1 more part: $\frac{13}{4}$ or $3\frac{1}{4}$

5. successful free throws → 11
total free throws → $\overline{12}$

$\frac{11}{12}$ of the free throws were made.

6. a. $131 - 23 = 108$
108 cars are not blue.

b. not blue → 108
total → $\overline{131}$

$\frac{108}{131}$ of the cars on the lot are not blue.

7. $\frac{3}{-3} = -1$

8. $\frac{-20}{-20} = 1$

Copyright © 2011 Pearson Education, Inc. Publishing as Prentice Hall.

9. $\frac{0}{-1} = 0$

10. $\frac{4}{0}$ is undefined

11.

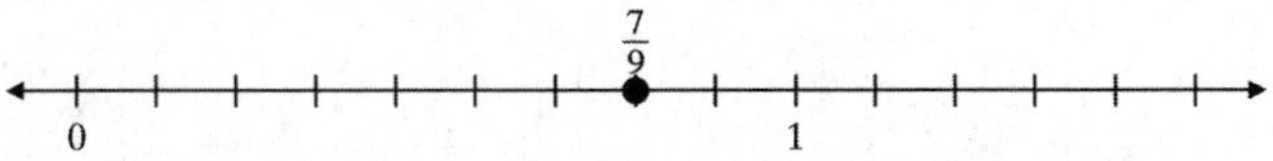

12.

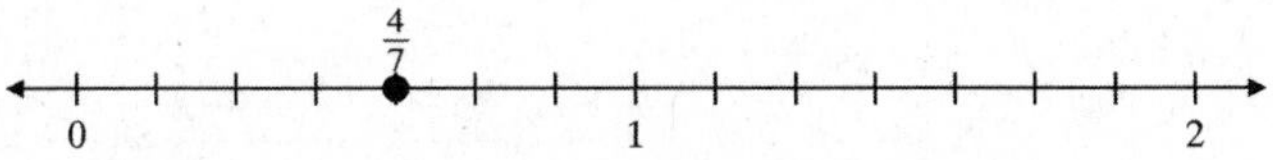

13.

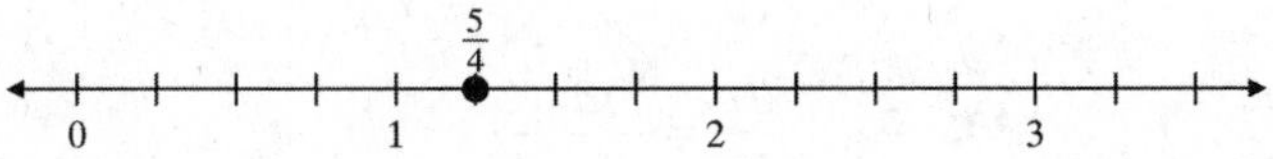

14.

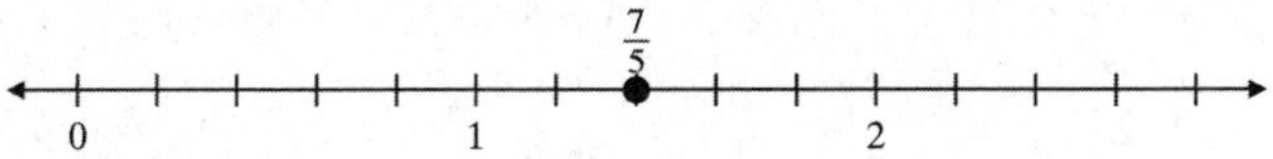

15.
$$\begin{array}{r} 3 \\ 4\overline{)15} \\ \underline{12} \\ 3 \end{array}$$

$\frac{15}{4} = 3\frac{3}{4}$

16.
$$\begin{array}{r} 3 \\ 13\overline{)39} \\ \underline{39} \\ 0 \end{array}$$

$\frac{39}{13} = 3$

17. $2\frac{1}{5} = \frac{5 \cdot 2 + 1}{5} = \frac{10 + 1}{5} = \frac{11}{5}$

18. $3\frac{8}{9} = \frac{9 \cdot 3 + 8}{9} = \frac{27 + 8}{9} = \frac{35}{9}$

19. $\frac{12}{28} = \frac{4 \cdot 3}{7 \cdot 4} = \frac{3}{7}$

20. $\frac{15}{27} = \frac{3 \cdot 5}{3 \cdot 9} = \frac{5}{9}$

21. $-\frac{25x}{75x^2} = -\frac{1 \cdot 25 \cdot x}{3 \cdot 25 \cdot x \cdot x} = -\frac{1}{3x}$

Copyright © 2011 Pearson Education, Inc. Publishing as Prentice Hall.

22. $-\frac{36y^3}{72y} = -\frac{1\cdot 36\cdot y\cdot y\cdot y}{2\cdot 36\cdot y} = -\frac{y^2}{2}$

23. $\frac{29ab}{32abc} = \frac{29\cdot a\cdot b}{32\cdot a\cdot b\cdot c} = \frac{29}{32c}$

24. $\frac{18xyz}{23xy} = \frac{18\cdot x\cdot y\cdot z}{23\cdot x\cdot y} = \frac{18z}{23}$

25. $\frac{45x^2y}{27xy^3} = \frac{9\cdot 5\cdot x\cdot x\cdot y}{9\cdot 3\cdot x\cdot y\cdot y\cdot y} = \frac{5x}{3y^2}$

26. $\frac{42ab^2c}{30abc^3} = \frac{6\cdot 7\cdot a\cdot b\cdot b\cdot c}{6\cdot 5\cdot a\cdot b\cdot c\cdot c\cdot c} = \frac{7b}{5c^2}$

27. $\frac{8 \text{ inches}}{12 \text{ inches}} = \frac{8}{12} = \frac{4\cdot 2}{4\cdot 3} = \frac{2}{3}$

8 inches represents $\frac{2}{3}$ of a foot.

28. $15 - 6 = 9$ cars are not white.

$\frac{9 \text{ non-white cars}}{15 \text{ total cars}} = \frac{9}{15} = \frac{3\cdot 3}{3\cdot 5} = \frac{3}{5}$

$\frac{3}{5}$ of the cars are not white.

29. Not equivalent, since the cross products are not equal: $34 \cdot 4 = 136$ and $10 \cdot 14 = 140$

30. Equivalent, since the cross products are equal: $30 \cdot 15 = 450$ and $50 \cdot 9 = 450$

31. $\frac{3}{5}\cdot\frac{1}{2} = \frac{3\cdot 1}{5\cdot 2} = \frac{3}{10}$

32. $-\frac{6}{7}\cdot\frac{5}{12} = -\frac{6\cdot 5}{7\cdot 12} = -\frac{6\cdot 5}{7\cdot 6\cdot 2} = -\frac{5}{7\cdot 2} = -\frac{5}{14}$

33.
$$-\frac{24x}{5}\cdot\left(-\frac{15}{8x^3}\right) = \frac{24x\cdot 15}{5\cdot 8x^3} = \frac{8\cdot 3\cdot x\cdot 5\cdot 3}{5\cdot 8\cdot x\cdot x\cdot x} = \frac{3\cdot 3}{x\cdot x} = \frac{9}{x^2}$$

34. $\frac{27y^3}{21}\cdot\frac{7}{18y^2} = \frac{27y^3\cdot 7}{21\cdot 18y^2} = \frac{3\cdot 9\cdot y\cdot y\cdot y\cdot 7}{3\cdot 7\cdot 9\cdot 2\cdot y\cdot y} = \frac{y}{2}$

35. $\left(-\frac{1}{3}\right)^3 = \left(-\frac{1}{3}\right)\left(-\frac{1}{3}\right)\left(-\frac{1}{3}\right) = -\frac{1\cdot 1\cdot 1}{3\cdot 3\cdot 3} = -\frac{1}{27}$

36. $\left(-\frac{5}{12}\right)^2 = \left(-\frac{5}{12}\right)\left(-\frac{5}{12}\right) = \frac{5\cdot 5}{12\cdot 12} = \frac{25}{144}$

37. $-\frac{3}{4}\div\frac{3}{8} = -\frac{3}{4}\cdot\frac{8}{3} = -\frac{3\cdot 8}{4\cdot 3} = -\frac{3\cdot 4\cdot 2}{1\cdot 4\cdot 3} = -\frac{2}{1} = -2$

38.
$$\frac{21a}{4}\div\frac{7a}{5} = \frac{21a}{4}\cdot\frac{5}{7a} = \frac{21a\cdot 5}{4\cdot 7a} = \frac{7\cdot 3\cdot a\cdot 5}{4\cdot 7\cdot a} = \frac{3\cdot 5}{4} = \frac{15}{4}$$

39. $-\frac{9}{2}\div -\frac{1}{3} = -\frac{9}{2}\cdot\left(-\frac{3}{1}\right) = \frac{9\cdot 3}{2\cdot 1} = \frac{27}{2}$

40. $-\frac{5}{3}\div 2y = -\frac{5}{3}\div\frac{2y}{1} = -\frac{5}{3}\cdot\frac{1}{2y} = -\frac{5\cdot 1}{3\cdot 2y} = -\frac{5}{6y}$

41.
$$x\div y = \frac{9}{7}\div\frac{3}{4} = \frac{9}{7}\cdot\frac{4}{3} = \frac{9\cdot 4}{7\cdot 3} = \frac{3\cdot 3\cdot 4}{7\cdot 3} = \frac{3\cdot 4}{7} = \frac{12}{7}$$

42. $ab = -7\cdot\frac{9}{10} = -\frac{7}{1}\cdot\frac{9}{10} = -\frac{7\cdot 9}{1\cdot 10} = -\frac{63}{10}$

Copyright © 2011 Pearson Education, Inc. Publishing as Prentice Hall.

43. area = length · width $= \frac{11}{6} \cdot \frac{7}{8} = \frac{11 \cdot 7}{6 \cdot 8} = \frac{77}{48}$

The area is $\frac{77}{48}$ square feet.

44. area = side · side $= \frac{2}{3} \cdot \frac{2}{3} = \frac{2 \cdot 2}{3 \cdot 3} = \frac{4}{9}$

The area is $\frac{4}{9}$ square meter.

45. $\frac{7}{11} + \frac{3}{11} = \frac{7+3}{11} = \frac{10}{11}$

46. $\frac{4}{9} + \frac{2}{9} = \frac{4+2}{9} = \frac{6}{9} = \frac{2 \cdot 3}{3 \cdot 3} = \frac{2}{3}$

47. $\frac{1}{12} - \frac{5}{12} = \frac{1-5}{12} = \frac{-4}{12} = -\frac{1 \cdot 4}{3 \cdot 4} = -\frac{1}{3}$

48. $\frac{11x}{15} + \frac{x}{15} = \frac{11x + x}{15} = \frac{12x}{15} = \frac{3 \cdot 4 \cdot x}{3 \cdot 5} = \frac{4x}{5}$

49. $\frac{4y}{21} - \frac{3}{21} = \frac{4y-3}{21}$

50. $\frac{4}{15} - \frac{3}{15} - \frac{2}{15} = \frac{4-3-2}{15} = \frac{-1}{15} = -\frac{1}{15}$

51. $3 = 3$
$x = x$
$\text{LCD} = 3 \cdot x = 3x$

52. $4 = 2 \cdot 2$
$8 = \boxed{2 \cdot 2 \cdot 2}$
$12 = 2 \cdot 2 \cdot \boxed{3}$
$\text{LCD} = 2 \cdot 2 \cdot 2 \cdot 3 = 24$

53. $\frac{2}{3} = \frac{2}{3} \cdot \frac{10}{10} = \frac{2 \cdot 10}{3 \cdot 10} = \frac{20}{30}$

54. $\frac{5}{8} = \frac{5}{8} \cdot \frac{7}{7} = \frac{5 \cdot 7}{8 \cdot 7} = \frac{35}{56}$

55. $\frac{7a}{6} = \frac{7a}{6} \cdot \frac{7}{7} = \frac{7a \cdot 7}{6 \cdot 7} = \frac{49a}{42}$

56. $\frac{9b}{4} = \frac{9b}{4} \cdot \frac{5}{5} = \frac{9b \cdot 5}{4 \cdot 5} = \frac{45b}{20}$

57. $\frac{4}{5x} = \frac{4}{5x} \cdot \frac{10}{10} = \frac{4 \cdot 10}{5x \cdot 10} = \frac{40}{50x}$

58. $\frac{5}{9y} = \frac{5}{9y} \cdot \frac{2}{2} = \frac{5 \cdot 2}{9y \cdot 2} = \frac{10}{18y}$

59. $\frac{3}{8} + \frac{2}{8} + \frac{1}{8} = \frac{3+2+1}{8} = \frac{6}{8} = \frac{2 \cdot 3}{2 \cdot 4} = \frac{3}{4}$

He did $\frac{3}{4}$ of his homework that evening.

60.
$$\frac{9}{16} + \frac{3}{16} + \frac{9}{16} + \frac{3}{16} = \frac{9+3+9+3}{16} = \frac{24}{16} = \frac{3 \cdot 8}{2 \cdot 8} = \frac{3}{2}$$

The perimeter is $\frac{3}{2}$ miles.

61. The LCD is 18.
$\frac{7}{18} + \frac{2}{9} = \frac{7}{18} + \frac{2 \cdot 2}{9 \cdot 2} = \frac{7}{18} + \frac{4}{18} = \frac{11}{18}$

62. The LCD is 26.
$\frac{4}{13} - \frac{1}{26} = \frac{4 \cdot 2}{13 \cdot 2} - \frac{1}{26} = \frac{8}{26} - \frac{1}{26} = \frac{7}{26}$

63. The LCD is 12.
$-\frac{1}{3} + \frac{1}{4} = -\frac{1 \cdot 4}{3 \cdot 4} + \frac{1 \cdot 3}{4 \cdot 3} = -\frac{4}{12} + \frac{3}{12} = -\frac{1}{12}$

64. The LCD is 12.
$-\frac{2}{3} + \frac{1}{4} = -\frac{2 \cdot 4}{3 \cdot 4} + \frac{1 \cdot 3}{4 \cdot 3} = -\frac{8}{12} + \frac{3}{12} = -\frac{5}{12}$

65. The LCD is 55.
$\frac{5x}{11} + \frac{2}{55} = \frac{5x \cdot 5}{11 \cdot 5} + \frac{2}{55} = \frac{25x}{55} + \frac{2}{55} = \frac{25x+2}{55}$

66. The LCD is 15.
$\frac{4}{15} + \frac{b}{5} = \frac{4}{15} + \frac{b \cdot 3}{5 \cdot 3} = \frac{4}{15} + \frac{3b}{15} = \frac{4+3b}{15}$

67. The LCD is 36.
$\frac{5y}{12} - \frac{2y}{9} = \frac{5y \cdot 3}{12 \cdot 3} - \frac{2y \cdot 4}{9 \cdot 4} = \frac{15y}{36} - \frac{8y}{36} = \frac{7y}{36}$

Copyright © 2011 Pearson Education, Inc. Publishing as Prentice Hall.

68. The LCD is 18.
$$\frac{7x}{18}+\frac{2x}{9}=\frac{7x}{18}+\frac{2x\cdot 2}{9\cdot 2}=\frac{7x}{18}+\frac{4x}{18}=\frac{11x}{18}$$

69. The LCD is 9*y*.
$$\frac{4}{9}+\frac{5}{y}=\frac{4\cdot y}{9\cdot y}+\frac{5\cdot 9}{y\cdot 9}=\frac{4y}{9y}+\frac{45}{9y}=\frac{4y+45}{9y}$$

70. The LCD is 14.
$$-\frac{9}{14}-\frac{3}{7}=-\frac{9}{14}-\frac{3\cdot 2}{7\cdot 2}=-\frac{9}{14}-\frac{6}{14}=-\frac{15}{14}$$

71. The LCD is 150.
$$\frac{4}{25}+\frac{23}{75}+\frac{7}{50}=\frac{4\cdot 6}{25\cdot 6}+\frac{23\cdot 2}{75\cdot 2}+\frac{7\cdot 3}{50\cdot 3}=\frac{24}{150}+\frac{46}{150}+\frac{21}{150}=\frac{91}{150}$$

72. The LCD is 18.
$$\frac{2}{3}-\frac{2}{9}-\frac{1}{6}=\frac{2\cdot 6}{3\cdot 6}-\frac{2\cdot 2}{9\cdot 2}-\frac{1\cdot 3}{6\cdot 3}=\frac{12}{18}-\frac{4}{18}-\frac{3}{18}=\frac{5}{18}$$

73. The LCD is 18.
$$\frac{2}{9}+\frac{5}{6}+\frac{2}{9}+\frac{5}{6}=\frac{2\cdot 2}{9\cdot 2}+\frac{5\cdot 3}{6\cdot 3}+\frac{2\cdot 2}{9\cdot 2}+\frac{5\cdot 3}{6\cdot 3}=\frac{4}{18}+\frac{15}{18}+\frac{4}{18}+\frac{15}{18}=\frac{38}{18}=\frac{2\cdot 19}{2\cdot 9}=\frac{19}{9}$$

The perimeter is $\frac{19}{9}$ meters.

74. The LCD is 10.
$$\frac{1}{5}+\frac{3}{5}+\frac{7}{10}=\frac{1\cdot 2}{5\cdot 2}+\frac{3\cdot 2}{5\cdot 2}+\frac{7}{10}=\frac{2}{10}+\frac{6}{10}+\frac{7}{10}=\frac{15}{10}=\frac{3\cdot 5}{2\cdot 5}=\frac{3}{2}$$

The perimeter is $\frac{3}{2}$ feet.

75. The LCD is 50.
$$\frac{9}{25}+\frac{3}{50}=\frac{9\cdot 2}{25\cdot 2}+\frac{3}{50}=\frac{18}{50}+\frac{3}{50}=\frac{21}{50}$$

$\frac{21}{50}$ of the donors have type A blood.

76. The LCD is 12.
$$\frac{2}{3}-\frac{5}{12}=\frac{2\cdot 4}{3\cdot 4}-\frac{5}{12}=\frac{8}{12}-\frac{5}{12}=\frac{3}{12}=\frac{1\cdot 3}{4\cdot 3}=\frac{1}{4}$$

The difference in length is $\frac{1}{4}$ yard.

77. $$\frac{\frac{2x}{5}}{\frac{7}{10}}=\frac{2x}{5}\div\frac{7}{10}=\frac{2x}{5}\cdot\frac{10}{7}=\frac{2x\cdot 10}{5\cdot 7}=\frac{2\cdot x\cdot 5\cdot 2}{5\cdot 7}=\frac{4x}{7}$$

78. $$\frac{\frac{3y}{7}}{\frac{11}{7}}=\frac{3y}{7}\div\frac{11}{7}=\frac{3y}{7}\cdot\frac{7}{11}=\frac{3y\cdot 7}{7\cdot 11}=\frac{3y}{11}$$

Copyright © 2011 Pearson Education, Inc. Publishing as Prentice Hall.

79. $\dfrac{\frac{2}{5}-\frac{1}{2}}{\frac{3}{4}-\frac{7}{10}}=\dfrac{20\left(\frac{2}{5}-\frac{1}{2}\right)}{20\left(\frac{3}{4}-\frac{7}{10}\right)}$
$=\dfrac{20\cdot\frac{2}{5}-20\cdot\frac{1}{2}}{20\cdot\frac{3}{4}-20\cdot\frac{7}{10}}$
$=\dfrac{8-10}{15-14}$
$=\dfrac{-2}{1}$
$=-2$

80. $\dfrac{\frac{5}{6}-\frac{1}{4}}{\frac{-1}{12y}}=\dfrac{12y\left(\frac{5}{6}-\frac{1}{4}\right)}{12y\left(\frac{-1}{12y}\right)}$
$=\dfrac{12y\cdot\frac{5}{6}-12y\cdot\frac{1}{4}}{-1}$
$=\dfrac{10y-3y}{-1}$
$=\dfrac{7y}{-1}$
$=-7y$

81. $\dfrac{x}{y+z}=\dfrac{\frac{1}{2}}{-\frac{2}{3}+\frac{4}{5}}$
$=\dfrac{30\cdot\frac{1}{2}}{30\left(-\frac{2}{3}+\frac{4}{5}\right)}$
$=\dfrac{15}{30\left(-\frac{2}{3}\right)+30\cdot\frac{4}{5}}$
$=\dfrac{15}{-20+24}$
$=\dfrac{15}{4}$

82. $\dfrac{x+y}{z}=\dfrac{\frac{1}{2}+\left(-\frac{2}{3}\right)}{\frac{4}{5}}$
$=\dfrac{30\left(\frac{1}{2}-\frac{2}{3}\right)}{30\cdot\frac{4}{5}}$
$=\dfrac{30\cdot\frac{1}{2}-30\cdot\frac{2}{3}}{6\cdot4}$
$=\dfrac{15-20}{24}$
$=-\dfrac{5}{24}$

83. $\dfrac{5}{13}\div\dfrac{1}{2}\cdot\dfrac{4}{5}=\dfrac{5}{13}\cdot\dfrac{2}{1}\cdot\dfrac{4}{5}=\dfrac{5\cdot2\cdot4}{13\cdot1\cdot5}=\dfrac{8}{13}$

84. $\dfrac{2}{27}-\left(\dfrac{1}{3}\right)^2=\dfrac{2}{27}-\dfrac{1}{9}$
$=\dfrac{2}{27}-\dfrac{1\cdot3}{9\cdot3}$
$=\dfrac{2}{27}-\dfrac{3}{27}$
$=-\dfrac{1}{27}$

85. $\dfrac{9}{10}\cdot\dfrac{1}{3}-\dfrac{2}{5}\cdot\dfrac{1}{11}=\dfrac{9\cdot1}{10\cdot3}-\dfrac{2\cdot1}{5\cdot11}$
$=\dfrac{3}{10}-\dfrac{2}{55}$
$=\dfrac{3\cdot11}{10\cdot11}-\dfrac{2\cdot2}{55\cdot2}$
$=\dfrac{33}{110}-\dfrac{4}{110}$
$=\dfrac{29}{110}$

86. $-\dfrac{2}{7}\cdot\left(\dfrac{1}{5}+\dfrac{3}{10}\right)=-\dfrac{2}{7}\cdot\left(\dfrac{1\cdot2}{5\cdot2}+\dfrac{3}{10}\right)$
$=-\dfrac{2}{7}\cdot\left(\dfrac{2}{10}+\dfrac{3}{10}\right)$
$=-\dfrac{2}{7}\cdot\dfrac{5}{10}$
$=-\dfrac{2\cdot5}{7\cdot2\cdot5}$
$=-\dfrac{1}{7}$

87.
$$\begin{array}{rr} 7\frac{3}{8} \\ 9\frac{5}{6} \\ +\,3\frac{1}{12} \\ \hline \end{array} \qquad \begin{array}{r} 7\frac{9}{24} \\ 9\frac{20}{24} \\ +\,3\frac{2}{24} \\ \hline 19\frac{31}{24} \end{array}$$
$19\dfrac{31}{24}=19+1\dfrac{7}{24}=20\dfrac{7}{24}$

Copyright © 2011 Pearson Education, Inc. Publishing as Prentice Hall.

88. $8\frac{1}{5}$ rounds to 8.

$5\frac{3}{11}$ rounds to 5.

An estimate is 8 − 5 = 3.

$$\begin{array}{r} 8\frac{1}{5} \\ -5\frac{3}{11} \\ \hline \end{array} \quad \begin{array}{r} 8\frac{11}{55} \\ -5\frac{15}{55} \\ \hline \end{array} \quad \begin{array}{r} 7\frac{66}{55} \\ -5\frac{15}{55} \\ \hline 2\frac{51}{55} \end{array}$$

89. $1\frac{5}{8}$ rounds to 2.

$3\frac{1}{5}$ rounds to 3.

An estimate is 2 · 3 = 6.

$$1\frac{5}{8}\cdot 3\frac{1}{5}=\frac{13}{8}\cdot\frac{16}{5}=\frac{13\cdot 16}{8\cdot 5}=\frac{13\cdot 8\cdot 2}{8\cdot 5}=\frac{26}{5}=5\frac{1}{5}$$

90. $6\frac{3}{4}\div 1\frac{2}{7}=\frac{27}{4}\div\frac{9}{7}=\frac{27}{4}\cdot\frac{7}{9}=\frac{9\cdot 3\cdot 7}{4\cdot 9}=\frac{21}{4}=5\frac{1}{4}$

91. $341\div 15\frac{1}{2}=\frac{341}{1}\div\frac{31}{2}=\frac{341}{1}\cdot\frac{2}{31}=\frac{11\cdot 31\cdot 2}{1\cdot 31}=22$

We would expect 22 miles on one gallon.

92. $7\frac{1}{3}\cdot 5=\frac{22}{3}\cdot\frac{5}{1}=\frac{22\cdot 5}{3\cdot 1}=\frac{110}{3}=36\frac{2}{3}$

There are $\frac{110}{3}$ or $36\frac{2}{3}$ grams of fat in a 5-ounce hamburger patty.

93. $18\frac{7}{8}-10\frac{3}{8}=8\frac{4}{8}=8\frac{1}{2}$

$8\frac{1}{2}\div 2=\frac{17}{2}\div\frac{2}{1}=\frac{17}{2}\cdot\frac{1}{2}=\frac{17}{4}=4\frac{1}{4}$

Each measurement is $4\frac{1}{4}$ inches.

94. $1\frac{3}{10}-\frac{3}{5}=\frac{13}{10}-\frac{3}{5}=\frac{13}{10}-\frac{3\cdot 2}{5\cdot 2}=\frac{13}{10}-\frac{6}{10}=\frac{7}{10}$

The unknown measurement is $\frac{7}{10}$ yard.

95. $-12\frac{1}{7}+\left(-15\frac{3}{14}\right)=-12\frac{2}{14}+\left(-15\frac{3}{14}\right)=-27\frac{5}{14}$

96. $23\frac{7}{8}-24\frac{7}{10}=-\left(24\frac{7}{10}-23\frac{7}{8}\right)$

$$\begin{array}{r} 24\frac{7}{10} \\ -23\frac{7}{8} \\ \hline \end{array} \quad \begin{array}{r} 24\frac{28}{40} \\ -23\frac{35}{40} \\ \hline \end{array} \quad \begin{array}{r} 23\frac{68}{40} \\ -23\frac{35}{40} \\ \hline \frac{33}{40} \end{array}$$

$23\frac{7}{8}-24\frac{7}{10}=-\frac{33}{40}$

97.

$$\begin{aligned} -3\frac{1}{5}\div\left(-2\frac{7}{10}\right) &= -\frac{16}{5}\div\left(-\frac{27}{10}\right) \\ &= -\frac{16}{5}\cdot\left(-\frac{10}{27}\right) \\ &= \frac{16\cdot 5\cdot 2}{5\cdot 27} \\ &= \frac{32}{27} \\ &= 1\frac{5}{27} \end{aligned}$$

98. $-2\frac{1}{4}\cdot 1\frac{3}{4}=-\frac{9}{4}\cdot\frac{7}{4}=-\frac{63}{16}=-3\frac{15}{16}$

99.

$$\begin{aligned} a-\frac{2}{3}&=\frac{1}{6} \\ a-\frac{2}{3}+\frac{2}{3}&=\frac{1}{6}+\frac{2}{3} \\ a&=\frac{1}{6}+\frac{2\cdot 2}{3\cdot 2} \\ a&=\frac{1}{6}+\frac{4}{6} \\ a&=\frac{5}{6} \end{aligned}$$

100.

$$\begin{aligned} 9x+\frac{1}{5}-8x&=-\frac{7}{10} \\ x+\frac{1}{5}&=-\frac{7}{10} \\ x+\frac{1}{5}-\frac{1}{5}&=-\frac{7}{10}-\frac{1}{5} \\ x&=-\frac{7}{10}-\frac{2}{10} \\ x&=-\frac{9}{10} \end{aligned}$$

Copyright © 2011 Pearson Education, Inc. Publishing as Prentice Hall.

101.
$$-\frac{3}{5}x=6$$
$$-\frac{5}{3}\cdot -\frac{3}{5}x=-\frac{5}{3}\cdot 6$$
$$x=-10$$

102.
$$\frac{2}{9}y=-\frac{4}{3}$$
$$\frac{9}{2}\cdot\frac{2}{9}y=\frac{9}{2}\cdot -\frac{4}{3}$$
$$y=-\frac{9\cdot 4}{2\cdot 3}$$
$$y=-\frac{3\cdot 3\cdot 2\cdot 2}{2\cdot 3}$$
$$y=-\frac{6}{1}$$
$$y=-6$$

103.
$$\frac{x}{7}-3=-\frac{6}{7}$$
$$7\left(\frac{x}{7}-3\right)=7\left(-\frac{6}{7}\right)$$
$$7\cdot\frac{x}{7}-7\cdot 3=-6$$
$$x-21=-6$$
$$x-21+21=-6+21$$
$$x=15$$

104.
$$\frac{y}{5}+2=\frac{11}{5}$$
$$5\left(\frac{y}{5}+2\right)=5\left(\frac{11}{5}\right)$$
$$5\cdot\frac{y}{5}+5\cdot 2=11$$
$$y+10=11$$
$$y+10-10=11-10$$
$$y=1$$

105.
$$\frac{1}{6}+\frac{x}{4}=\frac{17}{12}$$
$$12\left(\frac{1}{6}+\frac{x}{4}\right)=12\cdot\frac{17}{12}$$
$$12\cdot\frac{1}{6}+12\cdot\frac{x}{4}=17$$
$$2+3x=17$$
$$2+3x-2=17-2$$
$$3x=15$$
$$\frac{3x}{3}=\frac{15}{3}$$
$$x=5$$

106.
$$\frac{x}{5}-\frac{5}{4}=\frac{x}{2}-\frac{1}{20}$$
$$20\left(\frac{x}{5}-\frac{5}{4}\right)=20\left(\frac{x}{2}-\frac{1}{20}\right)$$
$$20\cdot\frac{x}{5}-20\cdot\frac{5}{4}=20\cdot\frac{x}{2}-20\cdot\frac{1}{20}$$
$$4x-25=10x-1$$
$$4x-25-10x=10x-1-10x$$
$$-25-6x=-1$$
$$-25-6x+25=-1+25$$
$$-6x=24$$
$$\frac{-6x}{-6}=\frac{24}{-6}$$
$$x=-4$$

107. $\frac{6}{15}\cdot\frac{5}{8}=\frac{6\cdot 5}{15\cdot 8}=\frac{3\cdot 2\cdot 5}{5\cdot 3\cdot 2\cdot 4}=\frac{1}{4}$

108.
$$\frac{5x^2}{y}\div\frac{10x^3}{y^3}=\frac{5x^2}{y}\cdot\frac{y^3}{10x^3}$$
$$=\frac{5x^2\cdot y^3}{y\cdot 10x^3}$$
$$=\frac{5\cdot x\cdot x\cdot y\cdot y\cdot y}{y\cdot 5\cdot 2\cdot x\cdot x\cdot x}$$
$$=\frac{y^2}{2x}$$

109. $\frac{3}{10}-\frac{1}{10}=\frac{3-1}{10}=\frac{2}{10}=\frac{1\cdot 2}{5\cdot 2}=\frac{1}{5}$

110. $\frac{7}{8x}\cdot -\frac{2}{3}=-\frac{7\cdot 2}{8x\cdot 3}=-\frac{7\cdot 2}{2\cdot 4\cdot x\cdot 3}=-\frac{7}{12x}$

111.
$$\frac{2x}{3}+\frac{x}{4}=\frac{2x}{3}\cdot\frac{4}{4}+\frac{x}{4}\cdot\frac{3}{3}$$
$$=\frac{8x}{12}+\frac{3x}{12}$$
$$=\frac{8x+3x}{12}$$
$$=\frac{11x}{12}$$

112. $-\frac{5}{11}+\frac{2}{55}=-\frac{5}{11}\cdot\frac{5}{5}+\frac{2}{55}=-\frac{25}{55}+\frac{2}{55}=-\frac{23}{55}$

Copyright © 2011 Pearson Education, Inc. Publishing as Prentice Hall.

113. $-1\frac{3}{5} \div \frac{1}{4} = -\frac{8}{5} \div \frac{1}{4}$
$= -\frac{8}{5} \cdot \frac{4}{1}$
$= -\frac{8 \cdot 4}{5 \cdot 1}$
$= -\frac{32}{5} \text{ or } -6\frac{2}{5}$

114. $2\frac{7}{8}$ rounds to 3.

$9\frac{1}{2}$ rounds to 10.

An estimate is 3 + 10 = 13.

$$\begin{array}{r} 2\frac{7}{8} \\ +\,9\frac{1}{2} \\ \hline \end{array} \qquad \begin{array}{r} 2\frac{7}{8} \\ +\,9\frac{4}{8} \\ \hline 11\frac{11}{8} \end{array} = 11 + 1\frac{3}{8} = 12\frac{3}{8}$$

115. $12\frac{1}{7}$ rounds to 12.

$9\frac{3}{5}$ rounds to 10.

An estimate is 12 − 10 = 2.

$$\begin{array}{r} 12\frac{1}{7} \\ -\;\;9\frac{3}{5} \\ \hline \end{array} \qquad \begin{array}{r} 12\frac{5}{35} \\ -\;\;9\frac{21}{35} \\ \hline \end{array} \qquad \begin{array}{r} 11\frac{40}{35} \\ -\;\;9\frac{21}{35} \\ \hline 2\frac{19}{35} \end{array}$$

116. $\frac{2+\frac{3}{4}}{1-\frac{1}{8}} = \frac{8\left(2+\frac{3}{4}\right)}{8\left(1-\frac{1}{8}\right)}$
$= \frac{8 \cdot 2 + 8 \cdot \frac{3}{4}}{8 \cdot 1 - 8 \cdot \frac{1}{8}}$
$= \frac{16+6}{8-1}$
$= \frac{22}{7} \text{ or } 3\frac{1}{7}$

117. $-\frac{3}{8} \cdot \left(\frac{2}{3} - \frac{4}{9}\right) = -\frac{3}{8}\left(\frac{2 \cdot 3}{3 \cdot 3} - \frac{4}{9}\right)$
$= -\frac{3}{8}\left(\frac{6}{9} - \frac{4}{9}\right)$
$= -\frac{3}{8}\left(\frac{2}{9}\right)$
$= -\frac{3 \cdot 2}{8 \cdot 9}$
$= -\frac{3 \cdot 2}{2 \cdot 4 \cdot 3 \cdot 3}$
$= -\frac{1}{12}$

118. $11x - \frac{2}{7} - 10x = -\frac{13}{14}$
$x - \frac{2}{7} = -\frac{13}{14}$
$x - \frac{2}{7} + \frac{2}{7} = -\frac{13}{14} + \frac{2}{7}$
$x = -\frac{13}{14} + \frac{2 \cdot 2}{7 \cdot 2}$
$x = -\frac{13}{14} + \frac{4}{14}$
$x = -\frac{9}{14}$

119. $-\frac{3}{5}x = \frac{4}{15}$
$-\frac{5}{3} \cdot -\frac{3}{5}x = -\frac{5}{3} \cdot \frac{4}{15}$
$x = -\frac{5 \cdot 4}{3 \cdot 15}$
$x = -\frac{5 \cdot 4}{3 \cdot 5 \cdot 3}$
$x = -\frac{4}{9}$

120. $\frac{x}{12} + \frac{5}{6} = -\frac{3}{4}$
$12\left(\frac{x}{12} + \frac{5}{6}\right) = 12\left(-\frac{3}{4}\right)$
$12 \cdot \frac{x}{12} + 12 \cdot \frac{5}{6} = -9$
$x + 10 = -9$
$x + 10 - 10 = -9 - 10$
$x = -19$

Copyright © 2011 Pearson Education, Inc. Publishing as Prentice Hall.

121. $50-5\frac{1}{2}=\frac{50}{1}-\frac{11}{2}$
$=\frac{50\cdot 2}{1\cdot 2}-\frac{11}{2}$
$=\frac{100}{2}-\frac{11}{2}$
$=\frac{89}{2}$
$=44\frac{1}{2}$

The length of the remaining piece is $44\frac{1}{2}$ yards.

122. $5\frac{1}{2}\cdot 7\frac{4}{11}=\frac{11}{2}\cdot\frac{81}{11}=\frac{11\cdot 81}{2\cdot 11}=\frac{81}{2}$ or $40\frac{1}{2}$

The area is $\frac{81}{2}$ or $40\frac{1}{2}$ square feet.

Chapter 4 Test

1. 7 of the 16 equal parts are shaded: $\frac{7}{16}$.

2. $7\frac{2}{3}=\frac{3\cdot 7+2}{3}=\frac{21+2}{3}=\frac{23}{3}$

3. $4\overline{)75}$ quotient 18
4
35
32
3

$\frac{75}{4}=18\frac{3}{4}$

4. $\frac{24}{210}=\frac{6\cdot 4}{6\cdot 7\cdot 5}=\frac{4}{35}$

5. $-\frac{42x}{70}=-\frac{3\cdot 14\cdot x}{5\cdot 14}=-\frac{3x}{5}$

6. Check the cross-products.
$5\cdot 11=55$
$7\cdot 8=56$
Since $55\neq 56$, the fractions are not equivalent.

7. Check the cross-products.
$6\cdot 63=378$
$27\cdot 14=378$
Since $378=378$, the fractions are equivalent.

8. $84=2\cdot 42=2\cdot 2\cdot 21=2\cdot 2\cdot 3\cdot 7=2^2\cdot 3\cdot 7$

9. $495=3\cdot 165=3\cdot 3\cdot 55=3\cdot 3\cdot 5\cdot 11=3^2\cdot 5\cdot 11$

10. $\frac{4}{4}\div\frac{3}{4}=\frac{4}{4}\cdot\frac{4}{3}=\frac{4\cdot 4}{4\cdot 3}=\frac{4}{3}$

11. $-\frac{4}{3}\cdot\frac{4}{4}=-\frac{4\cdot 4}{3\cdot 4}=-\frac{4}{3}$

12. $\frac{7x}{9}+\frac{x}{9}=\frac{7x+x}{9}=\frac{8x}{9}$

13. The LCD is $7x$.
$\frac{1}{7}-\frac{3}{x}=\frac{1}{7}\cdot\frac{x}{x}-\frac{3}{x}\cdot\frac{7}{7}=\frac{x}{7x}-\frac{21}{7x}=\frac{x-21}{7x}$

14. $\frac{xy^3}{z}\cdot\frac{z}{xy}=\frac{xy^3\cdot z}{z\cdot xy}=\frac{x\cdot y\cdot y\cdot y\cdot z}{x\cdot y\cdot z}=\frac{y\cdot y}{1}=y^2$

15. $-\frac{2}{3}\cdot -\frac{8}{15}=\frac{2\cdot 8}{3\cdot 15}=\frac{16}{45}$

16. $\frac{9a}{10}+\frac{2}{5}=\frac{9a}{10}+\frac{2\cdot 2}{5\cdot 2}=\frac{9a}{10}+\frac{4}{10}=\frac{9a+4}{10}$

17. $-\frac{8}{15y}-\frac{2}{15y}=\frac{-8-2}{15y}=\frac{-10}{15y}=-\frac{2\cdot 5}{3\cdot 5\cdot y}=-\frac{2}{3y}$

18. $\frac{3a}{8}\cdot\frac{16}{6a^3}=\frac{3a\cdot 16}{8\cdot 6a^3}=\frac{3\cdot a\cdot 8\cdot 2}{8\cdot 2\cdot 3\cdot a\cdot a\cdot a}=\frac{1}{a\cdot a}=\frac{1}{a^2}$

19. $\frac{11}{12}-\frac{3}{8}+\frac{5}{24}=\frac{11\cdot 2}{12\cdot 2}-\frac{3\cdot 3}{8\cdot 3}+\frac{5}{24}$
$=\frac{22}{24}-\frac{9}{24}+\frac{5}{24}$
$=\frac{22-9+5}{24}$
$=\frac{18}{24}$
$=\frac{3\cdot 6}{4\cdot 6}$
$=\frac{3}{4}$

Copyright © 2011 Pearson Education, Inc. Publishing as Prentice Hall.

20.
$$\begin{array}{r} 3\frac{7}{8} \\ 7\frac{2}{5} \\ +\ 2\frac{3}{4} \\ \hline \end{array} \qquad \begin{array}{r} 3\frac{35}{40} \\ 7\frac{16}{40} \\ +\ 2\frac{30}{40} \\ \hline 12\frac{81}{40} \end{array} = 12+2\frac{1}{40} = 14\frac{1}{40}$$

21.
$$\begin{array}{r} 19 \\ -\ 2\frac{3}{11} \\ \hline \end{array} \qquad \begin{array}{r} 18\frac{11}{11} \\ -\ 2\frac{3}{11} \\ \hline 16\frac{8}{11} \end{array}$$

22.
$$-\frac{16}{3} \div -\frac{3}{12} = -\frac{16}{3} \cdot -\frac{12}{3} = \frac{16 \cdot 12}{3 \cdot 3} = \frac{16 \cdot 3 \cdot 4}{3 \cdot 3} = \frac{64}{3} \text{ or } 21\frac{1}{3}$$

23.
$$3\frac{1}{3} \cdot 6\frac{3}{4} = \frac{10}{3} \cdot \frac{27}{4} = \frac{10 \cdot 27}{3 \cdot 4} = \frac{2 \cdot 5 \cdot 3 \cdot 9}{3 \cdot 2 \cdot 2} = \frac{5 \cdot 9}{2} = \frac{45}{2} \text{ or } 22\frac{1}{2}$$

24.
$$-\frac{2}{7} \cdot \left(6 - \frac{1}{6}\right) = -\frac{2}{7} \cdot \left(\frac{6}{1} - \frac{1}{6}\right) = -\frac{2}{7} \cdot \left(\frac{6 \cdot 6}{1 \cdot 6} - \frac{1}{6}\right) = -\frac{2}{7} \cdot \left(\frac{36}{6} - \frac{1}{6}\right) = -\frac{2}{7} \cdot \frac{35}{6} = -\frac{2 \cdot 35}{7 \cdot 6} = -\frac{2 \cdot 7 \cdot 5}{7 \cdot 2 \cdot 3} = -\frac{5}{3} \text{ or } -1\frac{2}{3}$$

25.
$$\frac{1}{2} \div \frac{2}{3} \cdot \frac{3}{4} = \frac{1}{2} \cdot \frac{3}{2} \cdot \frac{3}{4} = \frac{1 \cdot 3 \cdot 3}{2 \cdot 2 \cdot 4} = \frac{9}{16}$$

26.
$$\left(-\frac{3}{4}\right)^2 \div \left(\frac{2}{3} + \frac{5}{6}\right) = \left(-\frac{3}{4}\right)^2 \div \left(\frac{2}{3} \cdot \frac{2}{2} + \frac{5}{6}\right) = \left(-\frac{3}{4}\right)^2 \div \left(\frac{4}{6} + \frac{5}{6}\right) = \left(-\frac{3}{4}\right)^2 \div \frac{9}{6} = \frac{9}{16} \div \frac{9}{6} = \frac{9}{16} \cdot \frac{6}{9} = \frac{9 \cdot 2 \cdot 3}{2 \cdot 8 \cdot 9} = \frac{3}{8}$$

27.
$$\left(\frac{5}{6} + \frac{4}{3} + \frac{7}{12}\right) \div 3 = \left(\frac{5 \cdot 2}{6 \cdot 2} + \frac{4 \cdot 4}{3 \cdot 4} + \frac{7}{12}\right) \div 3 = \left(\frac{10}{12} + \frac{16}{12} + \frac{7}{12}\right) \div 3 = \frac{33}{12} \div \frac{3}{1} = \frac{33}{12} \cdot \frac{1}{3} = \frac{33 \cdot 1}{12 \cdot 3} = \frac{11 \cdot 3 \cdot 1}{12 \cdot 3} = \frac{11}{12}$$

28.
$$\frac{\frac{5x}{7}}{\frac{20x^2}{21}} = \frac{5x}{7} \div \frac{20x^2}{21} = \frac{5x}{7} \cdot \frac{21}{20x^2} = \frac{5 \cdot x \cdot 3 \cdot 7}{7 \cdot 4 \cdot 5 \cdot x \cdot x} = \frac{3}{4x}$$

29.
$$\frac{5+\frac{3}{7}}{2-\frac{1}{2}} = \frac{14\left(5+\frac{3}{7}\right)}{14\left(2-\frac{1}{2}\right)} = \frac{14 \cdot 5 + 14 \cdot \frac{3}{7}}{14 \cdot 2 - 14 \cdot \frac{1}{2}} = \frac{70+6}{28-7} = \frac{76}{21}$$

Copyright © 2011 Pearson Education, Inc. Publishing as Prentice Hall.

30.
$$-\frac{3}{8}x = \frac{3}{4}$$
$$-\frac{8}{3}\cdot -\frac{3}{8}x = -\frac{8}{3}\cdot\frac{3}{4}$$
$$x = -\frac{8\cdot 3}{3\cdot 4}$$
$$x = -\frac{4\cdot 2\cdot 3}{3\cdot 4}$$
$$x = -2$$

31.
$$\frac{x}{5}+x = -\frac{24}{5}$$
$$5\left(\frac{x}{5}+x\right) = 5\left(-\frac{24}{5}\right)$$
$$5\cdot\frac{x}{5}+5\cdot x = -24$$
$$x+5x = -24$$
$$6x = -24$$
$$\frac{6x}{6} = \frac{-24}{6}$$
$$x = -4$$

32.
$$\frac{2}{3}+\frac{x}{4} = \frac{5}{12}+\frac{x}{2}$$
$$12\left(\frac{2}{3}+\frac{x}{4}\right) = 12\left(\frac{5}{12}+\frac{x}{2}\right)$$
$$12\cdot\frac{2}{3}+12\cdot\frac{x}{4} = 12\cdot\frac{5}{12}+12\cdot\frac{x}{2}$$
$$8+3x = 5+6x$$
$$8+3x-6x = 5+6x-6x$$
$$8-3x = 5$$
$$8-8-3x = 5-8$$
$$-3x = -3$$
$$\frac{-3x}{-3} = \frac{-3}{-3}$$
$$x = 1$$

33. $-5x = -5\left(-\frac{1}{2}\right) = \frac{5}{1}\cdot\frac{1}{2} = \frac{5\cdot 1}{1\cdot 2} = \frac{5}{2}$

34.
$$x \div y = \frac{1}{2}\div 3\frac{7}{8}$$
$$= \frac{1}{2}\div\frac{31}{8}$$
$$= \frac{1}{2}\cdot\frac{8}{31}$$
$$= \frac{1\cdot 8}{2\cdot 31}$$
$$= \frac{1\cdot 2\cdot 4}{2\cdot 31}$$
$$= \frac{4}{31}$$

35.
$$\begin{array}{r} 6\frac{1}{2} \\ -\,2\frac{3}{4} \\ \hline \end{array} \qquad \begin{array}{r} 6\frac{2}{4} \\ -\,2\frac{3}{4} \\ \hline \end{array} \qquad \begin{array}{r} 5\frac{6}{4} \\ -\,2\frac{3}{4} \\ \hline 3\frac{3}{4} \end{array}$$

The remaining piece is $3\frac{3}{4}$ feet.

36. Housing: $\frac{8}{25}$

Food: $\frac{7}{50}$

$$\frac{8}{25}+\frac{7}{50} = \frac{8\cdot 2}{25\cdot 2}+\frac{7}{50} = \frac{16}{50}+\frac{7}{50} = \frac{16+7}{50} = \frac{23}{50}$$

$\frac{23}{50}$ of spending goes for housing and food combined.

37. Education: $\frac{1}{50}$

Transportation: $\frac{1}{5}$

Clothing: $\frac{1}{25}$

$$\frac{1}{50}+\frac{1}{5}+\frac{1}{25} = \frac{1}{50}+\frac{1}{5}\cdot\frac{10}{10}+\frac{1}{25}\cdot\frac{2}{2}$$
$$= \frac{1}{50}+\frac{10}{50}+\frac{2}{50}$$
$$= \frac{13}{50}$$

$\frac{13}{50}$ of spending goes for education, transportation, and clothing.

Copyright © 2011 Pearson Education, Inc. Publishing as Prentice Hall.

38. $\frac{3}{50}\cdot 47{,}000=\frac{3}{50}\cdot\frac{47{,}000}{1}=\frac{3\cdot 50\cdot 940}{50\cdot 1}=2820$
Expect to spend \$2820 on health care.

39. $\text{perimeter}=1+\frac{2}{3}+1+\frac{2}{3}$
$=\frac{3}{3}+\frac{2}{3}+\frac{3}{3}+\frac{2}{3}$
$=\frac{10}{3}$
$=3\frac{1}{3}$

$\text{area}=\text{length}\cdot\text{width}=1\cdot\frac{2}{3}=\frac{2}{3}$

The perimeter is $3\frac{1}{3}$ feet and the area is $\frac{2}{3}$ square foot.

40. $258\div 10\frac{3}{4}=\frac{258}{1}\div\frac{43}{4}=\frac{258}{1}\cdot\frac{4}{43}=\frac{43\cdot 6\cdot 4}{1\cdot 43}=24$
Expect to travel 24 miles on 1 gallon of gas.

Cumulative Review Chapters 1–4

1. 546 in words is five hundred forty-six.

2. 115 in words is one hundred fifteen.

3. 27,034 in words is twenty-seven thousand, thirty-four.

4. 6573 in words is six thousand, five hundred seventy-three.

5. $\begin{array}{r} 46 \\ +\ 713 \\ \hline 759 \end{array}$

6. $\begin{array}{r} 587 \\ +\ 44 \\ \hline 631 \end{array}$

7. $\begin{array}{r} 543 \\ -\ 29 \\ \hline 514 \end{array}$

Check: $\begin{array}{r} 514 \\ +\ 29 \\ \hline 543 \end{array}$

8. $\begin{array}{r} 995 \\ -\ 62 \\ \hline 933 \end{array}$

Check: $\begin{array}{r} 933 \\ +\ 62 \\ \hline 995 \end{array}$

9. To round 278,362 to the nearest thousand, observe that the digit in the hundreds place is 3. Since this digit is less than 5, we do not add 1 to the digit in the thousands place. 278,362 rounded to the nearest thousand is 278,000.

10. To round 1436 to the nearest ten, observe that the digit in the ones place is 6. Since this digit is at least 5, we add 1 to the digit in the tens place. 1436 rounded to the nearest ten is 1440.

11. $\begin{array}{r} 4800 \\ \times\ \ 12 \\ \hline 9\,600 \\ 48{,}000 \\ \hline 57{,}600 \end{array}$

Therefore, 12 DVDs can hold 57,600 megabytes of information.

12. $\begin{array}{r} 435 \\ \times\ 3 \\ \hline 1305 \end{array}$

He travels 1305 miles in 3 days.

13. $\begin{array}{r} 7089 \\ 8\overline{)56{,}717} \\ \underline{56} \\ 0\,7 \\ \underline{0} \\ 71 \\ \underline{64} \\ 77 \\ \underline{72} \\ 5 \end{array}$

$56{,}717\div 8=7089\text{ R }5$
Check: $7089\times 8+5=56{,}712+5=56{,}717$

Copyright © 2011 Pearson Education, Inc. Publishing as Prentice Hall.

14.
$$\begin{array}{r} 379 \\ 12\overline{)4558} \\ \underline{36} \\ 95 \\ \underline{84} \\ 118 \\ \underline{108} \\ 10 \end{array}$$

$4558 \div 12 = 379 \text{ R } 10$

Check: $379 \times 12 + 10 = 4548 + 10 = 4558$

15. $7 \cdot 7 \cdot 7 = 7^3$

16. $7 \cdot 7 = 7^2$

17. $3 \cdot 3 \cdot 3 \cdot 3 \cdot 9 \cdot 9 \cdot 9 = 3^4 \cdot 9^3$

18. $9 \cdot 9 \cdot 9 \cdot 9 \cdot 5 \cdot 5 = 9^4 \cdot 5^2$

19. $2(x - y) = 2(6 - 3) = 2(3) = 6$

20.
$$\begin{aligned} 8a + 3(b-5) &= 8 \cdot 5 + 3(9-5) \\ &= 8 \cdot 5 + 3 \cdot 4 \\ &= 40 + 12 \\ &= 52 \end{aligned}$$

21. Let 0 represent the surface of the earth. Then 6824 feet *below* the surface is represented as −6824.

22. Let 0 represent a temperature of 0°F. Then 21°F *below* zero is represented as −21.

23. $-7 + 3 = -4$

24. $-3 + 8 = 5$

25.
$$\begin{aligned} 7-8-(-5)-1 &= 7-8+5-1 \\ &= 7+(-8)+5+(-1) \\ &= -1+5+(-1) \\ &= 4+(-1) \\ &= 3 \end{aligned}$$

26.
$$\begin{aligned} 6+(-8)-(-9)+3 &= 6+(-8)+9+3 \\ &= -2+9+3 \\ &= 7+3 \\ &= 10 \end{aligned}$$

27. $(-5)^2 = (-5)(-5) = 25$

28. $-2^4 = -2 \cdot 2 \cdot 2 \cdot 2 = -16$

29.
$$\begin{aligned} 3(4-7)+(-2)-5 &= 3(-3)+(-2)-5 \\ &= -9+(-2)+(-5) \\ &= -11+(-5) \\ &= -16 \end{aligned}$$

30. $(20-5^2)^2 = (20-25)^2 = (-5)^2 = 25$

31. $2y-6+4y+8 = (2y+4y)+(-6+8) = 6y+2$

32. $5x - 1 + x + 10 = (5x + x) + (-1 + 10) = 6x + 9$

33.
$$\begin{aligned} 5x+2-4x &= 7-19 \\ x+2 &= -12 \\ x+2-2 &= -12-2 \\ x &= -14 \end{aligned}$$

34.
$$\begin{aligned} 9y+1-8y &= 3-20 \\ y+1 &= -17 \\ y+1-1 &= -17-1 \\ y &= -18 \end{aligned}$$

35.
$$\begin{aligned} 17-7x+3 &= -3x+21-3x \\ 20-7x &= 21-6x \\ 20-7x+7x &= 21-6x+7x \\ 20 &= 21+x \\ 20-21 &= 21-21+x \\ -1 &= x \text{ or } x = -1 \end{aligned}$$

36.
$$\begin{aligned} 9x-2 &= 7x-24 \\ 9x-7x-2 &= 7x-7x-24 \\ 2x-2 &= -24 \\ 2x-2+2 &= -24+2 \\ 2x &= -22 \\ \frac{2x}{2} &= \frac{-22}{2} \\ x &= -11 \end{aligned}$$

37. Two of five equal parts are shaded: $\frac{2}{5}$

38.
$$\begin{array}{l} 156 = 2 \cdot 78 \\ \quad\quad\; \downarrow \; \downarrow \searrow \\ \quad\quad\; 2 \cdot 2 \cdot 39 \\ \quad\quad\; \downarrow \; \downarrow \; \downarrow \searrow \\ \quad\quad\; 2 \cdot 2 \cdot 3 \cdot 13 \end{array}$$

$156 = 2^2 \cdot 3 \cdot 13$

39. a. $4\frac{2}{9} = \frac{9 \cdot 4 + 2}{9} = \frac{36+2}{9} = \frac{38}{9}$

Copyright © 2011 Pearson Education, Inc. Publishing as Prentice Hall.

b. $1\frac{8}{11}=\frac{11\cdot 1+8}{11}=\frac{11+8}{11}=\frac{19}{11}$

40. $\begin{array}{r} 7 \\ 5\overline{)\ 39} \\ \underline{-35} \\ 4 \end{array}$

$\frac{39}{5}=7\frac{4}{5}$

41. $\frac{42x}{66}=\frac{6\cdot 7\cdot x}{6\cdot 11}=\frac{7x}{11}$

42. $\frac{70}{105y}=\frac{35\cdot 2}{35\cdot 3\cdot y}=\frac{2}{3y}$

43. $3\frac{1}{3}\cdot\frac{7}{8}=\frac{10}{3}\cdot\frac{7}{8}$
$=\frac{10\cdot 7}{3\cdot 8}$
$=\frac{2\cdot 5\cdot 7}{3\cdot 2\cdot 4}$
$=\frac{5\cdot 7}{3\cdot 4}$
$=\frac{35}{12}$ or $2\frac{11}{12}$

44. $\frac{2}{3}\cdot 4=\frac{2}{3}\cdot\frac{4}{1}=\frac{2\cdot 4}{3\cdot 1}=\frac{8}{3}$ or $2\frac{2}{3}$

45. $\frac{5}{16}\div\frac{3}{4}=\frac{5}{16}\cdot\frac{4}{3}=\frac{5\cdot 4}{16\cdot 3}=\frac{5\cdot 4}{4\cdot 4\cdot 3}=\frac{5}{4\cdot 3}=\frac{5}{12}$

46. $1\frac{1}{10}\div 5\frac{3}{5}=\frac{11}{10}\div\frac{28}{5}=\frac{11}{10}\cdot\frac{5}{28}=\frac{11\cdot 5}{5\cdot 2\cdot 28}=\frac{11}{56}$

Copyright © 2011 Pearson Education, Inc. Publishing as Prentice Hall.

Chapter 5

Section 5.1

Practice Problems

1. **a.** 0.06 in words is six hundredths.

   b. −200.073 in words is negative two hundred and seventy-three thousandths.

   c. 0.0829 in words in eight hundred twenty-nine ten-thousandths.

2. 87.31 in words is eighty-seven and thirty-one hundredths.

3. 52.1085 in words is fifty-two and one thousand eighty-five ten-thousandths.

4. The check should be paid to "CLECO," for the amount of "207.40," which is written in words as "Two hundred seven and $\frac{40}{100}$."

5. Five hundred and ninety-six hundredths is 500.96.

6. Thirty-nine and forty-two thousandths is 39.042.

7. $0.051 = \frac{51}{1000}$

8. $29.97 = 29\frac{97}{100}$

9. $0.12 = \frac{12}{100} = \frac{3 \cdot 4}{25 \cdot 4} = \frac{3}{25}$

10. $64.8 = 64\frac{8}{10} = 64\frac{2 \cdot 4}{2 \cdot 5} = 64\frac{4}{5}$

11. $$\begin{aligned} -209.986 &= -209\frac{986}{1000} \\ &= -209\frac{2 \cdot 493}{2 \cdot 500} \\ &= -209\frac{493}{500} \end{aligned}$$

12. $$\begin{array}{ccc} 29.208 & & 26.28 \\ \uparrow & & \uparrow \\ 0 & < & 8 \end{array}$$
    so 26.208 < 26.28

13. $$\begin{array}{ccc} 0.12 & & 0.026 \\ \uparrow & & \uparrow \\ 1 & > & 0 \end{array}$$
    so 0.12 > 0.026

14. $$\begin{array}{ccc} 0.039 & & 0.0309 \\ \uparrow & & \uparrow \\ 9 & > & 0 \end{array}$$
    so 0.039 > 0.0309
    Thus, −0.039 < −0.0309.

15. To round 482.7817 to the nearest thousandth, observe that the digit in the ten-thousandths place is 7. Since this digit is at least 5, we add 1 to the digit in the thousandths place. The number 482.7817 rounded to the nearest thousandth is 482.782.

16. To round −0.032 to the nearest hundredth, observe that the digit in the thousandths place is 2. Since this digit is less than 5, we do not add 1 to the digit in the hundredths place. The number −0.032 rounded to the nearest hundredth is −0.03.

17. To round 3.14159265 to the nearest ten-thousandth, observe that the digit in the hundred-thousandths place is 9. Since this digit is at least 5, we add 1 to the digit in the ten-thousandths place. The number 3.14159265 rounded to the nearest the ten-thousandth is 3.1416, or $\pi \approx 3.1416$.

18. $24.62 rounded to the nearest dollar is $25, since $6 \geq 5$.

Copyright © 2011 Pearson Education, Inc. Publishing as Prentice Hall.

Vocabulary and Readiness Check

1. The number "twenty and eight hundredths" is written in words and "20.08" is written in standard form.

2. Another name for the distance around a circle is its circumference.

3. Like fractions, decimals are used to denote part of a whole.

4. When writing a decimal number in words, the decimal point is written as and.

5. The place value tenths is to the right of the decimal point while tens is to the left of the decimal point.

6. The decimal point in a whole number is after the last digit.

Exercise Set 5.1

1. 5.62 in words is five and sixty-two hundredths.

3. 16.23 in words is sixteen and twenty-three hundredths.

5. –0.205 in words is negative two hundred five thousandths.

7. 167.009 in words is one hundred sixty-seven and nine thousandths.

9. 3000.04 in words is three thousand and four hundredths.

11. 105.6 in words is one hundred five and six tenths.

13. 2.43 in words is two and forty-three hundredths.

15. The check should be paid to "R.W. Financial," for the amount "321.42," which is written in words as "Three hundred twenty-one and $\frac{42}{100}$."

Your Preprinted Name
Your Preprinted Address
60-8124/7233
1000613331
1407
DATE (Current date)
PAY TO THE ORDER OF R.W. Financial $ 321.42
Three hundred twenty-one and 42/100 — DOLLARS
FIRST STATE BANK
OF FARTHINGTON
FARTHINGTON, IL 64422
MEMO (Your signature)
⑆621497260⑆ 1000613331⑈ 1407

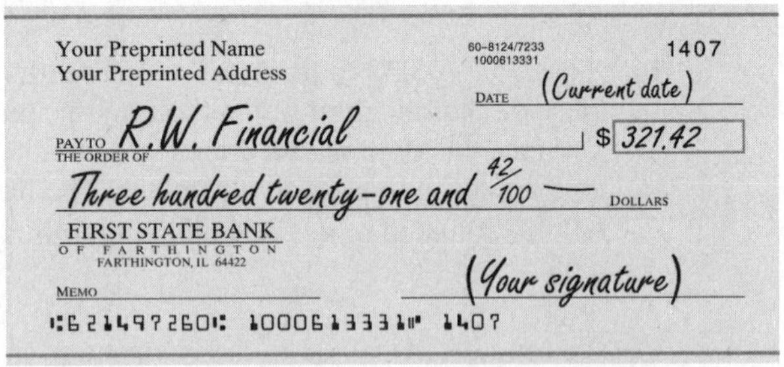

17. The check should be paid to "Bell South," for the amount of "59.68," which is written in words as "Fifty-nine and $\frac{68}{100}$."

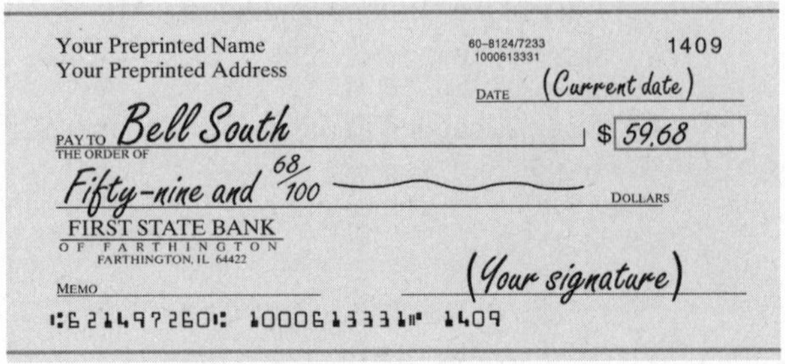

Your Preprinted Name
Your Preprinted Address
60-8124/7233
1000613331
1409
DATE (Current date)
PAY TO THE ORDER OF Bell South $ 59.68
Fifty-nine and 68/100 — DOLLARS
FIRST STATE BANK
OF FARTHINGTON
FARTHINGTON, IL 64422
MEMO (Your signature)
⑆621497260⑆ 1000613331⑈ 1409

Copyright © 2011 Pearson Education, Inc. Publishing as Prentice Hall.

19. Two and eight tenths is 2.8.

21. Nine and eight hundredths is 9.08.

23. Negative seven hundred five and six hundred twenty-five thousandths is −705.625.

25. Forty-six ten-thousandths is 0.0046.

27. $0.7 = \frac{7}{10}$

29. $0.27 = \frac{27}{100}$

31. $0.4 = \frac{4}{10} = \frac{2 \cdot 2}{2 \cdot 5} = \frac{2}{5}$

33. $5.4 = 5\frac{4}{10} = 5\frac{2 \cdot 2}{2 \cdot 5} = 5\frac{2}{5}$

35. $-0.058 = -\frac{58}{1000} = -\frac{2 \cdot 29}{2 \cdot 500} = -\frac{29}{500}$

37. $7.008 = 7\frac{8}{1000} = 7\frac{1 \cdot 8}{125 \cdot 8} = 7\frac{1}{125}$

39. $15.802 = 15\frac{802}{1000} = 15\frac{2 \cdot 401}{2 \cdot 500} = 15\frac{401}{500}$

41. $0.3005 = \frac{3005}{10{,}000} = \frac{601 \cdot 5}{2000 \cdot 5} = \frac{601}{2000}$

43. Eight tenths is 0.8 and as a fraction is $\frac{8}{10} = \frac{4}{5}$.

45. In words, 0.077 is seventy-seven thousandths. As a fraction, $0.077 = \frac{77}{1000}$.

47. 0.15 0.16
$$\begin{array}{ccc} \uparrow & & \uparrow \\ 5 & < & 6 \end{array}$$
so 0.15 < 0.16

49. 0.57 0.54
$$\begin{array}{ccc} \uparrow & & \uparrow \\ 7 & > & 4 \end{array}$$
so 0.57 > 0.54
Thus −0.57 < −0.54.

51. 0.098 0.1
$$\begin{array}{ccc} \uparrow & & \uparrow \\ 0 & < & 1 \end{array}$$
so 0.098 < 0.1

53. 0.54900 0.549
$$\begin{array}{ccc} \uparrow & & \uparrow \\ 9 & = & 9 \end{array}$$
so 0.54900 = 0.549

55. 167.908 167.980
$$\begin{array}{ccc} \uparrow & & \uparrow \\ 0 & < & 8 \end{array}$$
so 167.908 < 167.980

57. 1.062 1.07
$$\begin{array}{ccc} \uparrow & & \uparrow \\ 6 & < & 7 \end{array}$$
so 1.062 < 1.07
Thus, −1.062 > −1.07.

59. −7.052 7.0052
$$\begin{array}{ccc} \uparrow & & \uparrow \\ - & < & + \end{array}$$
so −7.052 < 7.0052

61. 0.023 0.024
$$\begin{array}{ccc} \uparrow & & \uparrow \\ 3 & < & 4 \end{array}$$
so 0.023 < 0.024
Thus, −0.023 > −0.024.

63. To round 0.57 to the nearest tenth, observe that the digit in the hundredths place is 7. Since this digit is at least 5, we add 1 to the digit in the tenths place. The number 0.57 rounded to the nearest tenth is 0.6.

65. To round 98,207.23 to the nearest ten, observe that the digit in the ones place is 7. Since this digit is at least 5, we add 1 to the digit in the tens place. The number 98,207.23 rounded to the nearest ten is 98,210.

67. To round −0.234 to the nearest hundredth, observe that the digit in the thousandths place is 4. Since this digit is less than 5, we do not add 1 to the digit in the hundredths place. The number −0.234 rounded to the nearest hundredth is −0.23.

69. To round 0.5942 to the nearest thousandth, observe that the digit in the ten-thousandths place is 2. Since this digit is less than 5, we do not add 1 to the digit in the thousandths place. The number 0.5942 rounded to the nearest thousandth is 0.594.

Copyright © 2011 Pearson Education, Inc. Publishing as Prentice Hall.

71. To round $\pi \approx 3.14159265$ to the nearest tenth, observe that the digit in the hundredths place is 4. Since this digit is less than 5, we do not add 1 to the digit in the tenths place. The number $\pi \approx 3.14159265$ rounded to the nearest tenth is 3.1.

73. To round $\pi \approx 3.14159265$ to the nearest thousandth, observe that the digit in the ten-thousandth place is 5. Since this digit is at least 5, we add 1 to the digit in the thousandths place. The number $\pi \approx 3.14159265$ rounded to the nearest thousandth is 3.142.

75. To round 26.95 to the nearest one, observe that the digit in the tenths place is 9. Since this digit is at least 5, we add 1 to the digit in the ones place. The number 26.95 rounded to the nearest one is 27. The amount is $27.

77. To round 0.1992 to the nearest hundredth, observe that the digit in the thousandths place is 9. Since this digit is at least 5, we add 1 to the digit in the hundredths place. The number 0.1992 rounded to the nearest hundredth is 0.2. The amount is $0.20.

79. To round 0.4064 to the nearest tenth, observe that the digit in the hundredths place is 0. Since this digit is less than 5, we do not add 1 to the digit in the tenths place. The number 0.4064 rounded to the nearest tenth is 0.4. The thickness is 0.4 centimeter.

81. To round 1.8672 to the nearest hundredth, observe that the digit in the thousandths place is 7. Since this digit is at least 5, we add 1 to the digit in the hundredths place. The number 1.8672 rounded to the nearest hundredth is 1.87. The time is 1.87 minutes.

83. To round 67.89 to the nearest one, observe that the digit in the tenths place is 8. Since this digit is at least 5, we add 1 to the digit in the ones place. The number 67.89 rounded to the nearest one is 68. The price is $68.

85. To round 224.695 to the nearest one, observe that the digit in the tenths place is 6. Since this digit is at least 5, we add 1 to the digit in the ones place. The number 224.695 rounded to the nearest one is 225. This is 225 days.

87.
$$\begin{array}{r} 3452 \\ +\ 2314 \\ \hline 5766 \end{array}$$

89.
$$\begin{array}{r} 82 \\ -\ 47 \\ \hline 35 \end{array}$$

91. To round 2849.1738 to the nearest hundred, observe that the digit in the tens place is 4. Since this digit is less than 5, we do not add 1 to the digit in the hundreds place. The number 2849.1738 rounded to the nearest hundred is 2800, which is choice b.

93. To round 2849.1738 to the nearest hundredth, observe that the digit in the thousandths place is 3. Since this digit is less than 5, we do not add 1 to the digit in the hundredths place. 2849.1738 rounded to the nearest hundredth is 2849.17, which is choice a.

95. answers may vary

97. $7\frac{12}{100} = 7.12$

99. $0.00026849577 = \frac{26,849,577}{100,000,000,000}$

101. answers may vary

103. answers may vary

105. 0.26499 and 0.25786 rounded to the nearest hundredth are 0.26. 0.26559 rounds to 0.27 and 0.25186 rounds to 0.25.

107. From smallest to largest, 0.10299, 0.1037, 0.1038, 0.9

109. Round to the nearest hundred million, then add.
$$\begin{array}{r} 600 \\ 500 \\ 500 \\ 400 \\ 400 \\ +\ 400 \\ \hline 2800 \end{array}$$
The total amount of money is estimated as $2800 million.

Copyright © 2011 Pearson Education, Inc. Publishing as Prentice Hall.

Section 5.2

Practice Problems

1. a.
$$\begin{array}{r} 19.520 \\ +\ 5.371 \\ \hline 24.891 \end{array}$$

b.
$$\begin{array}{r} 40.080 \\ +17.612 \\ \hline 57.692 \end{array}$$

c.
$$\begin{array}{r} 0.125 \\ +422.800 \\ \hline 422.925 \end{array}$$

2. a.
$$\begin{array}{r} 34.5670 \\ 129.4300 \\ +\ \ 2.8903 \\ \hline 166.8873 \end{array}$$

b.
$$\begin{array}{r} 11.210 \\ 46.013 \\ +362.526 \\ \hline 419.749 \end{array}$$

3.
$$\begin{array}{r} 19.000 \\ +26.072 \\ \hline 45.072 \end{array}$$

4. $7.12 + (-9.92)$
Subtract the absolute values.
$$\begin{array}{r} 9.92 \\ -7.12 \\ \hline 2.80 \end{array}$$
Attach the sign of the larger absolute value.
$7.12 + (-9.92) = -2.8$

5. a.
$$\begin{array}{r} 6.70 \\ -3.92 \\ \hline 2.78 \end{array} \qquad \textit{Check:} \begin{array}{r} 2.78 \\ +3.92 \\ \hline 6.70 \end{array}$$

b.
$$\begin{array}{r} 9.720 \\ -4.068 \\ \hline 5.652 \end{array} \qquad \textit{Check:} \begin{array}{r} 5.652 \\ +4.068 \\ \hline 9.720 \end{array}$$

6. a.
$$\begin{array}{r} 73.00 \\ -29.31 \\ \hline 43.69 \end{array} \qquad \textit{Check:} \begin{array}{r} 43.69 \\ +29.31 \\ \hline 73.00 \end{array}$$

b.
$$\begin{array}{r} 210.00 \\ -\ 68.22 \\ \hline 141.78 \end{array} \qquad \textit{Check:} \begin{array}{r} 141.78 \\ +\ 68.22 \\ \hline 210.00 \end{array}$$

7.
$$\begin{array}{r} 25.91 \\ -19.00 \\ \hline 6.91 \end{array}$$

8. $-5.4 - 9.6 = -5.4 + (-9.6)$
Add the absolute values.
$$\begin{array}{r} 5.4 \\ +9.6 \\ \hline 15.0 \end{array}$$
Attach the common sign.
$-5.4 - 9.6 = -15$

9. $-1.05 - (-7.23) = -1.05 + 7.23$
Subtract the absolute values.
$$\begin{array}{r} 7.23 \\ -1.05 \\ \hline 6.18 \end{array}$$
Attach the sign of the larger absolute value.
$-1.05 - (-7.23) = 6.18$

10. a.

Exact	Estimate 1	Estimate 2
$\begin{array}{r} 58.10 \\ +326.97 \\ \hline 385.07 \end{array}$	$\begin{array}{r} 60 \\ +300 \\ \hline 360 \end{array}$	$\begin{array}{r} 60 \\ +330 \\ \hline 390 \end{array}$

b.

Exact	Estimate 1	Estimate 2
$\begin{array}{r} 16.080 \\ -\ 0.925 \\ \hline 15.155 \end{array}$	$\begin{array}{r} 16 \\ -\ 1 \\ \hline 15 \end{array}$	$\begin{array}{r} 20 \\ -\ 1 \\ \hline 19 \end{array}$

11. $y - z = 11.6 - 10.8 = 0.8$

12.
$$\begin{aligned} y - 4.3 &= 7.8 \\ 12.1 - 4.3 &\stackrel{?}{=} 7.8 \\ 7.8 &= 7.8 \quad \text{True} \end{aligned}$$
Yes, 12.1 is a solution.

13.
$$\begin{aligned} &-4.3y + 7.8 - 20.1y + 14.6 \\ &= -4.3y - 20.1y + 7.8 + 14.6 \\ &= (-4.3 - 20.1)y + (7.8 + 14.6) \\ &= -24.4y + 22.4 \end{aligned}$$

14.
$$\begin{array}{r} 563.52 \\ 52.68 \\ +127.50 \\ \hline 743.70 \end{array}$$
The total cost is $743.70.

Copyright © 2011 Pearson Education, Inc. Publishing as Prentice Hall.

15. $$\begin{array}{r} 72.6 \\ -\ 70.8 \\ \hline 1.8 \end{array}$$

The average height in the Netherlands is 1.8 inches greater than the average height in Czechoslovakia.

Calculator Explorations

1. $315.782 + 12.96 = 328.742$

2. $29.68 + 85.902 = 115.582$

3. $6.249 - 1.0076 = 5.2414$

4. $5.238 - 0.682 = 4.556$

5. $$\begin{array}{r} 12.555 \\ 224.987 \\ 5.2 \\ +\ 622.65 \\ \hline 865.392 \end{array}$$

6. $$\begin{array}{r} 47.006 \\ 0.17 \\ 313.259 \\ +\ 139.088 \\ \hline 499.523 \end{array}$$

Vocabulary and Readiness Check

1. The decimal point in a whole number is positioned after the last digit.

2. In $89.2 - 14.9 = 74.3$, the number 74.3 is called the difference, 89.2 is the minuend, and 14.9 is the subtrahend.

3. To simplify an expression, we combine any like terms.

4. True or false: If we replace x with 11.2 and y with -8.6 in the expression $x - y$, we have $11.2 - 8.6$. false

5. To add or subtract decimals, we line up the decimal points vertically.

Exercise Set 5.2

1. $$\begin{array}{r} 5.6 \\ +\ 2.1 \\ \hline 7.7 \end{array}$$

3. $$\begin{array}{r} 8.20 \\ +\ 2.15 \\ \hline 10.35 \end{array}$$

5. $$\begin{array}{r} 24.6000 \\ 2.3900 \\ +\ \ 0.0678 \\ \hline 27.0578 \end{array}$$

7. $-2.6 + (-5.97)$

Add the absolute values.

$$\begin{array}{r} 2.60 \\ +\ 5.97 \\ \hline 8.57 \end{array}$$

Attach the common sign.

$-2.6 + (-5.97) = -8.57$

9. $18.56 + (-8.23)$

Subtract the absolute values.

$$\begin{array}{r} 18.56 \\ -\ 8.23 \\ \hline 10.33 \end{array}$$

Attach the sign of the larger absolute value.

$18.56 + (-8.23) = 10.33$

11. Exact: $$\begin{array}{r} 234.89 \\ +\ 230.67 \\ \hline 465.56 \end{array}$$

Estimate: $$\begin{array}{r} 230 \\ +\ 230 \\ \hline 460 \end{array}$$

13. Exact: $$\begin{array}{r} 100.009 \\ 6.080 \\ +\ \ 9.034 \\ \hline 115.123 \end{array}$$

Estimate: $$\begin{array}{r} 100 \\ 6 \\ +\ \ 9 \\ \hline 115 \end{array}$$

15. $$\begin{array}{r} 39.000 \\ 3.006 \\ +\ \ 8.403 \\ \hline 50.409 \end{array}$$

17. $$\begin{array}{r} 12.6 \\ -\ 8.2 \\ \hline 4.4 \end{array}$$

Check: $$\begin{array}{r} 4.4 \\ +\ 8.2 \\ \hline 12.6 \end{array}$$

Copyright © 2011 Pearson Education, Inc. Publishing as Prentice Hall.

19. $\begin{array}{r} 18.0 \\ -\ 2.7 \\ \hline 15.3 \end{array}$ *Check*: $\begin{array}{r} 15.3 \\ +\ 2.7 \\ \hline 18.0 \end{array}$

21. $\begin{array}{r} 654.90 \\ -\ 56.67 \\ \hline 598.23 \end{array}$ *Check:* $\begin{array}{r} 598.23 \\ +\ 56.67 \\ \hline 654.90 \end{array}$

23. Exact: $5.9 - 4.07 = 1.83$
Estimate: $6 - 4 = 2$
Check: $1.83 + 4.07 = 5.90$

25. Exact: $\begin{array}{r} 1000.0 \\ -\ 123.4 \\ \hline 876.6 \end{array}$ *Check:* $\begin{array}{r} 876.6 \\ +\ 123.4 \\ \hline 1000.0 \end{array}$
Estimate: $\begin{array}{r} 1000 \\ -\ 100 \\ \hline 900 \end{array}$

27. $\begin{array}{r} 200.0 \\ -\ 5.6 \\ \hline 194.4 \end{array}$ *Check:* $\begin{array}{r} 194.4 \\ +\ 5.6 \\ \hline 200.0 \end{array}$

29. $-1.12 - 5.2 = -1.12 + (-5.2)$
Add the absolute values.
$\begin{array}{r} 1.12 \\ +\ 5.20 \\ \hline 6.32 \end{array}$ *Check:* $\begin{array}{r} 6.32 \\ -\ 5.20 \\ \hline 1.12 \end{array}$
Attach the common sign.
$-1.12 - 5.2 = -6.32$

31. $5.21 - 11.36 = 5.21 + (-11.36)$
Subtract the absolute values.
$\begin{array}{r} 11.36 \\ -\ 5.21 \\ \hline 6.15 \end{array}$ *Check:* $\begin{array}{r} 6.15 \\ +\ 5.21 \\ \hline 11.36 \end{array}$
Attach the sign of the larger absolute value.
$5.21 - 11.36 = -6.15$

33. $-2.6 - (-5.7) = -2.6 + 5.7$
Subtract the absolute values.
$\begin{array}{r} 5.7 \\ -\ 2.6 \\ \hline 3.1 \end{array}$ *Check:* $\begin{array}{r} 3.1 \\ +\ 2.6 \\ \hline 5.7 \end{array}$
Attach the sign of the larger absolute value.
$-2.6 - (-5.7) = 3.1$

35. $\begin{array}{r} 3.0000 \\ -\ 0.0012 \\ \hline 2.9988 \end{array}$ *Check:* $\begin{array}{r} 2.9988 \\ +\ 0.0012 \\ \hline 3.0000 \end{array}$

37. $\begin{array}{r} 23.0 \\ -\ 6.7 \\ \hline 16.3 \end{array}$ *Check:* $\begin{array}{r} 16.3 \\ +\ 6.7 \\ \hline 23.0 \end{array}$

39. $\begin{array}{r} 0.9 \\ +\ 2.2 \\ \hline 3.1 \end{array}$

41. $-6.06 + 0.44$
Subtract the absolute values.
$\begin{array}{r} 6.06 \\ -\ 0.44 \\ \hline 5.62 \end{array}$
Attach the sign of the larger absolute value.
$-6.06 + 0.44 = -5.62$

43. $\begin{array}{r} 500.21 \\ -\ 136.85 \\ \hline 363.36 \end{array}$

45. $50.2 - 600 = 50.2 + (-600)$
Subtract the absolute values.
$\begin{array}{r} 600.0 \\ -\ 50.2 \\ \hline 549.8 \end{array}$
Attach the sign of the larger absolute value.
$50.2 - 600 = -549.8$

47. $\begin{array}{r} 923.5 \\ -\ 61.9 \\ \hline 861.6 \end{array}$

49. $\begin{array}{r} 100.009 \\ 6.080 \\ +\ 9.034 \\ \hline 115.123 \end{array}$

51. $-0.003 + 0.091$
Subtract the absolute values.
$\begin{array}{r} 0.091 \\ -\ 0.003 \\ \hline 0.088 \end{array}$
Attach the sign of the larger absolute value.
$-0.003 + 0.091 = 0.088$

53. $-102.4 - 78.04 = -102.4 + (-78.04)$
Add the absolute values.
$\begin{array}{r} 102.40 \\ +\ 78.04 \\ \hline 180.44 \end{array}$
Attach the common sign.
$-102.4 - 78.04 = -180.44$

Copyright © 2011 Pearson Education, Inc. Publishing as Prentice Hall.

55. $-2.9-(-1.8)=-2.9+1.8$
Subtract the absolute values.
$$\begin{array}{r} 2.9 \\ -\ 1.8 \\ \hline 1.1 \end{array}$$
Attach the sign of the larger absolute value.
$-2.9-(-1.8)=-1.1$

57. $x+z=3.6+0.21=3.81$

59. $x-z=3.6-0.21=3.39$

61.
$$\begin{aligned} y-x+z &= 5-3.6+0.21 \\ &= 5.00-3.60+0.21 \\ &= 1.40+0.21 \\ &= 1.61 \end{aligned}$$

63.
$$\begin{aligned} x+2.7 &= 9.3 \\ 7+2.7 &\stackrel{?}{=} 9.3 \\ 9.7 &= 9.3 \quad \text{False} \end{aligned}$$
No, 7 is not a solution.

65.
$$\begin{aligned} 27.4+y &= 16 \\ 27.4+(-11.4) &\stackrel{?}{=} 16 \\ 16 &= 16 \quad \text{True} \end{aligned}$$
Yes, −11.4 is a solution.

67.
$$\begin{aligned} 2.3+x &= 5.3-x \\ 2.3+1 &\stackrel{?}{=} 5.3-1 \\ 3.3 &= 4.3 \quad \text{False} \end{aligned}$$
No, 1 is not a solution.

69.
$$\begin{aligned} &30.7x+17.6-23.8x-10.7 \\ &= 30.7x-23.8x+17.6-10.7 \\ &= (30.7-23.8)x+(17.6-10.7) \\ &= 6.9x+6.9 \end{aligned}$$

71.
$$\begin{aligned} &-8.61+4.23y-2.36-0.76y \\ &= 4.23y-0.76y-8.61-2.36 \\ &= (4.23-0.76)y+(-8.61-2.36) \\ &= 3.47y-10.97 \end{aligned}$$

73. Change = $40 - \underline{32.48}$
$$\begin{array}{r} 40.00 \\ -\ 32.48 \\ \hline 7.52 \end{array}$$
If Ann-Margaret paid with two $20 bills, her change was $7.52.

75. Subtract the opening price from the closing price.
$$\begin{array}{r} 22.07 \\ -\ 21.90 \\ \hline 0.17 \end{array}$$
The price of each share increased by $0.17.

77.
$$\begin{aligned} \text{Perimeter} &= 7.14+7.14+7.14+7.14 \\ &= 28.56 \text{ meters} \end{aligned}$$

79. Perimeter = 4.5 + 2.4 + 4.5 + 2.4 = 13.8 inches

81. The phrase "How much faster" indicates that we should subtract the average wind speed from the record speed.
$$\begin{array}{r} 321.0 \\ -\ \ 35.2 \\ \hline 285.8 \end{array}$$
The highest wind speed is 285.8 miles per hour faster than the average wind speed.

83. To find the increase, subtract.
$$\begin{array}{r} 75.3 \\ -\ 62.9 \\ \hline 12.4 \end{array}$$
The increase was 12.4 million or 12,400,000 users.

85. To find the total, we add.
$$\begin{array}{r} 600.78 \\ 533.32 \\ +\ 460.99 \\ \hline 1595.09 \end{array}$$
The total ticket sales were $1595.09 million.

87. To find the amount of snow in Blue Canyon, add 111.6 to the amount in Marquette.
$$\begin{array}{r} 129.2 \\ +\ 111.6 \\ \hline 240.8 \end{array}$$
Blue Canyon receives on average 240.8 inches each year.

89. Add the lengths of the sides to get the perimeter.
$$\begin{array}{r} 12.40 \\ 29.34 \\ +\ 25.70 \\ \hline 67.44 \end{array}$$
67.44 feet of border material is needed.

Copyright © 2011 Pearson Education, Inc. Publishing as Prentice Hall.

91.
$$\begin{array}{r} 172.712 \\ -\ 152.672 \\ \hline 20.040 \end{array}$$
The difference is 20.04 miles per hour.

93. The tallest bar indicates the greatest chocolate consumption per person, so Switzerland has the greatest chocolate consumption per person.

95.
$$\begin{array}{r} 22.36 \\ -\ 17.93 \\ \hline 4.43 \end{array}$$
The difference in consumption is 4.43 pounds per year.

97.

Country	Pounds of Chocolate per Person
Switzerland	22.36
Austria	20.13
Ireland	19.47
Germany	18.04
Norway	17.93

99. $46 \cdot 3 = 138$

101. $\left(\frac{2}{3}\right)^2 = \frac{2}{3} \cdot \frac{2}{3} = \frac{2 \cdot 2}{3 \cdot 3} = \frac{4}{9}$

103. It is incorrect. Align the decimals.
$$\begin{array}{r} 9.200 \\ 8.630 \\ +\ 4.005 \\ \hline 21.835 \end{array}$$

105. $10.68 - (2.3 + 2.3) = 10.68 - 4.60 = 6.08$
The unknown length is 6.08 inches.

107. 3 nickels, 3 dimes, and 3 quarters:
$0.05 + 0.05 + 0.05 + 0.10 + 0.10 + 0.10 + 0.25 + 0.25 + 0.25 = 1.2$
The value of the coins shown is \$1.20.

109. 1 nickel, 1 dime, and 2 pennies:
$0.05 + 0.10 + 0.01 + 0.01 = 0.17$
3 nickels and 2 pennies:
$0.05 + 0.05 + 0.05 + 0.01 + 0.01 = 0.17$
1 dime and 7 pennies:
$0.10 + 0.01 + 0.01 + 0.01 + 0.01 + 0.01 + 0.01 + 0.01 = 0.17$
2 nickels and 7 pennies:
$0.05 + 0.05 + 0.01 + 0.01 + 0.01 + 0.01 + 0.01 + 0.01 + 0.01 = 0.17$

111. answers may vary

113. answers may vary

115.
$$\begin{aligned} &-8.689 + 4.286x - 14.295 - 12.966x + 30.861x \\ &= 4.286x - 12.966x + 30.861x - 8.689 - 14.295 \\ &= (4.286 - 12.966 + 30.861)x + (-8.689 - 14.295) \\ &= 22.181x - 22.984 \end{aligned}$$

Section 5.3

Practice Problems

1.
$$\begin{array}{rl} 34.8 & \text{1 decimal place} \\ \times\ 0.62 & \text{2 decimal places} \\ \hline 696 & \\ 20\ 880 & \\ \hline 21.576 & 1+2=3 \text{ decimal places} \end{array}$$

2.
$$\begin{array}{rl} 0.0641 & \text{4 decimal places} \\ \times\ 27 & \text{0 decimal places} \\ \hline 4487 & \\ 1\ 2820 & \\ \hline 1.7307 & 4+0=4 \text{ decimal places} \end{array}$$

3. $(7.3)(-0.9) = -6.57$ (Be sure to include the negative sign.)

4. Exact:
$$\begin{array}{r} 30.26 \\ \times\ 2.89 \\ \hline 2\ 7234 \\ 24\ 2080 \\ 60\ 5200 \\ \hline 87.4514 \end{array}$$
Estimate:
$$\begin{array}{r} 30 \\ \times\ 3 \\ \hline 90 \end{array}$$

5. $46.8 \times 10 = 468$

6. $203.004 \times 100 = 20{,}300.4$

7. $(-2.33)(1000) = -2330$

8. $6.94 \times 0.1 = 0.694$

9. $3.9 \times 0.01 = 0.039$

10. $(-7682)(-0.001) = 7.682$

11.
$$\begin{aligned} 60.7 \text{ million} &= 60.7 \times 1 \text{ million} \\ &= 60.7 \times 1{,}000{,}000 \\ &= 60{,}700{,}000 \end{aligned}$$

12. $7y = 7(-0.028) = -0.196$

Copyright © 2011 Pearson Education, Inc. Publishing as Prentice Hall.

13. $-6x = 33$
$-6(-5.5) \stackrel{?}{=} 33$
$33 = 33$ True
Yes, −5.5 is a solution.

14. $C = 2\pi r = 2\pi \cdot 11 = 22\pi \approx 22(3.14) = 69.08$
The circumference is 22π meters ≈ 69.08 meters.

15.
$$\begin{array}{r} 60.5 \\ \times\ \ 5.6 \\ \hline 36\ 30 \\ 302\ 50 \\ \hline 338.80 \end{array}$$
She needs 338.8 ounces of fertilizer.

Vocabulary and Readiness Check

1. When multiplying decimals, the number of decimal places in the product is equal to the sum of the number of decimal place in the factors.

2. In 8.6 × 5 = 43, the number 43 is called the product while 8.6 and 5 are each called a factor.

3. When multiplying a decimal number by powers of 10 such as 10, 100, 1000, and so on, we move the decimal point in the number to the right the same number of places as there are zeros in the power of 10.

4. When multiplying a decimal number by powers of 10 such as 0.1, 0.01, and so on, we move the decimal point in the number to the left the same number of places as there are decimal places in the power of 10.

5. The distance around a circle is called its circumference.

Exercise Set 5.3

1.
$$\begin{array}{rl} 0.17 & \text{2 decimal places} \\ \times\ \ \ 8 & \text{0 decimal places} \\ \hline 1.36 & 2+0=2 \text{ decimal places} \end{array}$$

3.
$$\begin{array}{rl} 1.2 & \text{1 decimal place} \\ \times 0.5 & \text{1 decimal place} \\ \hline 0.60 & 1+1=2 \text{ decimal places} \end{array}$$

5. The product (−2.3)(7.65) is negative.
$$\begin{array}{rl} 7.65 & \text{2 decimal places} \\ \times\ \ 2.3 & \text{1 decimal place} \\ \hline 2\ 295 & \\ 15\ 300 & \\ \hline -17.595 & 2+1=3 \text{ decimal places and include the negative sign} \end{array}$$

7. The product (−5.73)(−9.6) is positive.
$$\begin{array}{rl} 5.73 & \text{2 decimal places} \\ \times\ \ 9.6 & \text{1 decimal place} \\ \hline 3\ 438 & \\ 51\ 570 & \\ \hline 55.008 & 2+1=3 \text{ decimal places} \end{array}$$

9. Exact:
$$\begin{array}{r} 6.8 \\ \times 4.2 \\ \hline 1\ 36 \\ 27\ 20 \\ \hline 28.56 \end{array}$$
Estimate:
$$\begin{array}{r} 7 \\ \times 4 \\ \hline 28 \end{array}$$

11.
$$\begin{array}{rl} 0.347 & \text{3 decimal places} \\ \times\ \ 0.3 & \text{1 decimal place} \\ \hline 0.1041 & 3+1=4 \text{ decimal places} \end{array}$$

13. Exact:
$$\begin{array}{r} 1.0047 \\ \times\ \ \ \ 8.2 \\ \hline 20094 \\ 8\ 03760 \\ \hline 8.23854 \end{array}$$
Estimate:
$$\begin{array}{r} 1 \\ \times 8 \\ \hline 8 \end{array}$$

15.
$$\begin{array}{rl} 490.2 & \text{1 decimal place} \\ \times 0.023 & \text{3 decimal places} \\ \hline 1\ 4706 & \\ 9\ 8040 & \\ \hline 11.2746 & 1+3=4 \text{ decimal places} \end{array}$$

17. $6.5 \times 10 = 65$

19. $8.3 \times 0.1 = 0.83$

21. $(-7.093)(1000) = -7093$

23. $0.7 \times 100 = 70$

25. $(-9.83)(-0.01) = 0.0983$

27. $25.23 \times 0.001 = 0.02523$

Copyright © 2011 Pearson Education, Inc. Publishing as Prentice Hall.

29.
$$\begin{array}{r} 0.123 \\ \times\ \ 0.4 \\ \hline 0.0492 \end{array}$$

31. $(147.9)(100) = 14{,}790$

33.
$$\begin{array}{r} 8.6 \\ \times 0.15 \\ \hline 430 \\ 860 \\ \hline 1.290 \end{array} \text{ or } 1.29$$

35. $(937.62)(-0.01) = -9.3762$

37. $562.3 \times 0.001 = 0.5623$

39.
$$\begin{array}{r} 6.32 \\ \times\ \ 5.7 \\ \hline 4\ 424 \\ 31\ 600 \\ \hline 36.024 \end{array}$$

41.
$$\begin{aligned} 1.5 \text{ billion} &= 1.5 \times 1 \text{ billion} \\ &= 1.5 \times 1{,}000{,}000{,}000 \\ &= 1{,}500{,}000{,}000 \end{aligned}$$
The cost at launch was \$1,500,000,000.

43.
$$\begin{aligned} 49.8 \text{ million} &= 49.8 \times 1 \text{ million} \\ &= 49.8 \times 1{,}000{,}000 \\ &= 49{,}800{,}000 \end{aligned}$$
The roller coaster has given more than 49,800,000 rides.

45. $xy = 3(-0.2) = -0.6$

47.
$$\begin{aligned} xz - y &= 3(5.7) - (-0.2) \\ &= 17.1 - (-0.2) \\ &= 17.1 + 0.2 \\ &= 17.3 \end{aligned}$$

49.
$$\begin{aligned} 0.6x &= 4.92 \\ 0.6(14.2) &\stackrel{?}{=} 4.92 \\ 8.52 &= 4.92 \quad \text{False} \end{aligned}$$
No, 14.2 is not a solution.

51.
$$\begin{aligned} 3.5y &= -14 \\ 3.5(-4) &\stackrel{?}{=} -14 \\ -14 &= -14 \quad \text{True} \end{aligned}$$
Yes, –4 is a solution.

53. $C = \pi d$ is $\pi(10 \text{ cm}) = 10\pi$ cm
$C \approx 10(3.14)$ cm $= 31.4$ cm

55. $C = 2\pi r$ is $2\pi \cdot 9.1$ yards $= 18.2\pi$ yards
$C \approx 18.2(3.14)$ yards $= 57.148$ yards

57. pay before taxes = $\underline{17.88} \times \underline{40}$
$$\begin{array}{r} 17.88 \\ \times\ \ \ \ 40 \\ \hline 715.20 \end{array}$$
His pay for last week was \$715.20.

59. Multiply the number of ounces by the number of grams of saturated fat in 1 ounce.
$$\begin{array}{r} 6.2 \\ \times\ \ 4 \\ \hline 24.8 \end{array}$$
There are 24.8 grams of saturated fat in a 4-ounce serving of cream cheese.

61. Area = length · width
$$\begin{array}{r} 4.5 \\ \times 2.4 \\ \hline 1\ 80 \\ 9\ 00 \\ \hline 10.80 \end{array}$$
The area is 10.8 square inches.

63. Circumference = π · diameter
$C = \pi \cdot 250 = 250\pi$
$$\begin{array}{r} 250 \\ \times\ \ 3.14 \\ \hline 10\ 00 \\ 25\ 00 \\ 750\ 00 \\ \hline 785.00 \end{array}$$
The circumference is 250π feet, which is approximately 785 feet.

65. $C = \pi \cdot d$
$C = \pi \cdot 135 = 135\pi$
$$\begin{array}{r} 135 \\ \times 3.14 \\ \hline 5\ 40 \\ 13\ 50 \\ 405\ 00 \\ \hline 423.90 \end{array}$$
He travels 135π meters or approximately 423.9 meters.

Copyright © 2011 Pearson Education, Inc. Publishing as Prentice Hall.

67. Multiply her height in meters by the number of inches in 1 meter.

$$\begin{array}{r} 39.37 \\ \times\ \ 1.65 \\ \hline 1\,9685 \\ 23\,6220 \\ 39\,3700 \\ \hline 64.9605 \end{array}$$

She is approximately 64.9605 inches tall.

69. a. Circumference = $2 \cdot \pi \cdot$ radius

Smaller circle:

$C = 2 \cdot \pi \cdot 10 = 20\pi$

$C \approx 20(3.14) = 62.8$

The circumference of the smaller circle is approximately 62.8 meters.

Larger circle:

$C = 2 \cdot \pi \cdot 20 = 40\pi$

$C \approx 40(3.14) = 125.6$

The circumference of the larger circle is approximately 125.6 meters.

b. Yes, the circumference gets doubled when the radius is doubled.

71. $12.145 \times 100 = 1214.5$

The cost of 100 bushels of wheat was $1214.50.

73.

$$\begin{array}{r} 1.182 \\ \times\ \ \ 750 \\ \hline 59\,100 \\ 827\,400 \\ \hline 886.500 \end{array}$$

$750 U.S. is equivalent to 886.50 Canadian dollars.

75.

$$\begin{array}{r} 1.6252 \\ \times\ \ \ \ \ 800 \\ \hline 1300.1600 \end{array}$$

They can "buy" 1300.16 New Zealand dollars with 800 U.S. dollars.

77.

$$\begin{array}{r} 486 \\ 6\overline{)\,2916} \\ \underline{-24} \\ 51 \\ \underline{-48} \\ 36 \\ \underline{-36} \\ 0 \end{array}$$

79. $-\dfrac{24}{7} \div \dfrac{8}{21} = -\dfrac{24}{7} \cdot \dfrac{21}{8} = -\dfrac{8 \cdot 3 \cdot 7 \cdot 3}{7 \cdot 8} = -\dfrac{3 \cdot 3}{1} = -9$

81.

$$\begin{array}{r} 3.60 \\ +\ 0.04 \\ \hline 3.64 \end{array}$$

83.

$$\begin{array}{r} 3.60 \\ -\ 0.04 \\ \hline 3.56 \end{array}$$

85. The product of a negative number and a positive number is a negative number.

$$\begin{array}{r} 0.221 \\ \times\ \ \ 0.5 \\ \hline 0.1105 \end{array}$$

The product is –0.1105.

87.

$$\begin{array}{r} 20.6 \\ \times\ 1.86 \\ \hline 1\,236 \\ 16\,480 \\ 20\,600 \\ \hline 38.316 \end{array}$$

$38.316 \times 100{,}000 = 3{,}831{,}600$

The radio wave travels 3,831,600 miles in 20.6 seconds.

89. answers may vary

91. answers may vary

Section 5.4

Practice Problems

1.

$$\begin{array}{r} 46.3 \\ 8\overline{)\,370.4} \\ \underline{-32} \\ 50 \\ \underline{-48} \\ 2\,4 \\ \underline{-2\,4} \\ 0 \end{array}$$

Check:

$$\begin{array}{r} 46.3 \\ \times\ \ \ 8 \\ \hline 370.4 \end{array}$$

2.

$$\begin{array}{r} 0.71 \\ 48\overline{)\,34.08} \\ \underline{-33\,6} \\ 48 \\ \underline{-48} \\ 0 \end{array}$$

Check:

$$\begin{array}{r} 0.71 \\ \times\ 48 \\ \hline 5\,68 \\ 28\,40 \\ \hline 34.08 \end{array}$$

Copyright © 2011 Pearson Education, Inc. Publishing as Prentice Hall.

3. a.
$$\begin{array}{r} 1.135 \\ 14\overline{)\,15.890} \\ \underline{-14} \\ 1\,8 \\ \underline{-1\,4} \\ 49 \\ \underline{-42} \\ 70 \\ \underline{-70} \\ 0 \end{array}$$

Check:
$$\begin{array}{r} 1.135 \\ \underline{\times\ \ 14} \\ 4\,540 \\ \underline{11\,350} \\ 15.890 \end{array}$$

Thus, $-15.89 \div 14 = -1.135$.

b.
$$\begin{array}{r} 0.027 \\ 104\overline{)\,2.808} \\ \underline{-2\,08} \\ 728 \\ \underline{-728} \\ 0 \end{array}$$

Check:
$$\begin{array}{r} 0.027 \\ \underline{\times\ 104} \\ 108 \\ \underline{2\,700} \\ 2.808 \end{array}$$

Thus, $-2.808 \div (-104) = 0.027$

4. $5.6\overline{)166.88}$ becomes
$$\begin{array}{r} 29.8 \\ 56\overline{)\,1668.8} \\ \underline{-112} \\ 548 \\ \underline{-504} \\ 44\,8 \\ \underline{-44\,8} \\ 0 \end{array}$$

5. $0.16\overline{)1.976}$ becomes
$$\begin{array}{r} 12.35 \\ 16\overline{)\,197.60} \\ \underline{-16} \\ 37 \\ \underline{-32} \\ 5\,6 \\ \underline{-4\,8} \\ 80 \\ \underline{-80} \\ 0 \end{array}$$

6. $0.57\overline{)23.4}$ becomes
$$\begin{array}{r} 41.052 \approx 41.05 \\ 57\overline{)\,2340.000} \\ \underline{-228} \\ 60 \\ \underline{-57} \\ 3\,00 \\ \underline{-2\,85} \\ 150 \\ \underline{-114} \\ 36 \end{array}$$

7. $91.5\overline{)713.7}$ becomes
$$\begin{array}{r} 7.8 \\ 915\overline{)\,7137.0} \\ \underline{-6405} \\ 732\,0 \\ \underline{-732\,0} \\ 0 \end{array}$$

Estimate: $\begin{array}{r} 7 \\ 100\overline{)700} \end{array}$

8. $\dfrac{362.1}{1000} = 0.3621$

9. $-\dfrac{0.49}{10} = -0.049$

10. $x \div y = 0.035 \div 0.02$

$0.02\overline{)0.035}$ becomes
$$\begin{array}{r} 1.75 \\ 2\overline{)\,3.50} \\ \underline{-2} \\ 1\,5 \\ \underline{-1\,4} \\ 10 \\ \underline{-10} \\ 0 \end{array}$$

11. $\dfrac{x}{100} = 3.9$

$\dfrac{39}{100} \stackrel{?}{=} 3.9$

$0.39 = 3.9$ False

No, 39 is not a solution.

Copyright © 2011 Pearson Education, Inc. Publishing as Prentice Hall.

12. $$\begin{array}{r} 11.84 \\ 1250\overline{)\,14800.00} \\ \underline{-1250} \\ 2300 \\ \underline{-1250} \\ 10500 \\ \underline{-10000} \\ 5000 \\ \underline{-5000} \\ 0 \end{array}$$

He needs 11.84 bags or 12 whole bags.

Calculator Explorations

1. $102.62 \times 41.8 \approx 100 \times 40 = 4000$
Since 4000 is not close to 428.9516, it is not reasonable.

2. $174.835 \div 47.9 \approx 200 \div 50 = 4$
Since 4 is close to 3.65, it is reasonable.

3. $1025.68 - 125.42 \approx 1000 - 100 = 900$
Since 900 is close to 900.26, it is reasonable.

4. $562.781 + 2.96 \approx 563 + 3 = 566$
Since 566 is not close to 858.781, it is not reasonable.

Vocabulary and Readiness Check

1. In $6.5 \div 5 = 1.3$, the number 1.3 is called the quotient, 5 is the divisor, and 6.5 is the dividend.

2. To check a division exercise, we can perform the following multiplication:
quotient · divisor = dividend.

3. To divide a decimal number by a power of 10 such as 10, 100, 1000, or so on, we move the decimal point in the number to the left the same number of places as there are zeros in the power of 10.

4. True or false: If we replace x with -12.6 and y with 0.3 in the expression $y \div x$, we have $0.3 \div (-12.6)$ true

Exercise Set 5.4

1. $$\begin{array}{r} 4.6 \\ 6\overline{)\,27.6} \\ \underline{-24} \\ 3\,6 \\ \underline{-3\,6} \\ 0 \end{array}$$

3. $$\begin{array}{r} 0.094 \\ 5\overline{)0.470} \\ \underline{-45} \\ 20 \\ \underline{-20} \\ 0 \end{array}$$

5. $0.06\overline{)18}$ becomes $$\begin{array}{r} 300 \\ 6\overline{)\,1800} \\ \underline{-18} \\ 000 \end{array}$$

7. $0.82\overline{)4.756}$ becomes $$\begin{array}{r} 5.8 \\ 82\overline{)\,475.6} \\ \underline{-410} \\ 65\,6 \\ \underline{-65\,6} \\ 0 \end{array}$$

9. Exact: $5.5\overline{)36.3}$ becomes $$\begin{array}{r} 6.6 \\ 55\overline{)\,363.0} \\ \underline{-330} \\ 33\,0 \\ \underline{-33\,0} \\ 0 \end{array}$$

Estimate: $$\begin{array}{r} 6 \\ 6\overline{)36} \end{array}$$

11. $$\begin{array}{r} 0.413 \\ 18\overline{)\,7.434} \\ \underline{-7\,2} \\ 23 \\ \underline{-18} \\ 54 \\ \underline{-54} \\ 0 \end{array}$$

Copyright © 2011 Pearson Education, Inc. Publishing as Prentice Hall.

13. A positive number divided by a negative number is a negative number.

$$0.06\overline{)36} \text{ becomes } \begin{array}{r} 600 \\ 6\overline{)\ 3600} \\ \underline{-36} \\ 0 \end{array}$$

$36 \div (-0.06) = -600$

15. A negative number divided by a negative number is a positive number.

$$0.6\overline{)4.2} \text{ becomes } \begin{array}{r} 7 \\ 6\overline{)\ 42} \\ \underline{-42} \\ 0 \end{array}$$

$(-4.2) \div (-0.6) = 7$

17. $$0.27\overline{)1.296} \text{ becomes } \begin{array}{r} 4.8 \\ 27\overline{)\ 129.6} \\ \underline{-108} \\ 21\ 6 \\ \underline{-21\ 6} \\ 0 \end{array}$$

19. $$0.02\overline{)42} \text{ becomes } \begin{array}{r} 2100 \\ 2\overline{)\ 4200} \\ \underline{-4}\ \ \ \ \\ 02\ \ \\ \underline{-2}\ \ \\ 000 \end{array}$$

21. $$0.82\overline{)4.756} \text{ becomes } \begin{array}{r} 5.8 \\ 82\overline{)\ 475.6} \\ \underline{-410} \\ 65\ 6 \\ \underline{-65\ 6} \\ 0 \end{array}$$

23. A negative number divided by a negative number is a positive number.

$$6.6\overline{)36.3} \text{ becomes } \begin{array}{r} 5.5 \\ 66\overline{)\ 363.0} \\ \underline{-330} \\ 33\ 0 \\ \underline{-33\ 0} \\ 0 \end{array}$$

$-36.3 \div (-6.6) = 5.5$

25. Exact: $$7.2\overline{)70.56} \text{ becomes } \begin{array}{r} 9.8 \\ 72\overline{)\ 705.6} \\ \underline{-648} \\ 57\ 6 \\ \underline{-57\ 6} \\ 0 \end{array}$$

Estimate: $\begin{array}{r} 10 \\ 7\overline{)70} \end{array}$

27. $$5.4\overline{)51.84} \text{ becomes } \begin{array}{r} 9.6 \\ 54\overline{)\ 518.4} \\ \underline{-486} \\ 32\ 4 \\ \underline{-32\ 4} \\ 0 \end{array}$$

29. $$0.027\overline{)1.215} \text{ becomes } \begin{array}{r} 45 \\ 27\overline{)\ 1215} \\ \underline{-108} \\ 135 \\ \underline{-135} \\ 0 \end{array}$$

$\frac{1.215}{0.027} = 45$

31. $$0.25\overline{)13.648} \text{ becomes } \begin{array}{r} 54.592 \\ 25\overline{)\ 1364.800} \\ \underline{-125}\ \ \ \ \ \ \ \\ 114\ \ \ \ \ \ \\ \underline{-100}\ \ \ \ \ \ \\ 14\ 8\ \ \ \ \\ \underline{-12\ 5}\ \ \ \ \\ 2\ 30\ \ \\ \underline{-2\ 25}\ \ \\ 50 \\ \underline{-50} \\ 0 \end{array}$$

33. $$3.78\overline{)0.02079} \text{ becomes } \begin{array}{r} 0.0055 \\ 378\overline{)\ 2.0790} \\ \underline{-1\ 890}\ \\ 1890 \\ \underline{-1890} \\ 0 \end{array}$$

Copyright © 2011 Pearson Education, Inc. Publishing as Prentice Hall.

35. $0.023\overline{)0.549}$ becomes
$$\begin{array}{r} 23.869 \approx 23.87 \\ 23\overline{)\ 549.000} \\ \underline{-46} \\ 89 \\ \underline{-69} \\ 20\ 0 \\ \underline{-18\ 4} \\ 1\ 60 \\ \underline{-1\ 38} \\ 220 \\ \underline{-207} \\ 13 \end{array}$$

37. $0.6\overline{)68.39}$ becomes
$$\begin{array}{r} 113.98 \approx 114.0 \\ 6\overline{)\ 683.90} \\ \underline{-6} \\ 08 \\ \underline{-6} \\ 23 \\ \underline{-18} \\ 5\ 9 \\ \underline{-5\ 4} \\ 50 \\ \underline{-48} \\ 2 \end{array}$$

39. $\dfrac{83.397}{100} = 0.83397$

41. $\dfrac{26.87}{10} = 2.687$

43. $12.9 \div (-1000) = -0.0129$

45.
$$\begin{array}{r} 12.6 \\ 7\overline{)\ 88.2} \\ \underline{-7} \\ 18 \\ \underline{-14} \\ 4\ 2 \\ \underline{-4\ 2} \\ 0 \end{array}$$

47. $\dfrac{13.1}{10} = 1.31$

49. $\dfrac{456.25}{10,000} = 0.045625$

51.
$$\begin{array}{r} 0.413 \\ 3\overline{)\ 1.239} \\ \underline{-1\ 2} \\ 03 \\ \underline{-3} \\ 09 \\ \underline{-9} \\ 0 \end{array}$$

53. $0.6\overline{)4.8}$ becomes
$$\begin{array}{r} 8 \\ 6\overline{)\ 48} \\ \underline{-48} \\ 0 \end{array}$$

$4.8 \div (-0.6) = -8$

55. $0.17\overline{)1.224}$ becomes
$$\begin{array}{r} 7.2 \\ 17\overline{)\ 122.4} \\ \underline{-119} \\ 3\ 4 \\ \underline{-3\ 4} \\ 0 \end{array}$$

$-1.224 \div 0.17 = -7.2$

57. $0.03\overline{)42}$ becomes
$$\begin{array}{r} 1400 \\ 3\overline{)\ 4200} \\ \underline{-3} \\ 12 \\ \underline{-12} \\ 0 \end{array}$$

$42 \div 0.03 = 1400$

59. $0.6\overline{)18}$ becomes
$$\begin{array}{r} 30 \\ 6\overline{)\ 180} \\ \underline{-18} \\ 00 \end{array}$$

$-18 \div (-0.6) = 30$

61. $0.0015\overline{)87}$ becomes
$$\begin{array}{r} 58,000 \\ 15\overline{)\ 870,000} \\ \underline{-75} \\ 120 \\ \underline{-120} \\ 0000 \end{array}$$

$87 \div (-0.0015) = -58,000$

Copyright © 2011 Pearson Education, Inc. Publishing as Prentice Hall.

63. $1.6\overline{)1.104}$ becomes $16\overline{)11.04}$

$$\begin{array}{r} 0.69 \\ 16\overline{)11.04} \\ \underline{-9\,6} \\ 1\,44 \\ \underline{-1\,44} \\ 0 \end{array}$$

$-1.104 \div 1.6 = -0.69$

65. $-2.4 \div (-100) = \dfrac{-2.4}{-100} = 0.024$

67. $0.071\overline{)4.615}$ becomes $71\overline{)4615}$

$$\begin{array}{r} 65 \\ 71\overline{)\ 4615} \\ \underline{-426} \\ 355 \\ \underline{-355} \\ 0 \end{array}$$

$\dfrac{4.615}{0.071} = 65$

69. $z \div y = 4.52 \div (-0.8)$

$0.8\overline{)4.52}$ becomes $8\overline{)45.20}$

$$\begin{array}{r} 5.65 \\ 8\overline{)\ 45.20} \\ \underline{-40} \\ 5\,2 \\ \underline{-4\,8} \\ 40 \\ \underline{-40} \\ 0 \end{array}$$

$z \div y = 4.52 \div (-0.8) = -5.65$

71. $x \div y = 5.65 \div (-0.8)$

$0.8\overline{)5.65}$ becomes $8\overline{)56.5000}$

$$\begin{array}{r} 7.0625 \\ 8\overline{)\ 56.5000} \\ \underline{-56} \\ 0\,5 \\ \underline{-0} \\ 50 \\ \underline{-48} \\ 20 \\ \underline{-16} \\ 40 \\ \underline{-40} \\ 0 \end{array}$$

$x \div y = 5.65 \div (-0.8) = -7.0625$

73. $\dfrac{x}{4} = 3.04$

$\dfrac{12.16}{4} \stackrel{?}{=} 3.04$

$3.04 = 3.04$ True

Yes, 12.16 is a solution.

75. $\dfrac{z}{100} = 0.8$

$\dfrac{8}{100} \stackrel{?}{=} 0.8$

$0.08 = 0.8$ False

No, 8 is not a solution.

77. Number of quarts = <u>546</u> ÷ <u>52</u>

$$\begin{array}{r} 10.5 \approx 11 \\ 52\overline{)\ 546.0} \\ \underline{-52} \\ 26 \\ \underline{-0} \\ 26\,0 \\ \underline{-26\,0} \\ 0 \end{array}$$

Since he must buy whole quarts, 11 quarts are needed.

79. $39.37\overline{)200}$ becomes $3937\overline{)20000.00}$

$$\begin{array}{r} 5.08 \approx 5.1 \\ 3937\overline{)\ 20000.00} \\ \underline{-19685} \\ 315\,0 \\ \underline{-\quad 0} \\ 31500 \\ \underline{-31496} \\ 4 \end{array}$$

There are approximately 5.1 meters in 200 inches.

81. Divide the number of crayons by 64.

$$\begin{array}{r} 11.40 \approx 11.4 \\ 64\overline{)\ 730.00} \\ \underline{-64} \\ 90 \\ \underline{-64} \\ 26\,0 \\ \underline{-25\,6} \\ 40 \end{array}$$

740 crayons is approximately 11.4 boxes.

83. $6 \times 4 = 24$

There are 24 teaspoons in 4 fluid ounces.

Copyright © 2011 Pearson Education, Inc. Publishing as Prentice Hall.

85. From Exercise 83, we know that there are 24 teaspoons in 4 fluid ounces. Thus, there are 48 half teaspoons (0.5 tsp) or doses in 4 fluid ounces. To see how long the medicine will last, if a dose is taken every 4 hours, there are $24 \div 4 = 6$ doses taken per day.
48 (doses) $\div$ 6 (per day) = 8 days. The medicine will last 8 days.

87. There are 52 weeks in 1 year.

$$\begin{array}{r} 248.07 \approx 248.1 \\ 52\overline{)\,12{,}900.00} \\ \underline{-10\,4} \\ 2\,50 \\ \underline{-2\,08} \\ 420 \\ \underline{-416} \\ 4\,0 \\ \underline{-0} \\ 400 \\ \underline{-364} \\ 36 \end{array}$$

Americans aged 18–22 drive, on average, 248.1 miles per week.

89.
$$\begin{array}{r} 345.5 \\ 24\overline{)\,8292.0} \\ \underline{-72} \\ 109 \\ \underline{-96} \\ 132 \\ \underline{-120} \\ 12\,0 \\ \underline{-12\,0} \\ 0 \end{array}$$

There were 345.5 thousand books sold each hour.

91. $\frac{3}{5}\cdot\frac{7}{10}=\frac{3\cdot7}{5\cdot10}=\frac{21}{50}$

93. $\frac{3}{5}-\frac{7}{10}=\frac{3}{5}\cdot\frac{2}{2}-\frac{7}{10}=\frac{3\cdot2}{5\cdot2}-\frac{7}{10}=\frac{6}{10}-\frac{7}{10}=-\frac{1}{10}$

95. $0.3\overline{)1.278}$ becomes
$$\begin{array}{r} 4.26 \\ 3\overline{)\,12.78} \\ \underline{-12} \\ 0\,7 \\ \underline{-6} \\ 18 \\ \underline{-18} \\ 0 \end{array}$$

97.
$$\begin{array}{r} 1.278 \\ \underline{+\,0.300} \\ 1.578 \end{array}$$

99.
$$\begin{array}{rl} 8.6 & \text{1 decimal place} \\ \underline{\times\ 3.1} & \text{1 decimal place} \\ 86 & \\ \underline{25\,80} & \\ 26.66 & 1+1=2 \text{ decimal places} \end{array}$$

$(-8.6)(3.1) = -26.66$

101.
$$\begin{array}{r} 1000.00 \\ \underline{-\ \ 95.71} \\ 904.29 \end{array}$$

103. 8.62×41.7 is approximately $9 \times 40 = 360$, which is choice c.

105. $78.6 \div 97$ is approximately $78.6 \div 100 = 0.786$, which is choice b.

107. $\frac{86+78+91+87}{4}=\frac{342}{4}=85.5$

109. Area = (length)(width)

$4.5\overline{)38.7}$ becomes
$$\begin{array}{r} 8.6 \\ 45\overline{)\,387.0} \\ \underline{-360} \\ 27\,0 \\ \underline{-27\,0} \\ 0 \end{array}$$

The length is 8.6 feet.

111. answers may vary

Copyright © 2011 Pearson Education, Inc. Publishing as Prentice Hall.

113. $1.15\overline{)75}$ becomes $115\overline{)7500.00}$

$$\begin{array}{r} 65.21 \approx 65.2 \\ 115\overline{)\,7500.00} \\ \underline{-690} \\ 600 \\ \underline{-575} \\ 250 \\ \underline{-230} \\ 200 \\ \underline{-115} \\ 85 \end{array}$$

$1.15\overline{)95}$ becomes $115\overline{)9500.00}$

$$\begin{array}{r} 82.60 \approx 82.6 \\ 115\overline{)\,9500.00} \\ \underline{-920} \\ 300 \\ \underline{-230} \\ 700 \\ \underline{-690} \\ 100 \\ \underline{-0} \\ 100 \end{array}$$

The range of wind speeds is 65.2–82.6 knots.

115. First find the length for one round of wire. Then multiply by 4.

$$\begin{array}{r} 24.280 \\ 15.675 \\ 24.280 \\ \underline{+\,15.675} \\ 79.910 \end{array} \qquad \begin{array}{r} 79.91 \\ \underline{\times \quad\ 4} \\ 319.64 \end{array}$$

He will need 319.64 meters of wire.

Integrated Review

1. $$\begin{array}{r} 1.60 \\ \underline{+\,0.97} \\ 2.57 \end{array}$$

2. $$\begin{array}{r} 3.20 \\ \underline{+\,0.85} \\ 4.05 \end{array}$$

3. $$\begin{array}{r} 9.8 \\ \underline{-\,0.9} \\ 8.9 \end{array}$$

4. $$\begin{array}{r} 10.2 \\ \underline{-\ \ 6.7} \\ 3.5 \end{array}$$

5. $$\begin{array}{r} 0.8 \\ \underline{\times\ 0.2} \\ 0.16 \end{array}$$

6. $$\begin{array}{r} 0.6 \\ \underline{\times\ 0.4} \\ 0.24 \end{array}$$

7. $$\begin{array}{r} 0.27 \\ 8\overline{)\,2.16} \\ \underline{-1\,6} \\ 56 \\ \underline{-56} \\ 0 \end{array}$$

8. $$\begin{array}{r} 0.52 \\ 6\overline{)\,3.12} \\ \underline{-3\,0} \\ 12 \\ \underline{-12} \\ 0 \end{array}$$

9. $$\begin{array}{r} 9.6 \\ \underline{\times\ 0.5} \\ 4.80 \end{array}$$

$(9.6)(-0.5) = -4.8$

10. $$\begin{array}{r} 8.7 \\ \underline{\times\ 0.7} \\ 6.09 \end{array}$$

$(-8.7)(-0.7) = 6.09$

11. $$\begin{array}{r} 123.60 \\ \underline{-\ \ 48.04} \\ 75.56 \end{array}$$

12. $$\begin{array}{r} 325.20 \\ \underline{-\ \ 36.08} \\ 289.12 \end{array}$$

13. Subtract absolute values.

$$\begin{array}{r} 25.000 \\ \underline{-\ \ 0.026} \\ 24.974 \end{array}$$

Attach the sign of the larger absolute value.
$-25 + 0.026 = -24.974$

Copyright © 2011 Pearson Education, Inc. Publishing as Prentice Hall.

14. Subtract absolute values.

$$\begin{array}{r} 44.000 \\ -\ \ 0.125 \\ \hline 43.875 \end{array}$$

Attach the sign of the larger absolute value.
$0.125 + (-44) = -43.875$

15. $3.4\overline{)29.24}$ becomes $34\overline{)292.4}$

$$\begin{array}{r} 8.6 \\ 34\overline{)\,292.4} \\ \underline{-272} \\ 20\ 4 \\ \underline{-20\ 4} \\ 0 \end{array}$$

$29.24 \div (-3.4) = -8.6$

16. $1.9\overline{)10.26}$ becomes $19\overline{)102.6}$

$$\begin{array}{r} 5.4 \\ 19\overline{)\,102.6} \\ \underline{-95} \\ 7\ 6 \\ \underline{-7\ 6} \\ 0 \end{array}$$

$-10.26 \div (-1.9) = 5.4$

17. $-2.8 \times 100 = -280$

18. $1.6 \times 1000 = 1600$

19.

$$\begin{array}{r} 96.210 \\ 7.028 \\ +\ 121.700 \\ \hline 224.938 \end{array}$$

20.

$$\begin{array}{r} 0.268 \\ 1.930 \\ +\ 142.881 \\ \hline 145.079 \end{array}$$

21.

$$\begin{array}{r} 0.56 \\ 46\overline{)\,25.76} \\ \underline{-23\ 0} \\ 2\ 76 \\ \underline{-2\ 76} \\ 0 \end{array}$$

$-25.76 \div (-46) = 0.56$

22.

$$\begin{array}{r} 0.63 \\ 43\overline{)\,27.09} \\ \underline{-25\ 8} \\ 1\ 29 \\ \underline{-1\ 29} \\ 0 \end{array}$$

$-27.09 \div 43 = -0.63$

23.

$$\begin{array}{rl} 12.004 & \text{3 decimal places} \\ \times\ \ \ 2.3 & \text{1 decimal place} \\ \hline 3\ 6012 & \\ 24\ 0080 & \\ \hline 27.6092 & 3+1=4 \text{ decimal places} \end{array}$$

24.

$$\begin{array}{rl} 28.006 & \text{3 decimal places} \\ \times\ \ \ 5.2 & \text{1 decimal place} \\ \hline 5\ 6012 & \\ 140\ 0300 & \\ \hline 145.6312 & 3+1=4 \text{ decimal places} \end{array}$$

25.

$$\begin{array}{r} 10.0 \\ -\ \ 4.6 \\ \hline 5.4 \end{array}$$

26. Subtract absolute values.

$$\begin{array}{r} 18.00 \\ -\ \ 0.26 \\ \hline 17.74 \end{array}$$

Attach the sign of the greater absolute value.
$0.26 - 18 = -17.74$

27. $-268.19 - 146.25 = -268.19 + (-146.25)$
Add absolute values.

$$\begin{array}{r} 268.19 \\ +\ 146.25 \\ \hline 414.44 \end{array}$$

Attach the common sign.
$-268.19 - 146.25 = -414.44$

28. $-860.18 - 434.85 = -860.18 + (-434.85)$
Add absolute values.

$$\begin{array}{r} 860.18 \\ +\ \ 434.85 \\ \hline 1295.03 \end{array}$$

Attach the common sign.
$-860.18 - 434.85 = -1295.03$

Copyright © 2011 Pearson Education, Inc. Publishing as Prentice Hall.

29. $0.087\overline{)2.958}$ becomes $87\overline{)2958}$

$$\begin{array}{r} 34 \\ 87\overline{)2958} \\ -261 \\ \hline 348 \\ -348 \\ \hline 0 \end{array}$$

$$\frac{2.958}{-0.087} = -34$$

30. $0.061\overline{)1.708}$ becomes $61\overline{)1708}$

$$\begin{array}{r} 28 \\ 61\overline{)1708} \\ -122 \\ \hline 488 \\ -488 \\ \hline 0 \end{array}$$

$$\frac{-1.708}{0.061} = -28$$

31.
$$\begin{array}{r} 160.00 \\ -\ \ 43.19 \\ \hline 116.81 \end{array}$$

32.
$$\begin{array}{r} 120.00 \\ -101.21 \\ \hline 18.79 \end{array}$$

33. $15.62 \times 10 = 156.2$

34. $15.62 \div 10 = 1.562$

35.
$$\begin{array}{r} 15.62 \\ +10.00 \\ \hline 25.62 \end{array}$$

36.
$$\begin{array}{r} 15.62 \\ -10.00 \\ \hline 5.62 \end{array}$$

37.
$$\begin{array}{rlr} 53.7 & \text{rounds to} & 50 \\ 79.2 & \text{rounds to} & 80 \\ +71.2 & \text{rounds to} & +\ 70 \\ \hline & & 200 \end{array}$$

The estimated distance is 200 miles.

38.
$$\begin{array}{r} 4.80 \\ -4.11 \\ \hline 0.69 \end{array}$$

It costs \$0.69 more to send the package as Priority Mail.

39.
$$\begin{array}{r} 16.0 \\ +\ 7.5 \\ \hline 23.5 \end{array}$$

$$\begin{aligned} 23.5 \text{ billion} &= 23.5 \times 1 \text{ billion} \\ &= 23.5 \times 1{,}000{,}000{,}000 \\ &= 23{,}500{,}000{,}000 \end{aligned}$$

The total amount spent was \$23.5 billion or \$23,500,000,000.

Section 5.5

Practice Problems

1. a.
$$\begin{array}{r} 0.4 \\ 5\overline{)2.0} \\ -2.0 \\ \hline 0 \end{array}$$

$$\frac{2}{5} = 0.4$$

b.
$$\begin{array}{r} 0.225 \\ 40\overline{)9.000} \\ -8\ 0 \\ \hline 1\ 00 \\ -80 \\ \hline 200 \\ -200 \\ \hline 0 \end{array}$$

$$\frac{9}{40} = 0.225$$

2.
$$\begin{array}{r} 0.375 \\ 8\overline{)3.000} \\ -2\ 4 \\ \hline 60 \\ -56 \\ \hline 40 \\ -40 \\ \hline 0 \end{array}$$

$$-\frac{3}{8} = -0.375$$

3. a.
$$\begin{array}{r} 0.833... \\ 6\overline{)5.000} \\ -4\ 8 \\ \hline 20 \\ -18 \\ \hline 20 \\ -18 \\ \hline 2 \end{array}$$

$$\frac{5}{6} = 0.8\overline{3}$$

Copyright © 2011 Pearson Education, Inc. Publishing as Prentice Hall.

b.
$$\begin{array}{r} 0.22\ldots \\ 9\overline{)\,2.00} \\ \underline{-1\,8} \\ 20 \\ \underline{-18} \\ 2 \end{array}$$

$\frac{2}{9} = 0.\overline{2}$

4.
$$\begin{array}{r} 2.1538 \approx 2.154 \\ 13\overline{)\,28.0000} \\ \underline{-26} \\ 2\,0 \\ \underline{-1\,3} \\ 70 \\ \underline{-65} \\ 50 \\ \underline{-39} \\ 110 \\ \underline{-104} \\ 6 \end{array}$$

5. $3\frac{5}{16} = \frac{53}{16}$
$$\begin{array}{r} 3.3125 \\ 16\overline{)\,53.0000} \\ \underline{-48} \\ 5\,0 \\ \underline{-4\,8} \\ 20 \\ \underline{-16} \\ 40 \\ \underline{-32} \\ 80 \\ \underline{-80} \\ 0 \end{array}$$

Thus, $3\frac{5}{16} = 3.3125.$

6. $\frac{3}{5} = \frac{3}{5}\cdot\frac{2}{2} = \frac{6}{10} = 0.6$

7. $\frac{3}{50} = \frac{3}{50}\cdot\frac{2}{2} = \frac{6}{100} = 0.06$

8.
$$\begin{array}{r} 0.2 \\ 5\overline{)\,1.0} \\ \underline{-1\,0} \\ 0 \end{array}$$

Since $0.2 < 0.25$, then $\frac{1}{5} < 0.25$.

9. a. $\frac{1}{2} = 0.5$ and $0.5 < 0.54$, so $\frac{1}{2} < 0.54$.

b.
$$\begin{array}{r} 0.55\ldots \\ 9\overline{)\,5.00} \\ \underline{-4\,5} \\ 50 \\ \underline{-45} \\ 5 \end{array}$$

$0.\overline{5} = 0.55\ldots$, so $0.\overline{5} = \frac{5}{9}$.

c.
$$\begin{array}{r} 0.714 \\ 7\overline{)\,5.000} \\ \underline{-4\,9} \\ 10 \\ \underline{-7} \\ 30 \\ \underline{-28} \\ 2 \end{array}$$

$0.714 < 0.72$, so $\frac{5}{7} < 0.72$.

10. a. $\frac{1}{3} = 0.333\ldots$

$0.302 = 0.302$

$\frac{3}{8} = 0.375$

$0.302, \frac{1}{3}, \frac{3}{8}$

b. $1.26 = 1.26$

$1\frac{1}{4} = 1.25$

$1\frac{2}{5} = 1.40$

$1\frac{1}{4}, 1.26, 1\frac{2}{5}$

Copyright © 2011 Pearson Education, Inc. Publishing as Prentice Hall.

c. $0.4 = 0.40$
$0.41 = 0.41$
$\frac{3}{7} \approx 0.43$
$0.4, 0.41, \frac{3}{7}$

11. $897.8 \div 100 \times 10 = 8.978 \times 10 = 89.78$

12. $-8.69(3.2 - 1.8) = -8.69(1.4) = -12.166$

13. $(-0.7)^2 + 2.1 = 0.49 + 2.10 = 2.59$

14.
$$\begin{aligned}\frac{20.06-(1.2)^2 \div 10}{0.02} &= \frac{20.06 - 1.44 \div 10}{0.02} \\ &= \frac{20.06 - 0.144}{0.02} \\ &= \frac{19.916}{0.02} \\ &= 995.8\end{aligned}$$

15.
$$\begin{aligned}\text{Area} &= \frac{1}{2}\cdot \text{base}\cdot \text{height} \\ &= \frac{1}{2}\cdot 7\cdot 2.1 \\ &= 0.5\cdot 7\cdot 2.1 \\ &= 7.35\end{aligned}$$
The area of the triangle is 7.35 square meters.

16. $1.7y - 2 = 1.7(2.3) - 2 = 3.91 - 2 = 1.91$

Vocabulary and Readiness Check

1. The number $0.\overline{5}$ means 0.555. false

2. To write $\frac{9}{19}$ as a decimal, perform the division $19\overline{)9}$. true

3. $(-1.2)^2$ means $(-1.2)(-1.2)$ or -1.44. false

4. To simplify $8.6(4.8 - 9.6)$, we first subtract. true

Exercise Set 5.5

1.
$$\begin{array}{r} 0.2 \\ 5\overline{)\;1\,0} \\ -1.0 \\ \hline 0 \end{array}$$
$\frac{1}{5} = 0.2$

3.
$$\begin{array}{r} 0.68 \\ 25\overline{)\;17.00} \\ -15\,0 \\ \hline 2\,00 \\ -2\,00 \\ \hline 0 \end{array}$$
$\frac{17}{25} = 0.68$

5.
$$\begin{array}{r} 0.75 \\ 4\overline{)\;3.00} \\ -2\,8 \\ \hline 20 \\ -20 \\ \hline 0 \end{array}$$
$\frac{3}{4} = 0.75$

7.
$$\begin{array}{r} 0.08 \\ 25\overline{)\;2.00} \\ -2\,00 \\ \hline 0 \end{array}$$
$-\frac{2}{25} = -0.08$

9.
$$\begin{array}{r} 2.25 \\ 4\overline{)\;9.00} \\ -8 \\ \hline 1\,0 \\ -\,8 \\ \hline 20 \\ -20 \\ \hline 0 \end{array}$$
$\frac{9}{4} = 2.25$

11.
$$\begin{array}{r} 0.9166... \\ 12\overline{)\;11.0000} \\ -10\,8 \\ \hline 20 \\ -12 \\ \hline 80 \\ -72 \\ \hline 80 \\ -72 \\ \hline 8 \end{array}$$
$\frac{11}{12} = 0.91\overline{6}$

13.
$$\begin{array}{r} 0.425 \\ 40\overline{)\;17.000} \\ -16\,0 \\ \hline 1\,00 \\ -80 \\ \hline 200 \\ -200 \\ \hline 0 \end{array}$$
$\frac{17}{40} = 0.425$

Copyright © 2011 Pearson Education, Inc. Publishing as Prentice Hall.

15.
$$\begin{array}{r} 0.45 \\ 20\overline{)\ 9.00} \\ -8\ 0 \\ \hline 1\ 00 \\ -1\ 00 \\ \hline 0 \end{array}$$
$\frac{9}{20} = 0.45$

17.
$$\begin{array}{r} 0.333... \\ 3\overline{)1.000} \\ -9 \\ \hline 10 \\ -9 \\ \hline 10 \\ -9 \\ \hline 1 \end{array}$$
$-\frac{1}{3} = -0.\overline{3}$

19.
$$\begin{array}{r} 0.4375 \\ 16\overline{)\ 7.0000} \\ -6\ 4 \\ \hline 60 \\ -48 \\ \hline 120 \\ -112 \\ \hline 80 \\ -80 \\ \hline 0 \end{array}$$
$\frac{7}{16} = 0.4375$

21.
$$\begin{array}{r} 0.636363... \\ 11\overline{)\ 7.000000} \\ -6\ 6 \\ \hline 40 \\ -33 \\ \hline 70 \\ -66 \\ \hline 40 \\ -33 \\ \hline 70 \\ -66 \\ \hline 40 \\ -33 \\ \hline 7 \end{array}$$
$\frac{7}{11} = 0.\overline{63}$

23.
$$\begin{array}{r} 0.85 \\ 20\overline{)\ 17.00} \\ -16\ 0 \\ \hline 1\ 00 \\ -1\ 00 \\ \hline 0 \end{array}$$
$5\frac{17}{20} = 5.85$

25.
$$\begin{array}{r} 0.624 \\ 125\overline{)\ 78.000} \\ -75\ 0 \\ \hline 3\ 00 \\ -2\ 50 \\ \hline 500 \\ -500 \\ \hline 0 \end{array}$$
$\frac{78}{125} = 0.624$

27. $-\frac{1}{3} = -0.33\overline{3} \approx -0.33$

29. $\frac{7}{16} = 0.4375 \approx 0.44$

31. $\frac{7}{11} = 0.63\overline{63} \approx 0.6$

33.
$$\begin{array}{r} 0.615 \approx 0.62 \\ 91\overline{)\ 56.000} \\ -54\ 6 \\ \hline 1\ 40 \\ -91 \\ \hline 490 \\ -455 \\ \hline 35 \end{array}$$

35.
$$\begin{array}{r} 0.731 \approx 0.73 \\ 97\overline{)\ 71.000} \\ -\ 67\ 9 \\ \hline 3\ 10 \\ -2\ 91 \\ \hline 190 \\ -97 \\ \hline 93 \end{array}$$

37.
$$\begin{array}{r} 0.02 \\ 50\overline{)\ 1.00} \\ -1\ 00 \\ \hline 0 \end{array}$$

Copyright © 2011 Pearson Education, Inc. Publishing as Prentice Hall.

39. $0.562 \quad 0.569$

$$\begin{array}{ccc} \uparrow & & \uparrow \\ 2 & < & 9 \end{array}$$

so, $0.562 < 0.569$

41.
$$\begin{array}{r} 0.215 \\ 200\overline{)\ 43.000} \\ -40\ 0 \\ \hline 3\ 00 \\ -2\ 00 \\ \hline 1\ 000 \\ -1\ 000 \\ \hline 0 \end{array}$$

$0.215 = \frac{43}{200}$

43. $0.0932 \quad 0.0923$

$$\begin{array}{ccc} \uparrow & & \uparrow \\ 3 & > & 2 \end{array}$$

so, $0.0932 > 0.0923$

Thus, $-0.0932 < -0.0923$.

45.
$$\begin{array}{r} 0.833... \\ 6\overline{)\ 5.000} \\ -4\ 8 \\ \hline 20 \\ -18 \\ \hline 20 \\ -18 \\ \hline 2 \end{array}$$

$\frac{5}{6} = 0.8\overline{3}$ and $0.\overline{6} < 0.8\overline{3}$, so $0.\overline{6} < \frac{5}{6}$

47.
$$\begin{array}{r} 0.5604 \approx 0.560 \\ 91\overline{)\ 51.0000} \\ -45\ 5 \\ \hline 5\ 50 \\ -5\ 46 \\ \hline 40 \\ -0 \\ \hline 400 \\ -364 \\ \hline 36 \end{array}$$

$\frac{51}{91} \approx 0.560$ and $0.560 < 0.56\overline{4}$, so $\frac{51}{91} < 0.56\overline{4}$.

49.
$$\begin{array}{r} 0.571 \approx 0.57 \\ 7\overline{)\ 4.000} \\ -3\ 5 \\ \hline 50 \\ -49 \\ \hline 10 \\ -7 \\ \hline 3 \end{array}$$

$\frac{4}{7} \approx 0.57$ and $0.57 > 0.14$, so $\frac{4}{7} > 0.14$.

51.
$$\begin{array}{r} 1.3846 \approx 1.385 \\ 13\overline{)\ 18.0000} \\ -13 \\ \hline 5\ 0 \\ -3\ 9 \\ \hline 1\ 10 \\ -1\ 04 \\ \hline 60 \\ -52 \\ \hline 80 \\ -78 \\ \hline 2 \end{array}$$

$\frac{18}{13} \approx 1.385$ and $1.38 < 1.385$, so $1.38 < \frac{18}{13}$.

53.
$$\begin{array}{r} 7.125 \\ 64\overline{)\ 456.000} \\ -448 \\ \hline 8\ 0 \\ -6\ 4 \\ \hline 1\ 60 \\ -1\ 28 \\ \hline 320 \\ -320 \\ \hline 0 \end{array}$$

$\frac{456}{64} = 7.125$ and $7.123 < 7.125$, so

$7.123 < \frac{456}{64}$.

55. 0.32, 0.34, 0.35

57. $0.49 = 0.490$

0.49, 0.491, 0.498

Copyright © 2011 Pearson Education, Inc. Publishing as Prentice Hall.

59. $\frac{42}{8} = 5.25$

$5.23, \frac{42}{8}, 5.34$

61. $\frac{5}{8} = 0.625$

$0.612, \frac{5}{8}, 0.649$

63. $(0.3)^2 + 0.5 = 0.09 + 0.5 = 0.59$

65. $\frac{1+0.8}{-0.6} = \frac{1.8}{-0.6} = \frac{18}{-6} = -3$

67. $$\begin{aligned}(-2.3)^2(0.3+0.7) &= (-2.3)^2(1.0)\\ &= 5.29(1.0)\\ &= 5.29\end{aligned}$$

69. $(5.6 - 2.3)(2.4 + 0.4) = (3.3)(2.8) = 9.24$

71. $\frac{(4.5)^2}{100} = \frac{20.25}{100} = 0.2025$

73. $\frac{7+0.74}{-6} = \frac{7.74}{-6} = -1.29$

75. $\frac{1}{5} - 2(7.8) = \frac{1}{5} - 15.6 = 0.2 - 15.6 = -15.4$

77. $\frac{1}{4}(-9.6 - 5.2) = \frac{1}{4}(-14.8) = 0.25(-14.8) = -3.7$

79. $$\begin{aligned}\text{Area} &= \frac{1}{2}\cdot b\cdot h\\ &= \frac{1}{2}(5.7)(9)\\ &= 0.5(5.7)(9)\\ &= 25.65 \text{ square inches}\end{aligned}$$

81. $$\begin{aligned}\text{Area} &= l\cdot w\\ &= (0.62)\left(\frac{2}{5}\right)\\ &= (0.62)(0.4)\\ &= 0.248 \text{ square yard}\end{aligned}$$

83. $z^2 = (-2.4)^2 = 5.76$

85. $x - y = 6 - 0.3 = 5.7$

87. $4y - z = 4 \cdot 0.3 - (-2.4) = 1.2 + 2.4 = 3.6$

89. $\frac{9}{10} + \frac{16}{25} = \frac{9}{10}\cdot\frac{5}{5} + \frac{16}{25}\cdot\frac{2}{2} = \frac{45}{50} + \frac{32}{50} = \frac{77}{50}$

91. $\left(\frac{2}{5}\right)\left(\frac{5}{2}\right)^2 = \left(\frac{2}{5}\right)\left(\frac{5}{2}\right)\left(\frac{5}{2}\right) = \frac{2\cdot 5\cdot 5}{5\cdot 2\cdot 2} = \frac{5}{2}$

93. $1.0 = 1$

95. $1.00001 > 1$

97. $99 < 100$, so $\frac{99}{100} < 1$

99. $$\begin{array}{r} 0.0144 \approx 0.014 \\ 14{,}120\overline{)\,204.0000} \\ -141\ 20 \\ \hline 62\ 800 \\ -56\ 480 \\ \hline 6\ 3200 \\ -5\ 6480 \\ \hline 6720 \end{array}$$

Approximately 0.014 of radio stations had a hip hop music format.

101.

2092	rounds to	2100
1342	rounds to	1300
204	rounds to	200
455	rounds to	500
745	rounds to	700
+ 436	rounds to	+ 400
		5200

The total number of stations with the top six formats was about 5200 stations.

103. answers may vary

Section 5.6

Practice Problems

1. $$\begin{aligned} z + 0.9 &= 1.3\\ z + 0.9 - 0.9 &= 1.3 - 0.9\\ z &= 0.4\end{aligned}$$

2. $$\begin{aligned} 0.17x &= -0.34\\ \frac{0.17x}{0.17} &= \frac{-0.34}{0.17}\\ x &= -2\end{aligned}$$

Copyright © 2011 Pearson Education, Inc. Publishing as Prentice Hall.

3.
$$2.9 = 1.7 + 0.3x$$
$$2.9 - 1.7 = 1.7 + 0.3x - 1.7$$
$$1.2 = 0.3x$$
$$\frac{1.2}{0.3} = \frac{0.3x}{0.3}$$
$$4 = x$$

4.
$$8x + 4.2 = 10x + 11.6$$
$$8x + 4.2 - 4.2 = 10x + 11.6 - 4.2$$
$$8x = 10x + 7.4$$
$$8x - 10x = 10x - 10x + 7.4$$
$$-2x = 7.4$$
$$\frac{-2x}{-2} = \frac{7.4}{-2}$$
$$x = -3.7$$

5.
$$6.3 - 5x = 3(x + 2.9)$$
$$6.3 - 5x = 3x + 8.7$$
$$6.3 - 5x - 6.3 = 3x + 8.7 - 6.3$$
$$-5x = 3x + 2.4$$
$$-5x - 3x = 3x + 2.4 - 3x$$
$$-8x = 2.4$$
$$\frac{-8x}{-8} = \frac{2.4}{-8}$$
$$x = -0.3$$

6.
$$0.2y + 2.6 = 4$$
$$10(0.2y + 2.6) = 10(4)$$
$$10(0.2y) + 10(2.6) = 10(4)$$
$$2y + 26 = 40$$
$$2y + 26 - 26 = 40 - 26$$
$$2y = 14$$
$$\frac{2y}{2} = \frac{14}{2}$$
$$y = 7$$

Exercise Set 5.6

1.
$$x + 1.2 = 7.1$$
$$x + 1.2 - 1.2 = 7.1 - 1.2$$
$$x = 5.9$$

3.
$$-5y = 2.15$$
$$\frac{-5y}{-5} = \frac{2.15}{-5}$$
$$y = -0.43$$

5.
$$6.2 = y - 4$$
$$6.2 + 4 = y - 4 + 4$$
$$10.2 = y$$

7.
$$3.1x = -13.95$$
$$\frac{3.1x}{3.1} = \frac{-13.95}{3.1}$$
$$x = -4.5$$

9.
$$-3.5x + 2.8 = -11.2$$
$$-3.5x + 2.8 - 2.8 = -11.2 - 2.8$$
$$-3.5x = -14$$
$$\frac{-3.5x}{-3.5} = \frac{-14}{-3.5}$$
$$x = 4$$

11.
$$6x + 8.65 = 3x + 10$$
$$6x + 8.65 - 8.65 = 3x + 10 - 8.65$$
$$6x = 3x + 1.35$$
$$6x - 3x = 3x - 3x + 1.35$$
$$3x = 1.35$$
$$\frac{3x}{3} = \frac{1.35}{3}$$
$$x = 0.45$$

13.
$$2(x - 1.3) = 5.8$$
$$2x - 2.6 = 5.8$$
$$2x - 2.6 + 2.6 = 5.8 + 2.6$$
$$2x = 8.4$$
$$\frac{2x}{2} = \frac{8.4}{2}$$
$$x = 4.2$$

15.
$$0.4x + 0.7 = -0.9$$
$$10(0.4x + 0.7) = 10(-0.9)$$
$$4x + 7 = -9$$
$$4x + 7 - 7 = -9 - 7$$
$$4x = -16$$
$$\frac{4x}{4} = \frac{-16}{4}$$
$$x = -4$$

17.
$$7x - 10.8 = x$$
$$10(7x - 10.8) = 10 \cdot x$$
$$70x - 108 = 10x$$
$$70x - 70x - 108 = 10x - 70x$$
$$-108 = -60x$$
$$\frac{-108}{-60} = \frac{-60x}{-60}$$
$$1.8 = x$$

Copyright © 2011 Pearson Education, Inc. Publishing as Prentice Hall.

19.
$$\begin{aligned} 2.1x+5-1.6x &= 10 \\ 10(2.1x+5-1.6x) &= 10\cdot 10 \\ 21x+50-16x &= 100 \\ 5x+50 &= 100 \\ 5x+50-50 &= 100-50 \\ 5x &= 50 \\ \frac{5x}{5} &= \frac{50}{5} \\ x &= 10 \end{aligned}$$

21.
$$\begin{aligned} y-3.6 &= 4 \\ y-3.6+3.6 &= 4+3.6 \\ y &= 7.6 \end{aligned}$$

23.
$$\begin{aligned} -0.02x &= -1.2 \\ \frac{-0.02x}{-0.02} &= \frac{-1.2}{-0.02} \\ x &= 60 \end{aligned}$$

25.
$$\begin{aligned} 6.5 &= 10x+7.2 \\ 6.5-7.2 &= 10x+7.2-7.2 \\ -0.7 &= 10x \\ \frac{-0.7}{10} &= \frac{10x}{10} \\ -0.07 &= x \end{aligned}$$

27.
$$\begin{aligned} 2.7x-25 &= 1.2x+5 \\ 2.7x-25+25 &= 1.2x+5+25 \\ 2.7x &= 1.2x+30 \\ 2.7x-1.2x &= 1.2x-1.2x+30 \\ 1.5x &= 30 \\ \frac{1.5x}{1.5} &= \frac{30}{1.5} \\ x &= 20 \end{aligned}$$

29.
$$\begin{aligned} 200x-0.67 &= 100x+0.81 \\ 200x-0.67+0.67 &= 100x+0.81+0.67 \\ 200x &= 100x+1.48 \\ 200x-100x &= 100x-100x+1.48 \\ 100x &= 1.48 \\ \frac{100x}{100} &= \frac{1.48}{100} \\ x &= 0.0148 \end{aligned}$$

31.
$$\begin{aligned} 3(x+2.71) &= 2x \\ 3x+8.13 &= 2x \\ 3x-3x+8.13 &= 2x-3x \\ 8.13 &= -x \\ \frac{8.13}{-1} &= \frac{-x}{-1} \\ -8.13 &= x \end{aligned}$$

33.
$$\begin{aligned} 8x-5 &= 10x-8 \\ 8x-5+8 &= 10x-8+8 \\ 8x+3 &= 10x \\ 8x+3-8x &= 10x-8x \\ 3 &= 2x \\ \frac{3}{2} &= \frac{2x}{2} \\ 1.5 &= x \end{aligned}$$

35.
$$\begin{aligned} 1.2+0.3x &= 0.9 \\ 1.2+0.3x-1.2 &= 0.9-1.2 \\ 0.3x &= -0.3 \\ \frac{0.3x}{0.3} &= \frac{-0.3}{0.3} \\ x &= -1 \end{aligned}$$

37.
$$\begin{aligned} -0.9x+2.65 &= -0.5x+5.45 \\ 100(-0.9x+2.65) &= 100(-0.5x+5.45) \\ -90x+265 &= -50x+545 \\ -90x+265+90x &= -50x+545+90x \\ 265 &= 40x+545 \\ 265-545 &= 40x+545-545 \\ -280 &= 40x \\ \frac{-280}{40} &= \frac{40x}{40} \\ -7 &= x \end{aligned}$$

39.
$$\begin{aligned} 4x+7.6 &= 2(3x-3.2) \\ 4x+7.6 &= 6x-6.4 \\ 10(4x+7.6) &= 10(6x-6.4) \\ 40x+76 &= 60x-64 \\ 40x+76+64 &= 60x-64+64 \\ 40x+140 &= 60x \\ 40x-40x+140 &= 60x-40x \\ 140 &= 20x \\ \frac{140}{20} &= \frac{20x}{20} \\ 7 &= x \end{aligned}$$

41.
$$\begin{aligned} 0.7x+13.8 &= x-2.16 \\ 100(0.7x+13.8) &= 100(x-2.16) \\ 70x+1380 &= 100x-216 \\ 70x+1380+216 &= 100x-216+216 \\ 70x+1596 &= 100x \\ 70x+1596-70x &= 100x-70x \\ 1596 &= 30x \\ \frac{1596}{30} &= \frac{30x}{30} \\ 53.2 &= x \end{aligned}$$

43. $2x-7+x-9=(2x+x)+(-7-9)=3x-16$

Copyright © 2011 Pearson Education, Inc. Publishing as Prentice Hall.

45. $\frac{6x}{5}\cdot\frac{1}{2x^2}=\frac{6x\cdot 1}{5\cdot 2x^2}=\frac{2\cdot 3\cdot x}{5\cdot 2\cdot x\cdot x}=\frac{3}{5x}$

47. $$\begin{aligned}\frac{x}{3}+\frac{2x}{7}&=\frac{x}{3}\cdot\frac{7}{7}+\frac{2x}{7}\cdot\frac{3}{3}\\&=\frac{7x}{21}+\frac{6x}{21}\\&=\frac{7x+6x}{21}\\&=\frac{13x}{21}\end{aligned}$$

49. $$\begin{aligned}b+4.6&=8.3\\b+4.6-4.6&=8.3-4.6\\b&=3.7\end{aligned}$$

51. $$\begin{aligned}2x-0.6+4x-0.01&=2x+4x-0.6-0.01\\&=6x-0.61\end{aligned}$$

53. $$\begin{aligned}5y-1.2-7y+8&=5y-7y-1.2+8\\&=-2y+6.8\end{aligned}$$

55. $$\begin{aligned}2.8&=z-6.3\\2.8+6.3&=z-6.3+6.3\\9.1&=z\end{aligned}$$

57. $$\begin{aligned}4.7x+8.3&=-5.8\\4.7x+8.3-8.3&=-5.8-8.3\\4.7x&=-14.1\\\frac{4.7x}{4.7}&=\frac{-14.1}{4.7}\\x&=-3\end{aligned}$$

59. $$\begin{aligned}7.76+8z-12z+8.91&=8z-12z+7.76+8.91\\&=-4z+16.67\end{aligned}$$

61. $$\begin{aligned}5(x-3.14)&=4x\\5\cdot x-5\cdot 3.14&=4x\\5x-15.7&=4x\\5x-5x-15.7&=4x-5x\\-15.7&=-x\\15.7&=x\end{aligned}$$

63. $$\begin{aligned}2.6y+8.3&=4.6y-3.4\\10(2.6y+8.3)&=10(4.6y-3.4)\\26y+83&=46y-34\\26y+83-83&=46y-34-83\\26y&=46y-117\\26y-46y&=46y-46y-117\\-20y&=-117\\\frac{-20y}{-20}&=\frac{-117}{-20}\\y&=5.85\end{aligned}$$

65. $$\begin{aligned}9.6z-3.2-11.7z-6.9&=9.6z-11.7z-3.2-6.9\\&=-2.1z-10.1\end{aligned}$$

67. answers may vary

69. answers may vary

71. $$\begin{aligned}-5.25x&=-40.33575\\\frac{-5.25x}{-5.25}&=\frac{-40.33575}{-5.25}\\x&=7.683\end{aligned}$$

73. $$\begin{aligned}1.95y+6.834&=7.65y-19.8591\\1.95y+6.834-6.834&=7.65y-19.8591-6.834\\1.95y&=7.65y-26.6931\\1.95y-7.65y&=7.65y-7.65y-26.6931\\-5.7y&=-26.6931\\\frac{-5.7y}{-5.7}&=\frac{-26.6931}{-5.7}\\y&=4.683\end{aligned}$$

Section 5.7

Practice Problems

1. Mean $=\frac{87+75+96+91+78}{5}=\frac{427}{5}=85.4$

2. gpa $=\frac{4\cdot 2+3\cdot 4+2\cdot 5+1\cdot 2+4\cdot 2}{2+4+5+2+2}=\frac{40}{15}\approx 2.67$

3. Because the numbers are in numerical order, and there are an odd number of items, the median is the middle number, 24.

4. Write the numbers in numerical order:
36, 65, 71, 78, 88, 91, 95, 95
Since there are an even number of scores, the median is the mean of the two middle numbers.
median $=\frac{78+88}{2}=83$

Copyright © 2011 Pearson Education, Inc. Publishing as Prentice Hall.

5. Mode: 15 because it occurs most often, 3 times.

6. Median: Write the numbers in order.
15, 15, 15, 16, 18, 26, 26, 30, 31, 35
Median is mean of middle two numbers,
$\frac{18+26}{2}=22.$
Mode: 15 because it occurs most often, 3 times.

Vocabulary and Readiness Check

1. Another word for "mean" is <u>average</u>.

2. The number that occurs most often in a set of numbers is called the <u>mode</u>.

3. The <u>mean (or average)</u> of a set of number items is $\frac{\text{sum of items}}{\text{number of items}}$.

4. The <u>median</u> of a set of numbers is the middle number. If the number of numbers is even, it is the <u>mean (or average)</u> of the two middle numbers.

5. An example of weighted mean is a calculation of <u>grade point average</u>.

Exercise Set 5.7

1. Mean: $\frac{15+23+24+18+25}{5}=\frac{105}{5}=21$
Median: Write the numbers in order:
15, 18, 23, 24, 25
The middle number is 23.
Mode: There is no mode, since each number occurs once.

3. Mean:
$\frac{7.6+8.2+8.2+9.6+5.7+9.1}{6}=\frac{48.4}{6}\approx 8.1$
Median: Write the numbers in order:
5.7, 7.6, 8.2, 8.2, 9.1, 9.6
Median is mean of middle two: $\frac{8.2+8.2}{2}=8.2$
Mode: 8.2 since this number appears twice.

5. Mean:
$\frac{0.5+0.2+0.2+0.6+0.3+1.3+0.8+0.1+0.5}{9}$
$=\frac{4.5}{9}$
$=0.5$
Median: Write the numbers in order:
0.1, 0.2, 0.2, 0.3, 0.5, 0.5, 0.6, 0.8, 1.3
The middle number is 0.5.
Mode: Since 0.2 and 0.5 occur twice, there are two modes, 0.2 and 0.5.

7. Mean:
$\frac{231+543+601+293+588+109+334+268}{8}$
$=\frac{2967}{8}$
≈ 370.9
Median: Write the numbers in order:
109, 231, 268, 293, 334, 543, 588, 601
The mean of the middle two: $\frac{293+334}{2}=313.5$
Mode: There is no mode, since each number occurs once.

9. Mean:
$\frac{1670+1614+1483+1483+1451}{5}=\frac{7701}{5}$
$=1540.2$ feet

11. Because the numbers are in numerical order, the median is mean of the middle two (of the top 8),
$\frac{1483+1451}{2}=1467$ feet.

13. answers may vary

15. $\text{gpa}=\frac{3\cdot 3+2\cdot 3+4\cdot 4+2\cdot 4}{3+3+4+4}=\frac{39}{14}\approx 2.79$

17. $\text{gpa}=\frac{4\cdot 3+4\cdot 3+4\cdot 4+3\cdot 3+2\cdot 1}{3+3+4+3+1}$
$=\frac{51}{14}$
≈ 3.64

19. Mean:
$\frac{7.8+6.9+7.5+4.7+6.9+7.0}{6}=\frac{40.8}{6}=6.8$

21. Mode: 6.9, since this number appears twice.

23. Median: Write the numbers in order.
79, 85, 88, 89, 91, 93
The mean of the middle two: $\frac{88+89}{2}=88.5$

25. Mean: $\frac{\text{sum of 15 pulse rates}}{15}=\frac{1095}{15}=73$

Copyright © 2011 Pearson Education, Inc. Publishing as Prentice Hall.

27. Mode: Since 70 and 71 occur twice, there are two modes, 70 and 71.

29. There were 9 rates lower than the mean. They are 66, 68, 71, 64, 71, 70, 65, 70, and 72.

31. $\frac{6}{18}=\frac{1\cdot 6}{3\cdot 6}=\frac{1}{3}$

33. $\frac{18}{30y}=\frac{3\cdot 6}{5\cdot 6\cdot y}=\frac{3}{5y}$

35. $\frac{55y^2}{75y^2}=\frac{5\cdot 11\cdot y\cdot y}{5\cdot 15\cdot y\cdot y}=\frac{11}{15}$

37. Since the mode is 35, 35 must occur at least twice in the set.
Since there is an odd number of numbers in the set, the median, 37 is in the set.
Let n be the remaining unknown number.

$$\text{Mean: } \frac{35+35+37+40+n}{5}=38$$
$$\frac{147+n}{5}=38$$
$$5\cdot\frac{147+n}{5}=5\cdot 38$$
$$147+n=190$$
$$147-147+n=190-147$$
$$n=43$$

The missing numbers are 35, 35, 37, and 43.

39. yes; answers may vary

Chapter 5 Vocabulary Check

1. Like fractional notation, decimal notation is used to denote a part of a whole.

2. To write fractions as decimals, divide the numerator by the denominator.

3. To add or subtract decimals, write the decimals so that the decimal points line up vertically.

4. When writing decimals in words, write "and" for the decimal point.

5. When multiplying decimals, the decimal point in the product is placed so that the number of decimal places in the product is equal to the sum of the number of decimal places in the factors.

6. The mode of a set of numbers is the number that occurs most often.

7. The distance around a circle is called the circumference.

8. The median of a set of numbers in numerical order is the middle number. If there are an even number of numbers, the mode is the mean of the two middle numbers.

9. The mean of a list of numbers of items is $\frac{\text{sum of items}}{\text{number of items}}$.

10. When 2 million is written as 2,000,000, we say it is written in standard form.

Chapter 5 Review

1. In 23.45, the 4 is in the tenths place.

2. In 0.000345, the 4 is in the hundred-thousandths place.

3. –23.45 in words is negative twenty-three and forty-five hundredths.

4. 0.00345 in words is three hundred forty-five hundred-thousandths.

5. 109.23 in words is one hundred nine and twenty-three hundredths.

6. 200.000032 in words is two hundred and thirty-two millionths.

7. Eight and six hundredths is 8.06.

8. Negative five hundred three and one hundred two thousandths is –503.102.

9. Sixteen thousand twenty-five and fourteen ten-thousandths is 16,025.0014.

10. Fourteen and eleven thousandths is 14.011.

11. $0.16=\frac{16}{100}=\frac{4\cdot 4}{25\cdot 4}=\frac{4}{25}$

12. $-12.023=-12\frac{23}{1000}$

13. $\frac{231}{100,000}=0.00231$

14. $25\frac{1}{4}=25\frac{25}{100}=25.25$

Copyright © 2011 Pearson Education, Inc. Publishing as Prentice Hall.

15. 0.49 0.43
$\uparrow$ $\uparrow$
9 > 3
so 0.49 > 0.43

16. 0.973 = 0.9730

17. 38.0027 38.00056
$\uparrow$ $\uparrow$
2 > 0
so 38.0027 > 38.00056
Thus, −38.0027 < −38.00056.

18. 0.230505 0.23505
$\uparrow$ $\uparrow$
0 < 5
so 0.230505 < 0.23505
Thus, −0.230505 > −0.23505.

19. To round 0.623 to the nearest tenth, observe that the digit in the hundredths place is 2. Since this digit is less than 5, we do not add 1 to the digit in the tenths place. The number 0.623 rounded to the nearest tenth is 0.6.

20. To round 0.9384 to the nearest hundredth, observe that the digit in the thousandths place is 8. Since this digit is at least 5, we add 1 to the digit in the hundredths place. The number 0.9384 rounded to the nearest hundredth is 0.94.

21. To round −42.895 to the nearest hundredth, observe that the digit in the thousandths place is 5. Since this digit is at least 5, we add 1 to the digit in the hundredths place. The number −42.895 rounded to the nearest hundredth is −42.90.

22. To round 16.34925 to the nearest thousandth, observe that the digit in the ten-thousandths place is 2. Since this digit is less than 5, we do not add 1 to the digit in the thousandths place. The number 16.34925 rounded to the nearest thousandth is 16.349.

23. $$\begin{aligned}887 \text{ million} &= 887\times 1 \text{ million}\\ &= 887\times 1,000,000\\ &= 887,000,000\end{aligned}$$

24. $$\begin{aligned}600 \text{ thousand} &= 600\times 1 \text{ thousand}\\ &= 600\times 1000\\ &= 600,000\end{aligned}$$

25. $$\begin{array}{r} 8.6\\ +\,9.5\\ \hline 18.1\end{array}$$

26. $$\begin{array}{r} 3.9\\ +\,1.2\\ \hline 5.1\end{array}$$

27. Add the absolute values.
$$\begin{array}{r} 6.40\\ +\,0.88\\ \hline 7.28\end{array}$$
Attach the common sign.
−6.4 + (−0.88) = −7.28

28. Subtract the absolute values.
$$\begin{array}{r} 19.02\\ -\,6.98\\ \hline 12.04\end{array}$$
Attach the sign of the larger absolute value.
−19.02 + 6.98 = −12.04

29. $$\begin{array}{r} 200.490\\ 16.820\\ +\,103.002\\ \hline 320.312\end{array}$$

30. $$\begin{array}{r} 0.00236\\ 100.45000\\ +\,48.29000\\ \hline 148.74236\end{array}$$

31. $$\begin{array}{r} 4.9\\ -\,3.2\\ \hline 1.7\end{array}$$

32. $$\begin{array}{r} 5.23\\ -\,2.74\\ \hline 2.49\end{array}$$

33. −892.1 − 432.4 = −892.1 + (−432.4)
Add the absolute values.
$$\begin{array}{r} 892.1\\ +\,432.4\\ \hline 1324.5\end{array}$$
Attach the common sign.
−892.1 − 432.4 = −1324.5

Copyright © 2011 Pearson Education, Inc. Publishing as Prentice Hall.

34. $0.064 - 10.2 = 0.064 + (-10.2)$
Subtract the absolute values.
$$\begin{array}{r} 10.200 \\ -\ \ 0.064 \\ \hline 10.136 \end{array}$$
Attach the sign of the larger absolute value.
$0.064 - 10.2 = -10.136$

35.
$$\begin{array}{r} 100.00 \\ -\ \ 34.98 \\ \hline 65.02 \end{array}$$

36.
$$\begin{array}{r} 200.00000 \\ -\ \ \ \ 0.00198 \\ \hline 199.99802 \end{array}$$

37.
$$\begin{array}{r} 19.9 \\ 15.1 \\ 10.9 \\ +\ \ 6.7 \\ \hline 52.6 \end{array}$$
The total distance is 52.6 miles.

38. $x - y = 1.2 - 6.9 = -5.7$

39. Perimeter = 6.2 + 4.9 + 6.2 + 4.9 = 22.2
The perimeter is 22.2 inches.

40. Perimeter = 11.8 + 12.9 + 14.2 = 38.9
The perimeter is 38.9 feet.

41. $7.2 \times 10 = 72$

42. $9.345 \times 1000 = 9345$

43. A negative number multiplied by a positive number is a negative number.
$$\begin{array}{rl} 34.02 & \text{2 decimal places} \\ \times\ \ 2.3 & \text{1 decimal place} \\ \hline 10\ 206 & \\ 68\ 040 & \\ \hline 78.246 & 2+1=3 \text{ decimal places} \end{array}$$
$-34.02 \times 2.3 = -78.246$

44. A negative number multiplied by a negative number is a positive number.
$$\begin{array}{rl} 839.02 & \text{2 decimal places} \\ \times\ \ \ \ 87.3 & \text{1 decimal place} \\ \hline 251\ 706 & \\ 5873\ 140 & \\ 67121\ 600 & \\ \hline 73246.446 & 2+1=3 \text{ decimal places} \end{array}$$
$-839.02 \times (-87.3) = 73{,}246.446$

45. $C = 2\pi r = 2\pi \cdot 7 = 14\pi$ meters
$C \approx 14 \cdot 3.14 = 43.96$ meters

46. $C = \pi d = \pi \cdot 20 = 20\pi$ inches
$C \approx 20 \cdot 3.14 = 62.8$ inches

47.
$$\begin{array}{r} 0.0877 \\ 3\overline{)0.2631} \\ \underline{-24\ \ } \\ 23\ \ \\ \underline{-21\ \ } \\ 21 \\ \underline{-21} \\ 0 \end{array}$$

48.
$$\begin{array}{r} 15.825 \\ 20\overline{)\ 316.500} \\ \underline{-20} \\ 116 \\ \underline{-100} \\ 16\ 5 \\ \underline{-16\ 0} \\ 50 \\ \underline{-40} \\ 100 \\ \underline{-100} \\ 0 \end{array}$$

49. A negative number divided by a negative number is a positive number.
$0.3\overline{)21}$ becomes
$$\begin{array}{r} 70 \\ 3\overline{)\ 210} \\ \underline{-21\ } \\ 00 \end{array}$$
$-21 \div (-0.3) = 70$

Copyright © 2011 Pearson Education, Inc. Publishing as Prentice Hall.

50. A negative number divided by a positive number is a negative number.

$0.03\overline{)0.0063}$ becomes $3\overline{)0.63}$

0.21
−6
03
−3
0

$-0.0063 \div 0.03 = -0.21$

51. $0.34\overline{)2.74}$ becomes $34\overline{)274.0000}$

8.0588 ≈ 8.059
−272
2 0
− 0
200
−1 70
300
−272
280
−272
8

52. $19.8\overline{)601.92}$ becomes $198\overline{)6019.2}$

30.4
−594
79
−0
79 2
−79 2
0

53. $\frac{23.65}{1000} = 0.02365$

54. $\frac{93}{-10} = -9.3$

55. $3.28\overline{)24}$ becomes $328\overline{)2400.00}$

7.31 ≈ 7.3
−2296
104 0
−98 4
5 60
−3 28
2 32

There are approximately 7.3 meters in 24 feet.

56. $69.71\overline{)3136.95}$ becomes $6971\overline{)313695}$

45
−27884
34855
−34855
0

It will take him 45 months to pay off the loan.

57. $5\overline{)4.0}$

0.8
−4 0
0

$\frac{4}{5} = 0.8$

58. $13\overline{)12.0000}$

0.9230 ≈ 0.923
−11 7
30
−26
40
−39
10
−0
10

$-\frac{12}{13} \approx -0.923$

59. $3\overline{)1.0000}$

0.3333...
−9
10
−9
10
−9
10
−9
1

$2\frac{1}{3} = 2.\overline{3}$ or 2.333

Copyright © 2011 Pearson Education, Inc. Publishing as Prentice Hall.

60. $$\begin{array}{r} 0.2166\ldots \\ 60\overline{)\,13.0000} \\ \underline{-12\,0} \\ 1\,00 \\ \underline{-60} \\ 400 \\ \underline{-360} \\ 400 \\ \underline{-360} \\ 40 \end{array}$$

$\frac{13}{16} = 0.21\overline{6}$ or 0.217

61. $0.392 = 0.39200$

62. 0.0231 0.0221
↑ ↑
3 > 2
so $0.0231 > 0.0221$, thus $-0.0231 < -0.0221$.

63. $$\begin{array}{r} 0.571 \approx 0.57 \\ 7\overline{)\,4.000} \\ \underline{-3\,5} \\ 50 \\ \underline{-49} \\ 10 \\ \underline{-7} \\ 3 \end{array}$$

$\frac{4}{7} \approx 0.57$ and $0.57 < 0.625$, so $\frac{4}{7} < 0.625$.

64. $$\begin{array}{r} 0.2941 \approx 0.294 \\ 17\overline{)\,5.0000} \\ \underline{-3\,4} \\ 1\,60 \\ \underline{-1\,53} \\ 70 \\ \underline{-68} \\ 20 \\ \underline{-17} \\ 3 \end{array}$$

$\frac{5}{17} \approx 0.294$ and $0.293 < 0.294$, so $0.293 < \frac{5}{17}$.

65. 0.832, 0.837, 0.839

66. $\frac{5}{8} = 0.625$

$\frac{5}{8}$, 0.626, 0.685

67. $\frac{3}{7} \approx 0.428$

0.42, $\frac{3}{7}$, 0.43

68. $\frac{18}{11} = 1.6\overline{36}3$

$1.63 = 1.63$

$\frac{19}{12} = 1.58\overline{3}$

$\frac{19}{12}$, 1.63, $\frac{18}{11}$

69. $-7.6 \times 1.9 + 2.5 = -14.44 + 2.5 = -11.94$

70. $(-2.3)^2 - 1.4 = 5.29 - 1.4 = 3.89$

71. $0.0726 \div 10 \times 1000 = 0.00726 \times 1000 = 7.26$

72. $0.9(6.5 - 5.6) = 0.9(0.9) = 0.81$

73. $\frac{(1.5)^2 + 0.5}{0.05} = \frac{2.25 + 0.5}{0.05} = \frac{2.75}{0.05} = 55$

74. $\frac{7 + 0.74}{-0.06} = \frac{7.74}{-0.06} = -129$

75. $$\begin{aligned} \text{Area} &= \frac{1}{2} \cdot b \cdot h \\ &= \frac{1}{2}(4.6)(3) \\ &= 0.5(4.6)(3) \\ &= 6.9 \text{ square feet} \end{aligned}$$

76. $$\begin{aligned} \text{Area} &= \frac{1}{2} \cdot b \cdot h \\ &= \frac{1}{2}(5.2)(2.1) \\ &= 0.5(5.2)(2.1) \\ &= 5.46 \text{ square inches} \end{aligned}$$

77. $$\begin{aligned} x + 3.9 &= 4.2 \\ x + 3.9 - 3.9 &= 4.2 - 3.9 \\ x &= 0.3 \end{aligned}$$

Copyright © 2011 Pearson Education, Inc. Publishing as Prentice Hall.

78. $$\begin{aligned} 70 &= y - 22.81 \\ 70 + 22.81 &= y - 22.81 + 22.81 \\ 92.81 &= y \end{aligned}$$

79. $$\begin{aligned} 2x &= 17.2 \\ \frac{2x}{2} &= \frac{17.2}{2} \\ x &= 8.6 \end{aligned}$$

80. $$\begin{aligned} -1.1y &= 88 \\ \frac{-1.1y}{-1.1} &= \frac{88}{-1.1} \\ y &= -80 \end{aligned}$$

81. $$\begin{aligned} 3x - 0.78 &= 1.2 + 2x \\ 3x - 0.78 + 0.78 &= 1.2 + 2x + 0.78 \\ 3x &= 1.98 + 2x \\ 3x - 2x &= 1.98 + 2x - 2x \\ x &= 1.98 \end{aligned}$$

82. $$\begin{aligned} -x + 0.6 - 2x &= -4x - 0.9 \\ -3x + 0.6 &= -4x - 0.9 \\ -3x + 0.6 - 0.6 &= -4x - 0.9 - 0.6 \\ -3x &= -4x - 1.5 \\ -3x + 4x &= -4x + 4x - 1.5 \\ x &= -1.5 \end{aligned}$$

83. $$\begin{aligned} -1.3x - 9.4 &= -0.4x + 8.6 \\ 10(-1.3x - 9.4) &= 10(-0.4x + 8.6) \\ -13x - 94 &= -4x + 86 \\ -13x - 94 + 94 &= -4x + 86 + 94 \\ -13x &= -4x + 180 \\ -13x + 4x &= -4x + 4x + 180 \\ -9x &= 180 \\ \frac{-9x}{-9} &= \frac{180}{-9} \\ x &= -20 \end{aligned}$$

84. $$\begin{aligned} 3(x - 1.1) &= 5x - 5.3 \\ 3x - 3.3 &= 5x - 5.3 \\ 3x - 3.3 + 3.3 &= 5x - 5.3 + 3.3 \\ 3x &= 5x - 2 \\ 3x - 5x &= 5x - 5x - 2 \\ -2x &= -2 \\ \frac{-2x}{-2} &= \frac{-2}{-2} \\ x &= 1 \end{aligned}$$

Copyright © 2011 Pearson Education, Inc. Publishing as Prentice Hall.

85. Mean: $\frac{13+23+33+14+6}{5}=\frac{89}{5}=17.8$
Median: Write the numbers in order.
6, 13, 14, 23, 33
The middle number is 14.
Mode: There is no mode, since each number occurs once.

86. $\text{Mean} = \frac{45+86+21+60+86+64+45}{7}$
$= \frac{407}{7}$
≈ 58.1
Median: Write the numbers in order.
21, 45, 45, 60, 64, 86, 86
The middle number is 60.
Mode: There are 2 numbers that occur twice, so there are two modes, 45 and 86.

87. $\text{Mean} = \frac{14,000+20,000+12,000+20,000+36,000+45,000}{6}=\frac{147,000}{6}=24,500$
Median: Write the numbers in order.
12,000, 14,000, 20,000, 20,000, 36,000, 45,000
The mean of the middle two: $\frac{20,000+20,000}{2}=20,000$
Mode: 20,000 is the mode because it occurs twice.

88. $\text{Mean} = \frac{560+620+123+400+410+300+400+780+430+450}{10}=\frac{4473}{10}=447.3$
Median: Write the numbers in order.
123, 300, 400, 400, 410, 430, 450, 560, 620, 780
The mean of the two middle numbers: $\frac{410+430}{2}=420$
Mode: 400 is the mode because it occurs twice.

89. $\text{gpa} = \frac{4\cdot 3+4\cdot 3+2\cdot 2+3\cdot 3+2\cdot 1}{3+3+2+3+1}=\frac{39}{12}=3.25$

90. $\text{gpa} = \frac{3\cdot 3+3\cdot 4+2\cdot 2+1\cdot 2+3\cdot 3}{3+4+2+2+3}=\frac{36}{14}\approx 2.57$

91. 200.0032 in words is two hundred and thirty-two ten-thousandths.

92. Negative sixteen and nine hundredths is –16.09 in standard form.

93. $0.0847=\frac{847}{10,000}$

Copyright © 2011 Pearson Education, Inc. Publishing as Prentice Hall.

94. $\frac{6}{7} \approx 0.857$

$\frac{8}{9} \approx 0.889$

$0.75 = 0.750$

$0.75, \frac{6}{7}, \frac{8}{9}$

95. $-\frac{7}{100} = -0.07$

96.
$$\begin{array}{r} 0.1125 \\ 80\overline{)\ 9.0000} \\ -8\ 0 \\ \hline 1\ 00 \\ -80 \\ \hline 200 \\ -160 \\ \hline 400 \\ -400 \\ \hline 0 \end{array}$$

$\frac{9}{80} = 0.1125$

97.
$$\begin{array}{r} 51.0571 \approx 51.057 \\ 175\overline{)\ 8935.0000} \\ -875 \\ \hline 185 \\ -175 \\ \hline 100 \\ -0 \\ \hline 1000 \\ -875 \\ \hline 1\ 250 \\ -1\ 225 \\ \hline 250 \\ -175 \\ \hline 75 \end{array}$$

$\frac{8935}{175} \approx 51.057$

98. 402.000032 402.00032

↑ ↑

0 < 3

so $402.000032 < 402.00032$

Thus, $-402.000032 > -402.00032$.

99. $\frac{6}{11} = 0.\overline{54}$

$0.\overline{54} < 0.55$, so $\frac{6}{11} < 0.55$

100. To round 86.905 to the nearest hundredth, observe that the digit in the thousandths place is 5. Since this digit is at least 5, we add 1 to the digit in the hundredths place. The number 86.905 rounded to the nearest hundredth is 86.91.

101. To round 3.11526 to the nearest thousandth, observe that the digit in the ten-thousandths place is 2. Since this digit is less than 5, we do not add 1 to the digit in the thousandths place. The number 3.11526 rounded to the nearest thousandth is 3.115.

102. To round 123.46 to the nearest one, observe that the digit in the tenths place is 4. Since this digit is less than 5, we do not add 1 to the digit in the ones place. The number $123.46 rounded to the nearest dollar (or one) is $123.00.

103. To round 3645.52 to the nearest one, observe that the digit in the tenths place is 5. Since this digit is at least 5, we add 1 to the digit in the ones place. The number $3645.52 rounded to the nearest dollar (or one) is $3646.00.

104. Subtract absolute values.
$$\begin{array}{r} 4.9 \\ -\ 3.2 \\ \hline 1.7 \end{array}$$
Attach the sign of the larger absolute value.
$3.2 - 4.9 = -1.7$

105.
$$\begin{array}{r} 9.12 \\ -\ 3.86 \\ \hline 5.26 \end{array}$$

106. Subtract absolute values.
$$\begin{array}{r} 102.06 \\ -\ \ 89.30 \\ \hline 12.76 \end{array}$$
Attach the sign of the larger absolute value.
$-102.06 + 89.3 = -12.76$

107.
$$\begin{array}{r} -4.021 \\ -10.830 \\ (+)\ \ -0.056 \\ \hline -14.907 \end{array}$$

Copyright © 2011 Pearson Education, Inc. Publishing as Prentice Hall.

108.
$$\begin{array}{rl} 2.54 & \text{2 decimal places} \\ \times\ 3.2 & \text{1 decimal place} \\ \hline 508 & \\ 7\,620 & \\ \hline 8.128 & 2+1=3 \text{ decimal places} \end{array}$$

109. The product of a negative number and a positive number is a negative number.
$$\begin{array}{rl} 3.45 & \text{2 decimal places} \\ \times\ 2.1 & \text{1 decimal place} \\ \hline 345 & \\ 6\,900 & \\ \hline 7.245 & 2+1=3 \text{ decimal places} \end{array}$$
The product is −7.245.

110. $0.005\overline{)24.5}$ becomes $5\overline{)24500}$
$$\begin{array}{r} 4900 \\ 5\overline{)\ 24500} \\ \underline{-20} \\ 45 \\ \underline{-45} \\ 000 \end{array}$$

111. $2.3\overline{)54.98}$ becomes $23\overline{)549.8000}$
$$\begin{array}{r} 23.9043 \approx 23.904 \\ 23\overline{)\ 549.8000} \\ \underline{-46} \\ 89 \\ \underline{-69} \\ 20\,8 \\ \underline{-20\,7} \\ 10 \\ \underline{-0} \\ 100 \\ \underline{-92} \\ 80 \\ \underline{-69} \\ 11 \end{array}$$

112. length = 115.9 ≈ 120
width = 77.3 ≈ 80
$$\begin{aligned} \text{Area} &= \text{length} \cdot \text{width} \\ &= 120 \cdot 80 \\ &= 9600 \text{ square feet} \end{aligned}$$

113.
$$\begin{array}{lll} \$1.89 & \text{rounds to} & \$2 \\ \$1.07 & \text{rounds to} & \$1 \\ \underline{\$0.99} & \text{rounds to} & \underline{\$1} \\ & & \$4 \end{array}$$
Yes, the items can be purchased with a $5 bill.

114. $\dfrac{(3.2)^2}{100} = \dfrac{10.24}{100} = 0.1024$

115. $(2.6 + 1.4)(4.5 - 3.6) = (4)(0.9) = 3.6$

116. Mean: $\dfrac{73+82+95+68+54}{5} = \dfrac{372}{5} = 74.4$
Median: Write the numbers in order.
54, 68, 73, 82, 95
The median is the middle value: 73.
Mode: There is no mode, since each number occurs once.

117. Mean:
$$\frac{952+327+566+814+327+729}{6} = \frac{3715}{6} \approx 619.17$$
Median: Write the numbers in order.
327, 327, 566, 729, 814, 952
The mean of the two middle values:
$$\frac{566+729}{2} = 647.5$$
Mode: 327, since this number appears twice.

Chapter 5 Test

1. 45.092 in words is forty-five and ninety-two thousandths.

2. Three thousand and fifty-nine thousandths in standard form is 3000.059.

3.
$$\begin{array}{r} 2.893 \\ 4.210 \\ +\ 10.492 \\ \hline 17.595 \end{array}$$

4. Add the absolute values.
$$\begin{array}{r} 47.92 \\ +\ \ 3.28 \\ \hline 51.20 \end{array}$$
Attach the common sign.
$-47.92 - 3.28 = -51.20$

5. Subtract the absolute values.
$$\begin{array}{r} 30.25 \\ -\ \ 9.83 \\ \hline 20.42 \end{array}$$
Attach the sign of the larger absolute value.
$9.83 - 30.25 = -20.42$

Copyright © 2011 Pearson Education, Inc. Publishing as Prentice Hall.

6.
$$\begin{array}{rl} 10.2 & \text{1 decimal place} \\ \times\ 4.01 & \text{2 decimal places} \\ \hline 102 & \\ 40\,800 & \\ \hline 40.902 & 1+2=3 \text{ decimal places} \end{array}$$

7. $0.23\overline{)0.00843}$ becomes
$$\begin{array}{r} 0.0366 \approx 0.037 \\ 23\overline{)0.8430} \\ \underline{-69} \\ 153 \\ \underline{-138} \\ 150 \\ \underline{-138} \\ 12 \end{array}$$

$-0.00843 \div (-0.23) \approx 0.037$

8. To round 34.8923 to the nearest tenth, observe that the digit in the hundredths place is 9. Since this digit is at least 5, we add 1 to the digit in the tenths place. 34.8923 rounded to the nearest tenth is 34.9.

9. To round 0.8623 to the nearest thousandth, observe that the digit in the ten-thousandths place is 3. Since this digit is less than 5, we do not add 1 to the digit in the thousandths place. 0.8623 rounded to the nearest thousandth is 0.862.

10.
$$\begin{array}{cc} 25.0909 & 25.9090 \\ \uparrow & \uparrow \\ 0 \quad < & 9 \end{array}$$
so $25.0909 < 25.9090$

11.
$$\begin{array}{r} 0.444... \\ 9\overline{)\ 4.000} \\ \underline{-3\ 6} \\ 40 \\ \underline{-36} \\ 40 \\ \underline{-36} \\ 4 \end{array}$$

$0.44\bar{4} < 0.445$, so $\frac{4}{9} < 0.445$.

12. $0.345 = \frac{345}{1000} = \frac{5 \cdot 69}{5 \cdot 200} = \frac{69}{200}$

13. $-24.73 = -24\frac{73}{100}$

14. $-\frac{13}{26} = -\frac{1 \cdot 13}{2 \cdot 13} = -\frac{1}{2} = -\frac{1 \cdot 5}{2 \cdot 5} = -\frac{5}{10} = -0.5$

15.
$$\begin{array}{r} 0.9411 \approx 0.941 \\ 17\overline{)\ 16.0000} \\ \underline{-15\ 3} \\ 70 \\ \underline{-68} \\ 20 \\ \underline{-17} \\ 30 \\ \underline{-17} \\ 13 \end{array}$$

$\frac{16}{17} \approx 0.941$

16. $(-0.6)^2 + 1.57 = 0.36 + 1.57 = 1.93$

17. $\frac{0.23+1.63}{-0.3} = \frac{1.86}{-0.3} = -6.2$

18. $2.4x - 3.6 - 1.9x - 9.8$
$= (2.4x - 1.9x) + (-3.6 - 9.8)$
$= 0.5x - 13.4$

19.
$$\begin{aligned} 0.2x + 1.3 &= 0.7 \\ 0.2x + 1.3 - 1.3 &= 0.7 - 1.3 \\ 0.2x &= -0.6 \\ \frac{0.2x}{0.2} &= \frac{-0.6}{0.2} \\ x &= -3 \end{aligned}$$

20.
$$\begin{aligned} 2(x+5.7) &= 6x - 3.4 \\ 2x + 11.4 &= 6x - 3.4 \\ 2x + 11.4 - 11.4 &= 6x - 3.4 - 11.4 \\ 2x &= 6x - 14.8 \\ 2x - 6x &= 6x - 6x - 14.8 \\ -4x &= -14.8 \\ \frac{-4x}{-4} &= \frac{-14.8}{-4} \\ x &= 3.7 \end{aligned}$$

21. Mean: $\frac{26+32+42+43+49}{5} = \frac{192}{5} = 38.4$
Median: The numbers are listed in order. The middle number is 42.
Mode: There is no mode since each number occurs once.

Copyright © 2011 Pearson Education, Inc. Publishing as Prentice Hall.

22. Mean:

$$\frac{8+10+16+16+14+12+12+13}{8} = \frac{101}{8} = 12.625$$

Median: List the numbers in order.
8, 10, 12, 12, 13, 14, 16, 16

The mean of the middle two: $\frac{12+13}{2} = 12.5$

Mode: 12 and 16 each occur twice, so the modes are 12 and 16.

23. $\text{gpa} = \frac{4\cdot 3+3\cdot 3+2\cdot 3+3\cdot 4+4\cdot 1}{3+3+3+4+1} = \frac{43}{14} \approx 3.07$

24.
$$\begin{aligned} 4{,}583 \text{ million} &= 4583 \times 1 \text{ million} \\ &= 4583 \times 1{,}000{,}000 \\ &= 4{,}583{,}000{,}000 \end{aligned}$$

25.
$$\begin{aligned} \text{Area} &= \frac{1}{2}(4.2 \text{ miles})(1.1 \text{ miles}) \\ &= 0.5(4.2)(1.1) \text{ square miles} \\ &= 2.31 \text{ square miles} \end{aligned}$$

26. $C = 2\pi r = 2\pi \cdot 9 = 18\pi$ miles
$C \approx 18 \cdot 3.14 = 56.52$ miles

27. a.
$$\begin{aligned} \text{Area} &= \text{length} \cdot \text{width} \\ &= (123.8) \times (80) \\ &= 9904 \end{aligned}$$
The area is 9904 square feet.

b. $9904 \times 0.02 = 198.08$
Vivian needs to purchase 198.08 ounces.

28.
$$\begin{array}{r} 14.2 \\ 16.1 \\ +\ 23.7 \\ \hline 54.0 \end{array}$$
The total distance is 54 miles.

Cumulative Review Chapters 1–5

1. 72 in words is seventy-two.

2. 107 in words is one hundred seven.

3. 546 in words is five hundred forty-six.

4. 5026 in words is five thousand twenty-six.

5.
$$\begin{array}{r} 46 \\ +\ 713 \\ \hline 759 \end{array}$$

6. $3 + 7 + 9 = 19$
The perimeter is 19 inches.

7.
$$\begin{array}{r} 543 \\ -\ 29 \\ \hline 514 \end{array} \qquad \textit{Check: } \begin{array}{r} 514 \\ +\ 29 \\ \hline 543 \end{array}$$

8.
$$\begin{array}{r} 121 \text{ R } 1 \\ 27\overline{)\,3268} \\ -27 \\ \hline 56 \\ -54 \\ \hline 28 \\ -27 \\ \hline 1 \end{array}$$

9. To round 278,362 to the nearest thousand, observe that the digit in the hundreds place is 3. Since this digit is less than 5, we do not add 1 to the digit in the thousands place. The number 278,362 rounded to the nearest thousand is 278,000.

10. $30 = 2 \cdot 15 = 2 \cdot 3 \cdot 5$

11.
$$\begin{array}{r} 236 \\ \times\ 86 \\ \hline 1\,416 \\ 18\,880 \\ \hline 20{,}296 \end{array}$$

12. $236 \times 86 \times 0 = 0$

13. a. $1\overline{)7}$ with quotient 7 *Check:* $7 \times 1 = 7$

b. $12 \div 1 = 12$ *Check:* $12 \times 1 = 12$

c. $\frac{6}{6} = 1$ *Check:* $1 \times 6 = 6$

d. $9 \div 9 = 1$ *Check:* $1 \times 9 = 9$

e. $\frac{20}{1} = 20$ *Check:* $20 \times 1 = 20$

f. $18\overline{)18}$ with quotient 1 *Check:* $1 \times 18 = 18$

Copyright © 2011 Pearson Education, Inc. Publishing as Prentice Hall.

14. $\frac{25+17+19+39}{4}=\frac{100}{4}=5$
The average is 25.

15. $2\cdot 4-3\div 3=8-3\div 3=8-1=7$

16. $77\div 11\cdot 7=7\cdot 7=49$

17. $9^2=9\cdot 9=81$

18. $5^3=5\cdot 5\cdot 5=125$

19. $3^4=3\cdot 3\cdot 3\cdot 3=81$

20. $10^3=10\cdot 10\cdot 10=1000$

21. $\frac{x-5y}{y}=\frac{35-5\cdot 5}{5}=\frac{35-25}{5}=\frac{10}{5}=2$

22. $\frac{2a+4}{c}=\frac{2\cdot 7+4}{3}=\frac{14+4}{3}=\frac{18}{3}=6$

23. a. The opposite of 13 is −13.

b. The opposite of −2 is −(−2) = 2.

c. The opposite of 0 is 0.

24. a. The opposite of −7 is −(−7) = 7.

b. The opposite of 4 is −4.

c. The opposite of −1 is −(−1) = 1.

25. $-2+(-21)=-23$

26. $-7+(-15)=-22$

27. $5\cdot 6^2=5\cdot 36=180$

28. $4\cdot 2^3=4\cdot 8=32$

29. $-7^2=-(7\cdot 7)=-49$

30. $(-2)^5=(-2)(-2)(-2)(-2)(-2)=-32$

31. $(-5)^2=(-5)(-5)=25$

32. $-3^2=-(3\cdot 3)=-9$

33. Each part represents $\frac{1}{3}$ of a whole. Four parts are shaded, or 1 whole and 1 part.
$\frac{4}{3}$ or $1\frac{1}{3}$

34. Each part represents $\frac{1}{4}$ of a whole. Seven parts are shaded or 1 whole and 3 parts.
$\frac{7}{4}$ or $1\frac{3}{4}$

35. Each part represents $\frac{1}{4}$ of a whole. Eleven parts are shaded or 2 wholes and 3 parts.
$\frac{11}{4}$ or $2\frac{3}{4}$

36. Each part represents $\frac{1}{3}$ of a whole. Fourteen parts are shaded, or 4 wholes and 2 parts.
$\frac{14}{3}$ or $4\frac{2}{3}$

37.
$$
\begin{array}{l}
252=2\cdot 126 \\
\quad\downarrow\ \downarrow\searrow \\
\quad 2\cdot 2\cdot 63 \\
\quad\downarrow\ \downarrow\ \downarrow\searrow \\
\quad 2\cdot 2\cdot 3\cdot 21 \\
\quad\downarrow\ \downarrow\ \downarrow\ \downarrow\searrow \\
\quad 2\cdot 2\cdot 3\cdot 3\cdot 7
\end{array}
$$
$252=2^2\cdot 3^2\cdot 7$

38.
$$
\begin{array}{r}
87 \\
-\ 25 \\
\hline
62
\end{array}
$$

39. $-\frac{72}{26}=-\frac{36\cdot 2}{13\cdot 2}=-\frac{36}{13}$

40. $9\frac{7}{8}=\frac{8\cdot 9+7}{8}=\frac{72+7}{8}=\frac{79}{8}$

41. $\frac{16}{40}=\frac{2\cdot 8}{5\cdot 8}=\frac{2}{5}$
$\frac{10}{25}=\frac{2\cdot 5}{5\cdot 5}=\frac{2}{5}$
The fractions are equivalent.

Copyright © 2011 Pearson Education, Inc. Publishing as Prentice Hall.

42. $\frac{4}{7} \approx 0.571$

$\frac{5}{9} \approx 0.556$

Since $0.571 > 0.556$, then $\frac{4}{7} > \frac{5}{9}$.

or

$\frac{4}{7} \cdot \frac{9}{9} = \frac{36}{63}$ and $\frac{5}{9} \cdot \frac{7}{7} = \frac{35}{63}$

Since $36 > 35$, then $\frac{36}{63} > \frac{35}{63}$ and $\frac{4}{7} > \frac{5}{9}$.

43. $\frac{2}{3} \cdot \frac{5}{11} = \frac{2 \cdot 5}{3 \cdot 11} = \frac{10}{33}$

44. $2\frac{5}{8} \cdot \frac{4}{7} = \frac{21}{8} \cdot \frac{4}{7} = \frac{21 \cdot 4}{8 \cdot 7} = \frac{7 \cdot 3 \cdot 4}{4 \cdot 2 \cdot 7} = \frac{3}{2}$ or $1\frac{1}{2}$

45. $\frac{1}{4} \cdot \frac{1}{2} = \frac{1 \cdot 1}{4 \cdot 2} = \frac{1}{8}$

46. $7 \cdot 5\frac{2}{7} = \frac{7}{1} \cdot \frac{37}{7} = \frac{7 \cdot 37}{1 \cdot 7} = \frac{37}{1} = 37$

47.
$$\frac{z}{-4} = 11 - 5$$
$$\frac{z}{-4} = 6$$
$$-4 \cdot \frac{z}{-4} = -4 \cdot 6$$
$$z = -24$$

48.
$$6x - 12 - 5x = -20$$
$$x - 12 = -20$$
$$x - 12 + 12 = -20 + 12$$
$$x = -8$$

49.
$$\begin{array}{r} 763.7651 \\ 22.0010 \\ +\ 43.8900 \\ \hline 829.6561 \end{array}$$

50.
$$\begin{array}{r} 89.2700 \\ 14.3610 \\ +\ 127.2318 \\ \hline 230.8628 \end{array}$$

51.
$$\begin{array}{rl} 23.6 & \text{1 decimal place} \\ \times\ 0.78 & \text{2 decimal places} \\ \hline 1\,888 & \\ 16\,520 & \\ \hline 18.408 & 1+2=3 \text{ decimal places} \end{array}$$

52.
$$\begin{array}{rl} 43.8 & \text{1 decimal place} \\ \times\ 0.645 & \text{3 decimal places} \\ \hline 2190 & \\ 1\,7520 & \\ 26\,2800 & \\ \hline 28.2510 & 1+3=4 \text{ decimal places} \end{array}$$

Copyright © 2011 Pearson Education, Inc. Publishing as Prentice Hall.

Chapter 6

Section 6.1

Practice Problems

1. The ratio of 19 to 30 is $\frac{19}{30}$.

2. The ratio of \$16 to \$12 is $\frac{\$16}{\$12} = \frac{16}{12} = \frac{4 \cdot 4}{4 \cdot 3} = \frac{4}{3}$.

3. The ratio of 1.68 to 4.8 is
$\frac{1.68}{4.8} = \frac{1.68 \cdot 100}{4.8 \cdot 100} = \frac{168}{480} = \frac{7 \cdot 24}{20 \cdot 24} = \frac{7}{20}$.

4. The ratio of $2\frac{2}{3}$ to $1\frac{13}{15}$ is

$$\frac{2\frac{2}{3}}{1\frac{13}{15}} = 2\frac{2}{3} \div 1\frac{13}{15} = \frac{8}{3} \div \frac{28}{15} = \frac{8}{3} \cdot \frac{15}{28} = \frac{4 \cdot 2 \cdot 3 \cdot 5}{3 \cdot 7 \cdot 4} = \frac{10}{7}.$$

5. The ratio of work miles to total miles is
$\frac{4800 \text{ miles}}{15{,}000 \text{ miles}} = \frac{8 \cdot 600}{25 \cdot 600} = \frac{8}{25}$.

6. **a.** The ratio of the length of the shortest side to the length of the longest side is
$\frac{6 \text{ meters}}{10 \text{ meters}} = \frac{6}{10} = \frac{2 \cdot 3}{2 \cdot 5} = \frac{3}{5}$.

   b. The ratio of the length of the longest side to the perimeter is
$\frac{10 \text{ meters}}{24 \text{ meters}} = \frac{10}{24} = \frac{5 \cdot 2}{12 \cdot 2} = \frac{5}{12}$.

7. $\frac{\$1350}{6 \text{ weeks}} = \frac{\$225}{1 \text{ week}}$

8. $\frac{295 \text{ miles}}{15 \text{ gallons}} = \frac{59 \text{ miles}}{3 \text{ gallons}}$

9. $\frac{3200 \text{ feet}}{8 \text{ seconds}}$

$$\begin{array}{r} 400 \\ 8\overline{)\,3200} \\ \underline{-32} \\ 000 \end{array}$$

The unit rate is $\frac{400 \text{ feet}}{1 \text{ second}}$ or 400 feet/second.

10. $\frac{78 \text{ bushels}}{12 \text{ trees}}$

$$\begin{array}{r} 6.5 \\ 12\overline{)\,78.0} \\ \underline{-72} \\ 60 \\ \underline{-60} \\ 0 \end{array}$$

The unit rate is $\frac{6.5 \text{ bushels}}{1 \text{ tree}}$ or 6.5 bushels/ tree.

11. unit price $= \frac{\$170}{5 \text{ days}} = \frac{\$34}{1 \text{ day}}$ or \$34 per day

12. unit price $= \frac{\$2.32}{11 \text{ ounces}} \approx \frac{\$0.21}{1 \text{ ounce}}$
unit price $= \frac{\$3.59}{16 \text{ ounces}} \approx \frac{\$0.22}{1 \text{ ounce}}$
Since the 11-ounce bag has a cheaper price per ounce, it is the better buy.

Vocabulary and Readiness Check

1. A rate with a denominator of 1 is called a unit rate.
2. When a rate is written as money per item, a unit rate is called a unit price.
3. The word per translates to "division."
4. Rates are used to compare different types of quantities.
5. To write a rate as a unit rate, divide the numerator of the rate by the denominator.
6. The quotient of two quantities is called a ratio.
7. True or false: The ratio $\frac{7}{5}$ means the same as the ratio $\frac{5}{7}$. false

Copyright © 2011 Pearson Education, Inc. Publishing as Prentice Hall.

Exercise Set 6.1

1. The ratio of 16 to 24 is $\frac{16}{24}=\frac{8\cdot 2}{8\cdot 3}=\frac{2}{3}$.

3. The ratio of 7.7 to 10 is $\frac{7.7}{10}=\frac{7.7\cdot 10}{10\cdot 10}=\frac{77}{100}$.

5. The ratio of 4.63 to 8.21 is
$\frac{4.63}{8.21}=\frac{4.63\cdot 100}{8.21\cdot 100}=\frac{463}{821}$.

7. The ratio of 6 ounces to 16 ounces is
$\frac{6 \text{ ounces}}{16 \text{ ounces}}=\frac{3\cdot 2}{8\cdot 2}=\frac{3}{8}$.

9. The ratio of \$32 to \$100 is $\frac{\$32}{\$100}=\frac{4\cdot 8}{4\cdot 25}=\frac{8}{25}$.

11. The ratio of 24 days to 14 days is
$\frac{24 \text{ days}}{14 \text{ days}}=\frac{12\cdot 2}{7\cdot 2}=\frac{12}{7}$.

13. The ratio of $3\frac{1}{2}$ to $12\frac{1}{4}$ is

$$\begin{aligned}\frac{3\frac{1}{2}}{12\frac{1}{4}}&=3\frac{1}{2}\div 12\frac{1}{4}\\&=\frac{7}{2}\div\frac{49}{4}\\&=\frac{7}{2}\cdot\frac{4}{49}\\&=\frac{7\cdot 2\cdot 2}{2\cdot 7\cdot 7}\\&=\frac{2}{7}.\end{aligned}$$

15. The ratio of $7\frac{3}{5}$ hours to $1\frac{9}{10}$ hours is

$$\begin{aligned}\frac{7\frac{3}{5}\text{ hours}}{1\frac{9}{10}\text{ hours}}&=7\frac{3}{5}\div 1\frac{9}{10}\\&=\frac{38}{5}\div\frac{19}{10}\\&=\frac{38}{5}\cdot\frac{10}{19}\\&=\frac{19\cdot 2\cdot 2\cdot 5}{5\cdot 19\cdot 1}\\&=\frac{4}{1}.\end{aligned}$$

17. The ratio of the weight of an average mature Fin Whale to the weight of an average mature Blue Whale is $\frac{50 \text{ tons}}{145 \text{ tons}}=\frac{5\cdot 10}{5\cdot 29}=\frac{10}{29}$.

19. Perimeter = 32 + 19 + 32 + 19 = 102 feet
The ratio of width to perimeter is
$\frac{32 \text{ feet}}{102 \text{ feet}}=\frac{2\cdot 16}{2\cdot 51}=\frac{16}{51}$.

21. Perimeter = 94 + 50 + 94 + 50 = 288 feet
The ratio of width to perimeter is
$\frac{50 \text{ feet}}{288 \text{ feet}}=\frac{2\cdot 25}{2\cdot 144}=\frac{25}{144}$.

23. The ratio of women to men is
$\frac{125 \text{ women}}{100 \text{ men}}=\frac{5\cdot 25}{4\cdot 25}=\frac{5}{4}$.

25. The ratio of red blood cells to platelet cells is
$\frac{600 \text{ cells}}{40 \text{ cells}}=\frac{40\cdot 15}{40\cdot 1}=\frac{15}{1}$.

27. The ratio of the longest side to the perimeter is
$\frac{17 \text{ feet}}{(8+15+17)\text{ feet}}=\frac{17}{40}$.

29. The ratio of calories from fat to total calories is
$\frac{220 \text{ calories}}{500 \text{ calories}}=\frac{20\cdot 11}{20\cdot 25}=\frac{11}{25}$.

31. The ratio of the average amount of Coca-Cola beverages drunk by Mexicans to the average amount drunk by Americans is
$\frac{573 \text{ 8-oz beverages}}{423 \text{ 8-oz beverages}}=\frac{3\cdot 191}{3\cdot 141}=\frac{191}{141}$.

33. The rate of 5 shrubs every 15 feet is
$\frac{5 \text{ shrubs}}{15 \text{ feet}}=\frac{1 \text{ shrub}}{3 \text{ feet}}$.

35. The rate of 15 returns for 100 sales is
$\frac{15 \text{ returns}}{100 \text{ sales}}=\frac{3 \text{ returns}}{20 \text{ sales}}$.

37. The rate of 6 laser printers for 28 computers is
$\frac{6 \text{ laser printers}}{28 \text{ computers}}=\frac{3 \text{ laser printers}}{14 \text{ computers}}$.

Copyright © 2011 Pearson Education, Inc. Publishing as Prentice Hall.

39. The rate of 18 gallons of pesticide for 4 acres of crops is $\frac{18 \text{ gallons}}{4 \text{ acres}} = \frac{9 \text{ gallons}}{2 \text{ acres}}$.

41.
$$\begin{array}{r} 110 \\ 3\overline{)\,330} \\ \underline{-3} \\ 03 \\ \underline{-3} \\ 00 \end{array}$$

330 calories in a 3-ounce serving is $\frac{110 \text{ calories}}{1 \text{ ounce}}$ or 110 calories/ounce.

43.
$$\begin{array}{r} 75 \\ 5\overline{)\,375} \\ \underline{-35} \\ 25 \\ \underline{-25} \\ 0 \end{array}$$

375 riders in 5 subway cars is $\frac{75 \text{ riders}}{1 \text{ car}}$ or 75 riders/car.

45.
$$\begin{array}{r} 90 \\ 60\overline{)\,5400} \\ \underline{-540} \\ 00 \end{array}$$

5400 wingbeats per 60 seconds is $\frac{90 \text{ wingbeats}}{1 \text{ second}}$ or 90 wingbeats/second.

47.
$$\begin{array}{r} 50,000 \\ 20\overline{)\,1,000,000} \\ \underline{-1\,00} \\ 0 \end{array}$$

\$1,000,000 paid over 20 years is $\frac{\$50,000}{1 \text{ year}}$ or \$50,000/year.

49.
$$\begin{array}{r} 225,250 \\ 2\overline{)\,450,500} \\ \underline{-4} \\ 05 \\ \underline{-4} \\ 10 \\ \underline{-10} \\ 0\,5 \\ \underline{-4} \\ 10 \\ \underline{-10} \\ 0 \end{array}$$

450,500 voters for 2 senators is $\frac{225,250 \text{ voters}}{1 \text{ senator}}$ or 225,250 voters/senator.

51.
$$\begin{array}{r} 300 \\ 40\overline{)\,12,000} \\ \underline{-12\,0} \\ 0 \end{array}$$

12,000 good products to 40 defective is $\frac{300 \text{ good}}{1 \text{ defective}}$ or 300 good/defective.

53.
$$\begin{array}{r} 4,390,000 \\ 20\overline{)\,87,800,000} \\ \underline{-80} \\ 7\,8 \\ \underline{-6\,0} \\ 1\,80 \\ \underline{-1\,80} \\ 0 \end{array}$$

\$87,800,000 for 20 players is $\frac{\$4,390,000}{1 \text{ player}}$ or \$4,390,000/player.

55. a.
$$\begin{array}{r} 31.25 \\ 8\overline{)\,250.00} \\ \underline{-24} \\ 10 \\ \underline{-8} \\ 2\,0 \\ \underline{-1\,6} \\ 40 \\ \underline{-40} \\ 0 \end{array}$$

The unit rate for Charlie is 31.25 boards/hour.

Copyright © 2011 Pearson Education, Inc. Publishing as Prentice Hall.

b.
$$\begin{array}{r} 33.5 \\ 12\overline{)\ 402.0} \\ \underline{-36} \\ 42 \\ \underline{-36} \\ 6\,0 \\ \underline{-6\,0} \\ 0 \end{array}$$

The unit rate for Suellen is 33.5 boards/hour.

c. Suellen can assemble boards faster, since 33.5 > 31.25.

57. a. $14.5\overline{)400}$ becomes
$$\begin{array}{r} 27.58 \approx 27.6 \\ 145\overline{)\ 4000.00} \\ \underline{-290} \\ 1100 \\ \underline{-1015} \\ 850 \\ \underline{-725} \\ 1250 \\ \underline{-1160} \\ 90 \end{array}$$

The unit rate for the car is ≈27.6 miles/gallon.

b. $9.25\overline{)270}$ becomes
$$\begin{array}{r} 29.18 \approx 29.2 \\ 925\overline{)\ 27000.00} \\ \underline{-1850} \\ 8500 \\ \underline{-8325} \\ 175\,0 \\ \underline{-92\,5} \\ 82\,50 \\ \underline{-74\,00} \\ 8\,50 \end{array}$$

The unit rate for the truck is ≈29.2 miles/gallon.

c. The truck has the better gas mileage, since 29.2 > 27.6.

59.
$$\begin{array}{r} 11.50 \\ 5\overline{)\ 57.50} \\ \underline{-5} \\ 07 \\ \underline{-5} \\ 2\,5 \\ \underline{-2\,5} \\ 0 \end{array}$$

The unit price is $11.50 per compact disc.

61.
$$\begin{array}{r} 0.17 \\ 7\overline{)1.19} \\ \underline{-7} \\ 49 \\ \underline{-49} \\ 0 \end{array}$$

The unit price is $0.17 per banana.

63.
$$\begin{array}{r} 0.1487 \approx 0.149 \\ 8\overline{)1.1900} \\ \underline{-8} \\ 39 \\ \underline{-32} \\ 70 \\ \underline{-64} \\ 60 \\ \underline{-56} \\ 4 \end{array}$$

The 8-ounce size costs $0.149 per ounce.

$$\begin{array}{r} 0.1325 \approx 0.133 \\ 12\overline{)\ 1.5900} \\ \underline{-1\,2} \\ 39 \\ \underline{-36} \\ 30 \\ \underline{-24} \\ 60 \\ \underline{-60} \\ 0 \end{array}$$

The 12-ounce size costs $0.133 per ounce. The 12-ounce size is the better buy.

Copyright © 2011 Pearson Education, Inc. Publishing as Prentice Hall.

65. $$\begin{array}{r} 0.1056 \approx 0.106 \\ 16\overline{)\,1.6900} \\ \underline{-1\,6} \\ 090 \\ \underline{-80} \\ 100 \\ \underline{-96} \\ 4 \end{array}$$

The 16-ounce size costs \$0.106 per ounce.

$$\begin{array}{r} 0.115 \\ 6\overline{)0.690} \\ \underline{-6} \\ 09 \\ \underline{-6} \\ 30 \\ \underline{-30} \\ 0 \end{array}$$

The 6-ounce size costs \$0.115 per ounce.
The 16-ounce size is the better buy.

67. $$\begin{array}{r} 0.1908 \approx 0.191 \\ 12\overline{)\,2.2900} \\ \underline{-1\,2} \\ 1\,09 \\ \underline{-1\,08} \\ 100 \\ \underline{-96} \\ 4 \end{array}$$

The 12-ounce size costs \$0.191 per ounce.

$$\begin{array}{r} 0.1862 \approx 0.186 \\ 8\overline{)1.4900} \\ \underline{-8} \\ 69 \\ \underline{-64} \\ 50 \\ \underline{-48} \\ 20 \\ \underline{-16} \\ 4 \end{array}$$

The 8-ounce size costs \$0.186 per ounce.
The 8-ounce size is the better buy.

69. $$\begin{array}{r} 0.0059 \approx 0.006 \\ 100\overline{)0.5900} \\ \underline{-500} \\ 900 \\ \underline{-900} \\ 0 \end{array}$$

The 100-count size costs \$0.006 per napkin.

$$\begin{array}{r} 0.0051 \approx 0.005 \\ 180\overline{)0.9300} \\ \underline{-900} \\ 300 \\ \underline{-180} \\ 120 \end{array}$$

The 180-count size costs \$0.005 per napkin.
The 180-count size is the better buy.

71. $$\begin{array}{r} 2.3 \\ 9\overline{)\,20.7} \\ \underline{-18} \\ 2\,7 \\ \underline{-2\,7} \\ 0 \end{array}$$

73. $3.7\overline{)0.555}$ becomes $$\begin{array}{r} 0.15 \\ 37\overline{)\,5.55} \\ \underline{-3\,7} \\ 1\,85 \\ \underline{-1\,85} \\ 0 \end{array}$$

75. no; answers may vary

77. no; $\dfrac{6 \text{ inches}}{15 \text{ inches}} = \dfrac{2 \cdot 3}{5 \cdot 3} = \dfrac{2}{5}$

79. 10 defective to 200 good is $\dfrac{10 \text{ defective}}{200 \text{ good}} = \dfrac{1 \text{ defective}}{20 \text{ good}}$.
Yes, the machine should be repaired.

Copyright © 2011 Pearson Education, Inc. Publishing as Prentice Hall.

81. $29{,}543 - 29{,}286 = 257$

$$13.4\overline{)257} \text{ becomes } \begin{array}{r} 19.17 \approx 19.2 \\ 134\overline{)2570.00} \\ \underline{-134} \\ 1230 \\ \underline{-1206} \\ 24\;0 \\ \underline{-13\;4} \\ 10\;60 \\ \underline{-9\;38} \\ 1\;22 \end{array}$$

257 miles were driven, averaging approximately 19.2 miles per gallon.

83. $80{,}242 - 79{,}895 = 347$

$$16.1\overline{)347} \text{ becomes } \begin{array}{r} 21.55 \approx 21.6 \\ 161\overline{)3470.00} \\ \underline{-322} \\ 250 \\ \underline{-161} \\ 89\;0 \\ \underline{-80\;5} \\ 8\;50 \\ \underline{-8\;05} \\ 45 \end{array}$$

347 miles were driven, averaging approximately 21.6 miles per gallon.

85.

$$\begin{array}{r} 1.50 \approx 1.5 \\ 7759\overline{)11{,}674.00} \\ \underline{7\;759} \\ 3\;915\;0 \\ \underline{-3\;879\;5} \\ 35\;50 \end{array}$$

The unit rate is 1.5 steps/foot.

87. a. $\dfrac{20 \text{ states}}{50 \text{ states}} = \dfrac{20}{50} = \dfrac{2}{5}$

b. $\dfrac{20 \text{ states}}{(50-20) \text{ states}} = \dfrac{20}{30} = \dfrac{2}{3}$

c. no; answers may vary

89. answers may vary

91. no; answers may vary

Section 6.2

Practice Problems

1. a. $\dfrac{\text{cups} \rightarrow 24}{\text{cups} \rightarrow 6} = \dfrac{4 \leftarrow \text{cups}}{1 \leftarrow \text{cups}}$

b. $\dfrac{\text{students} \rightarrow 560}{\text{instructors} \rightarrow 25} = \dfrac{112 \leftarrow \text{students}}{5 \leftarrow \text{instructors}}$

2. $\dfrac{4}{8} \stackrel{?}{=} \dfrac{10}{20}$

$4 \cdot 20 \stackrel{?}{=} 8 \cdot 10$

$80 = 80$

The proportion is true.

3. $\dfrac{4.2}{6} \stackrel{?}{=} \dfrac{4.8}{8}$

$4.2 \cdot 8 \stackrel{?}{=} 6 \cdot 4.8$

$33.6 \neq 28.8$

The proportion is false.

4. $\dfrac{3\frac{3}{10}}{1\frac{5}{6}} \stackrel{?}{=} \dfrac{4\frac{1}{5}}{2\frac{1}{3}}$

$3\dfrac{3}{10} \cdot 2\dfrac{1}{3} \stackrel{?}{=} 1\dfrac{5}{6} \cdot 4\dfrac{1}{5}$

$\dfrac{33}{10} \cdot \dfrac{7}{3} \stackrel{?}{=} \dfrac{11}{6} \cdot \dfrac{21}{5}$

$\dfrac{231}{30} = \dfrac{231}{30}$

The proportion is true.

5. $\dfrac{2}{5} = \dfrac{x}{25}$

$2 \cdot 25 = 5 \cdot x$

$50 = 5x$

$\dfrac{50}{5} = \dfrac{5x}{5}$

$10 = x$

6. $\dfrac{-15}{2} = \dfrac{60}{x}$

$-15 \cdot x = 2 \cdot 60$

$-15x = 120$

$\dfrac{-15x}{-15} = \dfrac{120}{-15}$

$x = -8$

Copyright © 2011 Pearson Education, Inc. Publishing as Prentice Hall.

7. $\dfrac{\frac{7}{8}}{z} = \dfrac{\frac{2}{3}}{\frac{4}{7}}$

$\frac{7}{8} \cdot \frac{4}{7} = z \cdot \frac{2}{3}$

$\frac{1}{2} = z \cdot \frac{2}{3}$

$\frac{1}{2} \cdot \frac{3}{2} = z \cdot \frac{2}{3} \cdot \frac{3}{2}$

$\frac{3}{4} = z$

8. $\frac{y}{9} = \frac{0.6}{1.2}$

$y \cdot 1.2 = 9 \cdot 0.6$

$1.2y = 5.4$

$\frac{1.2y}{1.2} = \frac{5.4}{1.2}$

$y = 4.5$

9. $\frac{17}{z} = \frac{8}{10}$

$17 \cdot 10 = z \cdot 8$

$170 = 8z$

$\frac{170}{8} = \frac{8z}{8}$

$\frac{85}{4} = z$

$21.25 = z$

10. $\frac{4.5}{1.8} = \frac{y}{3}$

$4.5 \cdot 3 = 1.8 \cdot y$

$13.5 = 1.8y$

$\frac{13.5}{1.8} = \frac{1.8y}{1.8}$

$7.5 = y$

Vocabulary and Readiness Check

1. $\frac{4.2}{8.4} = \frac{1}{2}$ is called a proportion while $\frac{7}{8}$ is called a ratio.

2. In $\frac{a}{b} = \frac{c}{d}$, $a \cdot d$ and $b \cdot c$ are called cross products.

3. In a proportion, if cross products are equal, the proportion is true.

4. In a proportion, if cross products are not equal, the proportion is false.

Exercise Set 6.2

1. $\frac{10 \text{ diamonds}}{6 \text{ opals}} = \frac{5 \text{ diamonds}}{3 \text{ opals}}$

3. $\frac{20 \text{ students}}{5 \text{ microscopes}} = \frac{4 \text{ students}}{1 \text{ microscope}}$

5. $\frac{6 \text{ eagles}}{58 \text{ sparrows}} = \frac{3 \text{ eagles}}{29 \text{ sparrows}}$

7. $\frac{2\frac{1}{4} \text{ cups flour}}{24 \text{ cookies}} = \frac{6\frac{3}{4} \text{ cups flour}}{72 \text{ cookies}}$

9. $\frac{22 \text{ vanilla wafers}}{1 \text{ cup cookie crumbs}} = \frac{55 \text{ vanilla wafers}}{2.5 \text{ cups cookie crumbs}}$

11. $\frac{15}{9} \stackrel{?}{=} \frac{5}{3}$

$15 \cdot 3 \stackrel{?}{=} 9 \cdot 5$

$45 = 45$

The proportion is true.

13. $\frac{5}{8} \stackrel{?}{=} \frac{4}{7}$

$5 \cdot 7 \stackrel{?}{=} 8 \cdot 4$

$35 \neq 32$

The proportion is false.

15. $\frac{9}{36} \stackrel{?}{=} \frac{2}{8}$

$9 \cdot 8 \stackrel{?}{=} 36 \cdot 2$

$72 = 72$

The proportion is true.

17. $\frac{5}{8} \stackrel{?}{=} \frac{625}{1000}$

$5 \cdot 1000 \stackrel{?}{=} 8 \cdot 625$

$5000 = 5000$

The proportion is true.

19. $\frac{0.8}{0.3} \stackrel{?}{=} \frac{0.2}{0.6}$

$0.8 \cdot 0.6 \stackrel{?}{=} 0.3 \cdot 0.2$

$0.48 \neq 0.06$

The proportion is false.

Copyright © 2011 Pearson Education, Inc. Publishing as Prentice Hall.

21. $\frac{4.2}{8.4} \stackrel{?}{=} \frac{5}{10}$
$4.2 \cdot 10 \stackrel{?}{=} 8.4 \cdot 5$
$42 = 42$
The proportion is true.

23. $\frac{\frac{3}{4}}{\frac{4}{3}} \stackrel{?}{=} \frac{\frac{1}{2}}{\frac{8}{9}}$
$\frac{3}{4} \cdot \frac{8}{9} \stackrel{?}{=} \frac{4}{3} \cdot \frac{1}{2}$
$\frac{2}{3} = \frac{2}{3}$
The proportion is true.

25. $\frac{2\frac{2}{5}}{\frac{2}{3}} \stackrel{?}{=} \frac{\frac{10}{9}}{\frac{1}{4}}$
$2\frac{2}{5} \cdot \frac{1}{4} \stackrel{?}{=} \frac{2}{3} \cdot \frac{10}{9}$
$\frac{12}{5} \cdot \frac{1}{4} \stackrel{?}{=} \frac{20}{27}$
$\frac{12}{20} \neq \frac{20}{27}$
The proportion is false.

27. $\frac{\frac{4}{5}}{6} \stackrel{?}{=} \frac{\frac{6}{5}}{9}$
$\frac{4}{5} \cdot 9 \stackrel{?}{=} 6 \cdot \frac{6}{5}$
$\frac{36}{5} = \frac{36}{5}$
The proportion is true.

29. $\frac{10}{15} \stackrel{?}{=} \frac{4}{6}$
$10 \cdot 6 \stackrel{?}{=} 15 \cdot 4$
$60 = 60$
The proportion $\frac{10}{15} = \frac{4}{6}$ is true.

31. $\frac{11}{4} \stackrel{?}{=} \frac{5}{2}$
$11 \cdot 2 \stackrel{?}{=} 4 \cdot 5$
$22 \neq 20$
The proportion $\frac{11}{4} = \frac{5}{2}$ is false.

33. $\frac{0.15}{3} \stackrel{?}{=} \frac{0.35}{7}$
$0.15 \cdot 7 \stackrel{?}{=} 3 \cdot 0.35$
$1.05 = 1.05$
The proportion $\frac{0.15}{3} = \frac{0.35}{7}$ is true.

35. $\frac{\frac{2}{3}}{\frac{1}{5}} \stackrel{?}{=} \frac{\frac{2}{5}}{\frac{1}{9}}$
$\frac{2}{3} \cdot \frac{1}{9} \stackrel{?}{=} \frac{1}{5} \cdot \frac{2}{5}$
$\frac{2}{27} \neq \frac{2}{25}$
The proportion $\frac{\frac{2}{3}}{\frac{1}{5}} = \frac{\frac{2}{5}}{\frac{1}{9}}$ is false.

37. $\frac{x}{5} = \frac{6}{10}$
$x \cdot 10 = 5 \cdot 6$
$10x = 30$
$\frac{10x}{10} = \frac{30}{10}$
$x = 3$

39. $\frac{-18}{54} = \frac{3}{n}$
$-18 \cdot n = 54 \cdot 3$
$-18n = 162$
$\frac{-18n}{-18} = \frac{162}{-18}$
$n = -9$

41. $\frac{30}{10} = \frac{15}{y}$
$30 \cdot y = 10 \cdot 15$
$30y = 150$
$\frac{30y}{30} = \frac{150}{30}$
$y = 5$

43. $\frac{8}{15} = \frac{z}{6}$
$8 \cdot 6 = 15 \cdot z$
$48 = 15z$
$\frac{48}{15} = \frac{15z}{15}$
$3.2 = z$

Copyright © 2011 Pearson Education, Inc. Publishing as Prentice Hall.

45. $\frac{24}{x}=\frac{60}{96}$
$24\cdot 96=x\cdot 60$
$2304=60x$
$\frac{2304}{60}=\frac{60x}{60}$
$38.4=x$

47. $\frac{-3.5}{12.5}=\frac{-7}{n}$
$-3.5\cdot n=12.5\cdot(-7)$
$-3.5n=-87.5$
$\frac{-3.5n}{-3.5}=\frac{-87.5}{-3.5}$
$n=25$

49. $\frac{n}{0.6}=\frac{0.05}{12}$
$n\cdot 12=0.6\cdot 0.05$
$12n=0.030$
$\frac{12n}{12}=\frac{0.03}{12}$
$n=0.0025$

51. $\frac{8}{\frac{1}{3}}=\frac{24}{n}$
$8\cdot n=\frac{1}{3}\cdot 24$
$8n=8$
$\frac{8n}{8}=\frac{8}{8}$
$n=1$

53. $\frac{\frac{1}{3}}{\frac{3}{8}}=\frac{\frac{2}{5}}{n}$
$\frac{1}{3}\cdot n=\frac{3}{8}\cdot\frac{2}{5}$
$\frac{n}{3}=\frac{3}{20}$
$3\cdot\frac{n}{3}=3\cdot\frac{3}{20}$
$n=\frac{9}{20}$

55. $\frac{12}{n}=\frac{\frac{2}{3}}{\frac{6}{9}}$
$12\cdot\frac{6}{9}=n\cdot\frac{2}{3}$
$8=n\cdot\frac{2}{3}$
$\frac{3}{2}\cdot\frac{8}{1}=n\cdot\frac{2}{3}\cdot\frac{3}{2}$
$12=n$

57. $\frac{n}{1\frac{1}{5}}=\frac{4\frac{1}{6}}{6\frac{2}{3}}$
$n\cdot 6\frac{2}{3}=1\frac{1}{5}\cdot 4\frac{1}{6}$
$n\cdot\frac{20}{3}=\frac{6}{5}\cdot\frac{25}{6}$
$\frac{20}{3}n=5$
$\frac{3}{20}\cdot\frac{20}{3}n=\frac{3}{20}\cdot 5$
$n=\frac{3}{4}$

59. $\frac{25}{n}=\frac{3}{\frac{7}{30}}$
$25\cdot\frac{7}{30}=n\cdot 3$
$\frac{35}{6}=3n$
$\frac{1}{3}\cdot\frac{35}{6}=3n\cdot\frac{1}{3}$
$\frac{35}{18}=n$

61. $\frac{3.2}{0.3}=\frac{x}{1.4}$
$3.2\cdot 1.4=0.3\cdot x$
$4.48=0.3x$
$\frac{4.48}{0.3}=\frac{0.3x}{0.3}$
$14.9\approx x$

63. $\frac{z}{5.2}=\frac{0.08}{6}$
$z\cdot 6=5.2\cdot 0.08$
$6z=0.416$
$\frac{6z}{6}=\frac{0.416}{6}$
$z\approx 0.07$

Copyright © 2011 Pearson Education, Inc. Publishing as Prentice Hall.

65. $\frac{7}{18} = \frac{x}{5}$
$7 \cdot 5 = 18 \cdot x$
$35 = 18x$
$\frac{35}{18} = \frac{18x}{18}$
$1.9 \approx x$

67. $\frac{43}{17} = \frac{8}{z}$
$43 \cdot z = 17 \cdot 8$
$43z = 136$
$\frac{43z}{43} = \frac{136}{43}$
$z \approx 3.163$

69. 8.01 8.1
$\uparrow$ $\uparrow$
0 < 1
$8.01 < 8.1$

71. $\frac{1}{2} > \frac{1}{3}$

so $2\frac{1}{2} > 2\frac{1}{3}$

73. $\frac{75}{125} = \frac{3 \cdot 25}{5 \cdot 25} = \frac{3}{5}$

75. $\frac{12x}{42} = \frac{2 \cdot 6 \cdot x}{7 \cdot 6} = \frac{2x}{7}$

77. $\frac{9}{15} = \frac{3}{5}$
$\frac{9}{3} = \frac{15}{5}$
$\frac{5}{15} = \frac{3}{9}$
$\frac{15}{9} = \frac{5}{3}$

79. $\frac{6}{18} = \frac{1}{3}$
$\frac{6}{1} = \frac{18}{3}$
$\frac{3}{18} = \frac{1}{6}$
$\frac{18}{6} = \frac{3}{1}$

81. $\frac{a}{b} = \frac{c}{d}$
Possible answers include:
$\frac{d}{b} = \frac{c}{a}$
$\frac{a}{c} = \frac{b}{d}$
$\frac{b}{a} = \frac{d}{c}$

83. answers may vary

85. $\frac{x}{7} = \frac{0}{8}$
$x \cdot 8 = 7 \cdot 0$
$8x = 0$
$\frac{8x}{8} = \frac{0}{8}$
$x = 0$

87. $\frac{z}{1150} = \frac{588}{483}$
$483 \cdot z = 1150 \cdot 588$
$483z = 676,200$
$\frac{483z}{483} = \frac{676,200}{483}$
$z = 1400$

89. $\frac{222}{1515} = \frac{37}{y}$
$222 \cdot y = 1515 \cdot 37$
$222y = 56,055$
$\frac{222y}{222} = \frac{56,055}{222}$
$y = 252.5$

Integrated Review

1. The ratio of 27 to 30 is $\frac{27}{30} = \frac{9 \cdot 3}{10 \cdot 3} = \frac{9}{10}$.

2. The ratio of 18 to 50 is $\frac{18}{50} = \frac{9 \cdot 2}{25 \cdot 2} = \frac{9}{25}$.

3. The ratio of 9.4 to 10 is
$\frac{9.4}{10} = \frac{9.4 \times 10}{10 \times 10} = \frac{94}{100} = \frac{47 \cdot 2}{50 \cdot 2} = \frac{47}{50}$.

Copyright © 2011 Pearson Education, Inc. Publishing as Prentice Hall.

4. The ratio of 3.2 to 9.2 is
$\frac{3.2}{9.2}=\frac{3.2\times 10}{9.2\times 10}=\frac{32}{92}=\frac{4\cdot 8}{4\cdot 23}=\frac{8}{23}$.

5. The ratio of 8.65 to 6.95 is
$\frac{8.65}{6.95}=\frac{8.65\times 100}{6.95\times 100}=\frac{865}{695}=\frac{173\cdot 5}{139\cdot 5}=\frac{173}{139}$.

6. The ratio of 3.6 to 4.2 is
$\frac{3.6}{4.2}=\frac{3.6\times 10}{4.2\times 10}=\frac{36}{42}=\frac{6\cdot 6}{7\cdot 6}=\frac{6}{7}$.

7. The ratio of $\frac{7}{2}$ to 13 is
$\frac{\frac{7}{2}}{13}=\frac{7}{2}\div 13=\frac{7}{2}\cdot\frac{1}{13}=\frac{7\cdot 1}{2\cdot 13}=\frac{7}{26}$.

8. The ratio of $1\frac{2}{3}$ to $2\frac{3}{4}$ is
$\frac{1\frac{2}{3}}{2\frac{3}{4}}=1\frac{2}{3}\div 2\frac{3}{4}=\frac{5}{3}\div\frac{11}{4}=\frac{5}{3}\cdot\frac{4}{11}=\frac{5\cdot 4}{3\cdot 11}=\frac{20}{33}$.

9. The ratio of 16 inches to 24 inches is
$\frac{16 \text{ inches}}{24 \text{ inches}}=\frac{2\cdot 8}{3\cdot 8}=\frac{2}{3}$.

10. The ratio of 5 hours to 40 hours is
$\frac{5 \text{ hours}}{40 \text{ hours}}=\frac{5\cdot 1}{5\cdot 8}=\frac{1}{8}$.

11. $\frac{12 \text{ inches}}{18 \text{ inches}}=\frac{6\cdot 2}{6\cdot 3}=\frac{2}{3}$

12. a. 41 films were rated R.

b. $\frac{35 \text{ films}}{163 \text{ films}}=\frac{35}{163}$

13. $\frac{4 \text{ professors}}{20 \text{ graduate assistants}}=\frac{1 \text{ professor}}{5 \text{ graduate assistants}}$

14. $\frac{6 \text{ lights}}{20 \text{ feet}}=\frac{3 \text{ lights}}{10 \text{ feet}}$

15. $\frac{100 \text{ Senators}}{50 \text{ states}}=\frac{2 \text{ Senators}}{1 \text{ state}}$

16. $\frac{5 \text{ teachers}}{140 \text{ students}}=\frac{1 \text{ teacher}}{28 \text{ students}}$

17. $\frac{21 \text{ inches}}{7 \text{ seconds}}=\frac{3 \text{ inches}}{1 \text{ second}}$

18. $\frac{\$40}{5 \text{ hours}}=\frac{\$8}{1 \text{ hour}}$

19. $\frac{76 \text{ households with computers}}{100 \text{ households}}$
$=\frac{19 \text{ households with computers}}{25 \text{ households}}$

20. $\frac{538 \text{ electoral votes}}{50 \text{ states}}=\frac{269 \text{ electoral votes}}{25 \text{ states}}$

21. $\frac{560 \text{ feet}}{4 \text{ seconds}}$

$$\begin{array}{r} 140 \\ 4\overline{)\ 560} \\ \underline{-4} \\ 16 \\ \underline{-16} \\ 00 \end{array}$$

The unit rate is $\frac{140 \text{ feet}}{1 \text{ second}}$ or 140 feet/second

22. $\frac{195 \text{ miles}}{3 \text{ hours}}$

$$\begin{array}{r} 65 \\ 3\overline{)\ 195} \\ \underline{-18} \\ 15 \\ \underline{-15} \\ 0 \end{array}$$

The unit rate is $\frac{65 \text{ miles}}{1 \text{ hour}}$ or 65 miles/hour.

23. $\frac{63 \text{ employees}}{3 \text{ fax lines}}$

$$\begin{array}{r} 21 \\ 3\overline{)\ 63} \\ \underline{-6} \\ 03 \\ \underline{-3} \\ 0 \end{array}$$

The unit rate is $\frac{21 \text{ employees}}{1 \text{ fax line}}$ or 21 employees/fax line.

Copyright © 2011 Pearson Education, Inc. Publishing as Prentice Hall.

24. $\dfrac{85 \text{ phone calls}}{5 \text{ teenagers}}$

$$\begin{array}{r} 17 \\ 5\overline{)\;85} \\ \underline{-5} \\ 35 \\ \underline{-35} \\ 0 \end{array}$$

The unit rate is $\dfrac{17 \text{ phone calls}}{1 \text{ teenager}}$ or 17 phone calls/teenager.

25. $\dfrac{156 \text{ miles}}{6 \text{ gallons}}$

$$\begin{array}{r} 26 \\ 6\overline{)156} \\ \underline{12} \\ 36 \\ \underline{-36} \\ 0 \end{array}$$

The unit rate is $\dfrac{26 \text{ miles}}{1 \text{ gallon}}$ or 26 miles/gallon.

26. $\dfrac{112 \text{ teachers}}{7 \text{ computers}}$

$$\begin{array}{r} 16 \\ 7\overline{)112} \\ \underline{-7} \\ 42 \\ \underline{-42} \\ 0 \end{array}$$

The unit rate is $\dfrac{16 \text{ teachers}}{1 \text{ computer}}$ or 16 teachers/computer.

27. $\dfrac{8125 \text{ books}}{1250 \text{ college students}}$

$$\begin{array}{r} 6.5 \\ 1250\overline{)\;8125.0} \\ \underline{-7500} \\ 625\ 0 \\ \underline{-625\ 0} \\ 0 \end{array}$$

The unit rate is $\dfrac{6.5 \text{ books}}{1 \text{ college student}}$ or 6.5 books/student.

28. $\dfrac{2310 \text{ pounds}}{14 \text{ adults}}$

$$\begin{array}{r} 165 \\ 14\overline{)\;2310} \\ \underline{-14} \\ 91 \\ \underline{-84} \\ 70 \\ \underline{-70} \\ 0 \end{array}$$

The unit rate is $\dfrac{165 \text{ pounds}}{1 \text{ adult}}$ or 165 pounds/adult.

29.

$$\begin{array}{r} 0.27 \\ 8\overline{)\;2.16} \\ \underline{-1\ 6} \\ 56 \\ \underline{-56} \\ 0 \end{array}$$

The 8-pound size costs \$0.27 per pound.

$$\begin{array}{r} 0.2772 \approx 0.277 \\ 18\overline{)\;4.9900} \\ \underline{-3\ 6} \\ 1\ 39 \\ \underline{-1\ 26} \\ 130 \\ \underline{-126} \\ 40 \\ \underline{-36} \\ 4 \end{array}$$

The 18-pound size costs \$0.277 per pound.
The 8-pound size is the better buy.

30.

$$\begin{array}{r} 0.0198 \approx 0.020 \\ 100\overline{)\;1.9800} \\ \underline{-1\ 00} \\ 980 \\ \underline{-900} \\ 800 \\ \underline{-800} \\ 0 \end{array}$$

The 100-plate size costs \$0.020 per plate.

Copyright © 2011 Pearson Education, Inc. Publishing as Prentice Hall.

$$\begin{array}{r} 0.0179 \approx 0.018 \\ 500\overline{)\ 8.9900} \\ \underline{-5\ 00} \\ 3\ 990 \\ \underline{-3\ 500} \\ 4900 \\ \underline{-4500} \\ 400 \end{array}$$

The 500-plate size costs \$0.018 per plate.
The 500 paper plates is the better buy.

31.
$$\begin{array}{r} 0.7966\ldots \approx 0.797 \\ 3\overline{)\ 2.3900} \\ \underline{-2\ 1} \\ 29 \\ \underline{-27} \\ 20 \\ \underline{-18} \\ 20 \\ \underline{-18} \\ 2 \end{array}$$

The 3-pack size costs \$0.797 per pack.

$$\begin{array}{r} 0.7487 \approx 0.749 \\ 8\overline{)\ 5.9900} \\ \underline{-5\ 6} \\ 39 \\ \underline{-32} \\ 70 \\ \underline{-64} \\ 60 \\ \underline{-56} \\ 4 \end{array}$$

The 8-pack size costs \$0.749 per pack.
The 8 pack is the better buy.

32.
$$\begin{array}{r} 0.9225 \approx 0.923 \\ 4\overline{)\ 3.6900} \\ \underline{-3\ 6} \\ 09 \\ \underline{-8} \\ 10 \\ \underline{-8} \\ 20 \\ \underline{-20} \\ 0 \end{array}$$

The 4 batteries cost \$0.923 each.

$$\begin{array}{r} 0.989 \\ 10\overline{)\ 9.890} \\ \underline{-9\ 0} \\ 89 \\ \underline{-80} \\ 90 \\ \underline{-90} \\ 0 \end{array}$$

The 10 batteries cost \$0.989 each.
The 4 batteries are the better buy.

33.
$$\frac{7}{4} \stackrel{?}{=} \frac{5}{3}$$
$$7 \cdot 3 \stackrel{?}{=} 4 \cdot 5$$
$$21 \neq 20$$
The proportion is false.

34.
$$\frac{8.2}{2} \stackrel{?}{=} \frac{16.4}{4}$$
$$8.2 \cdot 4 \stackrel{?}{=} 2 \cdot 16.4$$
$$32.8 = 32.8$$
The proportion is true.

35.
$$\frac{5}{3} = \frac{40}{x}$$
$$5 \cdot x = 3 \cdot 40$$
$$5x = 120$$
$$\frac{5x}{5} = \frac{120}{5}$$
$$x = 24$$

36.
$$\frac{y}{10} = \frac{13}{4}$$
$$y \cdot 4 = 10 \cdot 13$$
$$4y = 130$$
$$\frac{4y}{4} = \frac{130}{4}$$
$$y = 32.5$$

37.
$$\frac{6}{11} = \frac{z}{5}$$
$$6 \cdot 5 = 11 \cdot z$$
$$30 = 11z$$
$$\frac{30}{11} = \frac{11z}{11}$$
$$2.\overline{72} = z \text{ or } z = 2\frac{8}{11}$$

Copyright © 2011 Pearson Education, Inc. Publishing as Prentice Hall.

38. $\frac{21}{x}=\frac{\frac{7}{2}}{3}$

$$21\cdot 3=x\cdot\frac{7}{2}$$
$$63=\frac{7}{2}\cdot x$$
$$\frac{2}{7}\cdot\frac{63}{1}=\frac{2}{7}\cdot\frac{7}{2}\cdot x$$
$$18=x$$

Section 6.3

Practice Problems

1. Let x be the length of the wall.

$$\begin{array}{l} \text{feet} \rightarrow \\ \text{inches} \rightarrow \end{array} \frac{4}{1}=\frac{x}{4\frac{1}{4}} \begin{array}{l} \leftarrow \text{feet} \\ \leftarrow \text{inches} \end{array}$$
$$4\cdot 4\frac{1}{4}=1\cdot x$$
$$\frac{4}{1}\cdot\frac{17}{4}=x$$
$$17=x$$

The wall is 17 feet long.

2. $$\begin{array}{l} \text{ounces} \rightarrow \\ \text{gallons} \rightarrow \end{array} \frac{5}{16}=\frac{8}{x} \begin{array}{l} \leftarrow \text{ounces} \\ \leftarrow \text{gallons} \end{array}$$
$$5\cdot x=16\cdot 8$$
$$5x=128$$
$$\frac{5x}{5}=\frac{128}{5}$$
$$x=25.6 \text{ or } 25\frac{3}{5}$$

Therefore, 25.6 or $25\frac{3}{5}$ gallons of gas can be treated with 8 ounces of alcohol.

3. Let x be the number of gallons.
$270 \times 11 = 2970$ square feet

$$\begin{array}{l} \text{gallons} \rightarrow \\ \text{square feet} \rightarrow \end{array} \frac{1}{450}=\frac{x}{2970} \begin{array}{l} \leftarrow \text{gallons} \\ \leftarrow \text{square feet} \end{array}$$
$$1\cdot 2970=450\cdot x$$
$$\frac{2970}{450}=\frac{450x}{450}$$
$$6.6=x$$
$$7\approx x$$

Therefore, 7 gallons are needed for the wall.

Exercise Set 6.3

1. Let x be the number of baskets made.

$$\begin{array}{l} \text{baskets} \rightarrow \\ \text{attempts} \rightarrow \end{array} \frac{45}{100}=\frac{x}{800} \begin{array}{l} \leftarrow \text{baskets} \\ \leftarrow \text{attempts} \end{array}$$
$$45\cdot 800=100\cdot x$$
$$36,000=100x$$
$$\frac{36,000}{100}=\frac{100x}{100}$$
$$360=x$$

He made 360 baskets.

3. Let x be the number of minutes.

$$\begin{array}{l} \text{minutes} \rightarrow \\ \text{pages} \rightarrow \end{array} \frac{30}{4}=\frac{x}{22} \begin{array}{l} \leftarrow \text{minutes} \\ \leftarrow \text{pages} \end{array}$$
$$30\cdot 22=4\cdot x$$
$$660=4x$$
$$\frac{660}{4}=\frac{4x}{4}$$
$$165=x$$

It will take her 165 minutes to word process and spell check 22 pages.

5. Let x be the number of applications received.

$$\begin{array}{l} \text{accepted} \rightarrow \\ \text{applied} \rightarrow \end{array} \frac{2}{7}=\frac{180}{x} \begin{array}{l} \leftarrow \text{accepted} \\ \leftarrow \text{applied} \end{array}$$
$$2\cdot x=7\cdot 180$$
$$2x=1260$$
$$\frac{2x}{2}=\frac{1260}{2}$$
$$x=630$$

The school received 630 applications.

7. Let x be the length of the wall.

$$\begin{array}{l} \text{inches} \rightarrow \\ \text{feet} \rightarrow \end{array} \frac{1}{8}=\frac{2\frac{7}{8}}{x} \begin{array}{l} \leftarrow \text{inches} \\ \leftarrow \text{feet} \end{array}$$
$$1\cdot x=8\cdot 2\frac{7}{8}$$
$$x=\frac{8}{1}\cdot\frac{23}{8}$$
$$x=23$$

The wall is 23 feet long.

9. Let x be the number of square feet required.

$$\begin{array}{l} \text{floor space} \rightarrow \\ \text{students} \rightarrow \end{array} \frac{9}{1}=\frac{x}{30} \begin{array}{l} \leftarrow \text{floor space} \\ \leftarrow \text{students} \end{array}$$
$$9\cdot 30=1\cdot x$$
$$270=x$$

30 students require 270 square feet of floor space.

Copyright © 2011 Pearson Education, Inc. Publishing as Prentice Hall.

11. Let x be the number of gallons.

$$\begin{matrix}\text{miles} \rightarrow \\ \text{gallons} \rightarrow\end{matrix} \frac{627}{12.3} = \frac{1250}{x} \begin{matrix}\leftarrow \text{miles} \\ \leftarrow \text{gallons}\end{matrix}$$

$$627 \cdot x = 12.3 \cdot 1250$$
$$627x = 15,375$$
$$\frac{627x}{627} = \frac{15,375}{627}$$
$$x \approx 25$$

He can expect to burn 25 gallons.

13. Let x be the distance between Milan and Rome.

$$\begin{matrix}\text{kilometers} \rightarrow \\ \text{cm on map} \rightarrow\end{matrix} \frac{30}{1} = \frac{x}{15} \begin{matrix}\leftarrow \text{kilometers} \\ \leftarrow \text{cm on map}\end{matrix}$$

$$30 \cdot 15 = 1 \cdot x$$
$$450 = x$$

Milan and Rome are 450 kilometers apart.

15. Let x be the number of bags.

$$\begin{matrix}\text{bags} \rightarrow \\ \text{square feet} \rightarrow\end{matrix} \frac{1}{3000} = \frac{x}{260 \cdot 180} \begin{matrix}\leftarrow \text{bags} \\ \leftarrow \text{square feet}\end{matrix}$$

$$1 \cdot 260 \cdot 180 = 3000 \cdot x$$
$$46,800 = 3000x$$
$$\frac{46,800}{3000} = \frac{3000x}{3000}$$
$$15.6 = x$$
$$16 \approx x$$

16 bags of fertilizer should be purchased.

17. Let x be the number of hits the player is expected to get.

$$\begin{matrix}\text{hits} \rightarrow \\ \text{at bats} \rightarrow\end{matrix} \frac{3}{8} = \frac{x}{40} \begin{matrix}\leftarrow \text{hits} \\ \leftarrow \text{at bats}\end{matrix}$$

$$3 \cdot 40 = 8 \cdot x$$
$$120 = 8x$$
$$\frac{120}{8} = \frac{8x}{8}$$
$$15 = x$$

The player would be expected to get 15 hits.

19. Let x be the number that prefer Coke.

$$\begin{matrix}\text{Coke} \rightarrow \\ \text{Total} \rightarrow\end{matrix} \frac{2}{3} = \frac{x}{40} \begin{matrix}\leftarrow \text{Coke} \\ \leftarrow \text{Total}\end{matrix}$$

$$2 \cdot 40 = 3 \cdot x$$
$$80 = 3x$$
$$\frac{80}{3} = \frac{3x}{3}$$
$$27 \approx x$$

About 27 people are likely to prefer Coke.

21. Let x be the number of applications she should expect.

$$\begin{matrix}\text{applications} \rightarrow \\ \text{ounces} \rightarrow\end{matrix} \frac{4}{3} = \frac{x}{14} \begin{matrix}\leftarrow \text{applications} \\ \leftarrow \text{ounces}\end{matrix}$$

$$4 \cdot 14 = 3 \cdot x$$
$$56 = 3x$$
$$\frac{56}{3} = \frac{3x}{3}$$
$$18\frac{2}{3} = x$$

She should expect 18 applications from the 14-ounce bottle.

23. Let x be the number of weeks.

$$\begin{matrix}\text{reams} \rightarrow \\ \text{weeks} \rightarrow\end{matrix} \frac{5}{3} = \frac{8}{x} \begin{matrix}\leftarrow \text{reams} \\ \leftarrow \text{weeks}\end{matrix}$$

$$5 \cdot x = 3 \cdot 8$$
$$5x = 24$$
$$\frac{5x}{5} = \frac{24}{5}$$
$$x = 4.8$$
$$x \approx 5$$

The case is likely to last 5 weeks.

25. Let x be the number of servings he can make.

$$\begin{matrix}\text{milk} \rightarrow \\ \text{servings} \rightarrow\end{matrix} \frac{1\frac{1}{2}}{4} = \frac{4}{x} \begin{matrix}\leftarrow \text{milk} \\ \leftarrow \text{servings}\end{matrix}$$

$$1\frac{1}{2} \cdot x = 4 \cdot 4$$
$$\frac{3}{2}x = 16$$
$$\frac{2}{3} \cdot \frac{3}{2}x = \frac{2}{3} \cdot 16$$
$$x = \frac{32}{3} = 10\frac{2}{3}$$

He can make $10\frac{2}{3}$ servings.

27. Let x be the time to reach the restaurant (in seconds).

$$\begin{matrix}\text{distance} \rightarrow \\ \text{time} \rightarrow\end{matrix} \frac{800}{60} = \frac{500}{x} \begin{matrix}\leftarrow \text{distance} \\ \leftarrow \text{time}\end{matrix}$$

$$800 \cdot x = 60 \cdot 500$$
$$800x = 30,000$$
$$\frac{800x}{800} = \frac{30,000}{800}$$
$$x = 37.5$$

It will take 37.5 seconds to reach the restaurant.

Copyright © 2011 Pearson Education, Inc. Publishing as Prentice Hall.

29. a. Let x be the number of teaspoons of granules needed.

$$\begin{array}{r} \text{water} \rightarrow \\ \text{granules} \rightarrow \end{array} \frac{25}{1} = \frac{450}{x} \begin{array}{l} \leftarrow \text{water} \\ \leftarrow \text{granules} \end{array}$$

$$25 \cdot x = 1 \cdot 450$$
$$25x = 450$$
$$\frac{25x}{25} = \frac{450}{25}$$
$$x = 18$$

18 teaspoons of granules are needed.

b. Let x be the number of tablespoons of granules needed.

$$\begin{array}{r} \text{tsp} \rightarrow \\ \text{tbsp} \rightarrow \end{array} \frac{3}{1} = \frac{18}{x} \begin{array}{l} \leftarrow \text{tsp} \\ \leftarrow \text{tbsp} \end{array}$$

$$3 \cdot x = 1 \cdot 18$$
$$3x = 18$$
$$\frac{3x}{3} = \frac{18}{3}$$
$$x = 6$$

6 tablespoons of granules are needed.

31. Let x be the number of people.

$$\begin{array}{r} \text{square feet} \rightarrow \\ \text{people} \rightarrow \end{array} \frac{625}{1} = \frac{3750}{x} \begin{array}{l} \leftarrow \text{square feet} \\ \leftarrow \text{people} \end{array}$$

$$625 \cdot x = 1 \cdot 3750$$
$$\frac{625x}{625} = \frac{3750}{625}$$
$$x = 6$$

It will provide enough oxygen for 6 people.

33. Let x be the estimated head-to-toe height of the Statue of Liberty.

$$\begin{array}{r} \text{height} \rightarrow \\ \text{arm length} \rightarrow \end{array} \frac{x}{42} = \frac{5\frac{1}{3}}{2} \begin{array}{l} \leftarrow \text{height} \\ \leftarrow \text{arm length} \end{array}$$

$$x \cdot 2 = 42 \cdot 5\frac{1}{3}$$
$$2x = 42 \cdot \frac{16}{3}$$
$$2x = 224$$
$$\frac{2x}{2} = \frac{224}{2}$$
$$x = 112$$

The estimated height is 112 feet.

$$112 - 111\frac{1}{12} = \frac{11}{12}$$

The difference is $\frac{11}{12}$ foot or 11 inches.

35. Let x be the number of milligrams.

$$\begin{array}{r} \text{milligrams} \rightarrow \\ \text{ounces} \rightarrow \end{array} \frac{72}{3.5} = \frac{x}{5} \begin{array}{l} \leftarrow \text{milligrams} \\ \leftarrow \text{ounces} \end{array}$$

$$72 \cdot 5 = 3.5 \cdot x$$
$$360 = 3.5x$$
$$\frac{360}{3.5} = \frac{3.5x}{3.5}$$
$$102.9 \approx x$$

There are about 102.9 milligrams of cholesterol in 5 ounces of lobster.

37. Let x be the estimated height of the Empire State Building.

$$\begin{array}{r} \text{height} \rightarrow \\ \text{stories} \rightarrow \end{array} \frac{x}{102} = \frac{881}{72} \begin{array}{l} \leftarrow \text{height} \\ \leftarrow \text{stories} \end{array}$$

$$x \cdot 72 = 102 \cdot 881$$
$$72x = 89,862$$
$$\frac{72x}{72} = \frac{89,862}{72}$$
$$x \approx 1248$$

The height of the Empire State Building is approximately 1248 feet.

39. Let x be the number of visits needing a prescription for medication.

$$\begin{array}{r} \text{medication} \rightarrow \\ \text{total} \rightarrow \end{array} \frac{7}{10} = \frac{x}{620} \begin{array}{l} \leftarrow \text{medication} \\ \leftarrow \text{total} \end{array}$$

$$7 \cdot 620 = 10 \cdot x$$
$$4340 = 10x$$
$$\frac{4340}{10} = \frac{10x}{10}$$
$$434 = x$$

Expect 434 emergency room visits to need a prescription for medication.

41. Let x be the number expected to have worked in the restaurant industry.

$$\begin{array}{r} \text{restaurant} \rightarrow \\ \text{workers} \rightarrow \end{array} \frac{x}{84} = \frac{1}{3} \begin{array}{l} \leftarrow \text{restaurant} \\ \leftarrow \text{workers} \end{array}$$

$$x \cdot 3 = 84 \cdot 1$$
$$3x = 84$$
$$\frac{3x}{3} = \frac{84}{3}$$
$$x = 28$$

You would expect 28 of the workers to have worked in the restaurant industry.

Copyright © 2011 Pearson Education, Inc. Publishing as Prentice Hall.

43. Let x be the cups of salt.

$$\begin{matrix}\text{ice} \rightarrow \\ \text{salt} \rightarrow\end{matrix}\frac{5}{1}=\frac{12}{x}\begin{matrix}\leftarrow \text{ice} \\ \leftarrow \text{salt}\end{matrix}$$

$$5 \cdot x = 1 \cdot 12$$

$$\frac{5x}{5}=\frac{12}{5}$$

$$x = 2.4$$

Mix 2.4 cups of salt with the ice.

45. a. Let x be the number of gallons of oil needed.

$$\begin{matrix}\text{oil} \rightarrow \\ \text{gas} \rightarrow\end{matrix}\frac{x}{5}=\frac{1}{50}\begin{matrix}\leftarrow \text{oil} \\ \leftarrow \text{gas}\end{matrix}$$

$$x \cdot 50 = 5 \cdot 1$$

$$50x = 5$$

$$\frac{50x}{50}=\frac{5}{50}$$

$$x=\frac{1}{10}=0.1$$

0.1 gallon of oil is needed.

b. Let x be the number of fluid ounces.

$$\begin{matrix}\text{gallons} \rightarrow \\ \text{fluid ounces} \rightarrow\end{matrix}\frac{1}{128}=\frac{0.1}{x}\begin{matrix}\leftarrow \text{gallons} \\ \leftarrow \text{fluid ounces}\end{matrix}$$

$$1 \cdot x = 128 \cdot 0.1$$

$$x = 12.8$$

0.1 gallon is approximately 13 fluid ounces

47. a. Let x be the milligrams of medicine.

$$\begin{matrix}\text{milligrams} \rightarrow \\ \text{pounds} \rightarrow\end{matrix}\frac{x}{275}=\frac{150}{20}\begin{matrix}\leftarrow \text{milligrams} \\ \leftarrow \text{pounds}\end{matrix}$$

$$x \cdot 20 = 275 \cdot 150$$

$$20x = 41,250$$

$$\frac{20x}{20}=\frac{41,250}{20}$$

$$x = 2062.5$$

The daily dose is 2062.5 milligrams.

b. $500 \times \frac{24}{8} = 500 \times 3 = 1500$

No, he is not receiving the proper dosage.

49.

$$\begin{aligned}200 &= 2 \cdot 100 \\ &\downarrow\ \ \downarrow\ \ \searrow \\ &= 2 \cdot 4 \cdot 25 \\ &\downarrow\ \swarrow\searrow\ \swarrow\searrow \\ &= 2 \cdot 2 \cdot 2 \cdot 5 \cdot 5 \\ &= 2^3 \cdot 5^2\end{aligned}$$

51.

$$\begin{aligned}32 &= 2 \cdot 16 \\ &\downarrow\ \downarrow\searrow \\ &2 \cdot 2 \cdot 8 \\ &\downarrow\ \downarrow\ \downarrow\searrow \\ &2 \cdot 2 \cdot 2 \cdot 4 \\ &\downarrow\ \downarrow\ \downarrow\ \downarrow\searrow \\ &2 \cdot 2 \cdot 2 \cdot 2 \cdot 2 \\ &= 2^5\end{aligned}$$

53. Let x be the number of ml.

$$\begin{matrix}\text{mg} \rightarrow \\ \text{ml} \rightarrow\end{matrix}\frac{15}{1}=\frac{12}{x}\begin{matrix}\leftarrow \text{mg} \\ \leftarrow \text{ml}\end{matrix}$$

$$15 \cdot x = 1 \cdot 12$$

$$15x = 12$$

$$\frac{15x}{15}=\frac{12}{15}$$

$$x=\frac{4}{5}=0.8$$

0.8 ml of the medicine should be administered.

55. Let x be the number of ml.

$$\begin{matrix}\text{mg} \rightarrow \\ \text{ml} \rightarrow\end{matrix}\frac{8}{1}=\frac{10}{x}\begin{matrix}\leftarrow \text{mg} \\ \leftarrow \text{ml}\end{matrix}$$

$$8 \cdot x = 1 \cdot 10$$

$$8x = 10$$

$$\frac{8x}{8}=\frac{10}{8}$$

$$x = 1.25$$

1.25 ml of the medicine should be administered.

57. 11 muffins are approximately 1 dozen (12) muffins.

$1.5 \cdot 8 = 12$

Approximately 12 cups of milk will be needed.

59.

$$\begin{matrix}\text{feet} \rightarrow \\ \text{pounds} \rightarrow\end{matrix}\frac{7}{60}=\frac{n}{40}\begin{matrix}\leftarrow \text{feet} \\ \leftarrow \text{pounds}\end{matrix}$$

$$7 \cdot 40 = 60 \cdot n$$

$$280 = 60n$$

$$\frac{280}{60}=\frac{60n}{60}$$

$$4\frac{2}{3}=n$$

The distance is $4\frac{2}{3}$ feet.

61. answers may vary

Copyright © 2011 Pearson Education, Inc. Publishing as Prentice Hall.

Section 6.4

Practice Problems

1. $\sqrt{100} = 10$ because $10^2 = 100$.

2. $\sqrt{64} = 8$ because $8^2 = 64$.

3. $\sqrt{169} = 13$ because $13^2 = 169$.

4. $\sqrt{0} = 0$ because $0^2 = 0$.

5. $\sqrt{\frac{1}{4}} = \frac{1}{2}$ because $\left(\frac{1}{2}\right)^2 = \frac{1}{4}$.

6. $\sqrt{\frac{9}{16}} = \frac{3}{4}$ because $\left(\frac{3}{4}\right)^2 = \frac{9}{16}$.

7. **a.** $\sqrt{10} \approx 3.162$

   b. $\sqrt{62} \approx 7.874$

8. Recall that $\sqrt{49} = 7$ and $\sqrt{64} = 8$. Since 62 is between 49 and 64, then $\sqrt{62}$ is between $\sqrt{49}$ and $\sqrt{64}$. Thus, $\sqrt{62}$ is between 7 and 8. Since 62 is closer to 64, then $\sqrt{62}$ is closer to $\sqrt{64}$, or 8.

9. Let $a = 12$ and $b = 16$.

$$\begin{aligned} a^2 + b^2 &= c^2 \\ 12^2 + 16^2 &= c^2 \\ 144 + 256 &= c^2 \\ 400 &= c^2 \\ \sqrt{400} &= c \\ 20 &= c \end{aligned}$$

The hypotenuse is 20 feet.

10. Let $a = 9$ and $b = 7$.

$$\begin{aligned} a^2 + b^2 &= c^2 \\ 9^2 + 7^2 &= c^2 \\ 81 + 49 &= c^2 \\ 130 &= c^2 \\ \sqrt{130} &= c \\ 11 &\approx c \end{aligned}$$

The hypotenuse is approximately 11 kilometers.

11. Let $a = 7$ and $c = 13$.

$$\begin{aligned} a^2 + b^2 &= c^2 \\ 7^2 + b^2 &= 13^2 \\ 49 + b^2 &= 169 \\ b^2 &= 120 \\ b &= \sqrt{120} \\ b &\approx 10.95 \end{aligned}$$

The length of the hypotenuse is exactly $\sqrt{120}$ feet or approximately 10.95 feet.

12.

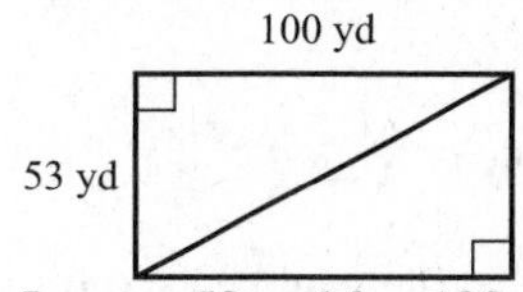

Let $a = 53$ and $b = 100$.

$$\begin{aligned} a^2 + b^2 &= c^2 \\ 53^2 + 100^2 &= c^2 \\ 2809 + 10{,}000 &= c^2 \\ 12{,}809 &= c^2 \\ \sqrt{12{,}809} &= c \\ 113 &\approx c \end{aligned}$$

The diagonal is approximately 113 yards.

Calculator Explorations

1. $\sqrt{1024} = 32$

2. $\sqrt{676} = 26$

3. $\sqrt{15} \approx 3.873$

4. $\sqrt{19} \approx 4.359$

5. $\sqrt{97} \approx 9.849$

6. $\sqrt{56} \approx 7.483$

Vocabulary and Readiness Check

1. The square roots of 100 are <u>10</u> and <u>−10</u> because $10 \cdot 10 = 100$ and $(-10)(-10) = 100$.

2. $\sqrt{100} = \underline{10}$ only because $10 \cdot 10 = 100$ and 10 is positive.

3. The <u>radical</u> sign is used to denote the positive square root of a nonnegative number.

Copyright © 2011 Pearson Education, Inc. Publishing as Prentice Hall.

4. The reverse process of squaring a number is finding a square root of a number.

5. The numbers 9, 1, and $\frac{1}{25}$ are called perfect squares.

6. Label the parts of the right triangle.

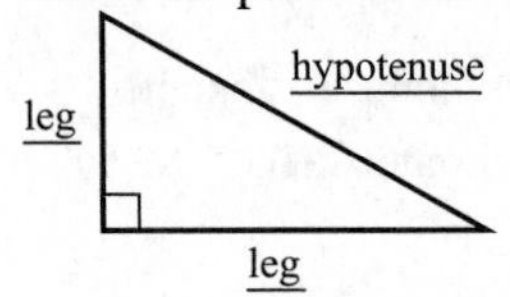

7. In the given triangle, $a^2 + \underline{c^2} = \underline{b^2}$.

8. The Pythagorean Theorem can be used for right triangles.

Exercise Set 6.4

1. $\sqrt{4} = 2$ because $2^2 = 4$.

3. $\sqrt{121} = 11$ because $11^2 = 121$.

5. $\sqrt{\frac{1}{81}} = \frac{1}{9}$ because $\left(\frac{1}{9}\right)^2 = \frac{1}{9} \cdot \frac{1}{9} = \frac{1}{81}$.

7. $\sqrt{\frac{16}{64}} = \frac{4}{8} = \frac{1}{2}$ because $\left(\frac{4}{8}\right)^2 = \frac{4}{8} \cdot \frac{4}{8} = \frac{16}{64}$.

9. $\sqrt{3} \approx 1.732$

11. $\sqrt{15} \approx 3.873$

13. $\sqrt{31} \approx 5.568$

15. $\sqrt{26} \approx 5.099$

17. Since 38 is between $36 = 6 \cdot 6$ and $49 = 7 \cdot 7$, $\sqrt{38}$ is between 6 and 7; $\sqrt{38} \approx 6.16$.

19. Since 101 is between $100 = 10 \cdot 10$ and $121 = 11 \cdot 11$, $\sqrt{101}$ is between 10 and 11; $\sqrt{101} \approx 10.05$.

21. $\sqrt{256} = 16$ because $16^2 = 256$.

23. $\sqrt{92} \approx 9.592$

25. $\sqrt{\frac{49}{144}} = \frac{7}{12}$ because $\left(\frac{7}{12}\right)^2 = \frac{7}{12} \cdot \frac{7}{12} = \frac{49}{144}$.

27. $\sqrt{71} \approx 8.426$

29. Let $a = 5$ and $b = 12$.

$$\begin{aligned} a^2 + b^2 &= c^2 \\ 5^2 + 12^2 &= c^2 \\ 25 + 144 &= c^2 \\ 169 &= c^2 \\ \sqrt{169} &= c \\ 13 &= c \end{aligned}$$

The missing length is 13 inches.

31. Let $a = 10$ and $c = 12$.

$$\begin{aligned} a^2 + b^2 &= c^2 \\ 10^2 + b^2 &= 12^2 \\ 100 + b^2 &= 144 \\ b^2 &= 44 \\ b &= \sqrt{44} \\ b &\approx 6.633 \end{aligned}$$

The missing length is approximately 6.633 centimeters.

33. Let $a = 22$ and $b = 48$.

$$\begin{aligned} c^2 &= a^2 + b^2 \\ c^2 &= 22^2 + 48^2 \\ c^2 &= 484 + 2304 \\ c^2 &= 2788 \\ c &= \sqrt{2788} \\ c &\approx 52.802 \end{aligned}$$

The missing length is approximately 52.802 meters.

35. Let $a = 108$ and $b = 45$.

$$\begin{aligned} a^2 + b^2 &= c^2 \\ 108^2 + 45^2 &= c^2 \\ 11,664 + 2025 &= c^2 \\ 13,689 &= c^2 \\ \sqrt{13,689} &= c \\ 117 &= c \end{aligned}$$

The missing length is 117 millimeters.

Copyright © 2011 Pearson Education, Inc. Publishing as Prentice Hall.

37.

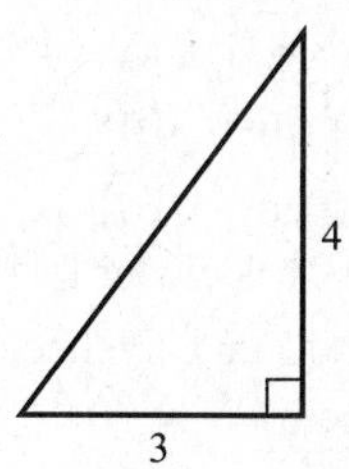

$$\begin{aligned}\text{hypotenuse} &= \sqrt{(\text{leg})^2 + (\text{other leg})^2} \\ &= \sqrt{(3)^2 + (4)^2} \\ &= \sqrt{9+16} \\ &= \sqrt{25} \\ &= 5\end{aligned}$$

The hypotenuse has length 5 units.

39.

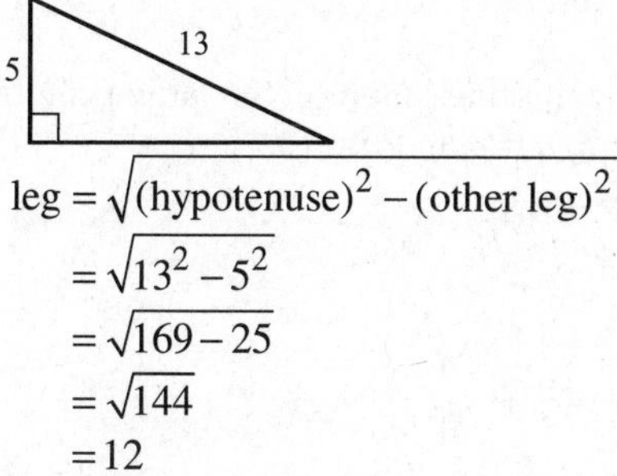

$$\begin{aligned}\text{leg} &= \sqrt{(\text{hypotenuse})^2 - (\text{other leg})^2} \\ &= \sqrt{13^2 - 5^2} \\ &= \sqrt{169-25} \\ &= \sqrt{144} \\ &= 12\end{aligned}$$

The leg has length 12 units.

41.

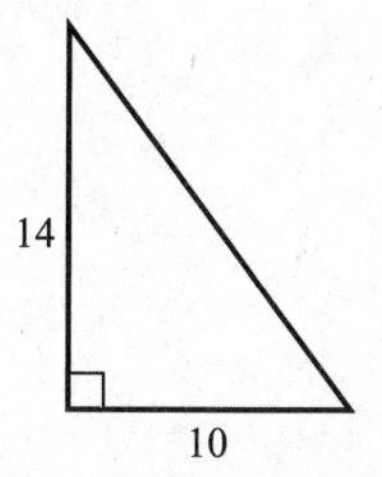

$$\begin{aligned}\text{hypotenuse} &= \sqrt{(\text{leg})^2 + (\text{other leg})^2} \\ &= \sqrt{(10)^2 + (14)^2} \\ &= \sqrt{100+196} \\ &= \sqrt{296} \\ &\approx 17.205\end{aligned}$$

The hypotenuse has length of about 17.205 units.

43.

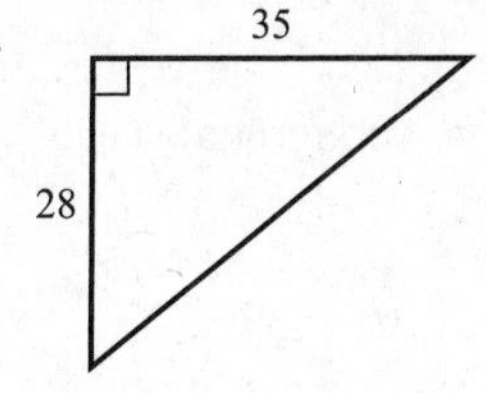

$$\begin{aligned}\text{hypotenuse} &= \sqrt{(\text{leg})^2 + (\text{other leg})^2} \\ &= \sqrt{35^2 + 28^2} \\ &= \sqrt{1225+784} \\ &= \sqrt{2009} \\ &\approx 44.822\end{aligned}$$

The hypotenuse has length of about 44.822 units.

45.

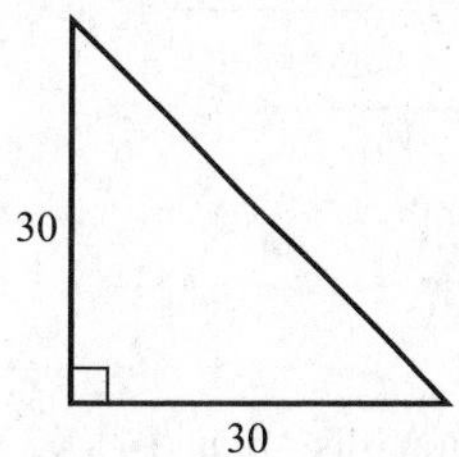

$$\begin{aligned}\text{hypotenuse} &= \sqrt{(\text{leg})^2 + (\text{other leg})^2} \\ &= \sqrt{(30)^2 + (30)^2} \\ &= \sqrt{900+900} \\ &= \sqrt{1800} \\ &\approx 42.426\end{aligned}$$

The hypotenuse has length of about 42.426 units.

47.

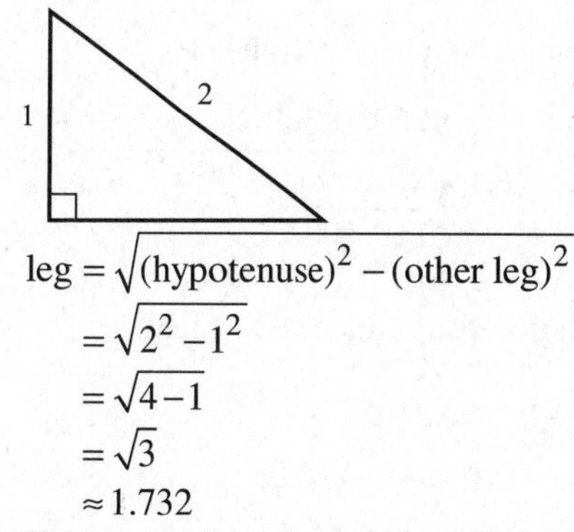

$$\begin{aligned}\text{leg} &= \sqrt{(\text{hypotenuse})^2 - (\text{other leg})^2} \\ &= \sqrt{2^2 - 1^2} \\ &= \sqrt{4-1} \\ &= \sqrt{3} \\ &\approx 1.732\end{aligned}$$

The leg has length of about 1.732 units.

49.

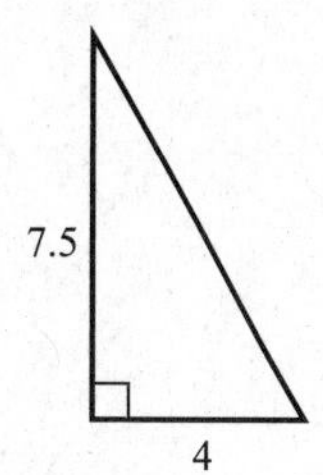

Copyright © 2011 Pearson Education, Inc. Publishing as Prentice Hall.

$$\begin{aligned}\text{hypotenuse} &= \sqrt{(\text{leg})^2 + (\text{other leg})^2}\\ &= \sqrt{(7.5)^2 + (4)^2}\\ &= \sqrt{56.25 + 16}\\ &= \sqrt{72.25}\\ &= 8.5\end{aligned}$$

The hypotenuse has length 8.5 units.

51. $$\begin{aligned}\text{hypotenuse} &= \sqrt{(\text{leg})^2 + (\text{other leg})^2}\\ &= \sqrt{100^2 + 100^2}\\ &= \sqrt{10,000 + 10,000}\\ &= \sqrt{20,000}\\ &\approx 141.42\end{aligned}$$

The length of the diagonal is about 141.42 yards.

53. $$\begin{aligned}\text{leg} &= \sqrt{(\text{hypotenuse})^2 - (\text{other leg})^2}\\ &= \sqrt{(32)^2 - (20)^2}\\ &= \sqrt{1024 - 400}\\ &= \sqrt{624}\\ &\approx 25.0\end{aligned}$$

The tree is about 25.0 feet tall.

55. $$\begin{aligned}\text{hypotenuse} &= \sqrt{(\text{leg})^2 + (\text{other leg})^2}\\ &= \sqrt{160^2 + 300^2}\\ &= \sqrt{25,600 + 90,000}\\ &= \sqrt{115,600}\\ &= 340\end{aligned}$$

The length of the run was 340 feet.

57. $\frac{10}{12} = \frac{5 \cdot 2}{6 \cdot 2} = \frac{5}{6}$

59. $\frac{2x}{60} = \frac{2 \cdot x}{2 \cdot 30} = \frac{x}{30}$

61. $\frac{9}{13y} + \frac{12}{13y} = \frac{9 + 12}{13y} = \frac{21}{13y}$

63. $\frac{9}{8} \cdot \frac{x}{8} = \frac{9 \cdot x}{8 \cdot 8} = \frac{9x}{64}$

65. Recall that $\sqrt{36} = 6$ and $\sqrt{49} = 7$. Since 38 is between 36 and 49, then $\sqrt{38}$ is between $\sqrt{36}$ and $\sqrt{49}$. Thus, $\sqrt{38}$ is between 6 and 7. Since 38 is closer to 36, then $\sqrt{38}$ is closer to $\sqrt{36}$, or 6. Check: $\sqrt{38} \approx 6.16$

67. Recall that $\sqrt{100} = 10$ and $\sqrt{121} = 11$. Since 101 is between 100 and 121, then $\sqrt{101}$ is between $\sqrt{100}$ and $\sqrt{121}$. Thus, $\sqrt{101}$ is between 10 and 11. Since 101 is closer to 100, then $\sqrt{101}$ is closer to $\sqrt{100}$, or 10. Check: $\sqrt{101} \approx 10.05$

69. answers may vary

71. $$\begin{aligned}a^2 + b^2 &= c^2\\ 25^2 + 60^2 &\stackrel{?}{=} 65^2\\ 625 + 3600 &\stackrel{?}{=} 4225\\ 4225 &= 4225\end{aligned}$$

Yes, the set forms the lengths of the sides of a right triangle.

73. Find the missing length in the large right triangle by letting $a = 8$ and $c = 12$.

$$\begin{aligned}a^2 + b^2 &= c^2\\ 8^2 + b^2 &= 12^2\\ 64 + b^2 &= 144\\ b^2 &= 80\\ b &= \sqrt{80}\end{aligned}$$

Find the unlabeled length by letting $a = 8$ and $c = 10$.

$$\begin{aligned}a^2 + b^2 &= c^2\\ 8^2 + b^2 &= 10^2\\ 64 + b^2 &= 100\\ b^2 &= 36\\ b^2 &= \sqrt{36}\\ b &= 6\end{aligned}$$

The unlabeled length is 6 inches. Thus $6 + x = \sqrt{80}$ or $x = \sqrt{80} - 6 \approx 2.94$ inches.

Section 6.5

Practice Problems

1. a. The triangles are congruent by Side-Angle-Side.

b. The triangles are not congruent.

2. $\frac{9 \text{ meters}}{13 \text{ meters}} = \frac{9}{13}$

The ratio of corresponding sides is $\frac{9}{13}$.

Copyright © 2011 Pearson Education, Inc. Publishing as Prentice Hall.

3. $\frac{x}{5} = \frac{6}{9}$
$x \cdot 9 = 5 \cdot 6$
$9x = 30$
$\frac{9x}{9} = \frac{30}{9}$
$x = \frac{10}{3}$ or $3\frac{1}{3}$

4. $\frac{5}{n} = \frac{8}{60}$
$5 \cdot 60 = n \cdot 8$
$300 = 8n$
$\frac{300}{8} = \frac{8n}{8}$
$37.5 = n$
The height of the building is approximately 37.5 feet.

Vocabulary and Readiness Check

1. Two triangles that have the same shape, but not necessarily the same size are congruent. false
2. Two triangles are congruent if they have the same shape and size. true
3. Congruent triangles are also similar. true
4. Similar triangles are also congruent. false
5. For the two similar triangles, the ratio of corresponding sides is $\frac{5}{6}$. false

Exercise Set 6.5

1. The triangles are congruent by Side-Side-Side.
3. The triangles are not congruent.
5. The triangles are congruent by Angle-Side-Angle.
7. The triangles are congruent by Side-Angle-Side.

9. $\frac{22}{11} = \frac{14}{7} = \frac{12}{6} = \frac{2}{1}$

The ratio of corresponding sides is $\frac{2}{1}$.

11. $\frac{10.5}{7} = \frac{9}{6} = \frac{12}{8} = \frac{3}{2}$

13. $\frac{x}{3} = \frac{9}{6}$
$x \cdot 6 = 3 \cdot 9$
$6x = 27$
$\frac{6x}{6} = \frac{27}{6}$
$x = 4.5$

15. $\frac{n}{18} = \frac{4}{12}$
$n \cdot 12 = 18 \cdot 4$
$12n = 72$
$\frac{12n}{12} = \frac{72}{12}$
$n = 6$

17. $\frac{y}{3.75} = \frac{12}{9}$
$y \cdot 9 = 12 \cdot 3.75$
$9y = 45$
$\frac{9y}{9} = \frac{45}{9}$
$y = 5$

19. $\frac{z}{18} = \frac{30}{40}$
$z \cdot 40 = 18 \cdot 30$
$40z = 540$
$\frac{40z}{40} = \frac{540}{40}$
$z = 13.5$

21. $\frac{x}{3.25} = \frac{17.5}{3.25}$
$x \cdot 3.25 = 3.25 \cdot 17.5$
$3.25x = 56.875$
$\frac{3.25x}{3.25} = \frac{56.875}{3.25}$
$x = 17.5$

Copyright © 2011 Pearson Education, Inc. Publishing as Prentice Hall.

23.
$$\frac{y}{2}=\frac{18\frac{1}{3}}{3\frac{2}{3}}$$
$$y\cdot 3\frac{2}{3}=2\cdot 18\frac{1}{3}$$
$$y\cdot\frac{11}{3}=2\cdot\frac{55}{3}$$
$$\frac{11}{3}\cdot y=\frac{110}{3}$$
$$\frac{3}{11}\cdot\frac{11}{3}y=\frac{3}{11}\cdot\frac{110}{3}$$
$$y=10$$

25.
$$\frac{z}{60}=\frac{15}{32}$$
$$z\cdot 32=60\cdot 15$$
$$32z=900$$
$$\frac{32z}{32}=\frac{900}{32}$$
$$z=28.125$$

27.
$$\frac{x}{7}=\frac{15}{10\frac{1}{2}}$$
$$x\cdot 10\frac{1}{2}=7\cdot 15$$
$$10.5x=105$$
$$\frac{10.5x}{10.5}=\frac{105}{10.5}$$
$$x=10$$

29.
$$\frac{x}{13}=\frac{80}{2}$$
$$x\cdot 2=13\cdot 80$$
$$2x=1040$$
$$\frac{2x}{2}=\frac{1040}{2}$$
$$x=520$$
The observation deck is 520 feet high.

31.
$$\frac{x}{25}=\frac{40}{2}$$
$$x\cdot 2=40\cdot 25$$
$$2x=1000$$
$$\frac{2x}{2}=\frac{1000}{2}$$
$$x=500$$
The building is 500 feet tall.

33.
$$\frac{x}{18}=\frac{24}{30}$$
$$x\cdot 30=18\cdot 24$$
$$30x=432$$
$$\frac{30x}{30}=\frac{432}{30}$$
$$x=14.4$$
The shadow of the tree is 14.4 feet long.

35.
$$\frac{x}{55}=\frac{19}{20}$$
$$x\cdot 20=55\cdot 19$$
$$20x=1045$$
$$\frac{20x}{20}=\frac{1045}{20}$$
$$x=52.25$$
$$x\approx 52$$
Pete can place 52 neon tetras in the tank.

37. Let $a = 200$ and $c = 430$.
$$a^2+b^2=c^2$$
$$200^2+b^2=430^2$$
$$40{,}000+b^2=184{,}900$$
$$b^2=144{,}900$$
$$b=\sqrt{144{,}900}$$
$$b\approx 381$$
The gantry is approximately 381 feet tall.

39.
$$\begin{array}{r} 3.60 \\ +\,0.41 \\ \hline 4.01 \end{array}$$

41. $(0.41)(-3) = -1.23$

43. Let x be the new width.
$$\frac{x}{7}=\frac{5}{9}$$
$$x\cdot 9=7\cdot 5$$
$$9x=35$$
$$\frac{9x}{9}=\frac{35}{9}$$
$$x=3\frac{8}{9}$$

The new width is $3\frac{8}{9}$ inches. No, it will not fit on a 3-by-5-inch card because $3\frac{8}{9}>3$.

Copyright © 2011 Pearson Education, Inc. Publishing as Prentice Hall.

45. $\frac{n}{5.2} = \frac{12.6}{7.8}$
$n \cdot 7.8 = 5.2 \cdot 12.6$
$7.8n = 65.52$
$\frac{7.8n}{7.8} = \frac{65.52}{7.8}$
$n = 8.4$

47. answers may vary

49. $\frac{x}{5} = \frac{10}{\frac{1}{4}}$
$x \cdot \frac{1}{4} = 5 \cdot 10$
$\frac{1}{4}x = 50$
$x = 200$
$\frac{y}{7\frac{1}{2}} = \frac{10}{\frac{1}{4}}$
$y \cdot \frac{1}{4} = 7\frac{1}{2} \cdot 10$
$\frac{1}{4}y = 75$
$y = 300$
$\frac{z}{10\frac{5}{8}} = \frac{10}{\frac{1}{4}}$
$z \cdot \frac{1}{4} = 10\frac{5}{8} \cdot 10$
$\frac{1}{4}z = 106.25$
$z = 425$
The actual proposed dimensions are 200 feet by 300 feet by 425 feet.

Chapter 6 Vocabulary Check

1. A ratio is the quotient of two numbers. It can be written as a fraction, using a colon, or using the word *to*.

2. $\frac{x}{2} = \frac{7}{16}$ is an example of a proportion.

3. A unit rate is a rate with a denominator of 1.

4. A unit price is a "money per item" unit rate.

5. A rate is used to compare different kinds of quantities.

6. In the proportion $\frac{x}{2} = \frac{7}{16}$, $x \cdot 16$ and $2 \cdot 7$ are called cross products.

7. If cross products are equal, the proportion is true.

8. If cross products are not equal, the proportion is false.

9. Congruent triangles have the same shape and the same size.

10. Similar triangles have exactly the same shape but not necessarily the same size.

11–13. Label the sides of the right triangle.

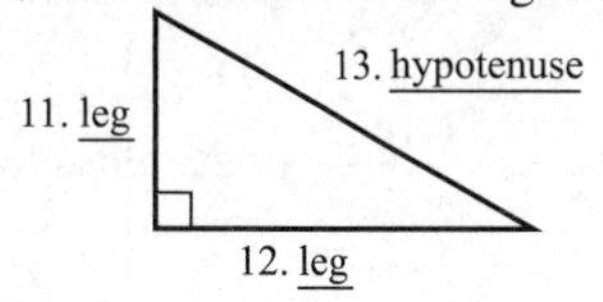

14. A triangle with one right angle is called a right triangle.

15. In the right triangle,

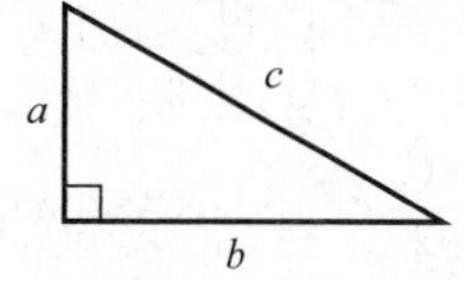

$a^2 + b^2 = c^2$ is called the Pythagorean theorem.

Chapter 6 Review

1. The ratio of 23 to 37 is $\frac{23}{37}$.

2. The ratio of \$121 to \$143 is $\frac{\$121}{\$143} = \frac{11 \cdot 11}{11 \cdot 13} = \frac{11}{13}$.

3. The ratio of 4.25 yards to 8.75 yards is $\frac{4.25 \text{ yards}}{8.75 \text{ yards}} = \frac{4.25 \times 100}{8.75 \times 100} = \frac{425}{875} = \frac{25 \cdot 17}{25 \cdot 35} = \frac{17}{35}$.

4. The ratio of $2\frac{1}{4}$ to $4\frac{3}{8}$ is
$\frac{2\frac{1}{4}}{4\frac{3}{8}} = 2\frac{1}{4} \div 4\frac{3}{8} = \frac{9}{4} \div \frac{35}{8} = \frac{9}{4} \cdot \frac{8}{35} = \frac{9 \cdot 4 \cdot 2}{4 \cdot 35} = \frac{18}{35}$.

5. $\frac{\text{length}}{\text{width}} = \frac{4.5 \text{ meters}}{2 \text{ meters}} = \frac{4.5 \times 10}{2 \times 10} = \frac{45}{20} = \frac{5 \cdot 9}{5 \cdot 4} = \frac{9}{4}$

Copyright © 2011 Pearson Education, Inc. Publishing as Prentice Hall.

6. $\dfrac{\text{width}}{\text{perimeter}} = \dfrac{2 \text{ meters}}{(4.5+2+4.5+2) \text{ meters}} = \dfrac{2}{13}$

7. $\dfrac{6000 \text{ people}}{2400 \text{ pets}} = \dfrac{1200 \cdot 5 \text{ people}}{1200 \cdot 2 \text{ pets}} = \dfrac{5 \text{ people}}{2 \text{ pets}}$

8. $\dfrac{15 \text{ pages}}{6 \text{ minutes}} = \dfrac{3 \cdot 5 \text{ pages}}{3 \cdot 2 \text{ minutes}} = \dfrac{5 \text{ pages}}{2 \text{ minutes}}$

9. $\dfrac{468 \text{ miles}}{9 \text{ hours}}$

$$\begin{array}{r} 52 \\ 9\overline{)\,468} \\ -45 \\ \hline 18 \\ -18 \\ \hline 0 \end{array}$$

The unit rate is $\dfrac{52 \text{ miles}}{1 \text{ hour}}$ or 52 miles/hour.

10. $\dfrac{180 \text{ feet}}{12 \text{ seconds}}$

$$\begin{array}{r} 15 \\ 12\overline{)\,180} \\ -12 \\ \hline 60 \\ -60 \\ \hline 0 \end{array}$$

The unit rate is $\dfrac{15 \text{ feet}}{1 \text{ second}}$ or 15 feet/second.

11. $\dfrac{\$6.96}{4 \text{ diskettes}}$

$$\begin{array}{r} 1.74 \\ 4\overline{)\,6.96} \\ -4 \\ \hline 29 \\ -28 \\ \hline 16 \\ -16 \\ \hline 0 \end{array}$$

The unit rate is $\dfrac{\$1.74}{1 \text{ diskette}}$ or \$1.74/diskette.

12. $\dfrac{104 \text{ bushels}}{8 \text{ trees}}$

$$\begin{array}{r} 13 \\ 8\overline{)104} \\ -8 \\ \hline 24 \\ -24 \\ \hline 0 \end{array}$$

The unit rate is $\dfrac{13 \text{ bushels}}{1 \text{ tree}}$ or 13 bushels/tree.

13.

$$\begin{array}{r} 0.1237 \approx 0.124 \\ 8\overline{)0.9900} \\ -8 \\ \hline 19 \\ -16 \\ \hline 30 \\ -24 \\ \hline 60 \\ -56 \\ \hline 4 \end{array}$$

The 8-ounce size costs \$0.124 per ounce.

$$\begin{array}{r} 0.1408 \approx 0.141 \\ 12\overline{)\,1.6900} \\ -1\,2 \\ \hline 49 \\ -48 \\ \hline 100 \\ -96 \\ \hline 4 \end{array}$$

The 12-ounce size costs \$0.141 per ounce.
The 8-ounce size is the better buy.

14.

$$\begin{array}{r} 0.0827 \approx 0.083 \\ 18\overline{)\,1.4900} \\ -1\,44 \\ \hline 50 \\ -36 \\ \hline 140 \\ -126 \\ \hline 14 \end{array}$$

The 18-ounce size costs \$0.083 per ounce.

$$\begin{array}{r} 0.0853 \approx 0.085 \\ 28\overline{)\,2.3900} \\ -2\,24 \\ \hline 150 \\ -140 \\ \hline 100 \\ -84 \\ \hline 16 \end{array}$$

The 28-ounce size costs \$0.085 per ounce.
The 18-ounce size is the better buy.

15. $\dfrac{24 \text{ uniforms}}{8 \text{ players}} = \dfrac{3 \text{ uniforms}}{1 \text{ player}}$

16. $\dfrac{12 \text{ tires}}{3 \text{ cars}} = \dfrac{4 \text{ tires}}{1 \text{ car}}$

Copyright © 2011 Pearson Education, Inc. Publishing as Prentice Hall.

17. $\frac{19}{8} \stackrel{?}{=} \frac{14}{6}$
$19 \cdot 6 \stackrel{?}{=} 8 \cdot 14$
$114 \neq 112$
The proportion is false.

18. $\frac{3.75}{3} \stackrel{?}{=} \frac{7.5}{6}$
$3.75 \cdot 6 \stackrel{?}{=} 3 \cdot 7.5$
$22.5 = 22.5$
The proportion is true.

19. $\frac{x}{3} = \frac{30}{18}$
$x \cdot 18 = 3 \cdot 30$
$18x = 90$
$\frac{18x}{18} = \frac{90}{18}$
$x = 5$

20. $\frac{x}{9} = \frac{7}{3}$
$x \cdot 3 = 9 \cdot 7$
$3x = 63$
$\frac{3x}{3} = \frac{63}{3}$
$x = 21$

21. $\frac{-8}{5} = \frac{9}{x}$
$-8 \cdot x = 5 \cdot 9$
$-8x = 45$
$\frac{-8x}{-8} = \frac{45}{-8}$
$x = -5.625$

22. $\frac{-27}{\frac{9}{4}} = \frac{x}{-5}$
$-27 \cdot (-5) = \frac{9}{4} \cdot x$
$135 = \frac{9}{4} \cdot x$
$\frac{4}{9} \cdot \frac{135}{1} = \frac{4}{9} \cdot \frac{9}{4} \cdot x$
$60 = x$

23. $\frac{0.4}{x} = \frac{2}{4.7}$
$0.4 \cdot 4.7 = x \cdot 2$
$1.88 = 2x$
$\frac{1.88}{2} = \frac{2x}{2}$
$0.94 = x$

24. $\frac{x}{4\frac{1}{2}} = \frac{2\frac{1}{10}}{8\frac{2}{5}}$
$x \cdot 8\frac{2}{5} = 4\frac{1}{2} \cdot 2\frac{1}{10}$
$x \cdot \frac{42}{5} = \frac{9}{2} \cdot \frac{21}{10}$
$\frac{42}{5} \cdot x = \frac{189}{20}$
$\frac{5}{42} \cdot \frac{42}{5} \cdot x = \frac{5}{42} \cdot \frac{189}{20}$
$x = \frac{9}{8}$ or $1\frac{1}{8}$

25. $\frac{x}{0.4} = \frac{4.7}{3}$
$x \cdot 3 = 0.4 \cdot 4.7$
$3x = 1.88$
$\frac{3x}{3} = \frac{1.88}{3}$
$x \approx 0.63$

26. $\frac{0.07}{0.3} = \frac{7.2}{n}$
$0.07 \cdot n = 0.3 \cdot 7.2$
$0.07n = 2.16$
$\frac{0.07n}{0.07} = \frac{2.16}{0.07}$
$n \approx 30.9$

27. Let x be the number of completed passes.
$\text{completed} \rightarrow \frac{3}{7} = \frac{x}{32} \leftarrow \text{completed}$
$\text{attempted} \rightarrow 7 \quad 32 \leftarrow \text{attempted}$
$3 \cdot 32 = 7 \cdot x$
$96 = 7x$
$\frac{96}{7} = \frac{7x}{7}$
$14 \approx x$
He completed 14 passes.

Copyright © 2011 Pearson Education, Inc. Publishing as Prentice Hall.

28. Let x be the number of attempted passes.

$$\begin{array}{l}\text{completed} \rightarrow \\ \text{attempted} \rightarrow\end{array} \frac{3}{7} = \frac{15}{x} \begin{array}{l}\leftarrow \text{completed} \\ \leftarrow \text{attempted}\end{array}$$

$$\begin{aligned} 3 \cdot x &= 7 \cdot 15 \\ 3x &= 105 \\ \frac{3x}{3} &= \frac{105}{3} \\ x &= 35 \end{aligned}$$

He attempted 35 passes.

29. Let x be the number of bags.

$$\begin{array}{l}\text{bags} \rightarrow \\ \text{square feet} \rightarrow\end{array} \frac{1}{4000} = \frac{x}{180 \cdot 175} \begin{array}{l}\leftarrow \text{bags} \\ \leftarrow \text{square feet}\end{array}$$

$$\begin{aligned} 1 \cdot 180 \cdot 175 &= 4000 \cdot x \\ 31{,}500 &= 4000x \\ \frac{31{,}500}{4000} &= \frac{4000x}{4000} \\ 7.875 &= x \end{aligned}$$

Purchase 8 bags of pesticide.

30. Let x be the number of bags.

$$\begin{array}{l}\text{bags} \rightarrow \\ \text{square feet} \rightarrow\end{array} \frac{1}{4000} = \frac{x}{250 \cdot 250} \begin{array}{l}\leftarrow \text{bags} \\ \leftarrow \text{square feet}\end{array}$$

$$\begin{aligned} 1 \cdot 250 \cdot 250 &= 4000 \cdot x \\ 62{,}500 &= 4000x \\ \frac{62{,}500}{4000} &= \frac{4000x}{4000} \\ 15.625 &= x \end{aligned}$$

Purchase 16 bags of pesticide.

31. Let x be the miles.

$$\begin{array}{l}\text{miles} \rightarrow \\ \text{inches} \rightarrow\end{array} \frac{x}{2} = \frac{80}{0.75} \begin{array}{l}\leftarrow \text{miles} \\ \leftarrow \text{inches}\end{array}$$

$$\begin{aligned} x \cdot 0.75 &= 2 \cdot 80 \\ 0.75x &= 160 \\ \frac{0.75x}{0.75} &= \frac{160}{0.75} \\ x &= \frac{640}{3} \text{ or } 213\frac{1}{3} \end{aligned}$$

The distance is $213\frac{1}{3}$ miles.

32. Let x be the inches.

$$\begin{array}{l}\text{miles} \rightarrow \\ \text{inches} \rightarrow\end{array} \frac{1025}{x} = \frac{80}{0.75} \begin{array}{l}\leftarrow \text{miles} \\ \leftarrow \text{inches}\end{array}$$

$$\begin{aligned} 1025 \cdot 0.75 &= x \cdot 80 \\ 768.75 &= 80x \\ \frac{768.75}{80} &= \frac{80x}{80} \\ 9.6 &\approx x \end{aligned}$$

The distance is about 9.6 inches.

33. $\sqrt{64} = 8$ because $8^2 = 64$.

34. $\sqrt{144} = 12$ because $12^2 = 144$.

35. $\sqrt{12} \approx 3.464$

36. $\sqrt{15} \approx 3.873$

37. $\sqrt{0} = 0$ because $0^2 = 0$.

38. $\sqrt{1} = 1$ because $1^2 = 1$.

39. $\sqrt{50} \approx 7.071$

40. $\sqrt{65} \approx 8.062$

41. $\sqrt{\frac{4}{25}} = \frac{2}{5}$ because $\left(\frac{2}{5}\right)^2 = \frac{2}{5} \cdot \frac{2}{5} = \frac{4}{25}$.

42. $\sqrt{\frac{1}{100}} = \frac{1}{10}$ because $\left(\frac{1}{10}\right)^2 = \frac{1}{10} \cdot \frac{1}{10} = \frac{1}{100}$.

43.

$$\begin{aligned} \text{hypotenuse} &= \sqrt{(\text{leg})^2 + (\text{other leg})^2} \\ &= \sqrt{12^2 + 5^2} \\ &= \sqrt{144 + 25} \\ &= \sqrt{169} \\ &= 13 \end{aligned}$$

The leg has length 13 units.

44.

$$\begin{aligned} \text{hypotenuse} &= \sqrt{(\text{leg})^2 + (\text{other leg})^2} \\ &= \sqrt{20^2 + 21^2} \\ &= \sqrt{400 + 441} \\ &= \sqrt{841} \\ &= 29 \end{aligned}$$

The leg has length 29 units.

45.

$$\begin{aligned} \text{leg} &= \sqrt{(\text{hypotenuse})^2 - (\text{other leg})^2} \\ &= \sqrt{14^2 - 9^2} \\ &= \sqrt{196 - 81} \\ &= \sqrt{115} \\ &\approx 10.7 \end{aligned}$$

The leg has length of about 10.7 units.

Copyright © 2011 Pearson Education, Inc. Publishing as Prentice Hall.

46. $\text{leg} = \sqrt{(\text{hypotenuse})^2 - (\text{other leg})^2}$
$= \sqrt{86^2 - 66^2}$
$= \sqrt{7396 - 4356}$
$= \sqrt{3040}$
≈ 55.1
The leg has length of about 55.1 units.

47. $\text{hypotenuse} = \sqrt{(\text{leg})^2 + (\text{other leg})^2}$
$= \sqrt{20^2 + 20^2}$
$= \sqrt{400 + 400}$
$= \sqrt{800}$
≈ 28.28
The diagonal is about 28.28 centimeters.

48. $\text{leg} = \sqrt{(\text{hypotenuse})^2 - (\text{other leg})^2}$
$= \sqrt{126^2 - 90^2}$
$= \sqrt{15{,}876 - 8100}$
$= \sqrt{7776}$
≈ 88.2
The height is about 88.2 feet.

49. The triangles are congruent by Angle-Side-Angle.

50. The triangles are not congruent.

51. $\frac{x}{20} = \frac{20}{30}$
$x \cdot 30 = 20 \cdot 20$
$30x = 400$
$\frac{30x}{30} = \frac{400}{30}$
$x = \frac{40}{3}$ or $13\frac{1}{3}$

52. $\frac{x}{5.8} = \frac{24}{8}$
$x \cdot 8 = 5.8 \cdot 24$
$8x = 139.2$
$\frac{8x}{8} = \frac{139.2}{8}$
$x = 17.4$

53. $\frac{x}{5.5} = \frac{42}{7}$
$x \cdot 7 = 5.5 \cdot 42$
$7x = 231$
$\frac{7x}{7} = \frac{231}{7}$
$x = 33$
The height of the building is approximately 33 feet.

54. $\frac{x}{10} = \frac{2}{24}$
$x \cdot 24 = 10 \cdot 2$
$24x = 20$
$\frac{24x}{24} = \frac{20}{24}$
$x = \frac{5}{6}$

$\frac{y}{26} = \frac{2}{24}$
$y \cdot 24 = 26 \cdot 2$
$24y = 52$
$\frac{24y}{24} = \frac{52}{24}$
$y = \frac{13}{6}$ or $2\frac{1}{6}$

The unknown lengths are $x = \frac{5}{6}$ inch and $y = 2\frac{1}{6}$ inches.

55. The ratio of 15 to 25 is $\frac{15}{25} = \frac{3 \cdot 5}{5 \cdot 5} = \frac{3}{5}$.

56. The ratio of 3 pints to 81 pints is $\frac{3 \text{ pints}}{81 \text{ pints}} = \frac{3}{81} = \frac{1}{27}$.

57. $\frac{2 \text{ teachers}}{18 \text{ students}} = \frac{2 \cdot 1 \text{ teachers}}{2 \cdot 9 \text{ students}} = \frac{1 \text{ teacher}}{9 \text{ students}}$

58. $\frac{6 \text{ nurses}}{24 \text{ patients}} = \frac{1 \cdot 6 \text{ nurses}}{4 \cdot 6 \text{ patients}} = \frac{1 \text{ nurse}}{4 \text{ patients}}$

59. $\frac{136 \text{ miles}}{4 \text{ hours}} = \frac{4 \cdot 34 \text{ miles}}{4 \cdot 1 \text{ hour}} = \frac{34 \text{ miles}}{1 \text{ hour}}$
The unit rate is $\frac{34 \text{ miles}}{1 \text{ hour}}$ or 34 miles/hour.

60. $\frac{12 \text{ gallons}}{6 \text{ cows}} = \frac{6 \cdot 2 \text{ gallons}}{6 \cdot 1 \text{ cow}} = \frac{2 \text{ gallons}}{1 \text{ cow}}$
The unit rate is $\frac{2 \text{ gallons}}{1 \text{ cow}}$ or 2 gallons/cow.

Copyright © 2011 Pearson Education, Inc. Publishing as Prentice Hall.

61. $\begin{array}{l} \text{The Netherlands} \rightarrow \\ \text{Total} \rightarrow \end{array} \dfrac{16 \text{ medals}}{958 \text{ medals}} = \dfrac{2 \cdot 8}{2 \cdot 479} = \dfrac{8}{479}$

62. $\dfrac{1576 \text{ steps}}{9.5 \text{ minutes}}$

$9.5\overline{)1576}$ becomes $95\overline{)15760.0}$ with quotient 165.8

$$\begin{array}{r} 165.8 \\ 95\overline{)15760.0} \\ -95 \\ \hline 626 \\ -570 \\ \hline 560 \\ -475 \\ \hline 850 \\ -760 \\ \hline 90 \end{array}$$

He ran 166 steps/minute.

63. 4 ounces:

$$\begin{array}{r} 1.235 \\ 4\overline{)\ 4.940} \\ -4 \\ \hline 0\,9 \\ -8 \\ \hline 14 \\ -12 \\ \hline 20 \\ -20 \\ \hline 0 \end{array}$$

The 4-ounce size costs $1.235 per ounce.
8 ounces:

$$\begin{array}{r} 1.2475 \approx 1.248 \\ 8\overline{)\ 9.9800} \\ -8 \\ \hline 1\,9 \\ -1\,6 \\ \hline 38 \\ -32 \\ \hline 60 \\ -56 \\ \hline 40 \\ -40 \\ \hline 0 \end{array}$$

The 8-ounce size costs $1.248 per ounce.
The 4-ounce size is the better buy.

64. 12 ounces:

$$\begin{array}{r} 0.0541 \approx 0.054 \\ 12\overline{)0.6500} \\ -60 \\ \hline 50 \\ -48 \\ \hline 20 \\ -12 \\ \hline 8 \end{array}$$

The 12-ounce size costs $0.054 per ounce.
64 ounces:

$$\begin{array}{r} 0.0465 \approx 0.047 \\ 64\overline{)\ 2.9800} \\ -2\,56 \\ \hline 420 \\ -384 \\ \hline 360 \\ -320 \\ \hline 40 \end{array}$$

The 64-ounce size costs $0.047 per ounce.
The 64-ounce size is the better buy.

65. $\dfrac{2 \text{ cups cookie dough}}{30 \text{ cookies}} = \dfrac{4 \text{ cups cookie dough}}{60 \text{ cookies}}$

66. $\dfrac{5 \text{ nickels}}{3 \text{ dollars}} = \dfrac{20 \text{ nickels}}{12 \text{ dollars}}$

67. $$\begin{aligned} \frac{3}{x} &= \frac{15}{8} \\ 3 \cdot 8 &= x \cdot 15 \\ 24 &= 15x \\ \frac{24}{15} &= \frac{15x}{15} \\ 1.6 &= x \end{aligned}$$

68. $$\begin{aligned} \frac{5}{4} &= \frac{x}{20} \\ 5 \cdot 20 &= 4 \cdot x \\ 100 &= 4x \\ \frac{100}{4} &= \frac{4x}{4} \\ 25 &= x \end{aligned}$$

Copyright © 2011 Pearson Education, Inc. Publishing as Prentice Hall.

69. $$\frac{x}{3}=\frac{7.5}{6}$$
$$x\cdot 6=3\cdot 7.5$$
$$6x=22.5$$
$$\frac{6x}{6}=\frac{22.5}{6}$$
$$x=3.75$$

70. $$\frac{\frac{1}{3}}{25}=\frac{x}{30}$$
$$\frac{1}{3}\cdot 30=25\cdot x$$
$$10=25x$$
$$\frac{10}{25}=\frac{25x}{25}$$
$$\frac{2}{5}=x$$

71. $\sqrt{36}=6$ because $6^2=36$.

72. $\sqrt{\frac{16}{81}}=\frac{4}{9}$ because $\left(\frac{4}{9}\right)^2=\frac{4}{9}\cdot\frac{4}{9}=\frac{16}{81}$.

73. $\sqrt{105}\approx 10.247$

74. $\sqrt{32}\approx 5.657$

75. $$\begin{aligned}\text{hypotenuse}&=\sqrt{(\text{leg})^2+(\text{other leg})^2}\\&=\sqrt{66^2+56^2}\\&=\sqrt{4356+3136}\\&=\sqrt{7492}\\&\approx 86.6\end{aligned}$$

76. $$\begin{aligned}\text{leg}&=\sqrt{(\text{hypotenuse})^2-(\text{other leg})^2}\\&=\sqrt{24^2-12^2}\\&=\sqrt{576-144}\\&=\sqrt{432}\\&\approx 20.8\end{aligned}$$

77. $$\frac{n}{6}=\frac{10}{5}$$
$$n\cdot 5=6\cdot 10$$
$$5n=60$$
$$\frac{5n}{5}=\frac{60}{5}$$
$$n=12$$

78. $$\frac{n}{8\frac{2}{3}}=\frac{9\frac{3}{8}}{12\frac{1}{2}}$$
$$n\cdot 12\frac{1}{2}=8\frac{2}{3}\cdot 9\frac{3}{8}$$
$$n\cdot\frac{25}{2}=\frac{26}{3}\cdot\frac{75}{8}$$
$$\frac{25}{2}n=\frac{325}{4}$$
$$\frac{2}{25}\cdot\frac{25}{2}n=\frac{2}{25}\cdot\frac{325}{4}$$
$$n=\frac{13}{2}\text{ or }6\frac{1}{2}$$

Chapter 6 Test

1. The ratio of 4500 trees to 6500 trees is
$\frac{4500\text{ trees}}{6500\text{ trees}}=\frac{9\cdot 500}{13\cdot 500}=\frac{9}{13}$.

2. The ratio of 9 inches of rain in 30 days is
$\frac{9\text{ inches}}{30\text{ days}}=\frac{3\cdot 3\text{ inches}}{3\cdot 10\text{ days}}=\frac{3\text{ inches}}{10\text{ days}}$.

3. The ratio of 8.6 to 10 is
$\frac{8.6}{10}=\frac{8.6\times 10}{10\times 10}=\frac{86}{100}=\frac{43}{50}$.

4. The ratio of $5\frac{7}{8}$ to $9\frac{3}{4}$ is

$\frac{5\frac{7}{8}}{9\frac{3}{4}}=5\frac{7}{8}\div 9\frac{3}{4}=\frac{47}{8}\div\frac{39}{4}=\frac{47}{8}\cdot\frac{4}{39}=\frac{47}{78}$.

5. The ratio 456 feet to 186 feet is
$\frac{456\text{ feet}}{186\text{ feet}}=\frac{6\cdot 76}{6\cdot 31}=\frac{76}{31}$.

Copyright © 2011 Pearson Education, Inc. Publishing as Prentice Hall.

6. $\dfrac{650 \text{ kilometers}}{8 \text{ hours}}$

$$\begin{array}{r} 81.25 \\ 8\overline{)\,650.00} \\ \underline{-64} \\ 10 \\ \underline{-8} \\ 2\,0 \\ \underline{-1\,6} \\ 40 \\ \underline{-40} \\ 0 \end{array}$$

The unit rate is $\dfrac{81.25 \text{ kilometers}}{1 \text{ hour}}$ or 81.25 kilometers/hour.

7. $\dfrac{140 \text{ students}}{5 \text{ teachers}}$

$$\begin{array}{r} 28 \\ 5\overline{)\,140} \\ \underline{-10} \\ 40 \\ \underline{-40} \\ 0 \end{array}$$

The unit rate is $\dfrac{28 \text{ students}}{1 \text{ teacher}}$ or 28 students/teacher.

8. $\dfrac{960 \text{ inches}}{60 \text{ minutes}}$

$$\begin{array}{r} 16 \\ 60\overline{)\,960} \\ \underline{-60} \\ 360 \\ \underline{-360} \\ 0 \end{array}$$

The unit rate is $\dfrac{16 \text{ inches}}{1 \text{ minute}}$ or 16 inches/minute.

9.

$$\begin{array}{r} 0.1487 \approx 0.149 \\ 8\overline{)1.1900} \\ \underline{-8} \\ 39 \\ \underline{-32} \\ 70 \\ \underline{-64} \\ 60 \\ \underline{-56} \\ 4 \end{array}$$

The 8-ounce size costs \$0.149/ounce.

$$\begin{array}{r} 0.1575 \approx 0.158 \\ 12\overline{)\,1.8900} \\ \underline{-1\,2} \\ 69 \\ \underline{-60} \\ 90 \\ \underline{-84} \\ 60 \\ \underline{-60} \\ 0 \end{array}$$

The 12-ounce size costs \$0.158/ounce.
The 8-ounce size is the better buy.

10.

$$\begin{array}{r} 0.09312 \approx 0.093 \\ 16\overline{)\,1.49000} \\ \underline{-1\,44} \\ 50 \\ \underline{-48} \\ 20 \\ \underline{-16} \\ 40 \\ \underline{-32} \\ 8 \end{array}$$

The 16-ounce size costs \$0.093 per ounce.

$$\begin{array}{r} 0.09958 \approx 0.100 \\ 24\overline{)\,2.39000} \\ \underline{-2\,16} \\ 230 \\ \underline{-216} \\ 140 \\ \underline{-120} \\ 200 \\ \underline{-192} \\ 8 \end{array}$$

The 24-ounce size costs \$0.100 per ounce.
The 16-ounce size is the better buy.

11.

$$\frac{28}{16} \stackrel{?}{=} \frac{14}{8}$$
$$28 \cdot 8 \stackrel{?}{=} 16 \cdot 14$$
$$224 = 224$$

The proportion is true.

12.

$$\frac{3.6}{2.2} \stackrel{?}{=} \frac{1.9}{1.2}$$
$$3.6 \cdot 1.2 \stackrel{?}{=} 2.2 \cdot 1.9$$
$$4.32 \neq 4.18$$

The proportion is false.

Copyright © 2011 Pearson Education, Inc. Publishing as Prentice Hall.

13. $\frac{n}{3}=\frac{15}{9}$
$n\cdot 9=3\cdot 15$
$9n=45$
$\frac{9n}{9}=\frac{45}{9}$
$n=5$

14. $\frac{8}{x}=\frac{11}{6}$
$8\cdot 6=x\cdot 11$
$48=11x$
$\frac{48}{11}=\frac{11x}{11}$
$4\frac{4}{11}=x$

15. $\frac{4}{\frac{3}{7}}=\frac{y}{\frac{1}{4}}$
$4\cdot\frac{1}{4}=\frac{3}{7}\cdot y$
$1=\frac{3}{7}y$
$\frac{7}{3}\cdot 1=\frac{7}{3}\cdot\frac{3}{7}y$
$\frac{7}{3}=y$

16. $\frac{1.5}{5}=\frac{2.4}{n}$
$1.5\cdot n=5\cdot 2.4$
$1.5n=12$
$\frac{1.5n}{1.5}=\frac{12}{1.5}$
$n=8$

17. Let x be the length of the home in feet.
$\frac{\text{feet} \to x}{\text{inches} \to 11}=\frac{9 \leftarrow \text{feet}}{2 \leftarrow \text{inches}}$
$x\cdot 2=11\cdot 9$
$2x=99$
$\frac{2x}{2}=\frac{99}{2}$
$x=49\frac{1}{2}$

The home is $49\frac{1}{2}$ feet long.

18. Let x be the number of hours.
$\frac{\text{miles} \to 80}{\text{hours} \to 3}=\frac{100 \leftarrow \text{miles}}{x \leftarrow \text{hours}}$
$80\cdot x=3\cdot 100$
$80x=300$
$\frac{80x}{80}=\frac{300}{80}$
$x=3\frac{3}{4}$

It will take $3\frac{3}{4}$ hours to travel 100 miles.

19. Let x be the number of grams.
$\frac{\text{grams} \to 10}{\text{pounds} \to 15}=\frac{x \leftarrow \text{grams}}{80 \leftarrow \text{pounds}}$
$10\cdot 80=15\cdot x$
$800=15x$
$\frac{800}{15}=\frac{15x}{15}$
$53\frac{1}{3}=x$

The standard dose for an 80-pound dog is $53\frac{1}{3}$ grams.

20. $\sqrt{49}=7$ because $7^2=49$.

21. $\sqrt{157}\approx 12.530$

22. $\sqrt{\frac{64}{100}}=\frac{8}{10}=\frac{4}{5}$ because $\left(\frac{8}{10}\right)^2=\frac{8}{10}\cdot\frac{8}{10}=\frac{64}{100}$.

23. hypotenuse $=\sqrt{(\text{leg})^2+(\text{other leg})^2}$
$=\sqrt{4^2+4^2}$
$=\sqrt{16+16}$
$=\sqrt{32}$
≈ 5.66

The hypotenuse is 5.66 centimeters.

24. $\frac{n}{12}=\frac{5}{8}$
$n\cdot 8=12\cdot 5$
$8n=60$
$\frac{8n}{8}=\frac{60}{8}$
$n=7.5$

Copyright © 2011 Pearson Education, Inc. Publishing as Prentice Hall.

25. Let x be the height of the tower.

$$\frac{x}{5\frac{3}{4}} = \frac{48}{4}$$
$$x \cdot 4 = 5\frac{3}{4} \cdot 48$$
$$4x = \frac{23}{4} \cdot 48$$
$$4x = 276$$
$$\frac{4x}{4} = \frac{276}{4}$$
$$x = 69$$

The tower is approximately 69 feet tall.

Cumulative Review Chapters 1–6

1. a. $12 - 9 = 3$ *Check*: $3 + 9 = 12$

b. $22 - 7 = 15$ *Check*: $15 + 7 = 22$

c. $35 - 35 = 0$ *Check*: $0 + 35 = 35$

d. $70 - 0 = 70$ *Check*: $70 + 0 = 70$

2. a. $20 \cdot 0 = 0$

b. $20 \cdot 1 = 20$

c. $0 \cdot 20 = 0$

d. $1 \cdot 20 = 20$

3. To round 248,982 to the nearest hundred, observe that the digit in the tens place is 8. Since this digit is at least 5, we add 1 to the digit in the hundreds place. The number 248,982 rounded to the nearest hundred is 249,000.

4. To round 248,982 to the nearest thousand, observe that the digit in the hundreds place is 9. Since this digit is at least 5, we add 1 to the digit in the thousands place. The number 248,982 rounded to the nearest thousand is 249,000.

5. a.
$$\begin{array}{r} \scriptstyle 4 \\ 25 \\ \times\ 8 \\ \hline 200 \end{array}$$

b.
$$\begin{array}{r} \scriptstyle 2\,3 \\ 246 \\ \times\ \ 5 \\ \hline 1230 \end{array}$$

6.
$$\begin{array}{r} 373 \text{ R } 24 \\ 28\overline{)10,468} \\ -8\,4 \\ \hline 2\,06 \\ -1\,96 \\ \hline 108 \\ -84 \\ \hline 24 \end{array}$$

7.
$$\begin{aligned} 1+(-10)+(-8)+9 &= -9+(-8)+9 \\ &= -17+9 \\ &= -8 \end{aligned}$$

8. $-12(7) = -84$

9.
$$\begin{aligned} 80 &= 2 \cdot 40 \\ &= 2 \cdot 2 \cdot 20 \\ &= 2 \cdot 2 \cdot 2 \cdot 10 \\ &= 2 \cdot 2 \cdot 2 \cdot 2 \cdot 5 \\ &= 2^4 \cdot 5 \end{aligned}$$

10. $3^2 - 1^2 = 9 - 1 = 8$

11. $\frac{12}{20} = \frac{4 \cdot 3}{4 \cdot 5} = \frac{3}{5}$

12. $9^2 \cdot 3 = 81 \cdot 3 = 243$

13. $\left(-\frac{6}{13}\right)\left(-\frac{26}{30}\right) = \frac{6 \cdot 26}{13 \cdot 30} = \frac{6 \cdot 13 \cdot 2}{13 \cdot 6 \cdot 5} = \frac{2}{5}$

14. $3\frac{3}{8} \cdot 4\frac{5}{9} = \frac{27}{8} \cdot \frac{41}{9} = \frac{9 \cdot 3 \cdot 41}{8 \cdot 9} = \frac{123}{8}$ or $15\frac{3}{8}$

15. $\frac{7}{8} + \frac{6}{8} + \frac{3}{8} = \frac{7+6+3}{8} = \frac{16}{8} = 2$

16. $\frac{7}{10} - \frac{3}{10} + \frac{4}{10} = \frac{7-3+4}{10} = \frac{8}{10} = \frac{4}{5}$

17. $7 = 7$

$14 = 2 \cdot 7$

$\text{LCD} = 2 \cdot 7 = 14$

Copyright © 2011 Pearson Education, Inc. Publishing as Prentice Hall.

18. $\frac{17}{25}+\frac{3}{10}=\frac{17}{25}\cdot\frac{2}{2}+\frac{3}{10}\cdot\frac{5}{5}$
$=\frac{17\cdot 2}{25\cdot 2}+\frac{3\cdot 5}{10\cdot 5}$
$=\frac{34}{50}+\frac{15}{50}$
$=\frac{34+15}{50}$
$=\frac{49}{50}$

19. $4\cdot 5=20$, so
$\frac{3}{4}=\frac{3\cdot 5}{4\cdot 5}=\frac{15}{20}$

20. $\frac{10}{55}=\frac{5\cdot 2}{5\cdot 11}=\frac{2}{11}$
$\frac{6}{33}=\frac{3\cdot 2}{3\cdot 11}=\frac{2}{11}$
Yes, they are equivalent.

21. $\frac{2}{3}-\frac{10}{11}=\frac{2}{3}\cdot\frac{11}{11}-\frac{10}{11}\cdot\frac{3}{3}$
$=\frac{22}{33}-\frac{30}{33}$
$=\frac{22-30}{33}$
$=-\frac{8}{33}$

22. $17\frac{5}{24} - 9\frac{5}{9}$ $17\frac{15}{72} - 9\frac{40}{72}$ $16\frac{87}{72} - 9\frac{40}{72} = 7\frac{47}{72}$

23. $\frac{5}{12}-\frac{1}{4}=\frac{5}{12}-\frac{3}{12}=\frac{5-3}{12}=\frac{2}{12}=\frac{1}{6}$
There is $\frac{1}{6}$ hour remaining.

24. $80\div 8\cdot 2+7=10\cdot 2+7=20+7=27$

25. $2\frac{1}{3} + 5\frac{3}{8}$ $2\frac{8}{24} + 5\frac{9}{24} = 7\frac{17}{24}$

26. $\frac{\frac{3}{5}+\frac{4}{9}+\frac{11}{15}}{3}=\left(\frac{27}{45}+\frac{20}{45}+\frac{33}{45}\right)\div 3$
$=\frac{80}{45}\cdot\frac{1}{3}$
$=\frac{16}{9}\cdot\frac{1}{3}$
$=\frac{16}{27}$

27. $\frac{3}{4}=\frac{3\cdot 11}{4\cdot 11}=\frac{33}{44}$
$\frac{9}{11}=\frac{9\cdot 4}{11\cdot 4}=\frac{36}{44}$
Since $33<36$, $\frac{33}{44}<\frac{36}{44}$ and $\frac{3}{4}<\frac{9}{11}$.

28. $5y-8y=24$
$-3y=24$
$\frac{-3y}{-3}=\frac{24}{-3}$
$y=-8$

29. $y-5=-2-6$
$y-5=-8$
$y-5+5=-8+5$
$y=-3$

30. $3y-6=7y-6$
$3y-6+6=7y-6+6$
$3y=7y$
$3y-7y=7y-7y$
$-4y=0$
$\frac{-4y}{-4}=\frac{0}{-4}$
$y=0$

31. $3a-6=a+4$
$3a-6-a=a+4-a$
$2a-6=4$
$2a-6+6=4+6$
$2a=10$
$\frac{2a}{2}=\frac{10}{2}$
$a=5$

Copyright © 2011 Pearson Education, Inc. Publishing as Prentice Hall.

32. $$\begin{aligned}4(y+1)-3&=21\\4y+4-3&=21\\4y+1&=21\\4y+1-1&=21-1\\4y&=20\\\frac{4y}{4}&=\frac{20}{4}\\y&=5\end{aligned}$$

33. $$\begin{aligned}3(2x-6)+6&=0\\6x-18+6&=0\\6x-12&=0\\6x-12+12&=0+12\\6x&=12\\\frac{6x}{6}&=\frac{12}{6}\\x&=2\end{aligned}$$

34. Seventy-five thousandths written in standard form is 0.075.

35. To round 736.2359 to the nearest tenth, observe that the digit in the hundredths place is 3. Since this digit is less than 5, we do not add 1 to the digit in the tenths place. The number 736.2359 rounded to the nearest tenth is 736.2.

36. To round 736.2359 to the nearest thousandth, observe that the digit in the ten-thousandths place is 9. Since this digit is at least 5, we add 1 to the digit in the thousandths place. The number 736.2359 rounded to the nearest thousandth is 736.236.

37. $$\begin{array}{r}23.850\\+\ 1.604\\\hline 25.454\end{array}$$

38. $$\begin{array}{r}700.00\\-\ 18.76\\\hline 681.24\end{array}$$

39. $$\begin{array}{r}0.0531\\\times\quad 16\\\hline 3186\\5310\\\hline 0.8496\end{array}$$

40. $$\begin{array}{r}0.375\\8\overline{)\ 3.000}\\-2\ 4\\\hline 60\\-56\\\hline 40\\-40\\\hline 0\end{array}$$

$\frac{3}{8}=0.375$

41. $$\begin{array}{r}0.052\\115\overline{)\ 5.980}\\-5\ 75\\\hline 230\\-230\\\hline 0\end{array}$$

$-5.98 \div 115 = -0.052$

42. $7.9=7\frac{9}{10}=\frac{10\cdot 7+9}{10}=\frac{79}{10}$

43. $-0.5(8.6-1.2)=-0.5(7.4)=-3.7$

44. $$\begin{aligned}\frac{n}{4}&=\frac{12}{16}\\n\cdot 16&=4\cdot 12\\16n&=48\\\frac{16n}{16}&=\frac{48}{16}\\n&=3\end{aligned}$$

45. $\frac{9}{20}=0.450$

$\frac{4}{9}=0.\overline{4}=0.444\ldots$

$0.456=0.456$

$\frac{4}{9}, \frac{9}{20}, 0.456$

46. $\frac{700\text{ meters}}{5\text{ seconds}}=\frac{5\cdot 140\text{ meters}}{5\cdot 1\text{ second}}=\frac{140\text{ meters}}{1\text{ second}}$

The unit rate is 140 meters/second.

47. The ratio of \$15 to \$10 is $\frac{\$15}{\$10}=\frac{15}{10}=\frac{5\cdot 3}{5\cdot 2}=\frac{3}{2}$.

48. The ratio of 7 to 21 is $\frac{7}{21}=\frac{1\cdot 7}{3\cdot 7}=\frac{1}{3}$.

Copyright © 2011 Pearson Education, Inc. Publishing as Prentice Hall.

49. The ratio of 2.5 to 3.15 is
$\frac{2.5}{3.15} = \frac{2.5 \cdot 100}{3.15 \cdot 100} = \frac{250}{315} = \frac{50 \cdot 5}{63 \cdot 5} = \frac{50}{63}$.

50. The ratio of 900 to 9000 is
$\frac{900}{9000} = \frac{900 \cdot 1}{900 \cdot 10} = \frac{1}{10}$.

Copyright © 2011 Pearson Education, Inc. Publishing as Prentice Hall.

Chapter 7

Section 7.1

Practice Problems

1. Since 27 students out of 100 students in a club are freshman, the fraction is $\frac{27}{100}$. Then $\frac{27}{100} = 27\%$.

2. $\frac{31}{100} = 31\%$

3. $49\% = 49(0.01) = 0.49$

4. $3.1\% = 3.1(0.01) = 0.031$

5. $175\% \ 175(0.01) = 1.75$

6. $0.46\% = 0.46(0.01) = 0.0046$

7. $600\% = 600(0.01) = 6.00$ or 6

8. $50\% = 50 \cdot \frac{1}{100} = \frac{50}{100} = \frac{1 \cdot 50}{2 \cdot 50} = \frac{1}{2}$

9. $2.3\% = 2.3 \cdot \frac{1}{100} = \frac{2.3}{100} = \frac{2.3 \cdot 10}{100 \cdot 10} = \frac{23}{1000}$

10. $150\% = 150 \cdot \frac{1}{100} = \frac{150}{100} = \frac{3 \cdot 50}{2 \cdot 50} = \frac{3}{2}$ or $1\frac{1}{2}$

11. $$\begin{aligned} 66\frac{2}{3}\% &= 66\frac{2}{3} \cdot \frac{1}{100} \\ &= \frac{200}{3} \cdot \frac{1}{100} \\ &= \frac{2 \cdot 100 \cdot 1}{3 \cdot 100} \\ &= \frac{2}{3} \end{aligned}$$

12. $12\% = 12 \cdot \frac{1}{100} = \frac{12}{100} = \frac{3 \cdot 4}{25 \cdot 4} = \frac{3}{25}$

13. $0.14 = 0.14(100\%) = 14.\%$ or 14%

14. $1.75 = 1.75(100\%) = 175.\%$ or 175%

15. $0.057 = 0.057(100\%) = 05.7\%$ or 5.7%

16. $0.5 = 0.5(100\%) = 050.\%$ or 50%

17. $\frac{3}{25} = \frac{3}{25} \cdot 100\% = \frac{3}{25} \cdot \frac{100}{1}\% = \frac{300}{25}\% = 12\%$

18. $$\begin{aligned} \frac{9}{40} &= \frac{9}{40} \cdot 100\% \\ &= \frac{9}{40} \cdot \frac{100}{1}\% \\ &= \frac{900}{40}\% \\ &= 22\frac{20}{40}\% \\ &= 22\frac{1}{2}\% \end{aligned}$$

19. $$\begin{aligned} 5\frac{1}{2} &= \frac{11}{2} \\ &= \frac{11}{2} \cdot 100\% \\ &= \frac{11}{2} \cdot \frac{100}{1}\% \\ &= \frac{1100}{2}\% \\ &= 550\% \end{aligned}$$

20. $\frac{3}{17} = \frac{3}{17} \cdot 100\% = \frac{3}{17} \cdot \frac{100\%}{1} = \frac{300}{17}\% \approx 17.65\%$

$$\begin{array}{r} 17.647 \approx 17.65 \\ 17 \overline{)\ 300.000} \\ \underline{-17} \\ 130 \\ \underline{-119} \\ 11\ 0 \\ \underline{-10\ 2} \\ 80 \\ \underline{-68} \\ 120 \\ \underline{-119} \\ 1 \end{array}$$

Thus, $\frac{3}{17}$ is approximately 17.65%.

Copyright © 2011 Pearson Education, Inc. Publishing as Prentice Hall.

21. As a decimal 27.5% = 27.5(0.01) = 0.275.

As a fraction $27.5\% = 27.5 \cdot \frac{1}{100}$
$$= \frac{27.5}{100}$$
$$= \frac{27.5}{100} \cdot \frac{10}{10}$$
$$= \frac{275}{1000}$$
$$= \frac{11 \cdot 25}{40 \cdot 25}$$
$$= \frac{11}{40}.$$

Thus, 27.5% written as a decimal is 0.275, and written as a fraction is $\frac{11}{40}$.

22. $1\frac{3}{4} = \frac{7}{4} = \frac{7}{4} \cdot 100\% = \frac{7}{4} \cdot \frac{100\%}{1} = \frac{700}{4}\% = 175\%$

Thus, a "$1\frac{3}{4}$ times" increase is the same as a "175% increase."

Vocabulary and Readiness Check

1. Percent means "per hundred."
2. 100% = 1.
3. The % symbol is read as percent.
4. To write a decimal or a fraction as a percent, multiply by 1 in the form of 100%.
5. To write a percent as a *decimal*, drop the % symbol and multiply by 0.01.
6. To write a percent as a *fraction*, drop the % symbol and multiply by $\frac{1}{100}$.

Exercise Set 7.1

1. $\frac{96}{100} = 96\%$

96% of these college students use the Internet.

3. 37 out of 100 adults preferred football.

$\frac{37}{100} = 37\%$

5. 37 of the adults preferred football, while 13 preferred soccer. Thus, 37 + 13 = 50 preferred football or soccer.

$\frac{50}{100} = 50\%$

7. 41% = 41(0.01) = 0.41

9. 6% = 6(0.01) = 0.06

11. 100% = 100(0.01) = 1.00 or 1

13. 73.6% = 73.6(0.01) = 0.736

15. 2.8% = 2.8(0.01) = 0.028

17. 0.6% = 0.6(0.01) = 0.006

19. 300% = 300(0.01) = 3.00 or 3

21. 32.58% = 32.58(0.01) = 0.3258

23. $8\% = 8 \cdot \frac{1}{100} = \frac{8}{100} = \frac{4 \cdot 2}{4 \cdot 25} = \frac{2}{25}$

25. $4\% = 4 \cdot \frac{1}{100} = \frac{4}{100} = \frac{1 \cdot 4}{4 \cdot 25} = \frac{1}{25}$

27. $4.5\% = 4.5 \cdot \frac{1}{100}$
$$= \frac{4.5}{100}$$
$$= \frac{4.5 \cdot 10}{100 \cdot 10}$$
$$= \frac{45}{1000}$$
$$= \frac{5 \cdot 9}{5 \cdot 200}$$
$$= \frac{9}{200}$$

29. $175\% = 175 \cdot \frac{1}{100} = \frac{175}{100} = \frac{7 \cdot 25}{4 \cdot 25} = \frac{7}{4}$ or $1\frac{3}{4}$

Copyright © 2011 Pearson Education, Inc. Publishing as Prentice Hall.

31. $6.25\% = 6.25 \cdot \frac{1}{100}$
$= \frac{6.25}{100}$
$= \frac{6.25 \cdot 100}{100 \cdot 100}$
$= \frac{625}{10{,}000}$
$= \frac{1 \cdot 625}{16 \cdot 625}$
$= \frac{1}{16}$

33. $10\frac{1}{3}\% = 10\frac{1}{3} \cdot \frac{1}{100} = \frac{31}{3} \cdot \frac{1}{100} = \frac{31}{300}$

35. $22\frac{3}{8}\% = 22\frac{3}{8} \cdot \frac{1}{100} = \frac{179}{8} \cdot \frac{1}{100} = \frac{179}{800}$

37. $0.22 = 0.22(100\%) = 22\%$

39. $0.006 = 0.006(100\%) = 0.6\%$

41. $5.3 = 5.3(100\%) = 530\%$

43. $0.056 = 0.056(100\%) = 5.6\%$

45. $0.2228 = 0.2228(100\%) = 22.28\%$

47. $3.00 = 3.00(100\%) = 300\%$

49. $0.7 = 0.7(100\%) = 70\%$

51. $\frac{7}{10} = \frac{7}{10} \cdot 100\% = \frac{700}{10}\% = 70\%$

53. $\frac{4}{5} = \frac{4}{5} \cdot 100\% = \frac{400}{5}\% = 80\%$

55. $\frac{34}{50} = \frac{34}{50} \cdot 100\% = \frac{3400}{50}\% = 68\%$

57. $\frac{3}{8} = \frac{3}{8} \cdot 100\% = \frac{300}{8}\% = \frac{75}{2}\% = 37\frac{1}{2}\%$

59. $\frac{1}{3} = \frac{1}{3} \cdot 100\% = \frac{100}{3}\% = 33\frac{1}{3}\%$

61. $4\frac{1}{2} = 4\frac{1}{2} \cdot 100\% = \frac{9}{2} \cdot 100\% = \frac{900}{2}\% = 450\%$

63. $1\frac{9}{10} = 1\frac{9}{10} \cdot 100\% = \frac{19}{10} \cdot 100\% = \frac{1900}{10}\% = 190\%$

65. $\frac{9}{11} = \frac{9}{11} \cdot 100\% = \frac{900}{11}\%$

$$\begin{array}{r} 81.818 \approx 81.82 \\ 11\overline{)\,900.000} \\ -88 \\ \hline 20 \\ -11 \\ \hline 90 \\ -88 \\ \hline 20 \\ -11 \\ \hline 90 \\ -88 \\ \hline 2 \end{array}$$

$\frac{9}{11}$ is approximately 81.82%.

67. $\frac{4}{15} = \frac{4}{15} \cdot 100\% = \frac{400}{15}\%$

$$\begin{array}{r} 26.666 \approx 26.67 \\ 15\overline{)\,400.000} \\ -30 \\ \hline 100 \\ -90 \\ \hline 100 \\ -90 \\ \hline 100 \\ -90 \\ \hline 100 \\ -90 \\ \hline 10 \end{array}$$

$\frac{4}{15}$ is approximately 26.67%.

Copyright © 2011 Pearson Education, Inc. Publishing as Prentice Hall.

69.

Percent	Decimal	Fraction
60%	0.6	$\frac{3}{5}$
$23\frac{1}{2}\%$	0.235	$\frac{47}{200}$
80%	0.8	$\frac{4}{5}$
$33\frac{1}{3}\%$	$0.333\overline{3}$	$\frac{1}{3}$
87.5%	0.875	$\frac{7}{8}$
7.5%	0.075	$\frac{3}{40}$

71.

Percent	Decimal	Fraction
200%	2	2
280%	2.8	$2\frac{4}{5}$
705%	7.05	$7\frac{1}{20}$
454%	4.54	$4\frac{27}{50}$

73. $38\% = 38(0.01) = 0.38$

$$38\% = 38\cdot\frac{1}{100} = \frac{38}{100} = \frac{19}{50}$$

75. $15.8\% = 15.8(0.01) = 0.158$

$$\begin{aligned}15.8\% &= 15.8\cdot\frac{1}{100}\\ &= \frac{15.8}{100}\\ &= \frac{15.8\cdot 10}{100\cdot 10}\\ &= \frac{158}{1000}\\ &= \frac{79}{500}\end{aligned}$$

77. $91\% = 91(0.01) = 0.91$

$$91\% = 91\cdot\frac{1}{100} = \frac{91}{100}$$

79. $0.5\% = 0.5(0.01) = 0.005$

$$\begin{aligned}0.5\% &= 0.5\cdot\frac{1}{100}\\ &= \frac{0.5}{100}\\ &= \frac{0.5}{100}\cdot\frac{10}{10}\\ &= \frac{5}{1000}\\ &= \frac{1}{200}\end{aligned}$$

81. $14.2\% = 14.2(0.01) = 0.142$

$$\begin{aligned}14.2\% &= 14.2\cdot\frac{1}{100}\\ &= \frac{14.2}{100}\\ &= \frac{14.2\cdot 10}{100\cdot 10}\\ &= \frac{142}{1000}\\ &= \frac{71}{500}\end{aligned}$$

83. $0.781 = 0.781(100\%) = 78.1\%$

85. $\frac{7}{1000} = \frac{7}{1000}(100\%) = \frac{700}{1000}\% = \frac{7}{10}\% = 0.7\%$

87. $46\% = 46(0.01) = 0.46$

89.

$$\begin{aligned}\frac{3}{4} - \frac{1}{2}\cdot\frac{8}{9} &= \frac{3}{4} - \frac{1\cdot 8}{2\cdot 9}\\ &= \frac{3}{4} - \frac{1\cdot 2\cdot 4}{2\cdot 9}\\ &= \frac{3}{4} - \frac{4}{9}\\ &= \frac{3}{4}\cdot\frac{9}{9} - \frac{4}{9}\cdot\frac{4}{4}\\ &= \frac{27}{36} - \frac{16}{36}\\ &= \frac{27-16}{36}\\ &= \frac{11}{36}\end{aligned}$$

Copyright © 2011 Pearson Education, Inc. Publishing as Prentice Hall.

91. $6\frac{2}{3} - 4\frac{5}{6} = \frac{20}{3} - \frac{29}{6}$
$= \frac{20}{3} \cdot \frac{2}{2} - \frac{29}{6}$
$= \frac{40}{6} - \frac{29}{6}$
$= \frac{11}{6}$
$= 1\frac{5}{6}$

93. a. 52.8647% rounded to the nearest tenth percent is 52.9%.

b. 52.8647% rounded to the nearest hundredth percent is 52.86%.

95. a. 6.5% = 6.5(0.01) = 0.065 INCORRECT

b. 7.8% = 7.8(0.01) = 0.078 CORRECT

c. 120% = 120(0.01) = 1.20 INCORRECT

d. 0.35% = 0.35(0.01) = 0.0035 CORRECT

b and d are correct.

97. 45% + 40% + 11% = 96%
100% − 96% = 4%
4% of the U.S. population have AB blood type.

99. 3 of 4 equal parts are shaded, so
$\frac{3}{4} = \frac{3}{4} \cdot 100\% = \frac{300}{4}\% = 75\%$

101. A fraction written as a percent is greater than 100% when the numerator is <u>greater</u> than the denominator.

103. $\frac{21}{79} \approx 0.2658 \approx 0.266$
$0.2658(100\%) = 26.58\% \approx 26.6\%$

105. The longest bar corresponds to network systems and data communication analysts, so that is predicted to be the fastest growing occupation.

107. The percent change for Veterinarians is 35%, so 35% = 35(0.01) = 0.35.

109. answers may vary

Section 7.2

Practice Problems

1. 8 is $\underbrace{\text{what percent}}$ of 48?
↓ ↓ ↓ ↓ ↓
$8 = x \cdot 48$

2. 2.6 is 40% of $\underbrace{\text{what number}}$?
↓ ↓ ↓ ↓ ↓
$2.6 = 40\% \cdot x$

3. $\underbrace{\text{What number}}$ is 90% of 0.045?
↓ ↓ ↓ ↓ ↓
$x = 90\% \cdot 0.045$

4. 56% of 180 is $\underbrace{\text{what number}}$?
↓ ↓ ↓ ↓ ↓
$56\% \cdot 180 = x$

5. 12% of $\underbrace{\text{what number}}$ is 21?
↓ ↓ ↓ ↓ ↓
$12\% \cdot x = 21$

6. $\underbrace{\text{What percent}}$ of 95 is 76?
↓ ↓ ↓ ↓ ↓
$x \cdot 95 = 76$

7. $\underbrace{\text{What number}}$ is 25% of 90?
↓ ↓ ↓ ↓ ↓
$x = 25\% \cdot 90$

$x = 25\% \cdot 90$
$x = 0.25 \cdot 90$
$x = 22.5$
Then 22.5 is 25% of 90.
Is this reasonable? To see, round 25% to 30%. Then 30% or 0.30(90) is 27. Our result is reasonable since 22.5 is close to 27.

Copyright © 2011 Pearson Education, Inc. Publishing as Prentice Hall.

8. 95% of 400 is $\underbrace{\text{what number}}$?

$$\begin{array}{ccccc} \downarrow & \downarrow & \downarrow & \downarrow & \downarrow \\ 95\% & \cdot & 400 & = & x \end{array}$$

$0.95 \cdot 400 = x$
$380 = x$

Then 95% of 400 is 380. Is this result reasonable? To see, round 95% to 100%. Then 100% of 400 or 1.0(400) = 400, which is close to 380.

9. 15% of $\underbrace{\text{what number}}$ is 2.4?

$$\begin{array}{ccccc} \downarrow & \downarrow & \downarrow & \downarrow & \downarrow \\ 15\% & \cdot & x & = & 2.4 \end{array}$$

$0.15 \cdot x = 2.4$
$\frac{0.15x}{0.15x} = \frac{2.4}{0.15}$
$x = 16$

Then 15% of 16 is 2.4. Is this result reasonable? To see, round 15% to 20%. Then 20% of 16 or 0.20(16) = 3.2, which is close to 2.4.

10. 18 is $4\frac{1}{2}$% of $\underbrace{\text{what number}}$?

$$\begin{array}{ccccc} \downarrow & \downarrow & \downarrow & \downarrow & \downarrow \\ 18 & = & 4\frac{1}{2}\% & \cdot & x \end{array}$$

$18 = 0.045x$
$\frac{18}{0.045} = \frac{0.045x}{0.045}$
$400 = x$

Then 18 is $4\frac{1}{2}$% of 400.

11. $\underbrace{\text{What percent}}$ of 90 is 27?

$$\begin{array}{ccccc} \downarrow & \downarrow & \downarrow & \downarrow & \downarrow \\ x & \cdot & 90 & = & 27 \end{array}$$

$90x = 27$
$\frac{90x}{90} = \frac{27}{90}$
$x = \frac{3}{10}$
or $x = 0.30$

Since we are looking for percent, we can write $\frac{3}{10}$ or 0.30 as a percent. $x = 30\%$

Then 30% of 90 is 27. To check, see that 30% · 90 = 27.

12. 63 is $\underbrace{\text{what percent}}$ of 45?

$$\begin{array}{ccccc} \downarrow & \downarrow & \downarrow & \downarrow & \downarrow \\ 63 & = & x & \cdot & 45 \end{array}$$

$63 = 45x$
$\frac{63}{45} = \frac{45x}{45}$
$\frac{7}{5} = x$
$1.4 = x$
$140\% = x$

Then 63 is 140% of 45.

Vocabulary and Readiness Check

1. The word is translates to "=."

2. The word of usually translates to "multiplication."

3. In the statement "10% of 90 is 9," the number 9 is called the amount, 90 is called the base, and 10 is called the percent.

4. 100% of a number = the number

5. Any "percent greater than 100%" of "a number" = "a number greater than the original number."

6. Any "percent less than 100%" of "a number" = "a number less than the original number."

Exercise Set 7.2

1. 18% of 81 is $\underbrace{\text{what number}}$?

$$\begin{array}{ccccc} \downarrow & \downarrow & \downarrow & \downarrow & \downarrow \\ 18\% & \cdot & 81 & = & x \end{array}$$

3. 20% of $\underbrace{\text{what number}}$ is 105?

$$\begin{array}{ccccc} \downarrow & \downarrow & \downarrow & \downarrow & \downarrow \\ 20\% & \cdot & x & = & 105 \end{array}$$

5. 0.6 is 40% of $\underbrace{\text{what number}}$?

$$\begin{array}{ccccc} \downarrow & \downarrow & \downarrow & \downarrow & \downarrow \\ 0.6 & = & 40\% & \cdot & x \end{array}$$

7. $\underbrace{\text{What percent}}$ of 80 is 3.8?

$$\begin{array}{ccccc} \downarrow & \downarrow & \downarrow & \downarrow & \downarrow \\ x & \cdot & 80 & = & 3.8 \end{array}$$

Copyright © 2011 Pearson Education, Inc. Publishing as Prentice Hall.

9. What number is 9% of 43?

$$\underbrace{\text{What number}}_{\downarrow\ x} \text{ is } \overset{}{\underset{=}{\downarrow}}\ \underset{9\%}{\downarrow}\ \underset{\cdot}{\downarrow}\ \underset{43}{\downarrow}$$

$x = 9\% \cdot 43$

11. What percent of 250 is 150?

$$\underbrace{\text{What percent}}_{\downarrow\ x} \text{ of 250 is 150?}$$

$x \cdot 250 = 150$

13. $10\% \cdot 35 = x$
$0.10 \cdot 35 = x$
$3.5 = x$
10% of 35 is 3.5.

15. $x = 14\% \cdot 205$
$x = 0.14 \cdot 205$
$x = 28.7$
28.7 is 14% of 205.

17. $1.2 = 12\% \cdot x$
$1.2 = 0.12x$
$\frac{1.2}{0.12} = \frac{0.12x}{0.12}$
$10 = x$
1.2 is 12% of 10.

19. $8\frac{1}{2}\% \cdot x = 51$
$0.085x = 51$
$\frac{0.085x}{0.085} = \frac{51}{0.085}$
$x = 600$
$8\frac{1}{2}\%$ of 600 is 51.

21. $x \cdot 80 = 88$
$\frac{x \cdot 80}{80} = \frac{88}{80}$
$x = 1.1$
$x = 110\%$
88 is 110% of 80.

23. $17 = x \cdot 50$
$\frac{17}{50} = \frac{x \cdot 50}{50}$
$0.34 = x$
$34\% = x$
17 is 34% of 50.

25. $0.1 = 10\% \cdot x$
$0.1 = 0.10x$
$\frac{0.1}{0.1} = \frac{0.1x}{0.1}$
$1 = x$
0.1 is 10% of 1.

27. $150\% \cdot 430 = x$
$1.5 \cdot 430 = x$
$645 = x$
150% of 430 is 645.

29. $82.5 = 16\frac{1}{2}\% \cdot x$
$82.5 = 0.165x$
$\frac{82.5}{0.165} = \frac{0.165x}{0.165}$
$500 = x$
82.5 is $16\frac{1}{2}\%$ of 500.

31. $2.58 = x \cdot 50$
$\frac{2.58}{50} = \frac{x \cdot 50}{50}$
$0.0516 = x$
$5.16\% = x$
2.58 is 5.16% of 50.

33. $x = 42\% \cdot 60$
$x = 0.42 \cdot 60$
$x = 25.2$
25.2 is 42% of 60.

35. $x \cdot 184 = 64.4$
$\frac{x \cdot 184}{184} = \frac{64.4}{184}$
$x = 0.35$
$x = 35\%$
35% of 184 is 64.4.

37. $120\% \cdot x = 42$
$1.20 \cdot x = 42$
$\frac{1.2x}{1.2} = \frac{42}{1.2}$
$x = 35$
120% of 35 is 42.

39. $2.4\% \cdot 26 = x$
$0.024 \cdot 26 = x$
$0.624 = x$
2.4% of 26 is 0.624.

Copyright © 2011 Pearson Education, Inc. Publishing as Prentice Hall.

41. $x \cdot 600 = 3$
$$\frac{x \cdot 600}{600} = \frac{3}{600}$$
$$x = 0.005$$
$$x = 0.5\%$$
0.5% of 600 is 3.

43. $6.67 = 4.6\% \cdot x$
$$6.67 = 0.046x$$
$$\frac{6.67}{0.046} = \frac{0.046x}{0.046}$$
$$145 = x$$
6.67 is 4.6% of 145.

45. $1575 = x \cdot 2500$
$$\frac{1575}{2500} = \frac{x \cdot 2500}{2500}$$
$$0.63 = x$$
$$63\% = x$$
1575 is 63% of 2500.

47. $2 = x \cdot 50$
$$\frac{2}{50} = \frac{x \cdot 50}{50}$$
$$0.04 = x$$
$$4\% = x$$
2 is 4% of 50.

49. $\frac{27}{x} = \frac{9}{10}$
$$27 \cdot 10 = x \cdot 9$$
$$270 = 9x$$
$$\frac{270}{9} = \frac{9x}{9}$$
$$30 = x$$

51. $\frac{x}{5} = \frac{8}{11}$
$$x \cdot 11 = 5 \cdot 8$$
$$11x = 40$$
$$\frac{11x}{11} = \frac{40}{11}$$
$$x = 3\frac{7}{11}$$

53. $\frac{17}{12} = \frac{x}{20}$

55. $\frac{8}{9} = \frac{14}{x}$

57. $5 \cdot n = 32$
$$\frac{5 \cdot n}{5} = \frac{32}{5}$$
$$n = \frac{32}{5}$$
Choice c is correct.

59. $0.06 = n \cdot 7$
$$\frac{0.06}{7} = \frac{n \cdot 7}{7}$$
$$\frac{0.06}{7} = n$$
Choice b is correct.

61. answers may vary

63. Since 100% of 20 is 20, and 30 is greater than 20, x must be greater than 100%; b.

65. Since 85 is less than 120, the percent is less than 100%; c.

67. Since 55% is less than 100%, which is 1, 55% of 45 is less than 45; c.

69. Since 100% is 1, 100% of 45 is equal to 45; a.

71. Since 100% is 1, 100% of 45 is equal to 45; a.

73. answers may vary

75. $1.5\% \cdot 45{,}775 = x$
$$0.015 \cdot 45{,}775 = x$$
$$686.625 = x$$
1.5% of 45,775 is 686.625.

77. $22{,}113 = 180\% \cdot x$
$$22{,}113 = 1.80 \cdot x$$
$$\frac{22{,}113}{1.8} = \frac{1.8 \cdot x}{1.8}$$
$$12{,}285 = x$$
22,113 is 180% of 12,285.

Copyright © 2011 Pearson Education, Inc. Publishing as Prentice Hall.

Section 7.3

Practice Problems

1. 27% of $\underbrace{\text{what number}}$ is 54?

↓ percent

↓ base — It appears after the word *of*.

↓ amount — It is the part compared to the whole

$$\frac{54}{b}=\frac{27}{100}$$

2. 30 is $\underbrace{\text{what percent}}$ of 90?

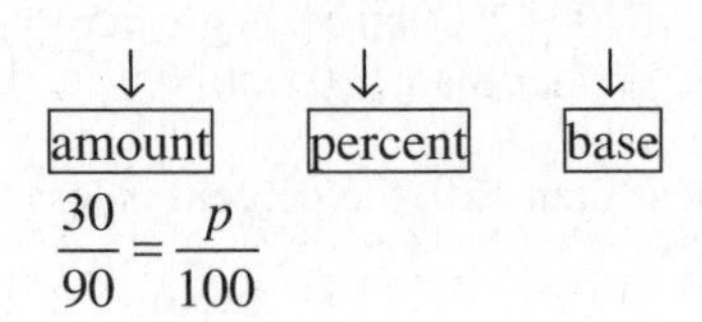

↓ amount ↓ percent ↓ base

$$\frac{30}{90}=\frac{p}{100}$$

3. $\underbrace{\text{What number}}$ is 25% of 116?

↓ amount ↓ percent ↓ base

$$\frac{a}{116}=\frac{25}{100}$$

4. 680 is 65% of $\underbrace{\text{what number}}$?

↓ amount ↓ percent ↓ base

$$\frac{680}{b}=\frac{65}{100}$$

5. $\underbrace{\text{What percent}}$ of 40 is 75?

↓ percent ↓ base ↓ amount

$$\frac{75}{40}=\frac{p}{100}$$

6. 46% of 80 is $\underbrace{\text{what number}}$?

↓ percent ↓ base ↓ amount

$$\frac{a}{80}=\frac{46}{100}$$

7. $\underbrace{\text{What number}}$ is 8% of 120?

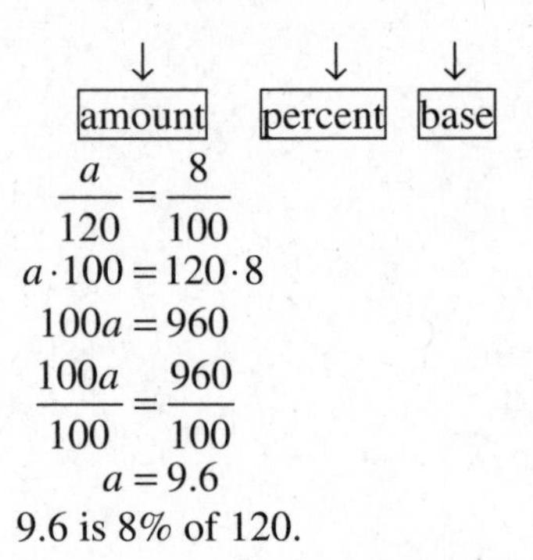

↓ amount ↓ percent ↓ base

$$\frac{a}{120}=\frac{8}{100}$$
$$a\cdot 100=120\cdot 8$$
$$100a=960$$
$$\frac{100a}{100}=\frac{960}{100}$$
$$a=9.6$$

9.6 is 8% of 120.

8. 65% of $\underbrace{\text{what number}}$ is 52?

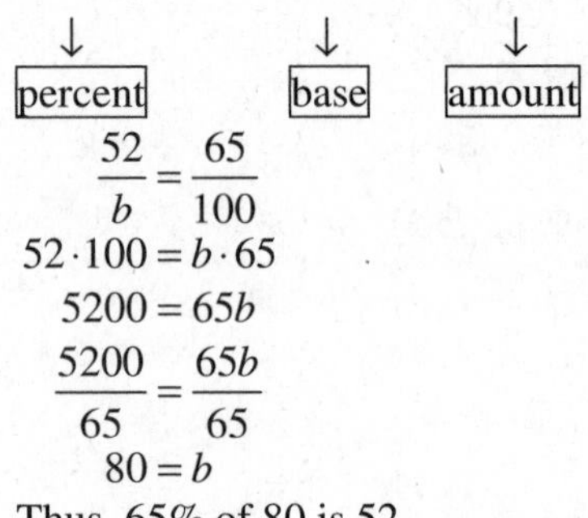

↓ percent ↓ base ↓ amount

$$\frac{52}{b}=\frac{65}{100}$$
$$52\cdot 100=b\cdot 65$$
$$5200=65b$$
$$\frac{5200}{65}=\frac{65b}{65}$$
$$80=b$$

Thus, 65% of 80 is 52.

9. 15.4 is 5% of $\underbrace{\text{what number}}$?

↓ amount ↓ percent ↓ base

$$\frac{15.4}{b}=\frac{5}{100} \text{ or } \frac{15.4}{b}=\frac{1}{20}$$
$$15.4\cdot 20=1\cdot b$$
$$308=b$$

So, 154 is 5% of 308.

10. $\underbrace{\text{What percent}}$ of 40 is 8?

↓ percent ↓ base ↓ amount

$$\frac{8}{40}=\frac{p}{100} \text{ or } \frac{1}{5}=\frac{p}{100}$$
$$1\cdot 100=5\cdot p$$
$$100=5p$$
$$\frac{100}{5}=\frac{5p}{5}$$
$$20=p$$

So, 20% of 40 is 8.

Copyright © 2011 Pearson Education, Inc. Publishing as Prentice Hall.

11. 414 is $\underbrace{\text{what percent}}$ of 180?

$\downarrow$ $\downarrow$ $\downarrow$

$\boxed{\text{amount}}$ $\boxed{\text{percent}}$ $\boxed{\text{base}}$

$$\frac{414}{180} = \frac{p}{100}$$
$$414 \cdot 100 = 180 \cdot p$$
$$41{,}400 = 180p$$
$$\frac{41{,}400}{180} = \frac{180p}{180}$$
$$230 = p$$

Then 414 is 230% of 180.

Vocabulary and Readiness Check

1. When translating the statement "20% of 15 is 3" to a proportion, the number 3 is called the amount, 15 is the base, and 20 is the percent.

2. In the question "50% of what number is 28?", which part of the percent proportion is unknown? base

3. In the question "What number is 25% of 200?", which part of the percent proportion is unknown? amount

4. In the question "38 is what percent of 380?", which part of the percent proportion is unknown? percent

Exercise Set 7.3

1. 98% of 45 is $\underbrace{\text{what number}}$?

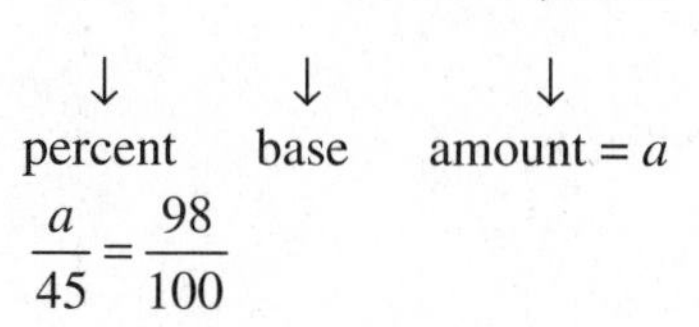

$\downarrow$ $\downarrow$ $\downarrow$

percent base amount $= a$

$$\frac{a}{45} = \frac{98}{100}$$

3. $\underbrace{\text{What number}}$ is 4% of 150?

$\downarrow$ $\downarrow$ $\downarrow$

amount $= a$ percent base

$$\frac{a}{150} = \frac{4}{100}$$

5. 14.3 is 26% of $\underbrace{\text{what number}}$?

$\downarrow$ $\downarrow$ $\downarrow$

amount percent base $= b$

$$\frac{14.3}{b} = \frac{26}{100}$$

7. 35% of $\underbrace{\text{what number}}$ is 84?

$\downarrow$ $\downarrow$ $\downarrow$

percent base $= b$ amount

$$\frac{84}{b} = \frac{35}{100}$$

9. $\underbrace{\text{What percent}}$ of 400 is 70?

$\downarrow$ $\downarrow$ $\downarrow$

percent $= p$ base amount

$$\frac{70}{400} = \frac{p}{100}$$

11. 8.2 is $\underbrace{\text{what percent}}$ of 82?

$\downarrow$ $\downarrow$ $\downarrow$

amount percent $= p$ base

$$\frac{8.2}{82} = \frac{p}{100}$$

13. $\frac{a}{65} = \frac{40}{100}$ or $\frac{a}{65} = \frac{2}{5}$

$$a \cdot 5 = 65 \cdot 2$$
$$5a = 130$$
$$\frac{5a}{5} = \frac{130}{5}$$
$$a = 26$$

40% of 65 is 26.

15.
$$\frac{a}{105} = \frac{18}{100}$$
$$a \cdot 100 = 105 \cdot 18$$
$$a \cdot 100 = 1890$$
$$\frac{a \cdot 100}{100} = \frac{1890}{100}$$
$$a = 18.9$$

18.9 is 18% of 105.

17. $\frac{90}{b} = \frac{15}{100}$ or $\frac{90}{b} = \frac{3}{20}$

$$90 \cdot 20 = b \cdot 3$$
$$1800 = 3b$$
$$\frac{1800}{3} = \frac{3b}{3}$$
$$600 = b$$

15% of 600 is 90.

Copyright © 2011 Pearson Education, Inc. Publishing as Prentice Hall.

19. $\frac{7.8}{b}=\frac{78}{100}$
$7.8\cdot 100=b\cdot 78$
$780=78\cdot b$
$\frac{780}{78}=\frac{78\cdot b}{78}$
$10=b$
7.8 is 78% of 10.

21. $\frac{42}{35}=\frac{p}{100}$ or $\frac{6}{5}=\frac{p}{100}$
$6\cdot 100=5\cdot p$
$600=5p$
$\frac{600}{5}=\frac{5p}{5}$
$120=p$
42 is 120% of 35.

23. $\frac{14}{50}=\frac{p}{100}$ or $\frac{7}{25}=\frac{p}{100}$
$7\cdot 100=25\cdot p$
$700=25p$
$\frac{700}{25}=\frac{25p}{25}$
$28=p$
14 is 28% of 50.

25. $\frac{3.7}{b}=\frac{10}{100}$ or $\frac{3.7}{b}=\frac{1}{10}$
$3.7\cdot 10=b\cdot 1$
$37=b$
3.7 is 10% of 37.

27. $\frac{a}{70}=\frac{2.4}{100}$
$a\cdot 100=70\cdot 2.4$
$100a=168$
$\frac{100a}{100}=\frac{168}{100}$
$a=1.68$
1.68 is 2.4% of 70.

29. $\frac{160}{b}=\frac{16}{100}$ or $\frac{160}{b}=\frac{4}{25}$
$160\cdot 25=b\cdot 4$
$4000=4b$
$\frac{4000}{4}=\frac{4b}{4}$
$1000=b$
160 is 16% of 1000.

31. $\frac{394.8}{188}=\frac{p}{100}$
$394.8\cdot 100=188\cdot p$
$39,480=188p$
$\frac{39,480}{188}=\frac{188p}{188}$
$210=p$
394.8 is 210% of 188.

33. $\frac{a}{62}=\frac{89}{100}$
$a\cdot 100=62\cdot 89$
$100a=5518$
$\frac{100a}{100}=\frac{5518}{100}$
$a=55.18$
55.18 is 89% of 62.

35. $\frac{2.7}{6}=\frac{p}{100}$
$2.7\cdot 100=6\cdot p$
$270=6p$
$\frac{270}{6}=\frac{6p}{6}$
$45=p$
45% of 6 is 2.7.

37. $\frac{105}{b}=\frac{140}{100}$ or $\frac{105}{b}=\frac{7}{5}$
$105\cdot 5=b\cdot 7$
$525=7b$
$\frac{525}{7}=\frac{7b}{7}$
$75=b$
140% of 75 is 105.

39. $\frac{a}{48}=\frac{1.8}{100}$
$a\cdot 100=48\cdot 1.8$
$100a=86.4$
$\frac{100a}{100}=\frac{86.4}{100}$
$a=0.864$
1.8% of 48 is 0.864.

Copyright © 2011 Pearson Education, Inc. Publishing as Prentice Hall.

41. $\frac{4}{800}=\frac{p}{100}$ or $\frac{1}{200}=\frac{p}{100}$

$$1\cdot 100=200\cdot p$$
$$100=200p$$
$$\frac{100}{200}=\frac{200p}{200}$$
$$0.5=p$$

0.5% of 800 is 4.

43. $\frac{3.5}{b}=\frac{2.5}{100}$

$$3.5\cdot 100=b\cdot 2.5$$
$$350=2.5b$$
$$\frac{350}{2.5}=\frac{2.5b}{2.5}$$
$$140=b$$

3.5 is 2.5% of 140.

45. $\frac{a}{48}=\frac{20}{100}$ or $\frac{a}{48}=\frac{1}{5}$

$$a\cdot 5=48\cdot 1$$
$$5a=48$$
$$\frac{5a}{5}=\frac{48}{5}$$
$$a=9.6$$

20% of 48 is 9.6.

47. $\frac{2486}{2200}=\frac{p}{100}$

$$2486\cdot 100=2200\cdot p$$
$$248,600=2200p$$
$$\frac{248,600}{2200}=\frac{2200p}{2200}$$
$$113=p$$

2486 is 113% of 2200.

49. $-\frac{11}{16}+\left(-\frac{3}{16}\right)=\frac{-11-3}{16}=\frac{-14}{16}=-\frac{2\cdot 7}{2\cdot 8}=-\frac{7}{8}$

51. $3\frac{1}{2}-\frac{11}{30}=\frac{7}{2}-\frac{11}{30}$

$$=\frac{7\cdot 15}{2\cdot 15}-\frac{11}{30}$$
$$=\frac{105}{30}-\frac{11}{30}$$
$$=\frac{105-11}{30}$$
$$=\frac{94}{30}$$
$$=3\frac{4}{30}$$
$$=3\frac{2}{15}$$

53.

$$\begin{array}{r} \scriptstyle 1 \\ 0.41 \\ +\ 0.29 \\ \hline 0.70 \end{array}$$

55.

$$\begin{array}{r} 2.38 \\ -\ 0.19 \\ \hline 2.19 \end{array}$$

57. answers may vary

59. $\frac{a}{64}=\frac{25}{100}$

$$\frac{17}{64}\stackrel{?}{=}\frac{25}{100}$$
$$17\cdot 100\stackrel{?}{=}64\cdot 25$$
$$1700=1600 \quad \text{False}$$

The amount is not 17.

$$\frac{a}{64}=\frac{25}{100}$$
$$a\cdot 100=64\cdot 25$$
$$100a=1600$$
$$\frac{100a}{100}=\frac{1600}{100}$$
$$a=16$$

25% of 64 is 16.

61. $\frac{p}{100}=\frac{13}{52}$

$$\frac{25}{100}\stackrel{?}{=}\frac{13}{52}$$
$$\frac{1}{4}=\frac{1}{4} \quad \text{True}$$

Yes, the percent is equal to 25 (25%).

63. answers may vary

Copyright © 2011 Pearson Education, Inc. Publishing as Prentice Hall.

65. $\frac{a}{53,862} = \frac{22.3}{100}$
$a \cdot 100 = 53,862 \cdot 22.3$
$100a = 1,201,122.6$
$\frac{100a}{100} = \frac{1,201,122.6}{100}$
$a \approx 12,011.2$
22.3% of 53,862 is 12,011.2.

67. $\frac{8652}{b} = \frac{119}{100}$
$8652 \cdot 100 = b \cdot 119$
$865,200 = 119b$
$\frac{865,200}{119} = \frac{119b}{119}$
$7270.6 \approx b$
8652 is 119% of 7270.6.

Integrated Review

1. $0.94 = 0.94(100\%) = 94\%$

2. $0.17 = 0.17(100\%) = 17\%$

3. $\frac{3}{8} = \frac{3}{8} \cdot 100\% = \frac{300}{8}\% = 37.5\%$

4. $\frac{7}{2} = \frac{7}{2} \cdot 100\% = \frac{700\%}{2} = 350\%$

5. $4.7 = 4.7(100\%) = 470\%$

6. $8 = 8(100\%) = 800\%$

7. $\frac{9}{20} = \frac{9}{20} \cdot 100\% = \frac{900}{20}\% = 45\%$

8. $\frac{53}{50} = \frac{53}{50} \cdot 100\% = \frac{5300}{50}\% = 106\%$

9. $6\frac{3}{4} = \frac{27}{4} = \frac{27}{4} \cdot 100\% = \frac{2700}{4}\% = 675\%$

10. $3\frac{1}{4} = \frac{13}{4} = \frac{13}{4} \cdot 100\% = \frac{1300}{4}\% = 325\%$

11. $0.02 = 0.02(100\%) = 2\%$

12. $0.06 = 0.06(100\%) = 6\%$

13. $71\% = 71(0.01) = 0.71$

14. $31\% = 31(0.01) = 0.31$

15. $3\% = 3(0.01) = 0.03$

16. $4\% = 4(0.01) = 0.04$

17. $224\% = 224(0.01) = 2.24$

18. $700\% = 700(0.01) = 7.0$ or 7

19. $2.9\% = 2.9(0.01) = 0.029$

20. $6.6\% = 6.6(0.01) = 0.066$

21. $7\% = 7(0.01) = 0.07$
$7\% = 7 \cdot \frac{1}{100} = \frac{7}{100}$

22. $5\% = 5(0.01) = 0.05$
$5\% = 5 \cdot \frac{1}{100} = \frac{5}{100} = \frac{1}{20}$

23. $6.8\% = 6.8(0.01) = 0.068$
$6.8\% = 6.8 \cdot \frac{1}{100} = \frac{6.8}{100} = \frac{6.8 \cdot 10}{100 \cdot 10} = \frac{68}{1000} = \frac{17}{250}$

24. $11.25\% = 11.25(0.01) = 0.1125$
$11.25\% = 11.25 \cdot \frac{1}{100}$
$= \frac{11.25}{100}$
$= \frac{11.25 \cdot 100}{100 \cdot 100}$
$= \frac{1125}{10,000}$
$= \frac{9}{80}$

25. $74\% = 74(0.01) = 0.74$
$74\% = 74 \cdot \frac{1}{100} = \frac{74}{100} = \frac{37}{50}$

26. $45\% = 45(0.01) = 0.45$
$45\% = 45 \cdot \frac{1}{100} = \frac{45}{100} = \frac{9}{20}$

27. $16\frac{1}{3}\% = 16.3\overline{3}(0.01) \approx 0.163$
$16\frac{1}{3}\% = \frac{49}{3}\% = \frac{49}{3} \cdot \frac{1}{100} = \frac{49}{300}$

Copyright © 2011 Pearson Education, Inc. Publishing as Prentice Hall.

28. $12\frac{2}{3}\% = 12.6\overline{6}(0.01) \approx 0.127$

$12\frac{2}{3}\% = \frac{38}{3}\% = \frac{38}{3} \cdot \frac{1}{100} = \frac{38}{300} = \frac{19}{150}$

29.
$$\frac{a}{90} = \frac{15}{100}$$
$$a \cdot 100 = 90 \cdot 15$$
$$100a = 1350$$
$$\frac{100a}{100} = \frac{1350}{100}$$
$$a = 13.5$$
15% of 90 is 13.5.

30.
$$\frac{78}{b} = \frac{78}{100}$$
$$78 \cdot 100 = b \cdot 78$$
$$7800 = 78b$$
$$\frac{7800}{78} = \frac{78b}{78}$$
$$100 = b$$
78% of 100 is 78.

31.
$$\frac{297.5}{b} = \frac{85}{100}$$
$$297.5 \cdot 100 = b \cdot 85$$
$$29,750 = 85b$$
$$\frac{29,750}{85} = \frac{85b}{85}$$
$$350 = b$$
297.5 is 85% of 350.

32.
$$\frac{78}{65} = \frac{p}{100}$$
$$78 \cdot 100 = 65 \cdot p$$
$$7800 = 65p$$
$$\frac{7800}{65} = \frac{65p}{65}$$
$$120 = p$$
78 is 120% of 65.

33.
$$\frac{23.8}{85} = \frac{p}{100}$$
$$23.8 \cdot 100 = 85 \cdot p$$
$$2380 = 85p$$
$$\frac{2380}{85} = \frac{85p}{85}$$
$$28 = p$$
23.8 is 28% of 85.

34.
$$\frac{a}{200} = \frac{38}{100}$$
$$a \cdot 100 = 200 \cdot 38$$
$$100a = 7600$$
$$\frac{100a}{100} = \frac{7600}{100}$$
$$a = 76$$
38% of 200 is 76.

35.
$$\frac{a}{85} = \frac{40}{100}$$
$$a \cdot 100 = 85 \cdot 40$$
$$100a = 3400$$
$$\frac{100a}{100} = \frac{3400}{100}$$
$$a = 34$$
34 is 40% of 85.

36.
$$\frac{128.7}{99} = \frac{p}{100}$$
$$128.7 \cdot 100 = 99 \cdot p$$
$$12,870 = 99p$$
$$\frac{12,870}{99} = \frac{99p}{99}$$
$$130 = p$$
130% of 99 is 128.7.

37.
$$\frac{115}{250} = \frac{p}{100}$$
$$115 \cdot 100 = 250 \cdot p$$
$$11,500 = 250p$$
$$\frac{11,500}{250} = \frac{250p}{250}$$
$$46 = p$$
46% of 250 is 115.

38.
$$\frac{a}{84} = \frac{45}{100}$$
$$a \cdot 100 = 84 \cdot 45$$
$$100a = 3780$$
$$\frac{100a}{100} = \frac{3780}{100}$$
$$a = 37.8$$
37.8 is 45% of 84.

Copyright © 2011 Pearson Education, Inc. Publishing as Prentice Hall.

39. $\frac{63}{b} = \frac{42}{100}$
$63 \cdot 100 = b \cdot 42$
$6300 = 42b$
$\frac{6300}{42} = \frac{42b}{42}$
$150 = b$
42% of 150 is 63.

40. $\frac{58.9}{b} = \frac{95}{100}$
$58.9 \cdot 100 = b \cdot 95$
$5890 = 95b$
$\frac{5890}{95} = \frac{95b}{95}$
$62 = b$
95% of 62 is 58.9.

Section 7.4

Practice Problems

1. *Method 1*:
What number is 25% of 2174?
$x = 25\% \cdot 2174$
$x = 0.25 \cdot 2174$
$x = 543.5$
We predict 543.5 miles of the trail resides in the state of Virginia.
Method 2:
What number is 25% of 2174?

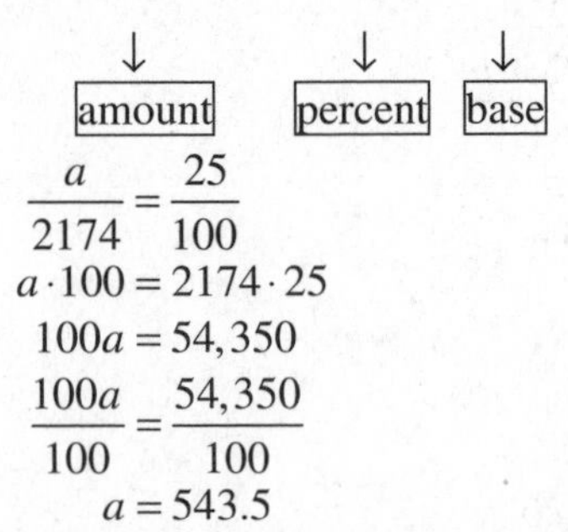

amount percent base
$\frac{a}{2174} = \frac{25}{100}$
$a \cdot 100 = 2174 \cdot 25$
$100a = 54,350$
$\frac{100a}{100} = \frac{54,350}{100}$
$a = 543.5$
We predict 543.5 miles of the trail resides in the state of Virginia.

2. *Method 1*:
34,000 is what percent of 130,000?
$34,000 = x \cdot 130,000$
$34,000 = 130,000x$
$\frac{34,000}{130,000} = \frac{130,000x}{130,000}$
$0.26 \approx x$
$26\% = x$
In Florida, 34,000 or 26% more new nurses were needed in 2006.
Method 2:
34,000 is what percent of 130,000?

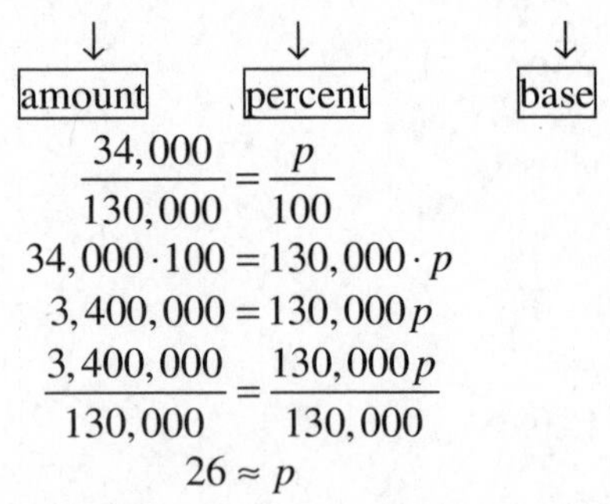

amount percent base
$\frac{34,000}{130,000} = \frac{p}{100}$
$34,000 \cdot 100 = 130,000 \cdot p$
$3,400,000 = 130,000p$
$\frac{3,400,000}{130,000} = \frac{130,000p}{130,000}$
$26 \approx p$
In Florida, 34,000 or 26% more new nurses were needed in 2006.

3. *Method 1*:
864 is 32% of what number?
$864 = 32\% \cdot x$
$864 = 0.32 \cdot x$
$\frac{864}{0.32} = \frac{0.32x}{0.32}$
$2700 = x$
There are 2700 students at Euclid University.
Method 2:
864 is 32% of what number?
amount percent base
$\frac{864}{b} = \frac{32}{100}$
$864 \cdot 100 = b \cdot 32$
$86,400 = 32b$
$\frac{86,400}{32} = \frac{32b}{32}$
$2700 = b$
There are 2700 students at Euclid University.

Copyright © 2011 Pearson Education, Inc. Publishing as Prentice Hall.

4. *Method 1*:

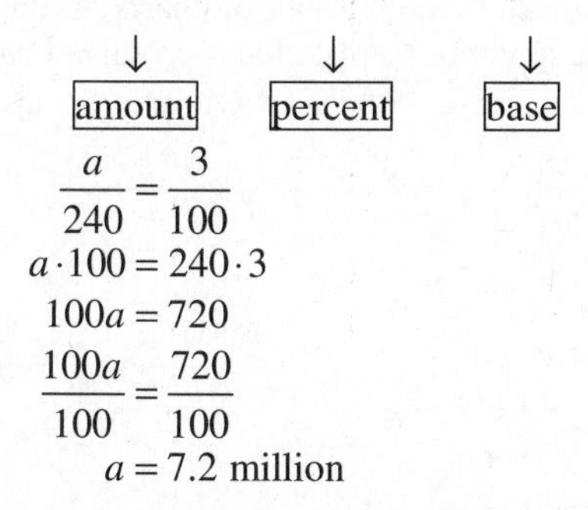

$x = 0.03 \cdot 240$
$x = 7.2$ million

a. The increase in the number of vehicles on the road in 2007 is 7.2 million.

b. The total number of registered vehicles on the road in 2007 was 240 million + 7.2 million = 247.2 million

Method 2:

What number is 3% of 240 million?

amount percent base

$$\frac{a}{240} = \frac{3}{100}$$
$$a \cdot 100 = 240 \cdot 3$$
$$100a = 720$$
$$\frac{100a}{100} = \frac{720}{100}$$
$$a = 7.2 \text{ million}$$

a. The increase in the number of vehicles on the road in 2007 is 7.2 million.

b. The total number of registered vehicles on the road in 2007 was 240 million + 7.2 million = 247.2 million

5. Find the amount of increase by subtracting the original number of attendants from the new number of attendants.
amount of increase = 333 − 285 = 48
The amount of increase is 48 attendants.

$$\text{percent of increase} = \frac{\text{amount of increase}}{\text{original amount}} = \frac{48}{285} \approx 0.168 = 16.8\%$$

The number of attendants to the local play, *Peter Pan*, increased by about 16.8%.

6. Find the amount of decrease by subtracting 18,483 from 20,200.
Amount of decrease = 20,200 − 18,483 = 1717
The amount of decrease is 1717.

$$\text{percent of decrease} = \frac{\text{amount of decrease}}{\text{original amount}} = \frac{1717}{20,200} = 0.085 = 8.5\%$$

The population decreased by 8.5%.

Exercise Set 7.4

1. 24 is 1.5% of what number?
Method 1:

$$24 = 1.5\% \cdot x$$
$$24 = 0.015x$$
$$\frac{24}{0.015} = \frac{0.015x}{0.015}$$
$$1600 = x$$

1600 bolts were inspected.
Method 2:

$$\frac{24}{b} = \frac{1.5}{100}$$
$$24 \cdot 100 = b \cdot 1.5$$
$$2400 = 1.5b$$
$$\frac{2400}{1.5} = \frac{1.5b}{1.5}$$
$$1600 = b$$

1600 bolts were inspected.

3. 4% of 220 is what number?
Method 1:

$$4\% \cdot 220 = x$$
$$0.04 \cdot 220 = x$$
$$8.8 = x$$

The minimum weight resistance is 8.8 pounds.
Method 2:

$$\frac{a}{220} = \frac{4}{100}$$
$$a \cdot 100 = 220 \cdot 4$$
$$100a = 880$$
$$\frac{100a}{100} = \frac{880}{100}$$
$$a = 8.8$$

The minimum weight resistance is 8.8 pounds.

Copyright © 2011 Pearson Education, Inc. Publishing as Prentice Hall.

5. 378 is what percent of 2700?
Method 1:
$378 = x \cdot 2700$
$\frac{378}{2700} = \frac{x \cdot 2700}{2700}$
$0.14 = x$
$14\% = x$
The student spent 14% of last semester's college cost on books.
Method 2:
$\frac{378}{2700} = \frac{p}{100}$
$378 \cdot 100 = 2700 \cdot p$
$37,800 = 2700p$
$\frac{37,800}{2700} = \frac{2700p}{2700}$
$14 = p$
The student spent 14% of last semester's college cost on books.

7. 32% of 725 is what number?
Method 1:
$32\% \cdot 725 = x$
$0.32 \cdot 725 = x$
$232 = x$
232 films were rated R.
Method 2:
$\frac{a}{725} = \frac{32}{100}$
$a \cdot 100 = 725 \cdot 32$
$100a = 23,200$
$\frac{100a}{100} = \frac{23,200}{100}$
$a = 232$
232 films were rated R.

9. 160,650 is what percent of 945,000?
Method 1:
$160,650 = x \cdot 945,000$
$\frac{160,650}{945,000} = \frac{x \cdot 945,000}{945,000}$
$0.17 = x$
$17\% = x$
17% of America's restaurants are pizza restaurants.
Method 2:
$\frac{160,650}{945,000} = \frac{p}{100}$
$160,650 \cdot 100 = 945,000 \cdot p$
$16,065,000 = 945,000p$
$\frac{16,065,000}{945,000} = \frac{945,000p}{945,000}$
$17 = p$
17% of America's restaurants are pizza restaurants.

11. 8% of 6200 is what number?
Method 1:
$8\% \cdot 6200 = x$
$0.08 \cdot 6200 = x$
$496 = x$
The decrease in the number of chairs produced is 496. The new number of chairs produced each month is 6200 – 496 = 5704 chairs.
Method 2:
$\frac{a}{6200} = \frac{8}{100}$
$100 \cdot a = 6200 \cdot 8$
$100a = 49,600$
$\frac{100a}{100} = \frac{49,600}{100}$
$a = 496$
The decrease in the number of chairs produced is 496. The new number of chairs produced each month is 6200 – 496 = 5704 chairs.

13. What number is 27% of 66,000?
Method 1:
$x = 27\% \cdot 66,000$
$x = 0.27 \cdot 66,000$
$x = 17,820$
The number of people employed as physician assistants is expected to be 66,000 + 17,820 =83,820.
Method 2:
$\frac{a}{66,000} = \frac{27}{100}$
$a \cdot 100 = 66,000 \cdot 27$
$100a = 1,782,000$
$\frac{100a}{100} = \frac{1,782,000}{100}$
$a = 17,820$
The number of people employed as physician assistants is expected to be 66,000 + 17,820 = 83,820.

Copyright © 2011 Pearson Education, Inc. Publishing as Prentice Hall.

15. 0.4% of 642,000 is what number?
Method 1:
$0.4\% \cdot 642,000 = x$
$0.004 \cdot 642,000 = x$
$2568 = x$
The population of North Dakota in 2007 was 642,000 – 2568 = 639,432.
Method 2:
$\frac{a}{642,000} = \frac{0.4}{100}$
$100 \cdot a = 642,000 \cdot 0.4$
$100a = 256,800$
$\frac{100a}{100} = \frac{256,800}{100}$
$a = 2568$
The population of North Dakota in 2007 was 642,000 – 2568 = 639,432.

17. 41 is what percent of 135?
Method 1:
$41 = x \cdot 135$
$\frac{41}{135} = \frac{x \cdot 135}{135}$
$0.30 \approx x$
$30\% \approx x$
30% of the runs are intermediate.
Method 2:
$\frac{41}{135} = \frac{p}{100}$
$41 \cdot 100 = 135 \cdot p$
$4100 = 135p$
$\frac{4100}{135} = \frac{135p}{135}$
$30 \approx p$
30% of the runs are intermediate.

19. 20 is what percent of 40?
Method 1:
$20 = x \cdot 40$
$\frac{20}{40} = \frac{40x}{40}$
$0.5 = x$
$50\% = x$
50% of the total calories come from fat.
Method 2:
$\frac{20}{40} = \frac{p}{100}$
$20 \cdot 100 = 40 \cdot p$
$2000 = 40p$
$\frac{2000}{40} = \frac{40p}{40}$
$50 = p$
50% of the total calories come from fat.

21. 10 is what percent of 80?
Method 1:
$10 = x \cdot 80$
$\frac{10}{80} = \frac{x \cdot 80}{80}$
$0.125 = x$
$12.5\% = x$
12.5% of the total calories come from fat.
Method 2:
$\frac{10}{80} = \frac{p}{100}$
$10 \cdot 100 = 80 \cdot p$
$1000 = 80p$
$\frac{1000}{80} = \frac{80p}{80}$
$12.5 = p$
12.5% of the total calories come from fat.

23. 35 is what percent of 120?
Method 1:
$35 = x \cdot 120$
$\frac{35}{120} = \frac{120x}{120}$
$0.292 \approx x$
$29.2\% \approx x$
29.2% of the total calories come from fat.
Method 2:
$\frac{35}{120} = \frac{p}{100}$
$35 \cdot 100 = 120 \cdot p$
$3500 = 120p$
$\frac{3500}{120} = \frac{120p}{120}$
$29.2 \approx p$
29.2% of the total calories come from fat.

Copyright © 2011 Pearson Education, Inc. Publishing as Prentice Hall.

25. 26,250 is 15% of what number?
Method 1:
$$26,250 = 15\% \cdot x$$
$$26,250 = 0.15 \cdot x$$
$$\frac{26,250}{0.15} = \frac{0.15 \cdot x}{0.15}$$
$$175,000 = x$$
The price of the home was $175,000.
Method 2:
$$\frac{26,250}{b} = \frac{15}{100}$$
$$26,250 \cdot 100 = b \cdot 15$$
$$2,625,000 = 15b$$
$$\frac{2,625,000}{15} = \frac{15b}{15}$$
$$175,000 = b$$
The price of the home was $175,000.

27. What number is 78% of 40?
Method 1:
$$x = 78\% \cdot 40$$
$$x = 0.78 \cdot 40$$
$$x = 31.2$$
The owner can bill 31.2 hours each week for a repairman.
Method 2:
$$\frac{a}{40} = \frac{78}{100}$$
$$a \cdot 100 = 40 \cdot 78$$
$$100a = 3120$$
$$\frac{100a}{100} = \frac{3120}{100}$$
$$a = 31.2$$
The owner can bill 31.2 hours each week for a repairman.

29. What number is 4.5% of 19,286?
Method 1:
$$x = 4.5\% \cdot 19,286$$
$$x = 0.045 \cdot 19,286$$
$$x = 867.87$$
The price of the car will increase by $867.87. The new price of that model will be $19,286 + $867.87 = $20,153.87.
Method 2:
$$\frac{a}{19,286} = \frac{4.5}{100}$$
$$a \cdot 100 = 19,286 \cdot 4.5$$
$$100a = 86,787$$
$$\frac{100a}{100} = \frac{86,787}{100}$$
$$a = 867.87$$
The price of the car will increase by $867.87. The new price of that model will be $19,286 + $867.87 = $20,153.87.

31. 21 is 60% of what number?
Method 1:
$$21 = 60\% \cdot x$$
$$21 = 0.60x$$
$$\frac{21}{0.60} = \frac{0.60x}{0.60}$$
$$35 = x$$
60% of the tower's total height is 35 feet.
Method 2:
$$\frac{21}{b} = \frac{60}{100}$$
$$21 \cdot 100 = b \cdot 60$$
$$2100 = 60b$$
$$\frac{2100}{60} = \frac{60b}{60}$$
$$35 = b$$
60% of the tower's total height is 35 feet.

33. 82.3% of 4761 is what number?
Method 1:
$$82.3\% \cdot 4761 = x$$
$$0.823 \cdot 4761 = x$$
$$3918 \approx x$$
4761 + 3918 = 8679
The increase is $3918 and the tuition in 2007–2008 was $8679.
Method 2:
$$\frac{a}{4761} = \frac{82.3}{100}$$
$$a \cdot 100 = 4761 \cdot 82.3$$
$$100a = 391,830.3$$
$$\frac{100a}{100} = \frac{391,830.3}{100}$$
$$a \approx 3918$$
4761 + 3918 = 8679
The increase is $3918 and the tuition in 2007–2008 was $8679.

Copyright © 2011 Pearson Education, Inc. Publishing as Prentice Hall.

35. What number is 5.7% of 731,000?

Method 1:

$x = 5.7\% \cdot 731{,}000$

$x = 0.057 \cdot 731{,}000$

$x = 41{,}667$

The increase is projected to be 41,667. The number of associate degrees awarded in 2017–2018 is projected to be 731,000 + 41,667 = 772,667.

Method 2:

$$\frac{a}{731{,}000} = \frac{5.7}{100}$$

$$a \cdot 100 = 731{,}000 \cdot 5.7$$

$$100a = 4{,}166{,}700$$

$$\frac{100a}{100} = \frac{4{,}166{,}700}{100}$$

$$a = 41{,}667$$

The increase is projected to be 41,667. The number of associate degrees awarded in 2017–2018 is projected to be 731,000 + 41,667 = 772,667.

	Original Amount	New Amount	Amount of Increase	Percent Increase
37.	50	80	80 − 50 = 30	$\frac{30}{50} = 0.6 = 60\%$
39.	65	117	117 − 65 = 52	$\frac{52}{65} = 0.8 = 80\%$

	Original Amount	New Amount	Amount of Decrease	Percent Decrease
41.	8	6	8 − 6 = 2	$\frac{2}{8} = 0.25 = 25\%$
43.	160	40	160 − 40 = 120	$\frac{120}{160} = 0.75 = 75\%$

45.
$$\text{percent decrease} = \frac{\text{amount of decrease}}{\text{original amount}} = \frac{150-84}{150} = \frac{66}{150} = 0.44$$

The decrease in calories is 44%.

Copyright © 2011 Pearson Education, Inc. Publishing as Prentice Hall.

47. $\text{percent decrease} = \dfrac{\text{amount of decrease}}{\text{original amount}}$
$= \dfrac{10,845 - 10,700}{10,845}$
$= \dfrac{145}{10,845}$
≈ 0.013
The decrease in cable TV systems was 1.3%.

49. $\text{percent increase} = \dfrac{\text{amount of increase}}{\text{original amount}}$
$= \dfrac{449 - 174}{174}$
$= \dfrac{275}{174}$
≈ 1.580
The increase in acres was 158%

51. $\text{percent decrease} = \dfrac{\text{amount of decrease}}{\text{original amount}}$
$= \dfrac{21.50 - 14.88}{21.50}$
$= \dfrac{6.62}{21.50}$
≈ 0.308
The decrease in the list price is 30.8%.

53. $\text{percent increase} = \dfrac{\text{amount of increase}}{\text{original amount}}$
$= \dfrac{3769 - 3570}{3570}$
$= \dfrac{199}{3570}$
≈ 0.056
The increase in elementary and secondary teachers is expected to be 5.6%.

55. $\text{percent decrease} = \dfrac{\text{amount of decrease}}{\text{original amount}}$
$= \dfrac{6903 - 5545}{6903}$
$= \dfrac{1358}{6903}$
≈ 0.197
The decrease in cinema sites was 19.7%

57. $\text{percent increase} = \dfrac{\text{amount of increase}}{\text{original amount}}$
$= \dfrac{19.9 - 13.1}{13.1}$
$= \dfrac{6.8}{13.1}$
≈ 0.519
The increase in soft drink size was 51.9%.

59. $\text{percent increase} = \dfrac{\text{amount of increase}}{\text{original amount}}$
$= \dfrac{220,472 - 178,025}{178,025}$
$= \dfrac{42,447}{178,025}$
≈ 0.24
The increase in cell sites was 24%.

61.
$$\begin{array}{r} 0.12 \\ \times\ 38 \\ \hline 96 \\ 360 \\ \hline 4.56 \end{array}$$

63.
$$\begin{array}{r} {}^{1\,1} \\ 9.20 \\ +\ 1.98 \\ \hline 11.18 \end{array}$$

65. $-\dfrac{3}{8} + \dfrac{5}{12} = -\dfrac{3}{8} \cdot \dfrac{3}{3} + \dfrac{5}{12} \cdot \dfrac{2}{2}$
$= -\dfrac{9}{24} + \dfrac{10}{24}$
$= \dfrac{1}{24}$

67. $2\dfrac{4}{5} \div 3\dfrac{9}{10} = \dfrac{14}{5} \div \dfrac{39}{10}$
$= \dfrac{14}{5} \cdot \dfrac{10}{39}$
$= \dfrac{14 \cdot 10}{5 \cdot 39}$
$= \dfrac{14 \cdot 2 \cdot 5}{5 \cdot 39}$
$= \dfrac{28}{39}$

69. The increased number is double the original number.

Copyright © 2011 Pearson Education, Inc. Publishing as Prentice Hall.

71. To find the percent increase, she should have divided by the original amount, which is 150.

$$\text{percent increase} = \frac{30}{150} = 0.2 = 20\%$$

73. False; the percents are different.
percent increase from 1980 to 1990 = 20%

$$\text{percent decrease from 1990 to 2000} = 16\frac{2}{3}\%$$

Section 7.5

Practice Problems

1. sales tax = tax rate · purchase price

$$\begin{aligned} &= 8.5\% \cdot \$59.90 \\ &= 0.085 \cdot \$59.90 \\ &\approx \$5.09 \end{aligned}$$

The sales tax is \$5.09.
Total Price = purchase price + sales tax

$$\begin{aligned} &= \$59.90 + \$5.09 \\ &= \$64.99 \end{aligned}$$

The sales tax on \$59.90 is \$5.09, and the total price is \$64.99.

2. sales tax = tax rate · purchase price

$$\begin{aligned} \$1665 &= r \cdot \$18,500 \\ \frac{1665}{18,500} &= \frac{r \cdot 18,500}{18,500} \\ 0.09 &= r \end{aligned}$$

The sales tax rate is 9%.

3. commission = commission rate · sales

$$\begin{aligned} &= 6.6\% \cdot \$47,632 \\ &= 0.066 \cdot \$47,632 \\ &\approx \$3143.712 \end{aligned}$$

The sales representative's commission for the month is \$3143.71.

4. commission = commission rate · sales

$$\begin{aligned} \$645 &= r \cdot \$4300 \\ \frac{645}{4300} &= r \\ 0.15 &= r \\ 15\% &= r \end{aligned}$$

The commission rate is 15%.

5. amount of discount
= discount rate · original price

$$\begin{aligned} &= 35\% \cdot \$700 \\ &= 0.35 \cdot \$700 \\ &= \$245 \end{aligned}$$

The discount is \$245.
sale price = original price − discount

$$\begin{aligned} &= \$700 - \$245 \\ &= \$455 \end{aligned}$$

The sale price is \$455.

Vocabulary and Readiness Check

1. sales tax = tax rate · purchase price.

2. total price = purchase price + sales tax.

3. commission = commission rate · sales.

4. amount of discount
= discount rate · original price.

5. sale price = original price − amount of discount.

Exercise Set 7.5

1. sales tax = 5% · \$150 = 0.05 · \$150 = \$7.50
The sales tax is \$7.50.

3. sales tax = 7.5% · \$799 = 0.075 · \$799 ≈ \$59.93
total price = \$799 + \$59.93 = \$858.93
The total price of the camcorder is \$858.93.

5. $\$335.30 = r \cdot \$4790$

$$\begin{aligned} \frac{335.30}{4790} &= r \\ 0.07 &= r \end{aligned}$$

The sales tax rate is 7%.

7. a. $\$10.20 = 8.5\% \cdot p$

$$\begin{aligned} \$10.2 &= 0.085p \\ \frac{\$10.2}{0.085} &= p \\ \$120 &= p \end{aligned}$$

The purchase price of the table saw is \$120.

b. total price = \$120 + \$10.20 = \$130.20
The total price of the table saw is \$130.20.

Copyright © 2011 Pearson Education, Inc. Publishing as Prentice Hall.

9. sales tax $= 6.5\% \cdot \$1800 = 0.065 \cdot \$1800 = \$117$
total price = \$1800 + \$117 = \$1917
The sales tax was \$117 and the total price of the bracelet is \$1917.

11. $\$24.25 = 5\% \cdot p$
$\$24.25 = 0.05p$
$\frac{\$24.25}{0.05} = p$
$\$485 = p$
The purchase price of the futon is \$485.

13. $\$98.70 = r \cdot \$1645$
$\frac{98.70}{1645} = r$
$0.06 = r$
The sales tax rate is 6%.

15. Purchase price = \$210 + \$15 + \$5 = \$230
Sales tax = 7% · \$230 = 0.07 · \$230 = \$16.10
Total price = \$230 + \$16.10 = \$246.10
The sales tax is \$16.10 and the total price of the items is \$246.10.

17. $\text{commission} = 4\% \cdot \$1,329,401$
$= 0.04 \cdot \$1,329,401$
$= \$53,176.04$
Her commission was \$53,176.04.

19. $\$1380.40 = r \cdot \$9860$
$\frac{\$1380.40}{\$9860} = r$
$0.14 = r$
The commission rate is 14%.

21. $\text{commission} = 1.5\% \cdot \$325,900$
$= 0.015 \cdot \$325,900$
$= \$4888.50$
His commission will be \$4888.50.

23. $\$5565 = 3\% \cdot \text{sales}$
$\$5565 = 0.03 \cdot s$
$\frac{\$5565}{0.03} = s$
$\$185,500 = s$
The selling price of the house was \$185,500.

	Original Price	Discount Rate	Amount of Discount	Sale Price
25.	\$89	10%	10% · \$89 = \$8.90	\$89 – \$8.90 = \$80.10
27.	\$196.50	50%	50% · \$196.50 = \$98.25	\$196.50 – \$98.25 = \$98.25
29.	\$410	35%	35% · \$410 = \$143.50	\$410 – \$143.50 = \$266.50
31.	\$21,700	15%	15% · \$21,700 = \$3255	\$21,700 – \$3255 = \$18,445

Copyright © 2011 Pearson Education, Inc. Publishing as Prentice Hall.

33. discount = 15% · \$300 = 0.15 · \$300 = \$45
sale price = \$300 − \$45 = \$255
The discount is \$45 and the sale price is \$255.

	Purchase Price	Tax Rate	Sales Tax	Total Price
35.	\$305	9%	9% · \$305 = \$27.45	\$305 + \$27.45 = \$332.45
37.	\$56	5.5%	5.5% · \$56 = \$3.08	\$56 + \$3.08 = \$59.08

	Sale	Commission Rate	Commission
39.	\$235,800	3%	\$235,800 · 3% = \$7074
41.	\$17,900	$\frac{\$1432}{\$17,900} = 0.08 = 8\%$	\$1432

43. $2000 \cdot \frac{3}{10} \cdot 2 = 600 \cdot 2 = 1200$

45. $400 \cdot \frac{3}{100} \cdot 11 = 12 \cdot 11 = 132$

47. $600 \cdot 0.04 \cdot \frac{2}{3} = 24 \cdot \frac{2}{3} = 16$

49. Round \$68 to \$70 and 9.5% to 10%.
10% · \$70 = 0.10 · \$70 = \$7
\$70 + \$7 = \$77
The best estimate of the total price is \$77; d.

	Bill Amount	10%	15%	20%
51.	$\$40.21 \approx \$40$	\$4.00	$\$4 + \frac{1}{2}(\$4) = \$4 + \$2 = \$6.00$	2(\$4) = \$8.00
53.	$\$72.17 \approx \$72.00$	\$7.20	$\$7.20 + \frac{1}{2}(\$7.20) = \$7.20 + \$3.60 = \$10.80$	2(\$7.20) = \$14.40

55. A discount of 60% is better; answers may vary.

57. 7.5% · \$24,966 = 0.075 · \$24,966 = \$1872.45
\$24,966 + \$1872.45 = \$26,838.45
The total price of the necklace is \$26,838.45.

Copyright © 2011 Pearson Education, Inc. Publishing as Prentice Hall.

Section 7.6

Practice Problems

1. $I = P \cdot R \cdot T$
$I = \$875 \cdot 7\% \cdot 5$
$= \$875 \cdot (0.07) \cdot 5$
$= \$306.25$
The simple interest is \$306.25.

2. $I = P \cdot R \cdot T$
$I = \$1500 \cdot 20\% \cdot \frac{9}{12}$
$= \$1500 \cdot (0.20) \cdot \frac{9}{12}$
$= \$225$
She paid \$225 in interest.

3. $I = P \cdot R \cdot T$
$= \$2100 \cdot 13\% \cdot \frac{6}{12}$
$= \$2100 \cdot (0.13) \cdot \frac{6}{12}$
$= \$136.50$
The interest is \$136.50.
total amount = principal + interest
$= \$2100 + \$136.50$
$= \$2236.50$
After 6 months, the total amount paid will be \$2236.50.

4. $P = \$3000$, $r = 4\% = 0.04$, $n = 1$, $t = 6$ years
$A = P\left(1+\frac{r}{n}\right)^{n \cdot t}$
$= \$3000\left(1+\frac{0.04}{1}\right)^{1 \cdot 6}$
$= \$3000(1.04)^6$
$\approx \$3795.96$
The total amount after 6 years is \$3795.96.

5. $P = \$5500,\ r = 6\frac{1}{4}\% = 0.0625,\ n = 365,\ t = 5$
$A = P\left(1+\frac{r}{n}\right)^{n \cdot t}$
$= \$5500\left(1+\frac{0.0625}{365}\right)^{365 \cdot 5}$
$\approx \$7517.41$
The total amount after 5 years is \$7517.41.

Calculator Explorations

1. $A = \$600\left(1+\frac{0.09}{4}\right)^{4 \cdot 5} \approx \$936.31$

2. $A = \$10,000\left(1+\frac{0.04}{365}\right)^{365 \cdot 15} \approx \$18,220.59$

3. $A = \$1200\left(1+\frac{0.11}{1}\right)^{1 \cdot 20} \approx \$9674.77$

4. $A = \$5800\left(1+\frac{0.07}{2}\right)^{2 \cdot 1} \approx \$6213.11$

5. $A = \$500\left(1+\frac{0.06}{4}\right)^{4 \cdot 4} \approx \$634.49$

6. $A = \$2500\left(1+\frac{0.05}{365}\right)^{365 \cdot 19} \approx \$6463.85$

Vocabulary and Readiness Check

1. To calculate simple interest, use $I = P \cdot R \cdot T$.

2. To calculate compound interest, use
$A = P\left(1+\frac{r}{n}\right)^{n \cdot t}$.

3. Compound interest is computed on not only the original principal, but on interest already earned in previous compounding periods.

4. When interest is computed on the original principal only, it is called simple interest.

5. Total amount (paid or received) = principal + interest.

6. The principal amount is the money borrowed, loaned, or invested.

Exercise Set 7.6

1. simple interest = principal · rate · time
$= \$200 \cdot 8\% \cdot 2$
$= \$200 \cdot 0.08 \cdot 2$
$= \$32$

Copyright © 2011 Pearson Education, Inc. Publishing as Prentice Hall.

3. simple interest = principal · rate · time
$$= \$160 \cdot 11.5\% \cdot 4$$
$$= \$160 \cdot 0.115 \cdot 4$$
$$= \$73.60$$

5. simple interest = principal · rate · time
$$= \$5000 \cdot 10\% \cdot 1\frac{1}{2}$$
$$= \$5000 \cdot 0.10 \cdot 1.5$$
$$= \$750$$

7. simple interest = principal · rate · time
$$= \$375 \cdot 18\% \cdot \frac{6}{12}$$
$$= \$375 \cdot 0.18 \cdot 0.5$$
$$= \$33.75$$

9. simple interest = principal · rate · time
$$= \$2500 \cdot 16\% \cdot \frac{21}{12}$$
$$= \$2500 \cdot 0.16 \cdot 1.75$$
$$= \$700$$

11. simple interest = principal · rate · time
$$= \$162{,}500 \cdot 12.5\% \cdot 5$$
$$= \$162{,}500 \cdot 0.125 \cdot 5$$
$$= \$101{,}562.50$$
$162,500 + $101,562.50 = $264,062.50
The amount of interest is $101,562.50.
The total amount paid back is $264,062.50.

13. simple interest = principal · rate · time
$$= \$5000 \cdot 9\% \cdot \frac{15}{12}$$
$$= \$5000 \cdot 0.09 \cdot 1.25$$
$$= \$562.50$$
Total = $5000 + $562.50 = $5562.50

15. Simple interest = principal · rate · time
$$= \$8500 \cdot 17\% \cdot 4$$
$$= \$8500 \cdot 0.17 \cdot 4$$
$$= \$5780$$
Total amount = $8500 + $5780 = $14,280

17. $A = P\left(1+\frac{r}{n}\right)^{n \cdot t}$
$$= 6150\left(1+\frac{0.14}{2}\right)^{2 \cdot 15}$$
$$= 6150(1.07)^{30}$$
$$\approx 46{,}815.37$$
The total amount is $46,815.37.

19. $A = P\left(1+\frac{r}{n}\right)^{n \cdot t}$
$$= 1560\left(1+\frac{0.08}{365}\right)^{365 \cdot 5}$$
$$= 1560\left(1+\frac{0.08}{365}\right)^{1825}$$
$$\approx 2327.14$$
The total amount is $2327.14.

21. $A = P\left(1+\frac{r}{n}\right)^{n \cdot t}$
$$= 10{,}000\left(1+\frac{0.09}{2}\right)^{2 \cdot 20}$$
$$= 10{,}000(1.045)^{40}$$
$$\approx 58{,}163.65$$
The total amount is $58,163.65.

23. $A = P\left(1+\frac{r}{n}\right)^{n \cdot t}$
$$= 2675\left(1+\frac{0.09}{1}\right)^{1 \cdot 1}$$
$$= 2675(1.09)$$
$$= 2915.75$$
The total amount is $2915.75.

25. $A = P\left(1+\frac{r}{n}\right)^{n \cdot t}$
$$= 2000\left(1+\frac{0.08}{1}\right)^{1 \cdot 5}$$
$$= 2000(1.08)^5$$
$$\approx 2938.66$$
The total amount is $2938.66.

27. $A = P\left(1+\frac{r}{n}\right)^{n \cdot t}$
$$= 2000\left(1+\frac{0.08}{4}\right)^{4 \cdot 5}$$
$$= 2000(1.02)^{20}$$
$$\approx 2971.89$$
The total amount is $2971.89.

29. perimeter = 10 + 6 + 10 + 6 = 32
The perimeter is 32 yards.

31. Perimeter = 7 + 7 + 7 + 7 + 7 = 35
The perimeter is 35 meters.

Copyright © 2011 Pearson Education, Inc. Publishing as Prentice Hall.

33. $\frac{x}{4}+\frac{x}{5}=\frac{x}{4}\cdot\frac{5}{5}+\frac{x}{5}\cdot\frac{4}{4}=\frac{5x}{20}+\frac{4x}{20}=\frac{9x}{20}$

35. $$\begin{aligned}\left(\frac{2}{3}\right)\left(-\frac{1}{3}\right)-\left(\frac{9}{10}\right)\left(\frac{2}{5}\right)&=-\frac{2}{9}-\frac{18}{50}\\&=-\frac{2}{9}\cdot\frac{50}{50}-\frac{18}{50}\cdot\frac{9}{9}\\&=-\frac{100}{450}-\frac{162}{450}\\&=-\frac{262}{450}\\&=-\frac{131}{225}\end{aligned}$$

37. answers may vary

39. answers may vary

Chapter 7 Vocabulary Check

1. In a mathematical statement, of usually means "multiplication."
2. In a mathematical statement, is means "equals."
3. Percent means "per hundred."
4. Compound interest is computed not only on the principal, but also on interest already earned in previous compounding periods.
5. In the percent proportion $\frac{\text{amount}}{\text{base}}=\frac{\text{percent}}{100}$.
6. To write a decimal or fraction as a percent, multiply by 100%.
7. The decimal equivalent of the % symbol is 0.01.
8. The fraction equivalent of the % symbol is $\frac{1}{100}$.
9. The percent equation is base · percent = amount.
10. Percent of decrease = $\frac{\text{amount of decrease}}{\text{original amount}}$.
11. Percent of increase = $\frac{\text{amount of increase}}{\text{original amount}}$.
12. Sales tax = tax rate · purchase price.
13. Total price = purchase price + sales tax.
14. Commission = commission rate · sales.
15. Amount of discount = discount rate · original price.
16. Sale price = original price − amount of discount.

Chapter 7 Review

1. $\frac{37}{100}=37\%$
37% of adults preferred pepperoni.
2. $\frac{77}{100}=77\%$
77% of free throws were made.
3. $26\%=26(0.01)=0.26$
4. $75\%=75(0.01)=0.75$
5. $3.5\%=3.5(0.01)=0.035$
6. $1.5\%=1.5(0.01)=0.015$
7. $275\%=275(0.01)=2.75$
8. $400\%=400(0.01)=4.00$ or 4
9. $47.85\%=47.85(0.01)=0.4785$
10. $85.34\%=85.34(0.01)=0.8534$
11. $1.6=1.6(100\%)=160\%$
12. $0.055=0.055(100\%)=5.5\%$
13. $0.076=0.076(100\%)=7.6\%$
14. $0.085=0.085(100\%)=8.5\%$
15. $0.71=0.71(100\%)=71\%$
16. $0.65=0.65(100\%)=65\%$
17. $6=6(100)\%=600\%$
18. $9=9(100\%)=900\%$
19. $7\%=7\left(\frac{1}{100}\right)=\frac{7}{100}$
20. $15\%=15\left(\frac{1}{100}\right)=\frac{15}{100}=\frac{3}{20}$

Copyright © 2011 Pearson Education, Inc. Publishing as Prentice Hall.

21. $25\% = 25\left(\frac{1}{100}\right) = \frac{25}{100} = \frac{1}{4}$

22. $$\begin{aligned} 8.5\% &= 8.5\left(\frac{1}{100}\right) \\ &= \frac{8.5}{100} \\ &= \frac{8.5 \cdot 10}{100 \cdot 10} \\ &= \frac{85}{1000} \\ &= \frac{17}{200} \end{aligned}$$

23. $$\begin{aligned} 10.2\% &= 10.2\left(\frac{1}{100}\right) \\ &= \frac{10.2}{100} \\ &= \frac{10.2 \cdot 10}{100 \cdot 10} \\ &= \frac{102}{1000} \\ &= \frac{51}{500} \end{aligned}$$

24. $16\frac{2}{3}\% = \frac{50}{3}\% = \frac{50}{3}\left(\frac{1}{100}\right) = \frac{50}{300} = \frac{1}{6}$

25. $33\frac{1}{3}\% = \frac{100}{3}\% = \frac{100}{3}\left(\frac{1}{100}\right) = \frac{100}{300} = \frac{1}{3}$

26. $110\% = 110\left(\frac{1}{100}\right) = \frac{110}{100} = 1\frac{10}{100} = 1\frac{1}{10}$

27. $\frac{2}{5} = \frac{2}{5} \cdot \frac{100}{1}\% = \frac{200}{5}\% = 40\%$

28. $\frac{7}{10} = \frac{7}{10} \cdot \frac{100}{1}\% = \frac{700}{10}\% = 70\%$

29. $\frac{7}{12} = \frac{7}{12} \cdot \frac{100}{1}\% = \frac{700}{12}\% = \frac{175}{3}\% = 58\frac{1}{3}\%$

30. $1\frac{2}{3} = \frac{5}{3} \cdot \frac{100}{1}\% = \frac{500}{3}\% = 166\frac{2}{3}\%$

31. $1\frac{1}{4} = \frac{5}{4} \cdot \frac{100}{1}\% = \frac{500}{4}\% = 125\%$

32. $\frac{3}{5} = \frac{3}{5} \cdot \frac{100}{1}\% = \frac{300}{5}\% = 60\%$

33. $\frac{1}{16} = \frac{1}{16} \cdot \frac{100}{1}\% = \frac{100}{16}\% = 6.25\%$

34. $\frac{5}{8} = \frac{5}{8} \cdot \frac{100}{1}\% = \frac{500}{8}\% = 62.5\%$

35. $$\begin{aligned} 1250 &= 1.25\% \cdot x \\ 1250 &= 0.0125x \\ \frac{1250}{0.0125} &= \frac{0.0125x}{0.0125} \\ 100,000 &= x \end{aligned}$$
1250 is 1.25% of 100,000.

36. $$\begin{aligned} x &= 33\frac{1}{3}\% \cdot 24,000 \\ x &= \frac{100}{3} \cdot \frac{1}{100} \cdot 24,000 \\ x &= \frac{1}{3} \cdot 24,000 \\ x &= 8000 \end{aligned}$$
8000 is $33\frac{1}{3}\%$ of 24,000.

37. $$\begin{aligned} 124.2 &= x \cdot 540 \\ \frac{124.2}{540} &= \frac{540x}{540} \\ 0.23 &= x \\ 23\% &= x \end{aligned}$$
124.2 is 23% of 540.

38. $$\begin{aligned} 22.9 &= 20\% \cdot x \\ 22.9 &= 0.20 \cdot x \\ \frac{22.9}{0.20} &= \frac{0.20x}{0.20} \\ 114.5 &= x \end{aligned}$$
22.9 is 20% of 114.5.

39. $$\begin{aligned} x &= 17\% \cdot 640 \\ x &= 0.17 \cdot 640 \\ x &= 108.8 \end{aligned}$$
108.8 is 17% of 640.

40. $$\begin{aligned} 693 &= x \cdot 462 \\ \frac{693}{462} &= \frac{462x}{462} \\ 1.5 &= x \\ 150\% &= x \end{aligned}$$
693 is 150% of 462.

Copyright © 2011 Pearson Education, Inc. Publishing as Prentice Hall.

41. $\frac{104.5}{b} = \frac{25}{100}$
$104.5 \cdot 100 = b \cdot 25$
$10,450 = 25b$
$\frac{10,450}{25} = \frac{25b}{25}$
$418 = b$
104.5 is 25% of 418.

42. $\frac{16.5}{b} = \frac{5.5}{100}$
$16.5 \cdot 100 = b \cdot 5.5$
$1650 = 5.5b$
$\frac{1650}{5.5} = \frac{5.5b}{5.5}$
$300 = b$
16.5 is 5.5% of 300.

43. $\frac{a}{532} = \frac{30}{100}$
$a \cdot 100 = 532 \cdot 30$
$100a = 15,960$
$\frac{100a}{100} = \frac{15,960}{100}$
$a = 159.6$
159.6 is 30% of 532.

44. $\frac{63}{35} = \frac{p}{100}$
$63 \cdot 100 = 35 \cdot p$
$6300 = 35p$
$\frac{6300}{35} = \frac{35p}{35}$
$180 = p$
63 is 180% of 35.

45. $\frac{93.5}{85} = \frac{p}{100}$
$93.5 \cdot 100 = 85 \cdot p$
$9350 = 85p$
$\frac{9350}{85} = \frac{85p}{85}$
$110 = p$
93.5 is 110% of 85.

46. $\frac{a}{500} = \frac{33}{100}$
$a \cdot 100 = 500 \cdot 33$
$100a = 16,500$
$\frac{100a}{100} = \frac{16,500}{100}$
$a = 165$
165 is 33% of 500.

47. 1320 is what percent of 2000?
Method 1:
$1320 = x \cdot 2000$
$\frac{1320}{2000} = \frac{2000x}{2000}$
$0.66 = x$
$66\% = x$
66% of people own microwaves.
Method 2:
$\frac{1320}{2000} = \frac{p}{100}$
$1320 \cdot 100 = 2000 \cdot p$
$132,000 = 2000p$
$\frac{132,000}{2000} = \frac{2000p}{2000}$
$66 = p$
66% of people own microwaves.

48. 2000 is what percent of 12,360?
Method 1:
$2000 = x \cdot 12,360$
$\frac{2000}{12,360} = \frac{12,360x}{12,360}$
$0.16 \approx x$
$16\% \approx x$
16% of freshmen are enrolled in prealgebra.
Method 2:
$\frac{200}{12,360} = \frac{p}{100}$
$2000 \cdot 100 = 12,360 \cdot p$
$200,000 = 12,360p$
$\frac{200,000}{12,360} = \frac{12,360p}{12,360}$
$16 \approx p$
16% of freshman are enrolled in prealgebra.

Copyright © 2011 Pearson Education, Inc. Publishing as Prentice Hall.

49. $\text{percent decrease} = \dfrac{\text{amount of decrease}}{\text{original amount}}$
$= \dfrac{675-534}{675}$
$= \dfrac{141}{675}$
≈ 0.209
$\approx 20.9\%$

Violent crime decreased 20.9%.

50. $\text{percent increase} = \dfrac{\text{amount of increase}}{\text{original amount}}$
$= \dfrac{33-16}{16}$
$= \dfrac{17}{16}$
$= 1.0625$
$= 106.25\%$

The charge will increase 106.25%.

51. Amount of increase
$= \text{percent increase} \cdot \text{original amount}$
$= 15\% \cdot \$11.50$
$= 0.15 \cdot \$11.50$
$\approx \$1.73$

Total amount = \$11.50 + \$1.73 = \$13.23

The new hourly rate is \$13.23.

52. Amount of decrease
$= \text{percent decrease} \cdot \text{original amount}$
$= 4\% \cdot \$215{,}000$
$= 0.04 \cdot \$215{,}000$
$= \$8600$

Total amount = \$215,000 – \$8600 = \$206,400

\$206,400 is expected to be collected next year.

53. $\text{sales tax} = \text{tax rate} \cdot \text{purchase price}$
$= 5.5\% \cdot \$250$
$= 0.055 \cdot \$250$
$= \$13.75$

Total amount = \$250 + \$13.75 = \$263.75

The total price for the coat is \$263.75.

54. $\text{sales tax} = \text{tax rate} \cdot \text{purchase price}$
$= 4.5\% \cdot \$25.50$
$= 0.045 \cdot \$25.50$
$\approx \$1.15$

The sales tax is \$1.15.

55. $\text{commission} = \text{commission rate} \cdot \text{sales}$
$= 5\% \cdot \$100{,}000$
$= 0.05 \cdot \$100{,}000$
$= \$5000$

His commission is \$5000.

56. $\text{commission} = \text{commission rate} \cdot \text{sales}$
$= 7.5\% \cdot \$4005$
$= 0.075 \cdot \$4005$
$\approx \$300.38$

Her commission is \$300.38.

57. $\text{Amount of discount} = \text{discount} \cdot \text{original price}$
$= 30\% \cdot \$3000$
$= 0.30 \cdot \$3000$
$= \$900$

Sale price = \$3000 – \$900 = \$2100

The amount of discount is \$900; the sale price is \$2100.

58. $\text{Amount of discount} = \text{discount} \cdot \text{original price}$
$= 10\% \cdot \$90$
$= 0.10 \cdot \$90$
$= \$9.00$

Sale price = \$90 – \$9 = \$81

The amount of discount is \$9; the sale price is \$81.

59. $I = P \cdot R \cdot T$
$= \$4000 \cdot 12\% \cdot \dfrac{4}{12}$
$= \$4000 \cdot 0.12 \cdot \dfrac{1}{3}$
$= \$160$

The simple interest is \$160.

60. $I = P \cdot R \cdot T$
$= \$6500 \cdot 20\% \cdot \dfrac{3}{12}$
$= \$6500 \cdot 0.20 \cdot 0.25$
$= \$325$

The simple interest is \$325.

61. $A = P\left(1+\dfrac{r}{n}\right)^{n \cdot t}$
$= 5500\left(1+\dfrac{0.12}{1}\right)^{1 \cdot 15}$
$= 5500(1.12)^{15}$
$\approx 30{,}104.61$

The total amount is \$30,104.61.

Copyright © 2011 Pearson Education, Inc. Publishing as Prentice Hall.

62. $A = P\left(1+\frac{r}{n}\right)^{n\cdot t}$
$= 6000\left(1+\frac{0.11}{2}\right)^{2\cdot 10}$
$= 6000(1.055)^{20}$
$\approx 17,506.54$
The total amount is $17,506.54.

63. $A = P\left(1+\frac{r}{n}\right)^{n\cdot t}$
$= 100\left(1+\frac{0.12}{4}\right)^{4\cdot 5}$
$= 100(1.03)^{20}$
≈ 180.61
The total amount is $180.61.

64. $A = P\left(1+\frac{r}{n}\right)^{n\cdot t}$
$= 1000\left(1+\frac{0.18}{4}\right)^{4\cdot 20}$
$= 1000(1.045)^{80}$
$\approx 33,830.10$
The total amount is $33,830.10.

65. $3.8\% = 3.8(0.01) = 0.038$

66. $124.5\% = 124.5(0.01) = 1.245$

67. $0.54 = 0.54(100\%) = 54\%$

68. $95.2 = 95.2(100\%) = 9520\%$

69. $47\% = 47\left(\frac{1}{100}\right) = \frac{47}{100}$

70. $5.6\% = 5.6\left(\frac{1}{100}\right)$
$= \frac{5.6}{100}$
$= \frac{5.6\cdot 10}{100\cdot 10}$
$= \frac{56}{1000}$
$= \frac{7}{125}$

71. $\frac{1}{8} = \frac{1}{8}\cdot\frac{100}{1}\% = \frac{100}{8}\% = 12\frac{1}{2}\%$ or 12.5%

72. $\frac{6}{5} = \frac{6}{5}\cdot\frac{100}{1}\% = \frac{600}{5}\% = 120\%$

73. $43 = 16\% \cdot x$
$43 = 0.16x$
$\frac{43}{0.16} = \frac{0.16x}{0.16}$
$268.75 = x$
43 is 16% of 268.75.

74. $27.5 = x \cdot 25$
$\frac{27.5}{25} = \frac{25x}{25}$
$1.1 = x$
$110\% = x$
27.5 is 110% of 25.

75. $x = 36\% \cdot 1968$
$x = 0.36 \cdot 1968$
$x = 708.48$
708.48 is 36% of 1968.

76. $67 = x \cdot 50$
$\frac{67}{50} = \frac{50x}{50}$
$1.34 = x$
$134\% = x$
67 is 134% of 50.

77. $\frac{75}{25} = \frac{p}{100}$
$75 \cdot 100 = 25 \cdot p$
$7500 = 25p$
$\frac{7500}{25} = \frac{25p}{25}$
$300 = p$
75 is 300% of 25.

78. $\frac{a}{240} = \frac{16}{100}$
$a \cdot 100 = 240 \cdot 16$
$100a = 3840$
$\frac{100a}{100} = \frac{3840}{100}$
$a = 38.4$
38.4 is 16% of 240.

Copyright © 2011 Pearson Education, Inc. Publishing as Prentice Hall.

79.
$$\frac{28}{b} = \frac{5}{100}$$
$$28 \cdot 100 = b \cdot 5$$
$$2800 = 5b$$
$$\frac{2800}{5} = \frac{5b}{5}$$
$$560 = b$$
28 is 5% of 560.

80.
$$\frac{52}{16} = \frac{p}{100}$$
$$52 \cdot 100 = 16 \cdot p$$
$$5200 = 16p$$
$$\frac{5200}{16} = \frac{16p}{16}$$
$$325 = p$$
52 is 325% of 16.

81. $\frac{78}{300} = 0.26 = 26\%$
26% of the soft drinks have been sold.

82. $\$96,950 \cdot 7\% = \$96,950 \cdot 0.07 = \$6786.50$
The house has lost \$6786.50 in value.

83. Sales tax $= 8.75\% \cdot \$568$
$= 0.0875 \cdot \$568$
$= \$49.70$
Total price = \$568 + \$49.70 = \$617.70
The total price is \$617.70.

84. Amount of discount $= 15\% \cdot \$23.00$
$= 0.15 \cdot \$23.00$
$= \$3.45$

85. commission = commission rate · sales
$$\$1.60 = r \cdot \$12.80$$
$$\frac{\$1.60}{\$12.80} = r$$
$$0.125 = r$$
$$12.5\% = r$$
His rate of commission is 12.5%.

86. Simple interest = principal · rate · time
$= \$1400 \cdot 13\% \cdot \frac{6}{12}$
$= \$1400 \cdot 0.13 \cdot 0.5$
$= \$91$
Total amount = \$1400 + \$91 = \$1491
The total amount is \$1491.

87. Simple interest = principal · rate · time
$= \$5500 \cdot 12.5\% \cdot 9$
$= \$5500 \cdot 0.125 \cdot 9$
$= \$6187.50$
Total amount = \$5500 + \$6187.50 = \$11,687.50
The total amount is \$11,687.50.

Chapter 7 Test

1. $85\% = 85(0.01) = 0.85$

2. $500\% = 500(0.01) = 5$

3. $0.6\% = 0.6(0.01) = 0.006$

4. $0.056 = 0.056(100\%) = 5.6\%$

5. $6.1 = 6.1(100\%) = 610\%$

6. $0.35 = 0.35(100\%) = 35\%$

7. $120\% = 120\left(\frac{1}{100}\right) = \frac{120}{100} = 1\frac{1}{5}$

8. $38.5\% = 38.5\left(\frac{1}{100}\right) = \frac{38.5}{100} = \frac{385}{1000} = \frac{77}{200}$

9. $0.2\% = 0.2\left(\frac{1}{100}\right) = \frac{0.2}{100} = \frac{2}{1000} = \frac{1}{500}$

10. $\frac{11}{20} = \frac{11}{20} \cdot \frac{100}{1}\% = \frac{1100}{20}\% = 55\%$

11. $\frac{3}{8} = \frac{3}{8} \cdot \frac{100}{1}\% = \frac{300}{8}\% = 37.5\%$

12. $1\frac{3}{4} = \frac{7}{4} \cdot \frac{100}{1}\% = \frac{700}{4}\% = 175\%$

13. $\frac{1}{5} = \frac{1}{5} \cdot \frac{100}{1}\% = \frac{100}{5}\% = 20\%$

14. $64\% = 64\left(\frac{1}{100}\right) = \frac{64}{100} = \frac{16}{25}$

15. *Method 1*:
$x = 42\% \cdot 80$
$x = 0.42 \cdot 80$
$x = 33.6$
33.6 is 42% of 80.

Copyright © 2011 Pearson Education, Inc. Publishing as Prentice Hall.

Method 2:

$$\frac{a}{80} = \frac{42}{100}$$
$$a \cdot 100 = 80 \cdot 42$$
$$100a = 3360$$
$$\frac{100a}{100} = \frac{3360}{100}$$
$$a = 33.6$$

33.6 is 42% of 80.

16. *Method 1*:

$$0.6\% \cdot x = 7.5$$
$$0.006x = 7.5$$
$$\frac{0.006x}{0.006} = \frac{7.5}{0.006}$$
$$x = 1250$$

0.6% of 1250 is 7.5.

Method 2:

$$\frac{7.5}{b} = \frac{0.6}{100}$$
$$7.5 \cdot 100 = b \cdot 0.6$$
$$750 = 0.6b$$
$$\frac{750}{0.6} = \frac{0.6b}{0.6}$$
$$1250 = b$$

0.6% of 1250 is 7.5.

17. *Method 1*:

$$567 = x \cdot 756$$
$$\frac{567}{756} = \frac{x \cdot 756}{756}$$
$$0.75 = x$$
$$75\% = x$$

567 is 75% of 756.

Method 2:

$$\frac{567}{756} = \frac{p}{100}$$
$$567 \cdot 100 = 756 \cdot p$$
$$56,700 = 756p$$
$$\frac{56,700}{756} = \frac{756p}{756}$$
$$75 = p$$

567 is 75% of 756.

18. 12% of 320 is what number?

$$12\% \cdot 320 = x$$
$$0.12 \cdot 320 = x$$
$$38.4 = x$$

There are 38.4 pounds of copper.

19. 20% of what number is \$11,350?

$$20\% \cdot x = \$11,350$$
$$0.20x = \$11,350$$
$$\frac{0.2x}{0.2} = \frac{\$11,350}{0.2}$$
$$x = \$56,750$$

The value of the potential crop is \$56,750.

20. tax = 1.25% · \$354 = 0.0125 · \$354 ≈ \$4.43
total amount = \$354 + \$4.43 = \$358.43
The total amount of the stereo is \$358.43.

21.

$$\text{percent increase} = \frac{\text{amount of increase}}{\text{original amount}}$$
$$= \frac{26,460 - 25,200}{25,200}$$
$$= \frac{1260}{25,200}$$
$$= 0.05$$

The increase in population was 5%.

22.

$$\text{Amount of discount} = 15\% \cdot \$120$$
$$= 0.15 \cdot \$120$$
$$= \$18$$

Sale price = \$120 − \$18 = \$102
The amount of the discount is \$18; the sale price is \$102.

23.

$$\text{commission} = 4\% \cdot \$9875$$
$$= 0.04 \cdot \$9875$$
$$= \$395$$

His commission was \$395.

24.

$$\$1.53 = \text{rate} \cdot \$152.99$$
$$\frac{\$1.53}{\$152.99} = r$$
$$0.01 \approx r$$
$$1\% \approx r$$

The sales tax rate is 1%.

25.

$$\text{simple interest} = \text{principal} \cdot \text{rate} \cdot \text{time}$$
$$= \$2000 \cdot 9.25\% \cdot 3\frac{1}{2}$$
$$= \$2000 \cdot 0.0925 \cdot 3.5$$
$$= \$647.5$$

Copyright © 2011 Pearson Education, Inc. Publishing as Prentice Hall.

26. $A = P\left(1+\frac{r}{n}\right)^{n\cdot t}$
$= 1365\left(1+\frac{0.08}{1}\right)^{1\cdot 5}$
$= 1365(1.08)^5$
≈ 2005.63
The total amount of \$2005.63.

27. Simple interest = principal · rate · time
$= \$400 \cdot 13.5\% \cdot \frac{6}{12}$
$= \$400 \cdot 0.135 \cdot 0.5$
$= \$27$
Total amount = \$400 + \$27 = \$427

28. percent decrease $= \frac{\text{amount of decrease}}{\text{original amount}}$
$= \frac{236{,}215 - 198{,}419}{236{,}215}$
$= \frac{37{,}796}{236{,}215}$
≈ 0.160
$\approx 16.0\%$
The number of crimes has decreased 16.0% or 16%.

Cumulative Review Chapters 1–7

1.
$$\begin{array}{r} 236 \\ \times \quad 86 \\ \hline 1\ 416 \\ 18\ 880 \\ \hline 20{,}296 \end{array}$$

2.
$$\begin{array}{r} 409 \\ \times \quad 76 \\ \hline 2\ 454 \\ 28\ 630 \\ \hline 31{,}084 \end{array}$$

3. −3 − 7 = −3 + (−7) = −10

4. 8 − (−2) = 8 + 2 = 10

5.
$x - 2 = -1$
$x - 2 + 2 = -1 + 2$
$x = 1$

6.
$x + 4 = 3$
$x + 4 - 4 = 3 - 4$
$x = -1$

7.
$3(2x - 6) + 6 = 0$
$3\cdot 2x - 3\cdot 6 + 6 = 0$
$6x - 18 + 6 = 0$
$6x - 12 = 0$
$6x - 12 + 12 = 0 + 12$
$6x = 12$
$\frac{6x}{6} = \frac{12}{6}$
$x = 2$

8.
$5(x - 2) = 3x$
$5\cdot x - 5\cdot 2 = 3x$
$5x - 10 = 3x$
$5x - 5x - 10 = 3x - 5x$
$-10 = -2x$
$\frac{-10}{-2} = \frac{-2x}{-2}$
$5 = x$

9. $3 = \frac{3}{1}\cdot\frac{7}{7} = \frac{3\cdot 7}{1\cdot 7} = \frac{21}{7}$

10. $8 = \frac{8}{1}\cdot\frac{5}{5} = \frac{8\cdot 5}{1\cdot 5} = \frac{40}{5}$

11. $-\frac{10}{27} = -\frac{2\cdot 5}{3\cdot 3\cdot 3}$

Since 10 and 27 have no common factors, $-\frac{10}{27}$ is already in simplest form.

12. $\frac{10y}{32} = \frac{5\cdot 2\cdot y}{16\cdot 2} = \frac{5y}{16}$

13. $-\frac{7}{12} \div -\frac{5}{6} = -\frac{7}{12}\cdot -\frac{6}{5} = \frac{7\cdot 6}{6\cdot 2\cdot 5} = \frac{7}{10}$

14. $\frac{-2}{5} \div \frac{7}{10} = -\frac{2}{5}\cdot\frac{10}{7} = -\frac{2\cdot 2\cdot 5}{5\cdot 7} = -\frac{4}{7}$

15. $y - x = -\frac{8}{10} - \left(-\frac{3}{10}\right) = -\frac{8}{10} + \frac{3}{10} = -\frac{5}{10} = -\frac{1}{2}$

16. $2x + 3y = 2\left(\frac{2}{5}\right) + 3\left(-\frac{1}{5}\right) = \frac{4}{5} + \left(-\frac{3}{5}\right) = \frac{1}{5}$

Copyright © 2011 Pearson Education, Inc. Publishing as Prentice Hall.

17. $-\frac{3}{4}-\frac{1}{14}+\frac{6}{7}=-\frac{3}{4}\cdot\frac{7}{7}-\frac{1}{14}\cdot\frac{2}{2}+\frac{6}{7}\cdot\frac{4}{4}$
$=-\frac{21}{28}-\frac{2}{28}+\frac{24}{28}$
$=\frac{1}{28}$

18. $\frac{2}{9}+\frac{7}{15}-\frac{1}{3}=\frac{2}{9}\cdot\frac{5}{5}+\frac{7}{15}\cdot\frac{3}{3}-\frac{1}{3}\cdot\frac{15}{15}$
$=\frac{10}{45}+\frac{21}{45}-\frac{15}{45}$
$=\frac{16}{45}$

19. $\frac{\frac{1}{2}+\frac{3}{8}}{\frac{3}{4}-\frac{1}{6}}=\frac{\frac{1}{2}\cdot\frac{4}{4}+\frac{3}{8}}{\frac{3}{4}\cdot\frac{3}{3}-\frac{1}{6}\cdot\frac{2}{2}}$
$=\frac{\frac{4}{8}+\frac{3}{8}}{\frac{9}{12}-\frac{2}{12}}$
$=\frac{\frac{7}{8}}{\frac{7}{12}}$
$=\frac{7}{8}\div\frac{7}{12}$
$=\frac{7}{8}\cdot\frac{12}{7}$
$=\frac{7\cdot 4\cdot 3}{4\cdot 2\cdot 7}$
$=\frac{3}{2}$

20. $\frac{\frac{2}{3}+\frac{1}{6}}{\frac{3}{4}-\frac{3}{5}}=\frac{\frac{2}{3}\cdot\frac{2}{2}+\frac{1}{6}}{\frac{3}{4}\cdot\frac{5}{5}-\frac{3}{5}\cdot\frac{4}{4}}$
$=\frac{\frac{4}{6}+\frac{1}{6}}{\frac{15}{20}-\frac{12}{20}}$
$=\frac{\frac{5}{6}}{\frac{3}{20}}$
$=\frac{5}{6}\div\frac{3}{20}$
$=\frac{5}{6}\cdot\frac{20}{3}$
$=\frac{5\cdot 2\cdot 10}{2\cdot 3\cdot 3}$
$=\frac{50}{9}$

21. $\frac{x}{2}=\frac{x}{3}+\frac{1}{2}$
$6\left(\frac{x}{2}\right)=6\left(\frac{x}{3}+\frac{1}{2}\right)$
$3x=6\cdot\frac{x}{3}+6\cdot\frac{1}{2}$
$3x=2x+3$
$3x-2x=2x+3-2x$
$x=3$

22. $\frac{x}{2}+\frac{1}{5}=3-\frac{x}{5}$
$10\left(\frac{x}{2}+\frac{1}{5}\right)=10\left(3-\frac{x}{5}\right)$
$10\cdot\frac{x}{2}+10\cdot\frac{1}{5}=10\cdot 3-10\cdot\frac{x}{5}$
$5x+2=30-2x$
$5x+2+2x=30-2x+2x$
$7x+2=30$
$7x+2-2=30-2$
$7x=28$
$\frac{7x}{7}=\frac{28}{7}$
$x=4$

23. a. $4\frac{2}{9}=\frac{9\cdot 4+2}{9}=\frac{36+2}{9}=\frac{38}{9}$

b. $1\frac{8}{11}=\frac{11\cdot 1+8}{11}=\frac{11+8}{11}=\frac{19}{11}$

24. a. $3\frac{2}{5}=\frac{5\cdot 3+2}{5}=\frac{15+2}{5}=\frac{17}{5}$

b. $6\frac{2}{7}=\frac{7\cdot 6+2}{7}=\frac{42+2}{7}=\frac{44}{7}$

25. $0.125=\frac{125}{1000}=\frac{1}{8}$

26. $0.85=\frac{85}{100}=\frac{17}{20}$

27. $-105.083=-105\frac{83}{1000}$

28. $17.015=17\frac{15}{1000}=17\frac{3}{200}$

Copyright © 2011 Pearson Education, Inc. Publishing as Prentice Hall.

29. $\begin{array}{r} 85.00 \\ -\ 17.31 \\ \hline 67.69 \end{array}$ *Check:* $\begin{array}{r} 67.69 \\ +\ 17.31 \\ \hline 85.00 \end{array}$

30. $\begin{array}{r} 38.00 \\ -\ 10.06 \\ \hline 27.94 \end{array}$ *Check:* $\begin{array}{r} 27.94 \\ +\ 10.06 \\ \hline 38.00 \end{array}$

31. $7.68 \times 10 = 76.8$

32. $12.483 \times 100 = 1248.3$

33. $(-76.3)(1000) = -76{,}300$

34. $-853.75 \times 10 = -8537.5$

35. $x \div y = 2.5 \div 0.05$

$0.05\overline{)2.5}$ becomes $\begin{array}{r} 50 \\ 5\overline{)\ 250} \\ \underline{-25} \\ 00 \end{array}$

36. $\dfrac{x}{100} = 4.75$

$\dfrac{470}{100} \stackrel{?}{=} 4.75$

$4.7 = 4.75$ False

No, 470 is not a solution.

37. 55, 67, 75, 86, 91, 91

$\text{median} = \dfrac{75+86}{2} = \dfrac{161}{2} = 80.5$

38. $\text{mean} = \dfrac{36+40+86+30}{4} = \dfrac{192}{4} = 48$

39. $\dfrac{2.5}{3.15} = \dfrac{2.5 \cdot 100}{3.15 \cdot 100} = \dfrac{250}{315} = \dfrac{50}{63}$

40. $\dfrac{5.8}{7.6} = \dfrac{5.8 \cdot 10}{7.6 \cdot 10} = \dfrac{58}{76} = \dfrac{29}{38}$

41. $\begin{array}{r} 0.21 \\ 16\overline{)\ 3.36} \\ \underline{-32} \\ 16 \\ \underline{-16} \\ 0 \end{array}$

The unit price is \$0.21/ounce.

42. $\begin{array}{r} 2.25 \\ 40\overline{)\ 90.00} \\ \underline{-80} \\ 10\ 0 \\ \underline{-8\ 0} \\ 2\ 00 \\ \underline{-2\ 00} \\ 0 \end{array}$

The price is \$2.25 per tile, or \$2.25 per square foot.

43. $\dfrac{4.1}{7} \stackrel{?}{=} \dfrac{2.9}{5}$

$4.1(5) \stackrel{?}{=} 7(2.9)$

$20.5 \neq 20.3$

No, it is not a true proportion.

44. $\dfrac{6.3}{9} \stackrel{?}{=} \dfrac{3.5}{5}$

$6.3(5) \stackrel{?}{=} 9(3.5)$

$31.5 = 31.5$

Yes, it is a true proportion.

45. $\dfrac{5 \text{ miles}}{2 \text{ inches}} = \dfrac{x \text{ miles}}{7 \text{ inches}}$

$5 \cdot 7 = 2 \cdot x$

$35 = 2x$

$\dfrac{35}{2} = \dfrac{2x}{2}$

$17.5 = x$

17.5 miles corresponds to 7 inches.

46. $\dfrac{7 \text{ problems}}{6 \text{ minutes}} = \dfrac{x \text{ problems}}{30 \text{ minutes}}$

$7 \cdot 30 = 6 \cdot x$

$210 = 6x$

$\dfrac{210}{6} = \dfrac{6x}{6}$

$35 = x$

The student can complete 35 problems in 30 minutes.

47. $1.9\% = 1.9\left(\dfrac{1}{100}\right) = \dfrac{1.9}{100} = \dfrac{19}{1000}$

48. $2.3\% = 2.3\left(\dfrac{1}{100}\right) = \dfrac{2.3}{100} = \dfrac{23}{1000}$

49. $33\frac{1}{3}\% = \dfrac{100}{3}\left(\dfrac{1}{100}\right) = \dfrac{100}{300} = \dfrac{1}{3}$

50. $108\% = 108\left(\dfrac{1}{100}\right) = \dfrac{108}{100} = 1\dfrac{2}{25}$

Copyright © 2011 Pearson Education, Inc. Publishing as Prentice Hall.

Chapter 8

Section 8.1

Practice Problems

1. a. English has 6 symbols and each symbol represents 50 million speakers, so English is spoken by $6(50) = 300$ million people.

b. Portuguese has 3.5 symbols, or $3.5(50) = 175$ million speakers. So, 300 million − 175 million = 125 million more people speak English than Portuguese.

2. a. The height of the bar for birds is 75, so approximately 75 endangered species are birds.

b. The shortest bar corresponds to arachnids, so arachnids have the fewest endangered species.

3.

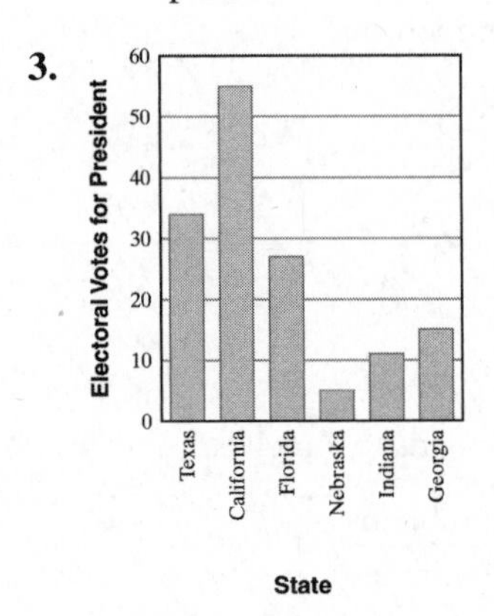

4. The height of the bar for 80–89 is 12, so 12 students scored 80–89 on the test.

5. The height of the bar for 40–49 is 1, for 50–59 is 3, for 60–69 is 2, for 70–79 is 10. So, $1 + 3 + 2 + 10 = 16$ students scored less than 80 on the test.

6.

Class Interval (Credit Card Balances)	Tally	Class Frequency (Number of Months)
\$0–\$49	|||	3
\$50–\$99	||||	4
\$100–\$149	||	2
\$150–\$199	|	1
\$200–\$249	|	1
\$250–\$299	|	1

7.

Number of Months
8
7
6
5
4
3
2
1
\$0–49
\$50–99
\$100–149
\$150–199
\$200–249
\$250–299
Unpaid Balance

8. a. The lowest point on the graph corresponds to January, so the average daily temperature is the lowest during January.

b. The point on the graph that corresponds to 25 is December, so the average daily temperature is 25°F in December.

c. The points on the graph that are greater than 70 are June, July, and August. So, the average daily temperature is greater than 70°F in June, July, and August.

Vocabulary and Readiness Check

1. A <u>bar</u> graph presents data using vertical or horizontal bars.

2. A <u>pictograph</u> is a graph in which pictures or symbols are used to visually present data.

3. A <u>line</u> graph displays information with a line that connects data points.

Copyright © 2011 Pearson Education, Inc. Publishing as Prentice Hall.

4. A histogram is a special bar graph in which the width of each bar represents a class interval and the height of each bar represents the class frequency.

Exercise Set 8.1

1. Kansas has the greatest number of wheat symbols, so the greatest quantity of acreage in wheat was planted by the state of Kansas.

3. Oklahoma is represented by 3.5 wheat symbols, and each symbol represents 1 million acres, so there were approximately 3.5(1 million) = 3.5 million or 3,500,000 acres of wheat planted.

5. Each wheat symbol represents 1 million acres, so find the state that has $\frac{5 \text{ million}}{1 \text{ million}} = 5$ wheat symbols. Montana has 5 wheat symbols, so it plants about 5,000,000 acres of wheat.

7. North Dakota is represented by 8 wheat symbols. From the pictograph, Montana, with 5 symbols, and South Dakota, with 3 symbols, together plant about the same acreage of wheat as North Dakota.

9. The year 2008 has 6.5 flames and each flame represents 12,000 wildfires, so there were approximately 6.5(12,000) = 78,000 wildfires in 2008.

11. The year with the most flames is 2006, so the most wildfires occurred in 2006.

13. 2004 has 5.5 flames and 2006 has 8 flames, which is 2.5 more. Thus, the increase in the number of wildfires from 2004 to 2006 was 2.5(12,000) or 30,000.

15. 2006 has 8 flames, 2007 has 7 flames, and 2008 has 6.5 flames. The average is $\frac{8+7+6.5}{3} = 7\frac{1}{6}$. Each flame represents 12,000 wildfires, so the average annual number of wildfires from 2006 to 2008 is $7\frac{1}{6}(12,000) = 86,000.$

17. The longest bar corresponds to September, so the month in which most hurricanes made landfall is September.

19. The length of the bar for August is 75, so approximately 75 hurricanes made landfall in August.

21. Two of the 76 hurricanes that made landfall in August did so in 2008. The fraction is $\frac{2}{76} = \frac{1}{38}$.

23. The longest bar corresponds to Tokyo, Japan, and the length of the bar is 33.8 million. So, the city with the largest population is Tokyo, Japan and its population is about 33.8 million or 33,800,000.

25. The longest bar corresponding to a city in the United States is the bar for New York. The population is approximately 21.9 million or 21,900,000.

27. The bar corresponding to Seoul, South Korea has length 23.8 and the bar corresponding to São Paolo, Brazil has length 20.9. Thus, Seoul, South Korea is about $23.8 - 20.9 = 2.9 \approx 3$ million larger than São Paolo, Brazil.

Copyright © 2011 Pearson Education, Inc. Publishing as Prentice Hall.

29.

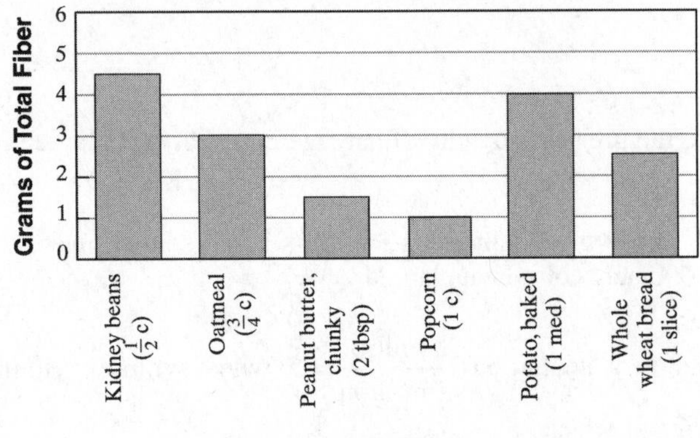

31.

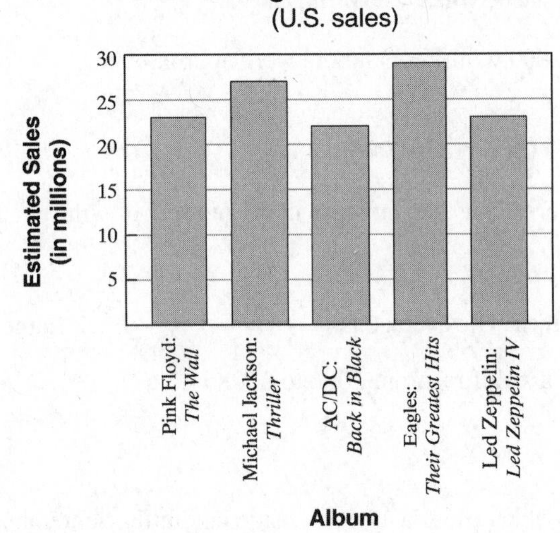

33. The height of the bar for 100–149 miles per week is 15, so 15 of the adults drive 100–149 miles per week.

35. 29 of the adults drive 0–49 miles per week, 17 of the adults drive 50–99 miles per week, and 15 of the adults drive 100–149 miles per week, so $29 + 17 + 15 = 61$ of the adults drive fewer than 150 miles per week.

37. 15 of the adults drive 100–149 miles per week and 9 of the adults drive 150–199 miles per week, so $15 + 9 = 24$ of the adults drive 100–199 miles per week.

39. 21 of the adults drive 250–299 miles per week and 9 of the adults drive 200–249 miles per week, so $21 - 9 = 12$ more adults drive 250–299 miles per week than 200–249 miles per week.

41. 9 of the 100 adults surveyed drive 150–199 miles per week, so the ratio is $\frac{9}{100}$.

43. The tallest bar corresponds to 45–54, so the most householders are in the 45–54 age range.

45. According to the bar graph, approximately 21 million householders will be 55–64 years old.

47. The sum of the first three bars is $21 + 17 + 6 = 44$, so approximately 44 million householders will be 44 years old or younger.

Copyright © 2011 Pearson Education, Inc. Publishing as Prentice Hall.

49. The height of the bar for 45–54 is 25 and the height of the bar for 55–64 is 21. So approximately 25 − 21 = 4 million more householders will be 45–54 years old than 55–64 years old.

	Class Interval (Scores)	Tally	Class Frequency (Number of Games)
51.	70–79	|	1
53.	90–99	~~||||~~ |||	8

	Class Interval (Account Balances)	Tally	Class Frequency (Number of People)
55.	\$0–\$99	~~||||~~ |	6
57.	\$200–\$299	~~||||~~ |	6
59.	\$400–\$499	||	2

61.

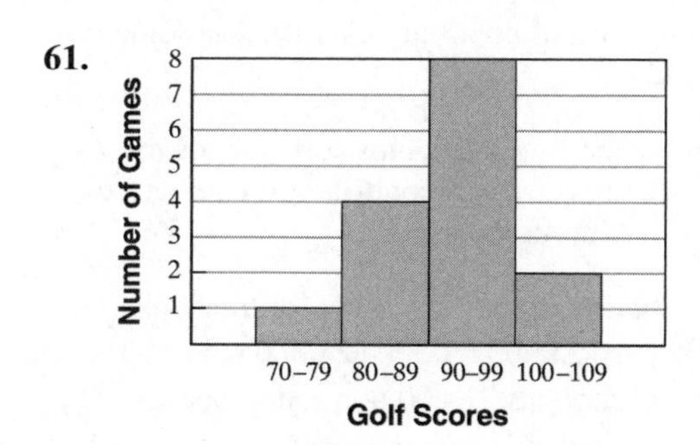

63. The point on the graph corresponding to 2004 is 7.8, so the average number of goals per game in 2004 was 7.8.

65. The highest point on the graph corresponds to 2003, so the average number of goals per game was the greatest in 2003.

67. The graph increases between 2004 and 2006, so the average number of goals per game increased from 2004 to 2006.

69. The dots for 2001, 2004, and 2007 are below the 8-level, so the average number of goals per game was less than 8 in 2001, 2004, and 2007.

71. 30% of 12 is $0.30 \cdot 12 = 3.6$.

73. 10% of 62 is $0.10 \cdot 62 = 6.2$

75. $\frac{1}{4} = \frac{1}{4} \cdot 100\% = \frac{25 \cdot 4}{4}\% = 25\%$

77. $\frac{17}{50} = \frac{17}{50} \cdot 100\% = \frac{17 \cdot 2 \cdot 50}{50}\% = 34\%$

79. The point on the high temperature graph corresponding to Thursday is 83, so the high temperature reading on Thursday was 83°F.

81. The lowest point on the graph of low temperatures corresponds to Sunday. The low temperature on Sunday was 68°F.

83. The difference between the graphs is the greatest for Tuesday. The high temperature was 86°F and the low temperature was 73°F, so the difference is 86 − 73 = 13°F.

85. answers may vary

Section 8.2

Practice Problems

1. Eight of the 100 adults prefer golf. The ratio is
$$\frac{\text{adults preferring golf}}{\text{total adults}} = \frac{8}{100} = \frac{2}{25}$$

2. Add the percents corresponding to Europe, Asia, and South America.
20% + 11% + 4% = 35%

3.
$$\begin{aligned}\text{amount} &= \text{percent} \cdot \text{base} \\ &= 0.25 \cdot 61{,}000{,}000 \\ &= 15{,}250{,}000\end{aligned}$$
Thus, 15,250,000 tourists might come from Mexico in 2011.

Copyright © 2011 Pearson Education, Inc. Publishing as Prentice Hall.

4.

Year	Percent	Degrees in Sector
Freshmen	30%	30% of 360° = 0.30(360°) = 108°
Sophomores	27%	27% of 360° = 0.27(360°) = 97.2°
Juniors	25%	25% of 360° = 0.25(360°) = 90°
Seniors	18%	18% of 360° = 0.18(360°) = 64.8°

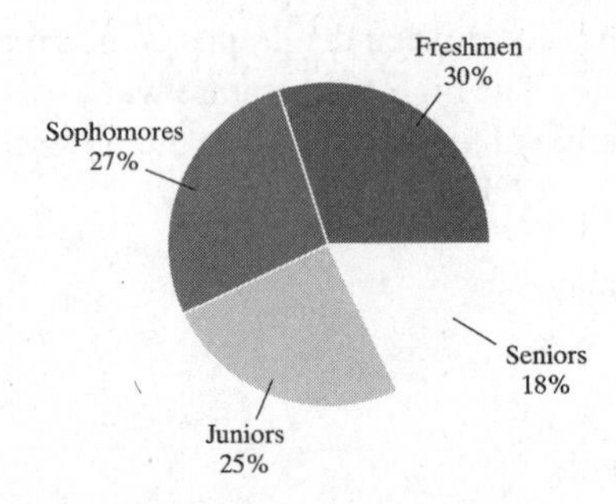

Vocabulary and Readiness Check

1. In a circle graph, each section (shaped like a piece of pie) shows a category and the relative size of the category.

2. A circle graph contains pie-shaped sections, each called a sector.

3. The number of degrees in a whole circle is 360.

4. If a circle graph has percent labels, the percents should add up to 100.

Exercise Set 8.2

1. The largest sector corresponds to the category "parent or guardian's home," thus most of the students live in a parent or guardian's home.

3. 180 of the 700 total students live in campus housing.
$\frac{180}{700} = \frac{9}{35}$
The ratio is $\frac{9}{35}$.

5. 180 of the students live in campus housing while 320 live in a parent or guardian's home.
$\frac{180}{320} = \frac{9}{16}$
The ratio is $\frac{9}{16}$.

7. The largest sector corresponds to Asia. Thus, the largest continent is Asia.

9. 30% + 7% = 37%
37% of the land on Earth is accounted for by Europe and Asia.

11. Asia accounts for 30% of the land on Earth.
$30\% \text{ of } 57{,}000{,}000 = 0.30 \cdot 57{,}000{,}000$
$= 17{,}100{,}000$
Asia is 17,100,000 square miles.

13. Australia accounts for 5% of the land on Earth.
$5\% \text{ of } 57{,}000{,}000 = 0.05 \cdot 57{,}000{,}000$
$= 2{,}850{,}000$
Australia is 2,850,000 square miles.

15. Add the percent for adult's fiction (33%) to the percent for children's fiction (22%).
33% + 22% = 55%
Thus, 55% of books are classified as some type of fiction.

17. The second-largest sector corresponds to nonfiction, so the second-largest category of books is nonfiction.

19. Nonfiction accounts for 25% of the books.
$25\% \text{ of } 125{,}600 = 0.25 \cdot 125{,}600 = 31{,}400$
The library has 31,400 nonfiction books.

21. Children's fiction accounts for 22% of the books.
$22\% \text{ of } 125{,}600 = 0.22 \cdot 125{,}600 = 27{,}632$
The library has 27,632 children's fiction books.

23. Reference or other accounts for 17% + 3% = 20% of the books.
$20\% \text{ of } 125{,}600 = 0.20 \cdot 125{,}600 = 25{,}120$
The library has 25,120 reference or other books.

Copyright © 2011 Pearson Education, Inc. Publishing as Prentice Hall.

25.

Type of Apple	Percent	Degrees in Sector
Red Delicious	37%	37% of 360° $= 0.37(360°) \approx 133°$
Golden Delicious	13%	13% of 360° $= 0.13(360°) \approx 47°$
Fuji	14%	14% of 360° $= 0.14(360°) \approx 50°$
Gala	15%	15% of 360° $= 0.15(360°) = 54°$
Granny Smith	12%	12% of 360° $= 0.12(360°) \approx 43°$
Other varieties	6%	6% of 360° $= 0.06(360°) \approx 22°$
Braeburn	3%	3% of 360° $= 0.03(360°) \approx 11°$

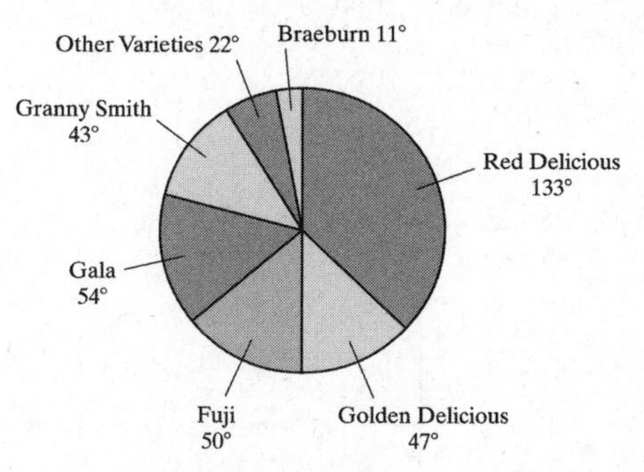

27.

Distribution of Large Dams by Continent		
Continent	**Percent**	**Degrees in Sector**
Europe	19%	19% of 360° $= 0.19(360°) \approx 68°$
North America	32%	32% of 360° $= 0.32(360°) \approx 115°$
South America	3%	3% of 360° $= 0.03(360°) \approx 11°$
Asia	39%	39% of 360° $= 0.39(360°) \approx 140°$
Africa	5%	5% of 360° $= 0.05(360°) = 18°$
Australia	2%	2% of 360° $= 0.02(360°) \approx 7°$

Copyright © 2011 Pearson Education, Inc. Publishing as Prentice Hall.

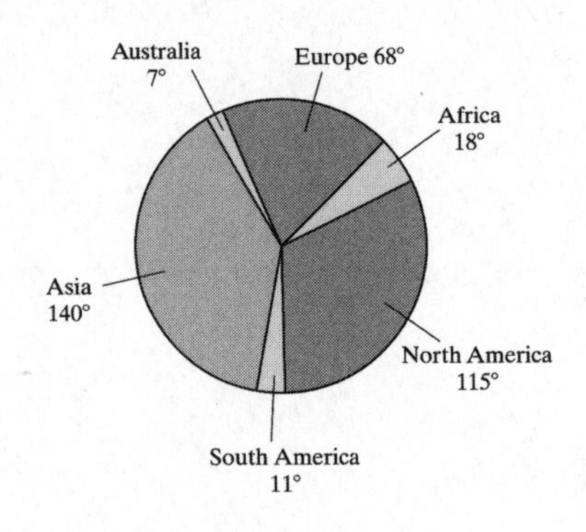

29. $20 = 2 \cdot 10 = 2 \cdot 2 \cdot 5 = 2^2 \cdot 5$

31. $40 = 2 \cdot 20 = 2 \cdot 2 \cdot 10 = 2 \cdot 2 \cdot 2 \cdot 5 = 2^3 \cdot 5$

33. $85 = 5 \cdot 17$

35. answers may vary

37. Pacific Ocean:
$$49\% \cdot 264,489,800 = 0.49 \cdot 264,489,800 = 129,600,002 \text{ square kilometers}$$

39. Indian Ocean:
$$21\% \cdot 264,489,800 = 0.21 \cdot 264,489,800 = 55,542,858 \text{ square kilometers}$$

41. $21.5\% \cdot 2800 = 0.215 \cdot 2800 = 602$ respondents

43. $21.5\% + 59.8\% = 81.3\%$
$0.813 \cdot 2800 \approx 2276$ respondents

45.
$$\frac{\text{number of respondents who spend \$0} - \text{\$15}}{\text{number of respondents who spend \$15} - \text{\$175}} = \frac{602}{1674} = \frac{2 \cdot 301}{2 \cdot 837} = \frac{301}{837}$$

47. no; answers may vary

Section 8.3

Practice Problems

1.

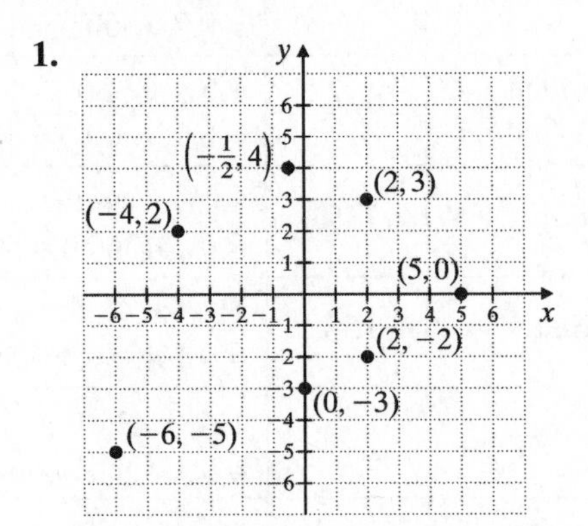

2. Point *A* has coordinates (5, 0).
Point *B* has coordinates (–5, 4)
Point *C* has coordinates (–3, –4).
Point *D* has coordinates (0, –1).
Point *E* has coordinates (2, 1).

3.
$$x + 3y = -12$$
$$0 + 3(-4) \stackrel{?}{=} -12$$
$$-12 = -12 \quad \text{True}$$
Yes, (0, –4) is a solution of $x + 3y = -12$.

4.

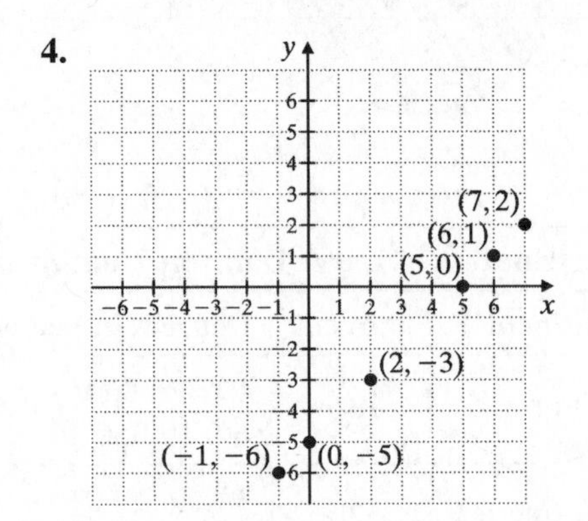

5. a.
$y = -5x$
$y = -5(5)$
$y = -25$
The ordered pair solution is (5, –25).

b.
$y = -5x$
$0 = -5x$
$\frac{0}{-5} = \frac{-5x}{-5}$
$0 = x$
The ordered pair solution is (0, 0).

Copyright © 2011 Pearson Education, Inc. Publishing as Prentice Hall.

c. $y = -5x$
$y = -5(-3)$
$y = 15$
The ordered pair solution is (−3, 15).

6. a. $y = 5x + 2$
$y = 5(0) + 2$
$y = 0 + 2$
$y = 2$
The ordered pair solution is (0, 2).

b.
$$y = 5x + 2$$
$$-3 = 5x + 2$$
$$-3 - 2 = 5x + 2 - 2$$
$$-5 = 5x$$
$$\frac{-5}{5} = \frac{5x}{5}$$
$$-1 = x$$
The ordered pair solution is (−1, −3).

Vocabulary and Readiness Check

1. In the ordered pair (−1, 2), the x-value is −1 and the y-value is 2.

2. In the rectangular coordinate system, the point of intersection of the x-axis and the y-axis is called the origin.

3. The axes divide the plane into four regions, called quadrants.

4. Every ordered pair of numbers, such as $\left(\frac{1}{2}, -3\right)$, corresponds to how many points in the plane? one

5. The process of locating a point on the rectangular coordinate system is called plotting the point.

6. The origin corresponds to the ordered pair (0, 0).

7. A plane is a flat surface that extends indefinitely in all directions.

8. Every point in the rectangular coordinate system corresponds to an ordered pair of numbers.

Exercise Set 8.3

1.

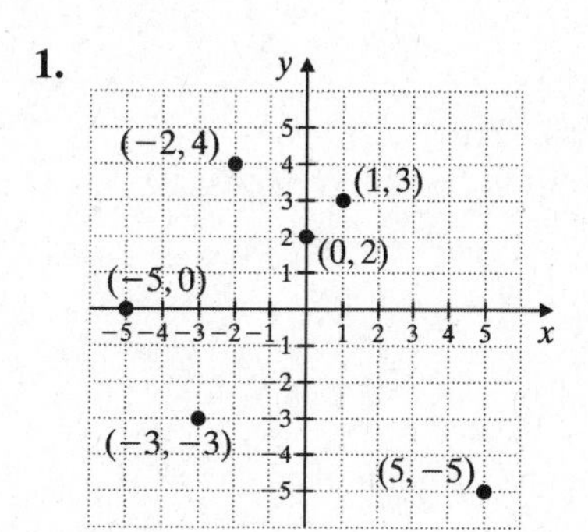

3.

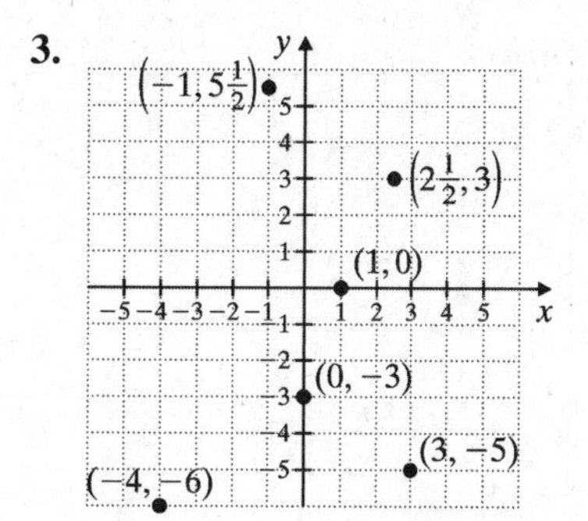

5. Point A has coordinates (0, 0).
Point B has coordinates $\left(3\frac{1}{2}, 0\right)$.
Point C has coordinates (3, 2).
Point D has coordinates (−1, 3).
Point E has coordinates (−2, −2).
Point F has coordinates (0, −1).
Point G has coordinates (2, −1).

7. $y = -20x$
$0 \stackrel{?}{=} -20(0)$
$0 = 0$ True
Yes, (0, 0) is a solution of $y = -20x$.

9. $x - y = 3$
$1 - 2 \stackrel{?}{=} 3$
$-1 = 3$ False
No, (1, 2) is not a solution of $x - y = 3$.

11. $y = 2x + 1$
$-3 \stackrel{?}{=} 2(-2) + 1$
$-3 \stackrel{?}{=} -4 + 1$
$-3 = -3$ True
Yes, (−2, −3) is a solution of $y = 2x + 1$.

13. $x = -3y$
$6 \stackrel{?}{=} -3(-2)$
$6 = 6$ True
Yes, (6, −2) is a solution of $x = -3y$.

Copyright © 2011 Pearson Education, Inc. Publishing as Prentice Hall.

15.
$$3y + 2x = 10$$
$$3 \cdot 0 + 2 \cdot 5 \stackrel{?}{=} 10$$
$$0 + 10 \stackrel{?}{=} 10$$
$$10 = 10 \quad \text{True}$$
Yes, (5, 0) is a solution of $3y + 2x = 10$.

17.
$$x - 5y = -1$$
$$3 - 5(1) \stackrel{?}{=} -1$$
$$3 - 5 \stackrel{?}{=} -1$$
$$-2 = -1 \quad \text{False}$$
No, (3, 1) is not a solution of $x - 5y = -1$.

19.

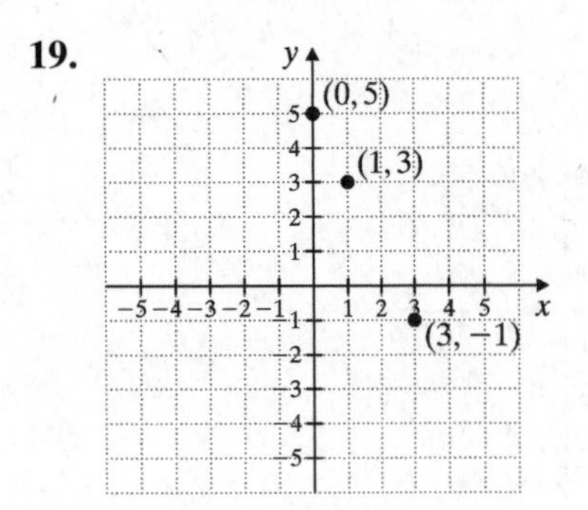

21.

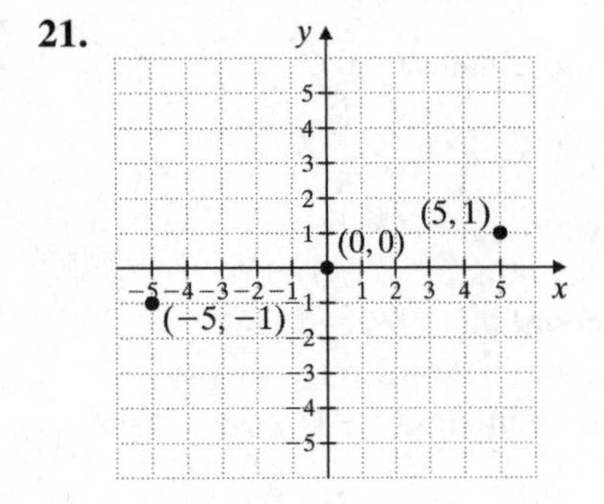

23.

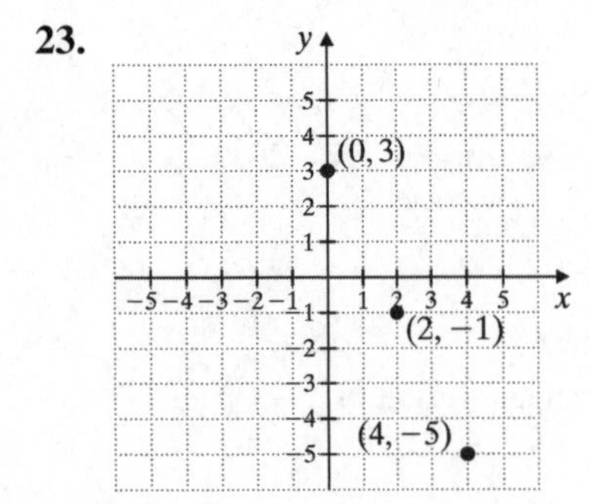

25. $y = -9x$
$y = -9(1)$
$y = -9$
The solution is (1, −9).
$y = -9x$
$y = -9(0)$
$y = 0$
The solution is (0, 0).
$$y = -9x$$
$$-18 = -9x$$
$$\frac{-18}{-9} = \frac{-9x}{-9}$$
$$2 = x$$
The solution is (2, −18).

27.
$$x - y = 14$$
$$2 - y = 14$$
$$-2 + 2 - y = -2 + 14$$
$$-y = 12$$
$$y = -12$$
The solution is (2, −12).
$$x - y = 14$$
$$x - (-8) = 14$$
$$x + 8 = 14$$
$$x + 8 - 8 = 14 - 8$$
$$x = 6$$
The solution is (6, −8).
$$x - y = 14$$
$$0 - y = 14$$
$$-y = 14$$
$$y = -14$$
The solution is (0, −14).

29.
$$x + y = -2$$
$$-2 + y = -2$$
$$-2 + 2 + y = -2 + 2$$
$$y = 0$$
The solution is (−2, 0).
$$x + y = -2$$
$$1 + y = -2$$
$$1 - 1 + y = -2 - 1$$
$$y = -3$$
The solution is (1, −3).
$$x + y = -2$$
$$x + 5 = -2$$
$$x + 5 - 5 = -2 - 5$$
$$x = -7$$
The solution is (−7, 5).

31. $x = y - 4$
$x = -12 - 4$
$x = -16$
The solution is (−16, −12).
$x = y - 4$
$x = 3 - 4$
$x = -1$
The solution is (−1, 3).

Copyright © 2011 Pearson Education, Inc. Publishing as Prentice Hall.

$x = y - 4$
$100 = y - 4$
$100 + 4 = y - 4 + 4$
$104 = y$
The solution is (100, 104).

33. $y = 3x - 5$
$y = 3 \cdot 1 - 5$
$y = 3 - 5$
$y = -2$
The solution is (1, –2).
$y = 3x - 5$
$y = 3 \cdot 2 - 5$
$y = 6 - 5$
$y = 1$
The solution is (2, 1).
$y = 3x - 5$
$4 = 3x - 5$
$4 + 5 = 3x - 5 + 5$
$9 = 3x$
$\frac{9}{3} = \frac{3x}{3}$
$3 = x$
The solution is (3, 4).

35. $x = -y$
$x = -0$
$x = 0$
The solution is (0, 0).
$x = -y$
$3 = -y$
$-3 = y$
The solution is (3, –3).
$x = -y$
$x = -(-9)$
$x = 9$
The solution is (9, –9).

37. $x + 2y = -8$
$4 + 2y = -8$
$4 - 4 + 2y = -8 - 4$
$2y = -12$
$\frac{2y}{2} = \frac{-12}{2}$
$y = -6$
The solution is (4, –6).
$x + 2y = -8$
$x + 2(-3) = -8$
$x - 6 = -8$
$x - 6 + 6 = -8 + 6$
$x = -2$
The solution is (–2, –3).
$x + 2y = -8$
$0 + 2y = -8$
$2y = -8$
$\frac{2y}{2} = \frac{-8}{2}$
$y = -4$
The solution is (0, –4).

39.
$$\begin{array}{r} 5.6 \\ -\ 3.9 \\ \hline 1.7 \end{array}$$

41.
$$\begin{array}{r} 5.6 \\ \times\ 3.9 \\ \hline 5\ 04 \\ 16\ 80 \\ \hline 21.84 \end{array}$$

43. (0.236)(–100) = –23.6

45. To plot (*a*, *b*), start at the origin and move *a* units to the right and *b* units up. The point will be in quadrant I. Thus, (*a*, *b*) is in quadrant I is a true statement.

47. To plot (0, *b*), start at the origin and move 0 units to the right and *b* units up. The point will lie on the *y*-axis. Thus, (0, *b*) lies on the *y*-axis is a true statement.

49. To plot (0, –*b*), start at the origin and move 0 units to the right and *b* units down. The point will be on the *y*-axis. Thus, (0, –*b*) lies on the *x*-axis is a false statement.

51. To plot (–*a*, *b*), start at the origin and move *a* units to the left and *b* units up. The point will lie in quadrant II. Thus, (–*a*, *b*) lies in quadrant III is a false statement.

53. To plot (4, –3), start at the origin and move 4 units to the right, so the point (4, –3) is plotted to the right of the *y*-axis.

Copyright © 2011 Pearson Education, Inc. Publishing as Prentice Hall.

55.

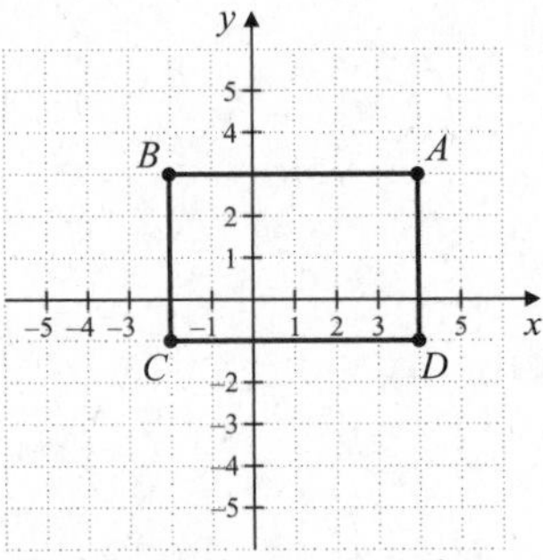

The figure is a rectangle.

57. $P = 2l + 2w$
Let $l = 6$ and $w = 4$.
$P = 2 \cdot 6 + 2 \cdot 4 = 12 + 8 = 20$
The perimeter is 20 units.

Integrated Review

1. Customer service representatives has 10 figures and each figure represents 50,000 workers. Thus, the increase in the number of customer service representatives is approximately $10 \cdot 50{,}000 = 500{,}000$.

2. Post-secondary teachers has 11 figures and each figure represents 50,000 workers. Thus, the increase in the number of post-secondary teachers is approximately $11 \cdot 50{,}000 = 550{,}000$.

3. Retail salespeople has the greatest number of figures, so the greatest increase is expected for retail salespeople.

4. Waitstaff has the least number of figures, so the least increase is expected for waitstaff.

5. The tallest bar corresponds to Oroville, CA. Thus, the U.S. dam with the greatest height is the Oroville Dam, which is approximately 755 feet.

6. The bar whose height is between 625 and 650 feet corresponds to New Bullards Bar, CA. Thus, the U.S. dam with a height between 625 and 650 feet is the New Bullards Bar Dam, which is approximately 635 feet.

7. From the graph, the Hoover Dam is approximately 725 feet and the Glen Canyon Dam is approximately 710 feet. The Hoover Dam is $725 - 710 = 15$ feet higher than the Glen Canyon Dam.

8. There are 4 bars that are taller than the 700-feet level, so there are 4 dams with heights over 700 feet.

9. The highest points on the graph correspond to Thursday and Saturday. Thus, the highest temperature occurs on Thursday and Saturday and is 100°F.

10. The lowest point on the graph corresponds to Monday. Thus, the lowest temperature occurs on Monday, and is 82°F.

11. The points on the graph that are lower than the 90 level correspond to Sunday, Monday, and Tuesday. Thus, the temperature was less than 90°F on Sunday, Monday, and Tuesday.

12. The points on the graph that are higher than the 90 level correspond to Wednesday, Thursday, Friday, and Saturday. Thus, the temperature was greater than 90°F on Wednesday, Thursday, Friday, and Saturday.

13. The sector corresponding to whole milk is 35%.
35% of 200 = $0.35 \cdot 200 = 70$
Thus, 70 quart containers of whole milk are sold.

14. The sector corresponding to skim milk is 26%.
26% of 200 = $0.26 \cdot 200 = 52$
Thus, 52 quart containers of skim milk are sold.

15. The sector corresponding to buttermilk is 1%.
1% of 200 = $0.01 \cdot 200 = 2$
Thus, 2 quart containers of buttermilk are sold.

16. The sector corresponding to Flavored reduced fat and skim milk is 3%.
3% of 200 = $0.03 \cdot 200 = 6$
Thus, 6 quart containers of flavored reduced fat and skim milk are sold.

	Class Intervals (Scores)	Tally	Class Frequency (Number of Quizzes)
17.	50–59	||	2
18.	60–69	|	1
19.	70–79	|||	3
20.	80–89	~~||||~~ |	6
21.	90–99	~~||||~~	5

Copyright © 2011 Pearson Education, Inc. Publishing as Prentice Hall.

22.

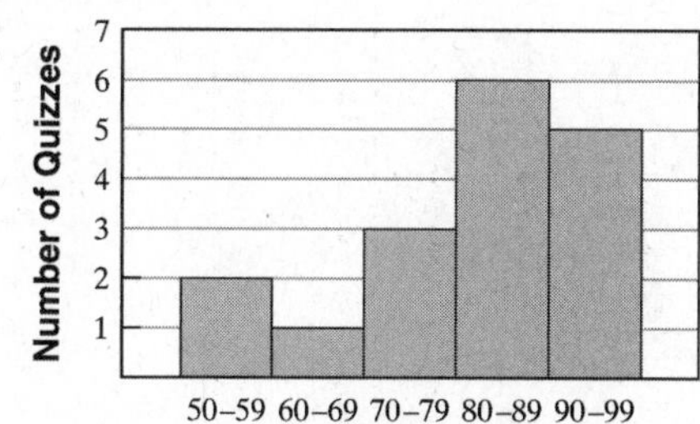

23.

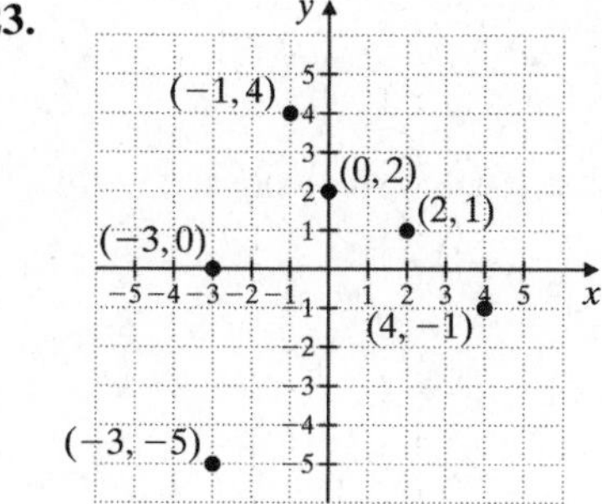

24. $x = 3y$

$1 \stackrel{?}{=} 3 \cdot 3$

$1 = 9$ False

No, (1, 3) is not a solution of $x = 3y$.

25. $x + y = -6$

$-2 + (-4) \stackrel{?}{=} -6$

$-6 = -6$ True

Yes, (–2, –4) is a solution of $x + y = -6$.

26. $x - y = 6$

$0 - y = 6$

$-y = 6$

$y = -6$

(0, –6) is a solution.

$x - y = 6$

$x - 0 = 6$

$x = 6$

(6, 0) is a solution.

$x - y = 6$

$2 - y = 6$

$2 - 2 - y = 6 - 2$

$-y = 4$

$y = -4$

(2, –4) is a solution.

Section 8.4

Practice Problems

1.

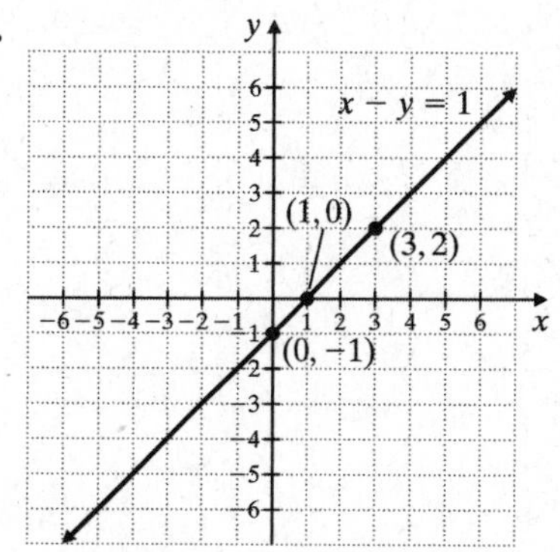

2. $y - x = 4$

Find any 3 ordered pair solutions.

Let $x = 0$.

$y - x = 4$

$y - 0 = 4$

$y = 4$

(0, 4)

Let $x = 2$.

$y - x = 4$

$y - 2 = 4$

$y - 2 + 2 = 4 + 2$

$y = 6$

(2, 6)

Let $y = 0$.

$y - x = 4$

$0 - x = 4$

$x = -4$

(–4, 0)

Plot (0, 4), (2, 6), and (–4, 0). Then draw a line through them.

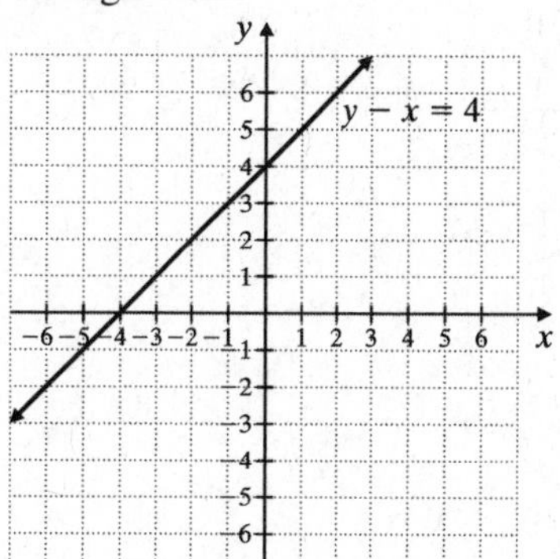

Copyright © 2011 Pearson Education, Inc. Publishing as Prentice Hall.

3. $y = -3x + 2$
Find any 3 ordered pair solutions.
Let $x = 0$.
$y = -3x + 2$
$y = -3 \cdot 0 + 2$
$y = 2$
$(0, 2)$
Let $x = 1$.
$y = -3x + 2$
$y = -3(1) + 2$
$y = -3 + 2$
$y = -1$
$(1, -1)$
Let $x = -1$.
$y = -3x + 2$
$y = -3(-1) + 2$
$y = 3 + 2$
$y = 5$
$(-1, 5)$
Plot $(0, 2)$, $(1, -1)$, and $(-1, 5)$. Then draw a line through them.

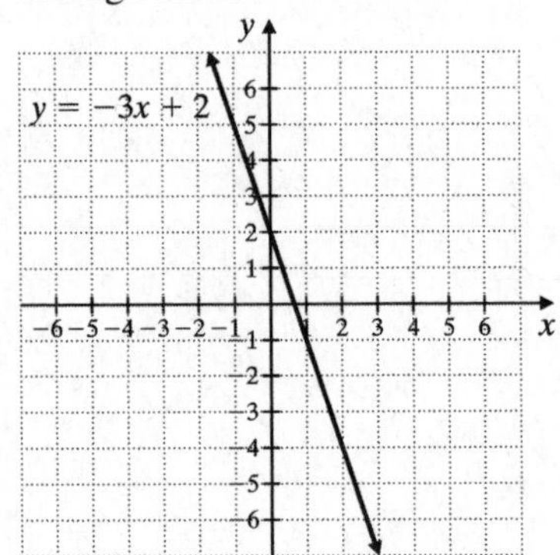

4. $y = -4$
No matter what x-value we choose, y is always -4.

x	y
−2	−4
0	−4
2	−4

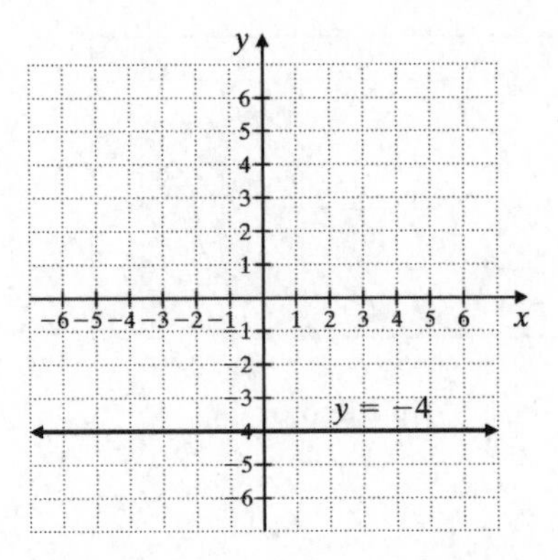

5. $x = 5$
No matter what y-value we choose, x is always 5.

x	y
5	−2
5	0
5	2

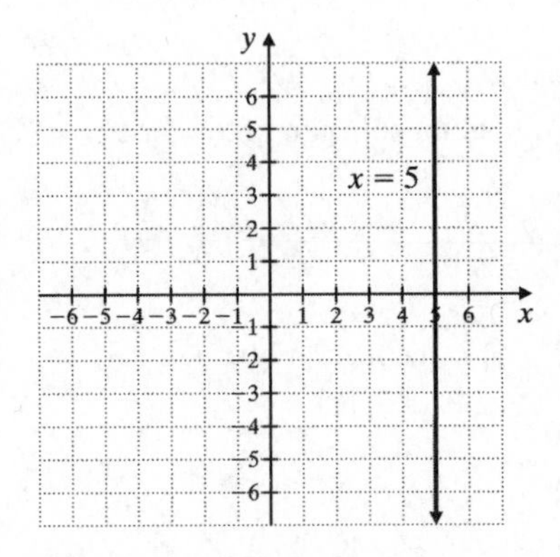

Vocabulary and Readiness Check

1. A linear equation in two variables can be written in the form $ax + by = c$.

2. The graph of the equation $x = 5$ is a vertical line.

3. The graph of the equation $y = -2$ is a horizontal line.

4. The graph of $3x - 2y = 6$ is a line that is not vertical or horizontal.

Copyright © 2011 Pearson Education, Inc. Publishing as Prentice Hall.

Exercise Set 8.4

1. $x + y = 4$
   Find any 3 ordered pair solutions.
   Let $x = 0$.
   $x + y = 4$
   $0 + y = 4$
   $y = 4$
   (0, 4)
   Let $x = 2$.
   $x + y = 4$
   $2 + y = 4$
   $2 - 2 + y = 4 - 2$
   $y = 2$
   (2, 2)
   Let $x = 4$.
   $x + y = 4$
   $4 + y = 4$
   $4 - 4 + y = 4 - 4$
   $y = 0$
   (4, 0)
   Plot (0, 4), (2, 2), and (4, 0). Then draw a line through them.

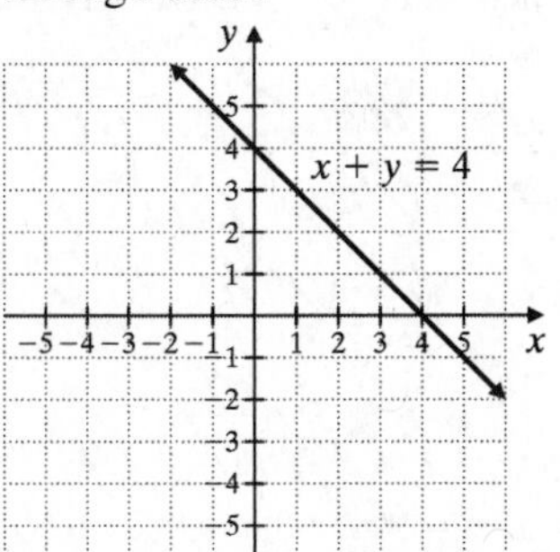

3. $x - y = -6$
   Find any 3 ordered pair solutions.
   Let $x = 0$.
   $x - y = -6$
   $0 - y = -6$
   $y = 6$
   (0, 6)
   Let $x = -2$.
   $x - y = -6$
   $-2 - y = -6$
   $2 - 2 - y = 2 - 6$
   $-y = -4$
   $y = 4$
   (−2, 4)
   Let $y = 0$.
   $x - y = -6$
   $x - 0 = -6$
   $x = -6$
   (−6, 0)
   Plot (0, 6), (−2, 4), and (−6, 0). Then draw a line through them.

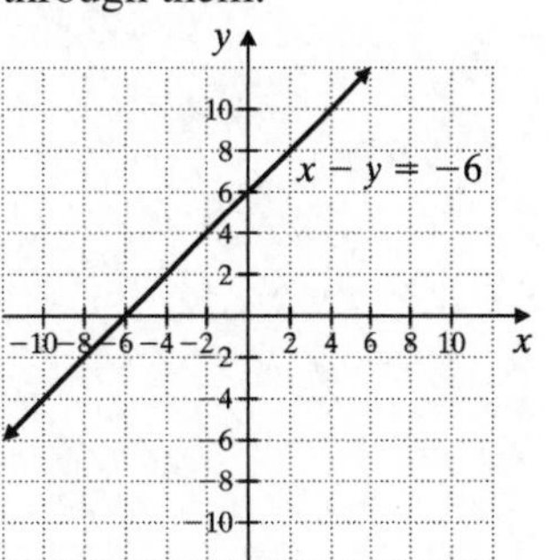

5. $y = 4x$
   Find any 3 ordered pair solutions.
   Let $x = 0$.
   $y = 4x$
   $y = 4(0)$
   $y = 0$
   (0, 0)
   Let $x = 1$.
   $y = 4x$
   $y = 4(1)$
   $y = 4$
   (1, 4)
   Let $x = -1$.
   $y = 4x$
   $y = 4(-1)$
   $y = -4$
   (−1, −4)
   Plot (0, 0), (1, 4) and (−1, −4). Then draw a line through them.

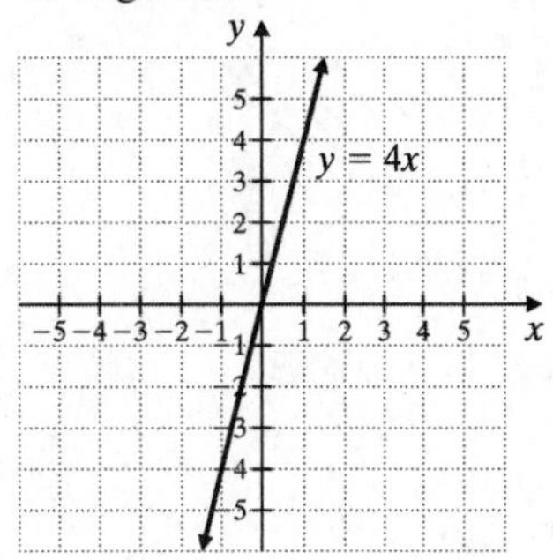

Copyright © 2011 Pearson Education, Inc. Publishing as Prentice Hall.

7. $y = 2x - 1$
Find any 3 ordered pair solutions.
Let $x = 0$.
$y = 2x - 1$
$y = 2 \cdot 0 - 1$
$y = -1$
$(0, -1)$
Let $x = 1$.
$y = 2x - 1$
$y = 2 \cdot 1 - 1$
$y = 2 - 1$
$y = 1$
$(1, 1)$
Let $x = 2$.
$y = 2x - 1$
$y = 2 \cdot 2 - 1$
$y = 4 - 1$
$y = 3$
$(2, 3)$
Plot $(0, -1)$, $(1, 1)$, and $(2, 3)$. Then draw a line through them.

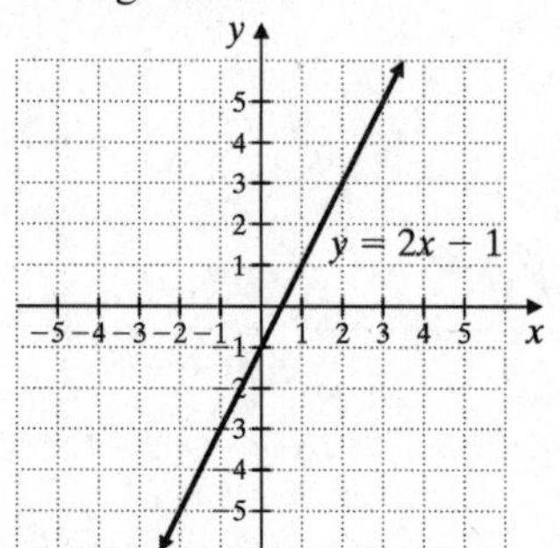

9. $x = -3$
No matter what y-value we choose, x is always -3.

x	y
−3	−4
−3	0
−3	4

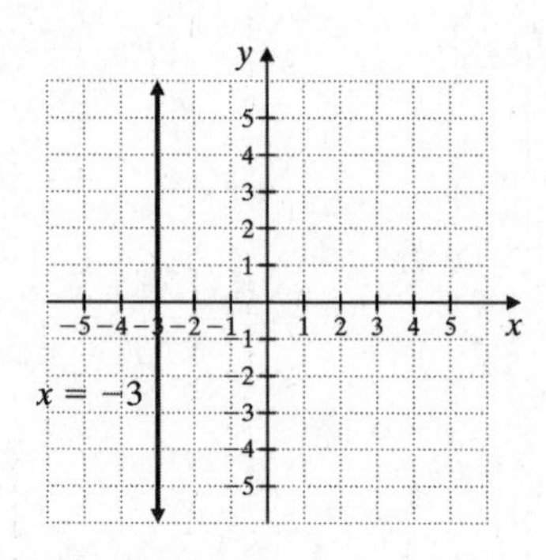

11. $y = -3$
No matter what x-value we choose, y is always -3.

x	y
−2	−3
0	−3
2	−3

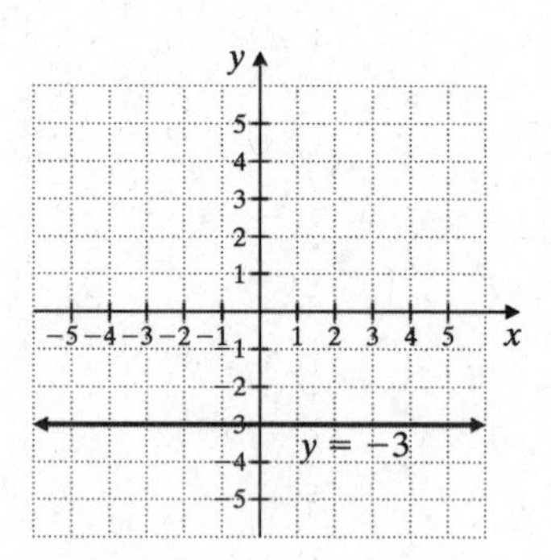

13. $x = 0$
No matter what y-value we choose, x is always 0.

x	y
0	−3
0	0
0	3

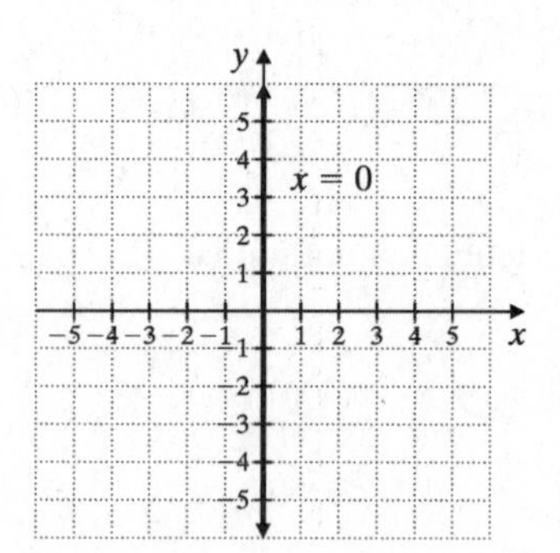

15. $y = -2x$
Find any 3 ordered pair solutions.
Let $x = 0$.
$y = -2 \cdot 0$
$y = 0$
$(0, 0)$
Let $x = 2$.
$y = -2x$
$y = -2 \cdot 2$
$y = -4$
$(2, -4)$

Copyright © 2011 Pearson Education, Inc. Publishing as Prentice Hall.

Let $x = -2$.

$y = -2x$

$y = -2(-2)$

$y = 4$

$(-2, 4)$

Plot (0, 0), (2, –4), and (–2, 4). Then draw a line through them.

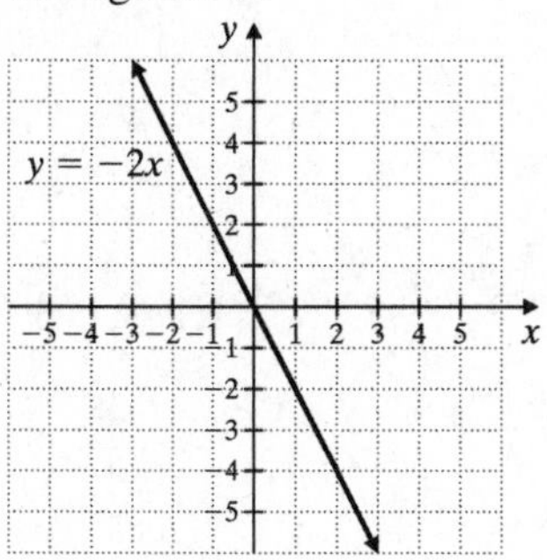

17. $y = -2$

No matter what x-value we choose, y is always –2.

x	y
–4	–2
0	–2
4	–2

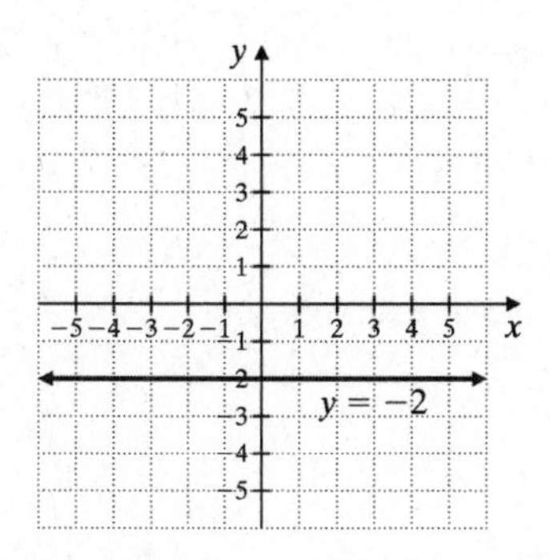

19. $x + 2y = 12$

Find any 3 ordered pair solutions.

Let $x = 0$.

$$x + 2y = 12$$
$$0 + 2y = 12$$
$$\frac{2y}{2} = \frac{12}{2}$$
$$y = 6$$

$(0, 6)$

Let $y = 0$.

$$x + 2y = 12$$
$$x + 2 \cdot 0 = 12$$
$$x + 0 = 12$$
$$x = 12$$

$(12, 0)$

Let $x = 6$.

$$x + 2y = 12$$
$$6 + 2y = 12$$
$$6 - 6 + 2y = 12 - 6$$
$$2y = 6$$
$$\frac{2y}{2} = \frac{6}{2}$$
$$y = 3$$

$(6, 3)$

Plot (0, 6), (12, 0), and (6, 3). Then draw a line through them.

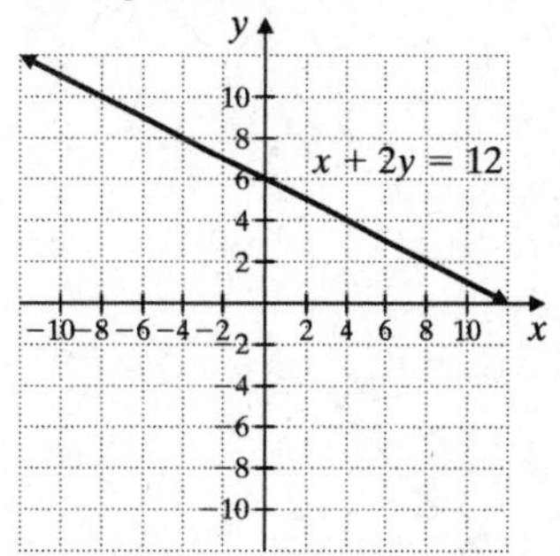

21. $x = 6$

No matter what y-value we choose, x is always 6.

x	y
6	–5
6	0
6	5

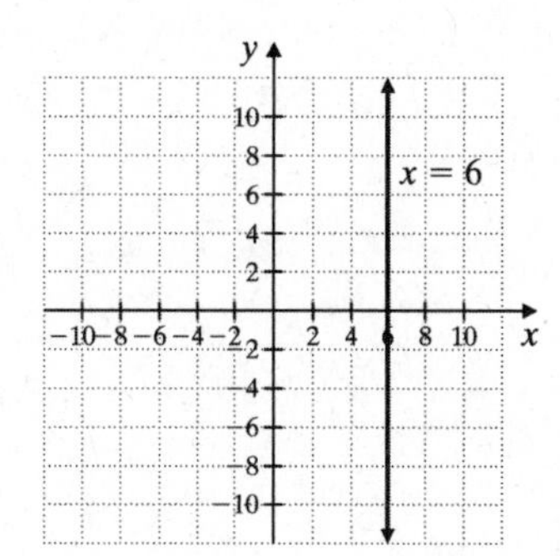

Copyright © 2011 Pearson Education, Inc. Publishing as Prentice Hall.

23. $y = x - 3$
Find any 3 ordered pair solutions.
Let $x = 0$.
$y = x - 3$
$y = 0 - 3$
$y = -3$
$(0, -3)$
Let $x = -3$.
$y = x - 3$
$y = -3 - 3$
$y = -6$
$(-3, -6)$
Let $y = 0$.
$y = x - 3$
$0 = x - 3$
$0 + 3 = x - 3 + 3$
$3 = x$
$(3, 0)$
Plot $(0, -3)$, $(-3, -6)$, and $(3, 0)$. Then draw a line through them.

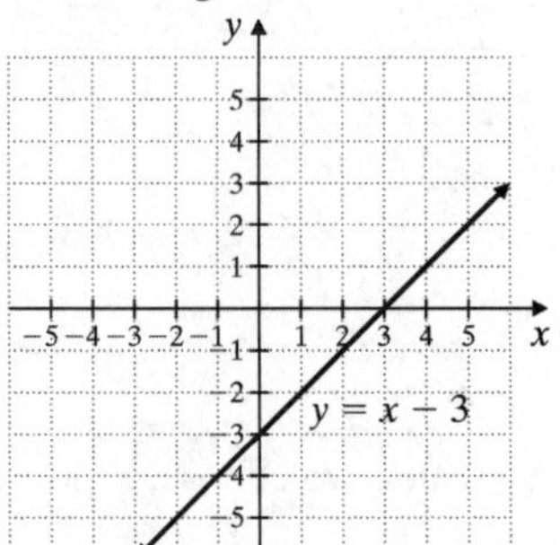

25. $x = y - 4$
Find any 3 ordered pair solutions.
Let $y = 0$.
$x = y - 4$
$x = 0 - 4$
$x = -4$
$(-4, 0)$
Let $y = 4$.
$x = y - 4$
$x = 4 - 4$
$x = 0$
$(0, 4)$
Let $y = 6$.
$x = y - 4$
$x = 6 - 4$
$x = 2$
$(2, 6)$
Plot $(-4, 0)$, $(0, 4)$, and $(2, 6)$. Then draw a line through them.

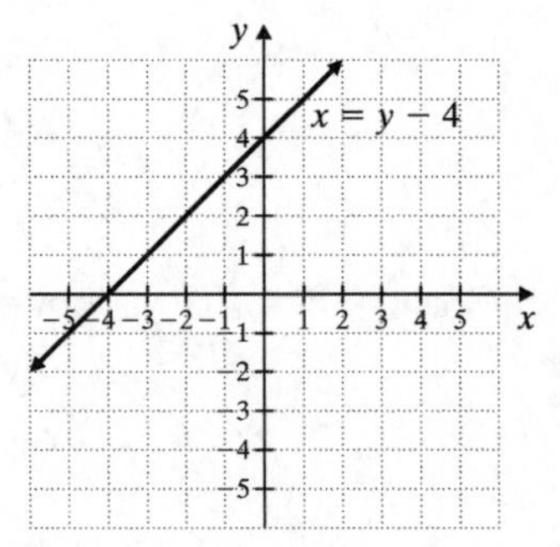

27. $x + 3 = 0$ or $x = -3$
No matter what y-value we choose, x is always -3.

x	y
-3	-2
-3	0
-3	2

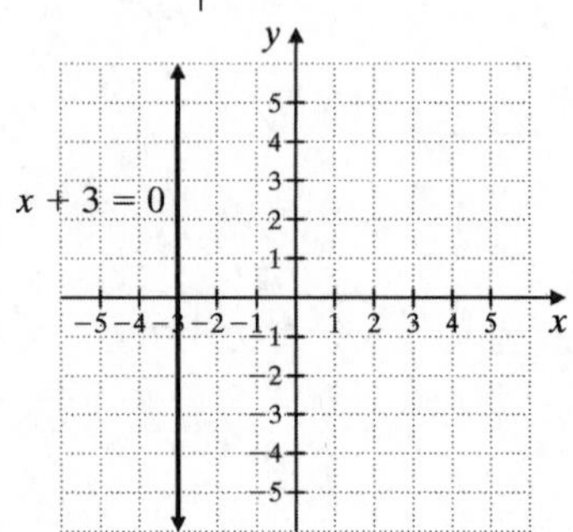

29. $y = -\frac{1}{4}x$
Find any 3 ordered pair solutions.
Let $x = -4$.
$y = -\frac{1}{4}x$
$y = -\frac{1}{4}(-4)$
$y = 1$
$(-4, 1)$
Let $x = 0$.
$y = -\frac{1}{4}x$
$y = -\frac{1}{4} \cdot 0$
$y = 0$
$(0, 0)$
Let $x = 4$.

Copyright © 2011 Pearson Education, Inc. Publishing as Prentice Hall.

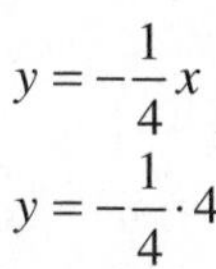

$y = -\frac{1}{4}x$

$y = -\frac{1}{4} \cdot 4$

$y = -1$

$(4, -1)$

Plot $(-4, 1)$, $(0, 0)$, and $(4, -1)$. Then draw a line through them.

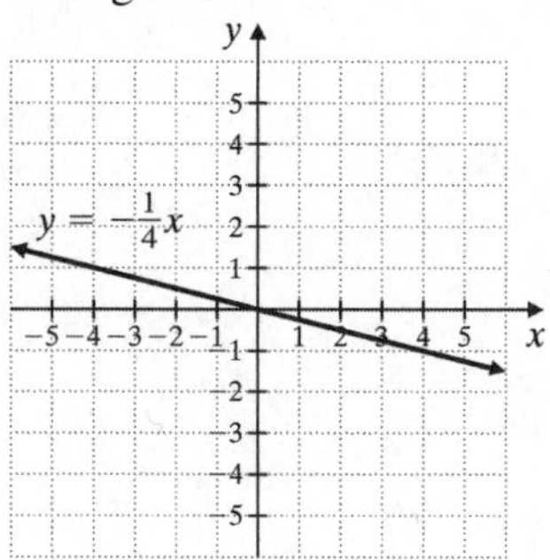

31. $y = \frac{1}{3}x$

Find any 3 ordered pair solutions.

Let $x = -3$.

$y = \frac{1}{3}x$

$y = \frac{1}{3}(-3)$

$y = -1$

$(-3, -1)$

Let $x = 0$.

$y = \frac{1}{3}x$

$y = \frac{1}{3} \cdot 0$

$y = 0$

$(0, 0)$

Let $x = 3$.

$y = \frac{1}{3}x$

$y = \frac{1}{3} \cdot 3$

$y = 1$

$(3, 1)$

Plot $(-3, -1)$, $(0, 0)$, and $(3, 1)$. Then draw a line through them.

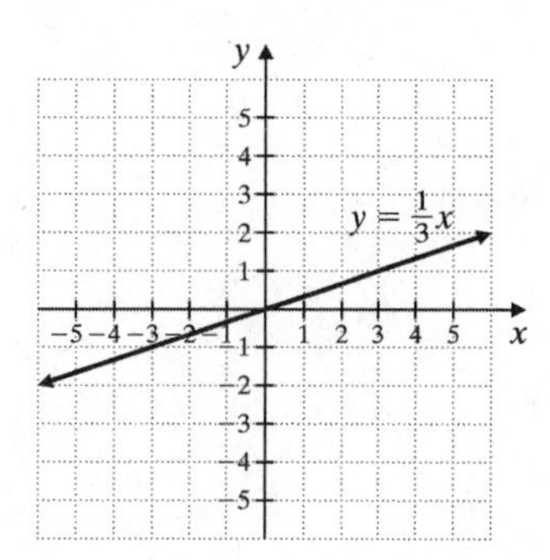

33. $y = 4x + 2$

Find any 3 ordered pair solutions.

Let $x = -1$.

$y = 4x + 2$

$y = 4(-1) + 2$

$y = -4 + 2$

$y = -2$

$(-1, -2)$

Let $x = 0$.

$y = 4x + 2$

$y = 4(0) + 2$

$y = 0 + 2$

$y = 2$

$(0, 2)$

Let $x = 1$.

$y = 4x + 2$

$y = 4(1) + 2$

$y = 4 + 2$

$y = 6$

$(1, 6)$

Plot $(-1, -2)$, $(0, 2)$ and $(1, 6)$. Then draw a line through them.

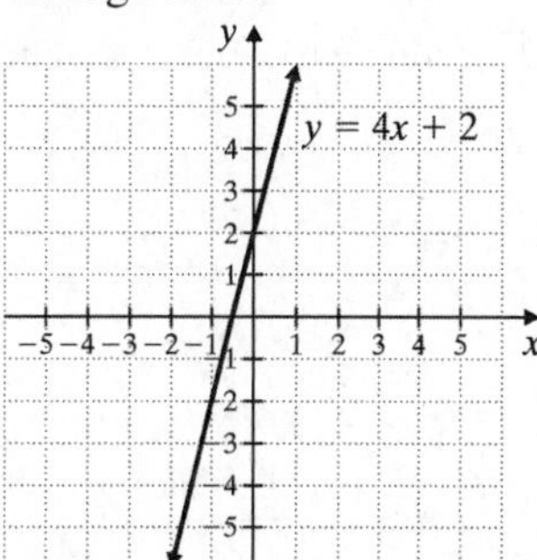

35. $2x + 3y = 6$

Find any 3 ordered pair solutions.

Let $x = 0$.

$$2x + 3y = 6$$
$$2 \cdot 0 + 3y = 6$$
$$3y = 6$$
$$\frac{3y}{3} = \frac{6}{3}$$
$$y = 2$$

$(0, 2)$

Copyright © 2011 Pearson Education, Inc. Publishing as Prentice Hall.

Let $y = 0$.

$$\begin{aligned} 2x+3y &= 6 \\ 2x+3\cdot 0 &= 6 \\ 2x &= 6 \\ \frac{2x}{2} &= \frac{6}{2} \\ x &= 3 \end{aligned}$$

(3, 0)

Let $x = 6$.

$$\begin{aligned} 2x+3y &= 6 \\ 2\cdot 6+3y &= 6 \\ 12+3y &= 6 \\ 12-12+3y &= 6-12 \\ 3y &= -6 \\ \frac{3y}{3} &= \frac{-6}{3} \\ y &= -2 \end{aligned}$$

(6, −2)

Plot (0, 2), (3, 0), and (6, −2). Then draw a line through them.

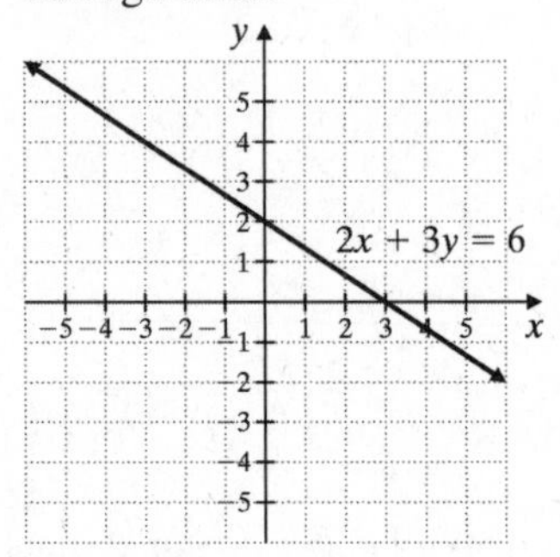

37. $x = -3.5$

No matter what y-value we choose, x is always −3.5.

x	y
−3.5	−3
−3.5	0
−3.5	3

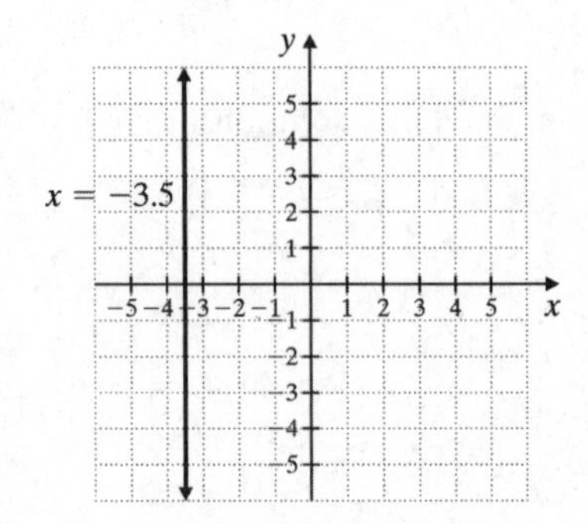

39. $3x - 4y = 24$

Find any 3 ordered pair solutions.

Let $x = 0$.

$$\begin{aligned} 3x-4y &= 24 \\ 3\cdot 0-4y &= 24 \\ -4y &= 24 \\ \frac{-4y}{-4} &= \frac{24}{-4} \\ y &= -6 \end{aligned}$$

(0, −6)

Let $y = 0$.

$$\begin{aligned} 3x-4y &= 24 \\ 3x-4\cdot 0 &= 24 \\ 3x &= 24 \\ \frac{3x}{3} &= \frac{24}{3} \\ x &= 8 \end{aligned}$$

(8, 0)

Let $x = 4$.

$$\begin{aligned} 3x-4y &= 24 \\ 3\cdot 4-4y &= 24 \\ 12-4y &= 24 \\ 12-12-4y &= 24-12 \\ -4y &= 12 \\ \frac{-4y}{-4} &= \frac{12}{-4} \\ y &= -3 \end{aligned}$$

(4, −3)

Plot (0, −6), (8, 0), and (4, −3). Then draw a line through them.

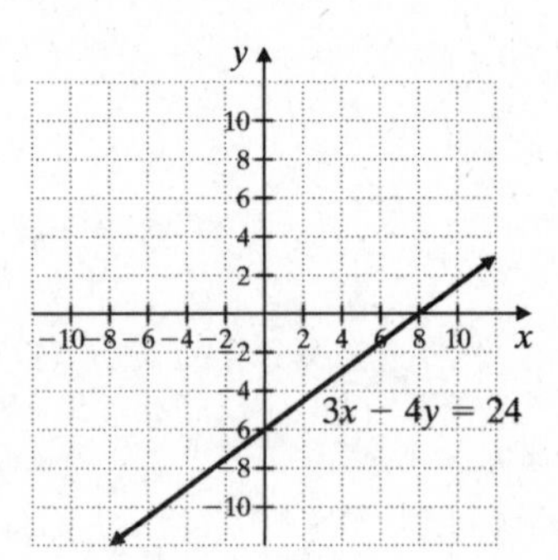

Copyright © 2011 Pearson Education, Inc. Publishing as Prentice Hall.

41. $y = \frac{1}{2}$

No matter what x-value we choose, y is always $\frac{1}{2}$.

x	y
−2	$\frac{1}{2}$
0	$\frac{1}{2}$
2	$\frac{1}{2}$

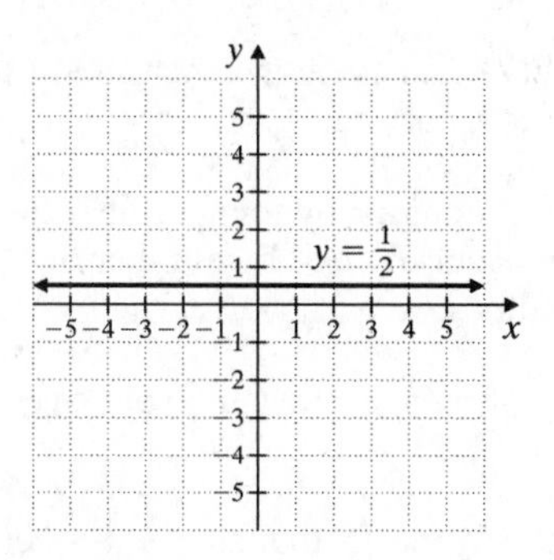

43. $y = \frac{1}{3}x - 2$

Find any 3 ordered pair solutions.

Let $x = -3$.

$y = \frac{1}{3}x - 2$

$y = \frac{1}{3}(-3) - 2$

$y = -1 - 2$

$y = -3$

$(-3, -3)$

Let $x = 0$.

$y = \frac{1}{3}x - 2$

$y = \frac{1}{3}(0) - 2$

$y = -2$

$(0, -2)$

Let $x = 3$.

$y = \frac{1}{3}x - 2$

$y = \frac{1}{3}(3) - 2$

$y = 1 - 2$

$y = -1$

$(3, -1)$

Plot $(-3, -3)$, $(0, -2)$, and $(3, -1)$. Then draw a line through them.

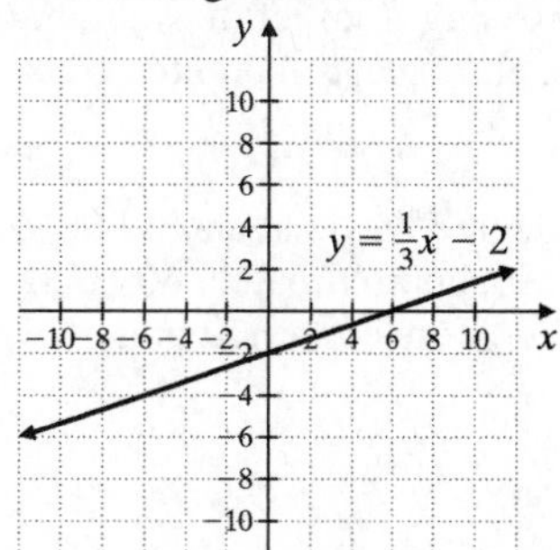

45. $(0.5)\left(-\frac{1}{8}\right) = \left(\frac{1}{2}\right)\left(-\frac{1}{8}\right) = -\frac{1 \cdot 1}{2 \cdot 8} = -\frac{1}{16}$

47. $\frac{3}{4} \div \left(-\frac{19}{20}\right) = \frac{3}{4} \cdot \left(-\frac{20}{19}\right)$

$= -\frac{3 \cdot 20}{4 \cdot 19}$

$= -\frac{3 \cdot 4 \cdot 5}{4 \cdot 19}$

$= -\frac{15}{19}$

49. $\frac{2}{11} - \frac{x}{11} = \frac{2 - x}{11}$

51.

x	$y = \lvert x \rvert$
−3	$\lvert -3 \rvert = 3$
−2	$\lvert -2 \rvert = 2$
−1	$\lvert -1 \rvert = 1$
0	$\lvert 0 \rvert = 0$
1	$\lvert 1 \rvert = 1$
2	$\lvert 2 \rvert = 2$
3	$\lvert 3 \rvert = 3$

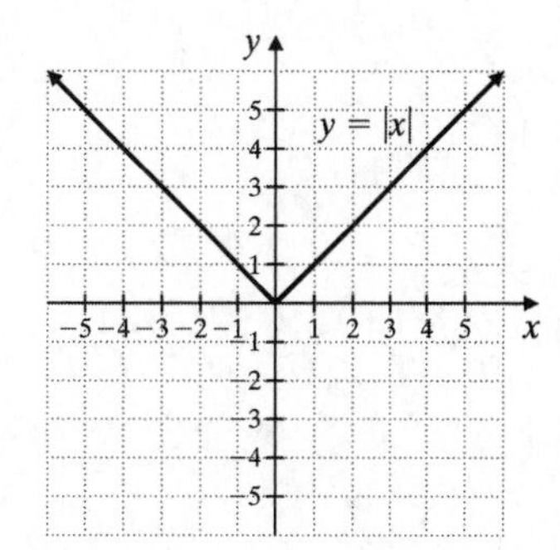

Copyright © 2011 Pearson Education, Inc. Publishing as Prentice Hall.

53. answers may vary

55. Since the line corresponding to Bachelor's Degrees is increasing from left to right, the number of bachelor's degrees conferred is increasing.

57. The associate degree line is at about 600 above the year 2007. So, approximately 600,000 associate degrees were conferred in the year 2007.

59. answers may vary

61. Since the line corresponding to cars is decreasing from left to right between 2000 to 2004, the passenger car sales are decreasing from 2000 to 2004.

63. The truck line is about halfway between 7 and 8 above the year 2000. So, approximately 7.5 million light trucks were sold in the United States in 2000.

65. answers may vary

Section 8.5

Practice Problems

1.

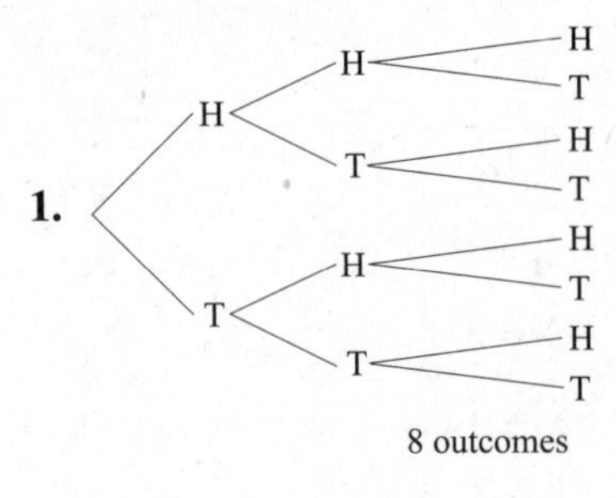

8 outcomes

2.

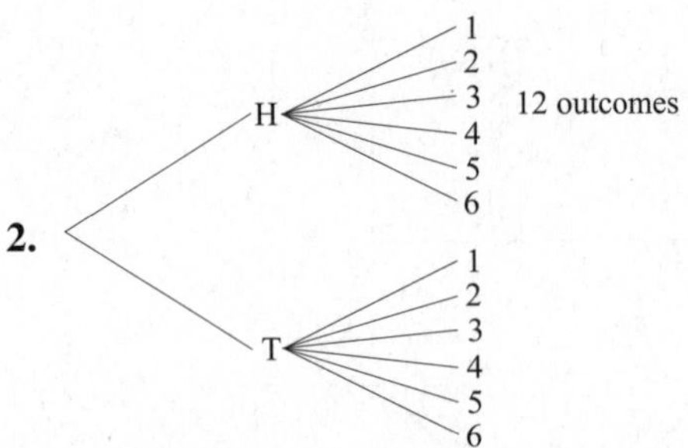

12 outcomes

3. The possibilities are:
H, H, H, H, H, T, H, T, H, H, T, T,
T, H, H, T, H, T, T, T, H, T, T, T
T, H, T is one of the 8 possible outcomes, so the probability is $\frac{1}{8}$.

4. A 2 or a 5 are two of the six possible outcomes. The probability is $\frac{2}{6} = \frac{1}{3}$.

5. A blue is 2 out of the 4 possible marbles. The probability is $\frac{2}{4} = \frac{1}{2}$.

Vocabulary and Readiness Check

1. A possible result of an experiment is called an outcome.

2. A tree diagram shows each outcome of an experiment as a separate branch.

3. The probability of an event is a measure of the likelihood of it occurring.

4. Probability is calculated by number of ways that the event can occur divided by number of possible outcomes.

5. A probability of 0 means that an event won't occur.

6. A probability of 1 means that an event is certain to occur.

Exercise Set 8.5

1.

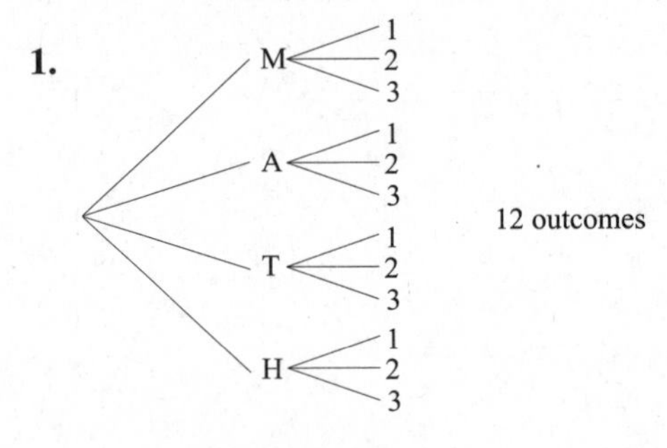

12 outcomes

3. Red, Blue, Yellow — 3 outcomes

Copyright © 2011 Pearson Education, Inc. Publishing as Prentice Hall.

5.

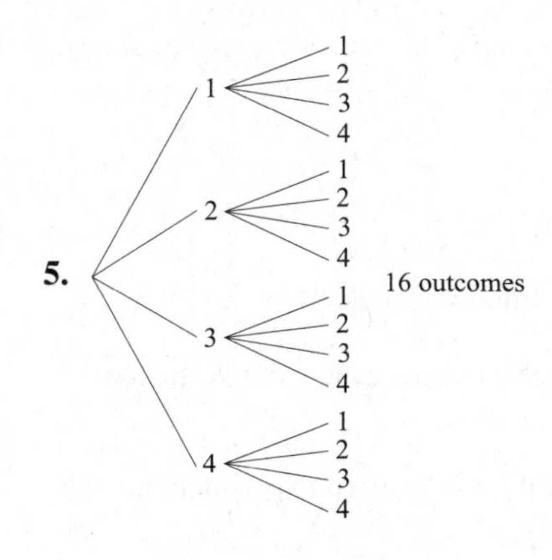

7.

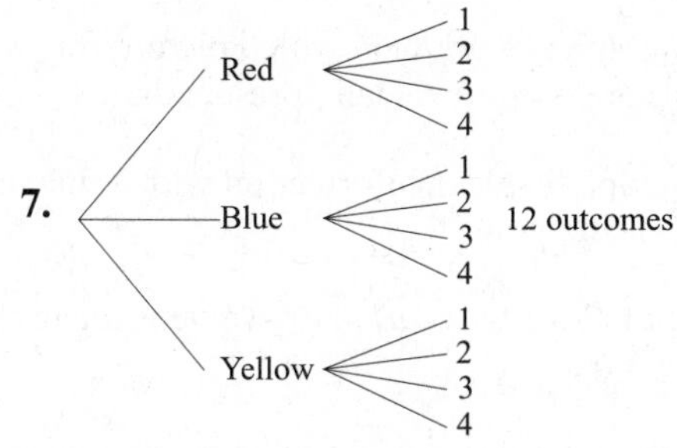

9.

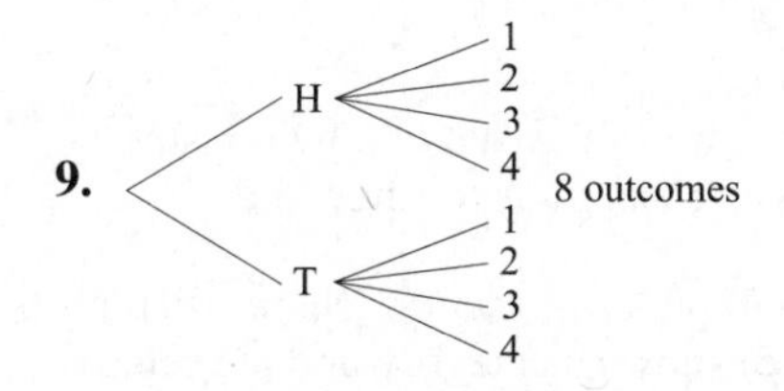

11. A 5 is one of the six possible outcomes. The probability is $\frac{1}{6}$.

13. A 1 or a 6 are two of the six possible outcomes. The probability is $\frac{2}{6} = \frac{1}{3}$.

15. Three of the six possible outcomes are even. The probability is $\frac{3}{6} = \frac{1}{2}$.

17. Four of the six possible outcomes are numbers greater than 2. The probability is $\frac{4}{6} = \frac{2}{3}$.

19. A 2 is one of three possible outcomes. The probability is $\frac{1}{3}$.

21. A 1, a 2, or a 3 are three of three possible outcomes. The probability is $\frac{3}{3} = 1$.

23. An odd number is a 1 or a 3, which are two of three possible outcomes. The probability is $\frac{2}{3}$.

25. One of the seven marbles is red. The probability is $\frac{1}{7}$.

27. Two of the seven marbles are yellow. The probability is $\frac{2}{7}$.

29. Four of the seven marbles are either green or red. The probability is $\frac{4}{7}$.

31. The blood pressure was higher for 38 of the 200 people. The probability is $\frac{38}{200} = \frac{19}{100}$.

33. The blood pressure did not change for 10 of the 200 people. The probability is $\frac{10}{200} = \frac{1}{20}$.

35. $\frac{1}{2} + \frac{1}{3} = \frac{1}{2} \cdot \frac{3}{3} + \frac{1}{3} \cdot \frac{2}{2} = \frac{3}{6} + \frac{2}{6} = \frac{3+2}{6} = \frac{5}{6}$

37. $\frac{1}{2} \cdot \frac{1}{3} = \frac{1 \cdot 1}{2 \cdot 3} = \frac{1}{6}$

39. $5 \div \frac{3}{4} = \frac{5}{1} \div \frac{3}{4} = \frac{5}{1} \cdot \frac{4}{3} = \frac{5 \cdot 4}{1 \cdot 3} = \frac{20}{3}$ or $6\frac{2}{3}$

41. One of the 52 cards is the king of hearts. The probability is $\frac{1}{52}$.

43. Four of the 52 cards are kings. The probability is $\frac{4}{52} = \frac{1}{13}$.

45. Thirteen of the 52 cards are hearts. The probability is $\frac{13}{52} = \frac{1}{4}$.

Copyright © 2011 Pearson Education, Inc. Publishing as Prentice Hall.

47. Twenty six of the cards are in black ink. The probability is $\frac{26}{52}=\frac{1}{2}$.

Tree diagram for 49.–51.

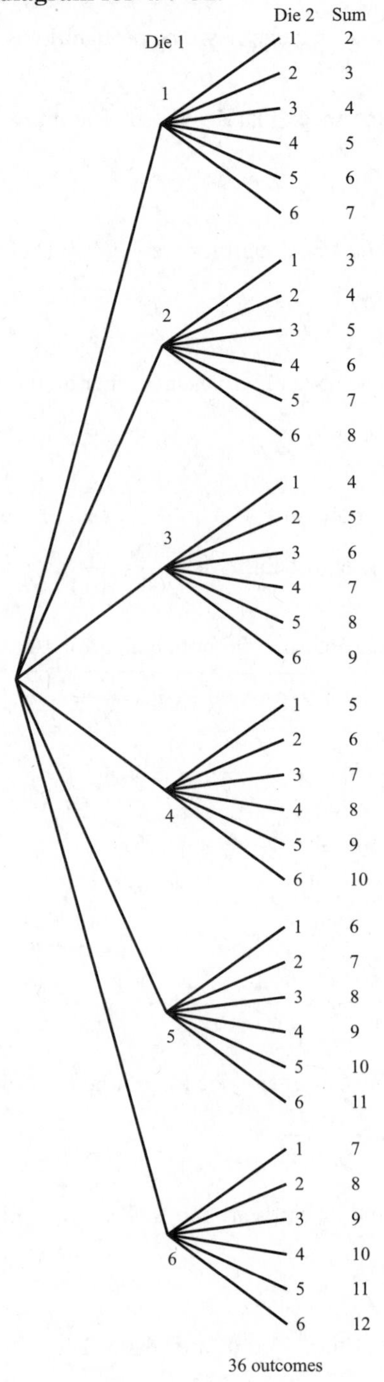

49. Five of the 36 sums are 6. The probability is $\frac{5}{36}$.

51. None of the 36 sums are 13. The probability is $\frac{0}{36}=0$.

53. answers may vary

Chapter 8 Vocabulary Check

1. A bar graph presents data using vertical or horizontal bars.

2. The possible results of an experiment are the outcomes.

3. A pictograph is a graph in which pictures or symbols are used to visually present data.

4. A line graph displays information with a line that connects data points.

5. In the ordered pair (a, b), the x-value is a and the y-value is b.

6. A tree diagram is one way to picture and count outcomes.

7. An experiment is an activity being considered, such as tossing a coin or rolling a die.

8. In a circle graph, each section (shaped like a piece of pie) shows a category and the relative size of the category.

9. The probability of an event is $\frac{\text{number of ways that event can occur}}{\text{number of possible outcomes}}$.

10. A histogram is a special bar graph in which the width of each bar represents a class interval and the height of each bar represents the class frequency.

11. The point of intersection of the x-axis and the y-axis in the rectangular coordinate system is called the origin.

12. The axes divide the plane into 4 regions, called quadrants.

13. The process of locating a point on the rectangular coordinate system is called plotting the point.

14. A linear equation in two variables can be written in the form $ax + by = c$.

Copyright © 2011 Pearson Education, Inc. Publishing as Prentice Hall.

Chapter 8 Review

1. Midwest has 4 houses, and each house represents 500,000 homes, so there were 4(500,000) = 2,000,000 new homes constructed in the Midwest.

2. Northeast has 3.5 houses, and each house represents 500,000 homes, so there were 3.5(500,000) = 1,750,000 new homes constructed in the Northeast.

3. South has the greatest number of houses, so the most new homes constructed were in the South.

4. Northeast has the least number of houses, so the fewest new homes constructed were in the Northeast.

5. Each house represents 500,000 homes, so look for the regions with $\frac{3,000,000}{500,000} = 6$ or more houses. The South and West had 3,000,000 or more new homes constructed.

6. Each house represents 500,000 homes, so look for the regions with fewer than $\frac{3,000,000}{500,000} = 8$ houses. The Northeast and Midwest had fewer than 3,000,000 new homes constructed.

7. The height of the bar representing 1970 is 11. Thus, approximately 11% of persons completed four or more years of college in 1970.

8. The tallest bar corresponds to 2006. Thus, the greatest percent of persons completing four or more years of college was in 2006.

9. The bars whose height is at a level of 20 or more are 1990, 2000, and 2006. Thus, 20% or more persons completed four or more years of college in 1990, 2000, and 2006.

10. answers may vary

11. The point on the graph corresponding to 2008 is about 960. Thus, there were approximately 960 medals awarded at the Summer Olympics in 2008.

12. The point on the graph corresponding to 2000 is about 920. Thus, there were approximately 920 medals awarded at the Summer Olympics in 2000.

13. The point on the graph corresponding to 2004 is about 930. Thus, there were approximately 930 medals awarded at the Summer Olympics in 2004.

14. The point on the graph corresponding to 1992 is about 815. Thus, there were approximately 815 medals awarded at the Summer Olympics in 1992.

15. The points on the graph corresponding to 1996 and 1992 are 840 and 815, respectively. Thus, there were 840 − 815 = 25 more medals awarded in 1996 than in 1992.

16. The points on the graph corresponding to 2008 and 1992 are 960 and 815, respectively. Thus, there were 960 − 815 = 145 more medals awarded in 2008 than in 1992.

17. The height of the bar corresponding to 41–45 is 1. Thus, 1 employee works 41–45 hours per week.

18. The height of the bar corresponding to 21–25 is 4. Thus, 4 employees work 21–25 hours per week.

19. Add the heights of the bars corresponding to 16–20, 21–25, and 26–30. Thus, 6 + 4 + 8 = 18 employees work 30 hours or less per week.

20. Add the heights of the bars corresponding to 36–40 and 41–45. Thus, 8 + 1 = 9 employees work 36 or more hours per week.

	Class Interval (Temperatures)	Tally	Class Frequency (Number of Months)
21.	80°–89°	𝍸	5
22.	90°–99°	|||	3
23.	100°–109°	||||	4

Copyright © 2011 Pearson Education, Inc. Publishing as Prentice Hall.

24.

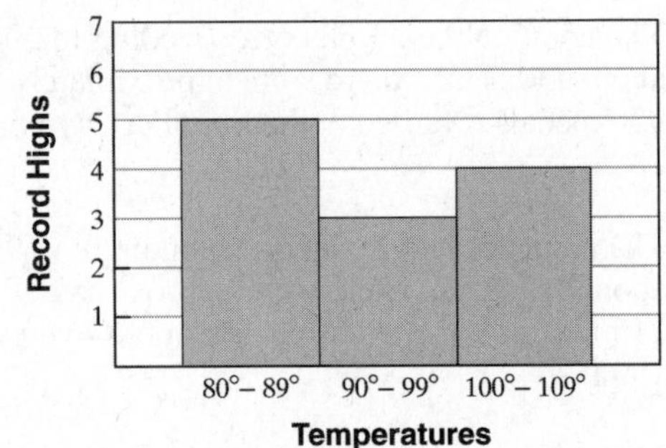

25. The largest sector corresponds to the category "Mortgage payment," thus the largest budget item is mortgage payment.

26. The smallest sector corresponds to the category "Utilities," thus the smallest budget item is utilities.

27. Add the amounts for mortgage payment and utilities. Thus, \$975 + \$250 = \$1225 is budgeted for the mortgage payment and utilities.

28. Add the amounts for savings and contributions. Thus, \$400 + \$300 = \$700 is budgeted for savings and contributions.

29. $\frac{\text{mortgage payment}}{\text{total}} = \frac{\$975}{\$4000} = \frac{39 \cdot 25}{160 \cdot 25} = \frac{39}{160}$

The ratio is $\frac{39}{160}$.

30. $\frac{\text{food}}{\text{total}} = \frac{\$700}{\$4000} = \frac{7 \cdot 100}{40 \cdot 100} = \frac{7}{40}$

The ratio is $\frac{7}{40}$.

31. The sector corresponding to Asia is 62%.
62% of 61 = $0.62 \cdot 61 \approx 38$
Thus, 38 tall buildings are located in Asia.

32. The sector corresponding to North America is 29.5%.
29.5% of 61 = $0.295 \cdot 61 \approx 18$
Thus, 18 tall buildings are located in North America.

33. The sector corresponding to Oceania is 1.6%.
1.6% of 61 = $0.016 \cdot 61 \approx 1$
Thus, 1 tall building is located in Oceania.

34. The sector corresponding to Europe is 6.6%.
6.6% of 61 = $0.066 \cdot 61 \approx 4$
Thus, 4 tall buildings are located in Europe.

35. $x = -6y$
$0 = -6y$
$\frac{0}{-6} = \frac{-6y}{-6}$
$0 = y$
The solution is (0, 0).
$x = -6y$
$x = -6(-1)$
$x = 6$
The solution is (6, −1).
$x = -6y$
$-6 = -6y$
$\frac{-6}{-6} = \frac{-6y}{-6}$
$1 = y$
The solution is (−6, 1).

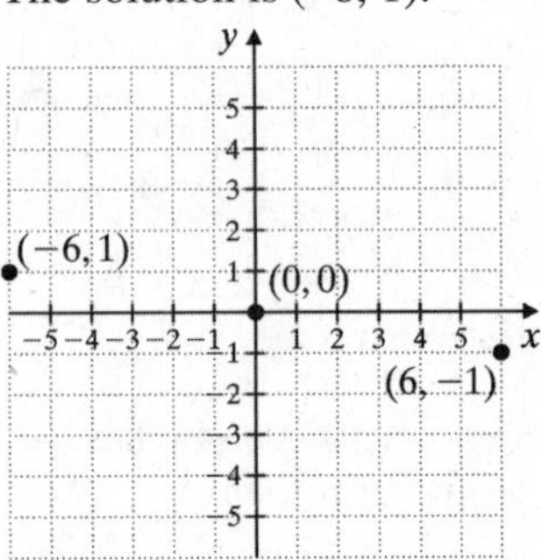

36. $y = 3x - 2$
$y = 3 \cdot 0 - 2$
$y = 0 - 2$
$y = -2$
The solution is (0, −2).
$y = 3x - 2$
$y = 3(-1) - 2$
$y = -3 - 2$
$y = -5$
The solution is (−1, −5).
$y = 3x - 2$
$y = 3 \cdot 2 - 2$
$y = 6 - 2$
$y = 4$
The solution is (2, 4).

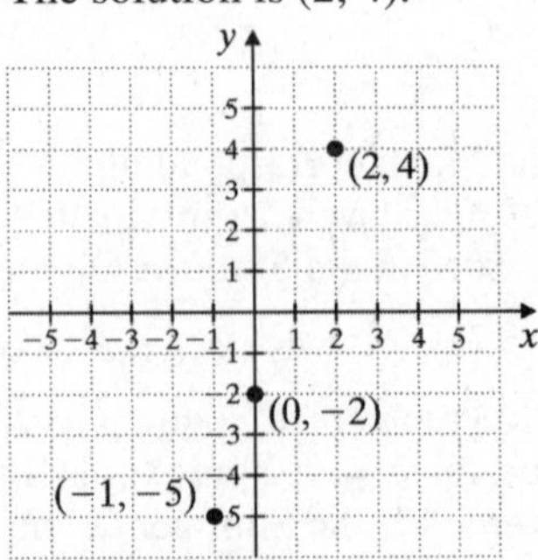

Copyright © 2011 Pearson Education, Inc. Publishing as Prentice Hall.

37. $$\begin{aligned} x+y&=-4\\ -1+y&=-4\\ 1-1+y&=1-4\\ y&=-3 \end{aligned}$$
The solution is (–1, –3).
$$\begin{aligned} x+y&=-4\\ x+0&=-4\\ x&=-4 \end{aligned}$$
The solution is (–4, 0).
$$\begin{aligned} x+y&=-4\\ -5+y&=-4\\ 5-5+y&=5-4\\ y&=1 \end{aligned}$$
The solution is (–5, 1).

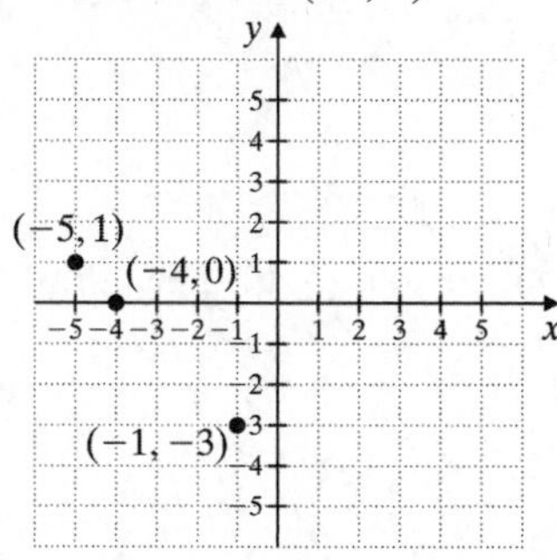

38. $$\begin{aligned} x-y&=3\\ 4-y&=3\\ -4+4-y&=-4+3\\ -y&=-1\\ y&=1 \end{aligned}$$
The solution is (4, 1).
$$\begin{aligned} x-y&=3\\ 0-y&=3\\ -y&=3\\ y&=-3 \end{aligned}$$
The solution is (0, –3).
$$\begin{aligned} x-y&=3\\ x-3&=3\\ x-3+3&=3+3\\ x&=6 \end{aligned}$$
The solution is (6, 3).

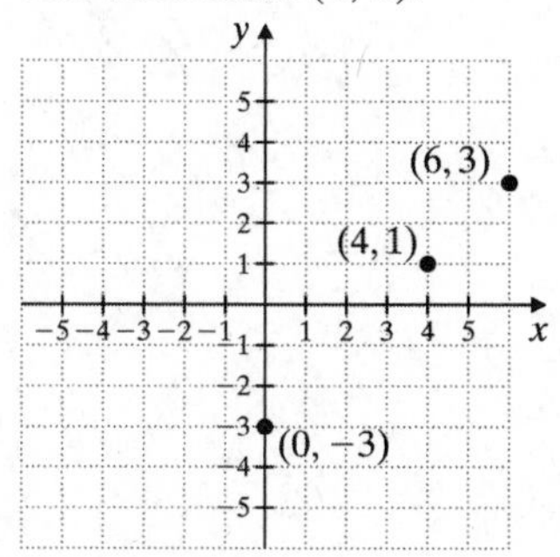

39. $$\begin{aligned} y&=3x\\ y&=3\cdot 1\\ y&=3 \end{aligned}$$
The solution is (1, 3).
$$\begin{aligned} y&=3x\\ y&=3(-2)\\ y&=-6 \end{aligned}$$
The solution is (–2, –6).
$$\begin{aligned} y&=3x\\ 0&=3x\\ \frac{0}{3}&=\frac{3x}{3}\\ 0&=x \end{aligned}$$
The solution is (0, 0).

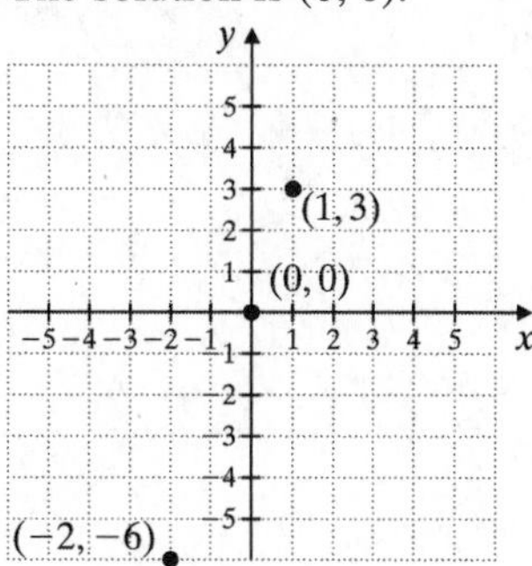

40. $$\begin{aligned} x&=y+6\\ 1&=y+6\\ 1-6&=y+6-6\\ -5&=y \end{aligned}$$
The solution is (1, –5).
$$\begin{aligned} x&=y+6\\ 6&=y+6\\ 6-6&=y+6-6\\ 0&=y \end{aligned}$$
The solution is (6, 0).
$$\begin{aligned} x&=y+6\\ x&=-4+6\\ x&=2 \end{aligned}$$
The solution is (2, –4).

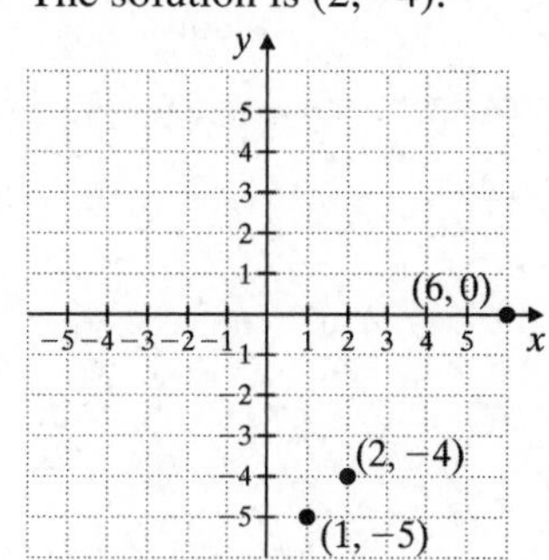

Copyright © 2011 Pearson Education, Inc. Publishing as Prentice Hall.

41. $x = -5$
No matter what *y*-value we choose, *x* is always −5.

x	y
−5	−2
−5	0
−5	2

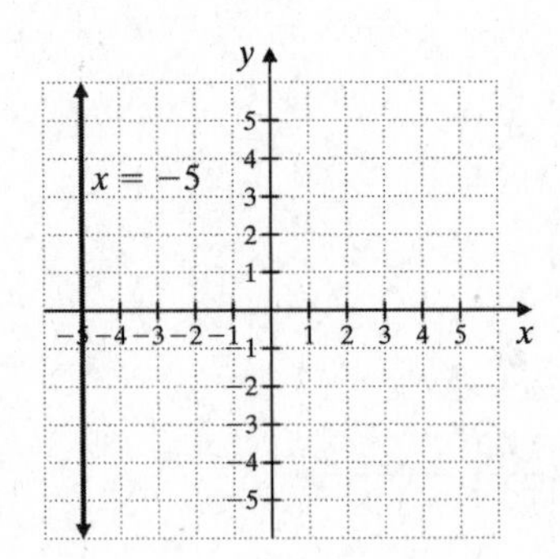

42. $y = \frac{3}{2}$
No matter what *x*-value we choose, *y* is always $\frac{3}{2}$.

x	y
−2	$\frac{3}{2}$
0	$\frac{3}{2}$
2	$\frac{3}{2}$

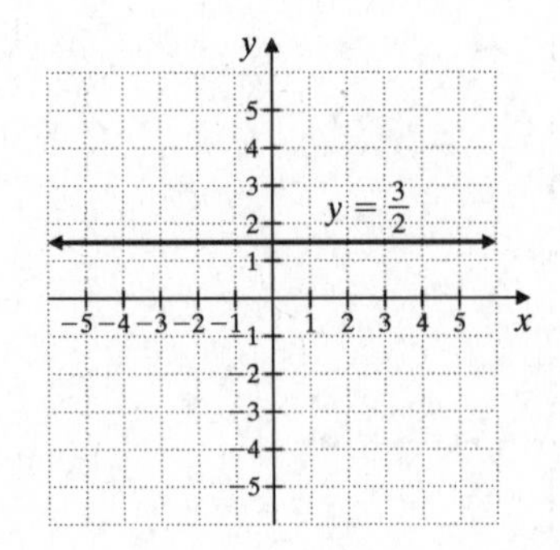

43. $x + y = 11$
Find any 3 ordered pair solutions.
Let $x = 0$.

$$x + y = 11$$
$$0 + y = 11$$
$$y = 11$$

(0, 11)
Let $y = 0$.

$$x + y = 11$$
$$x + 0 = 11$$
$$x = 11$$

(11, 0)
Let $x = 5$.

$$x + y = 11$$
$$5 + y = 11$$
$$5 + y = 11$$
$$5 - 5 + y = 11 - 5$$
$$y = 6$$

(5, 6)
Plot (0, 11), (11, 0), and (5, 6). Then draw a line through them.

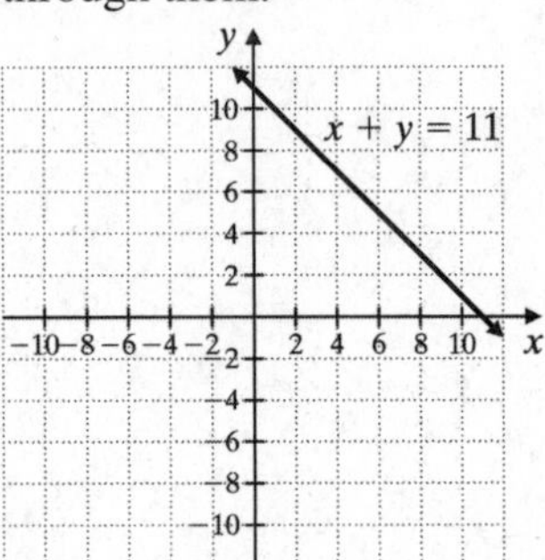

44. $x - y = 11$
Find any 3 ordered pair solutions.
Let $x = 0$.

$$x - y = 11$$
$$0 - y = 11$$
$$-y = 11$$
$$y = -11$$

(0, −11)
Let $y = 0$.

$$x - y = 11$$
$$x - 0 = 11$$
$$x = 11$$

(11, 0)
Let $x = 5$.

$$x - y = 11$$
$$5 - y = 11$$
$$5 - 5 - y = 11 - 5$$
$$-y = 6$$
$$y = -6$$

(5, −6)

Copyright © 2011 Pearson Education, Inc. Publishing as Prentice Hall.

Plot (0, −11), (11, 0), and (5, −6). Then draw a line through them.

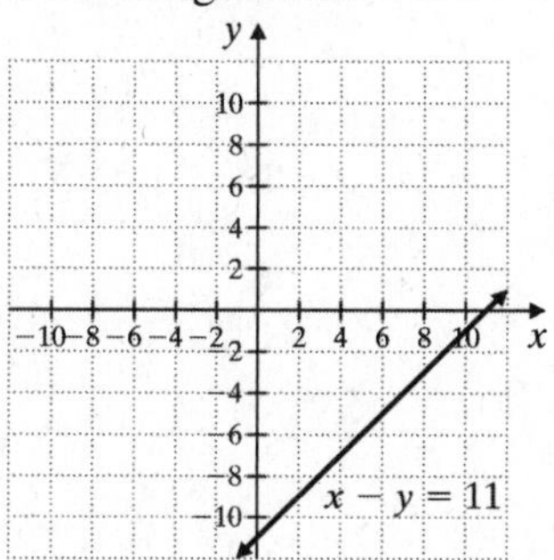

45. $y = 4x - 2$
Find any 3 ordered pair solutions.
Let $x = 0$.
$y = 4x - 2$
$y = 4 \cdot 0 - 2$
$y = 0 - 2$
$y = -2$
(0, −2)
Let $y = 0$.

$$\begin{aligned} y &= 4x - 2 \\ 0 &= 4x - 2 \\ 0 + 2 &= 4x - 2 + 2 \\ 2 &= 4x \\ \frac{2}{4} &= \frac{4x}{4} \\ \frac{1}{2} &= x \end{aligned}$$

$\left(\frac{1}{2}, 0\right)$
Let $x = 1$.
$y = 4x - 2$
$y = 4 \cdot 1 - 2$
$y = 4 - 2$
$y = 2$
(1, 2)

Plot (0, −2), $\left(\frac{1}{2}, 0\right)$, and (1, 2). Then draw a line through them.

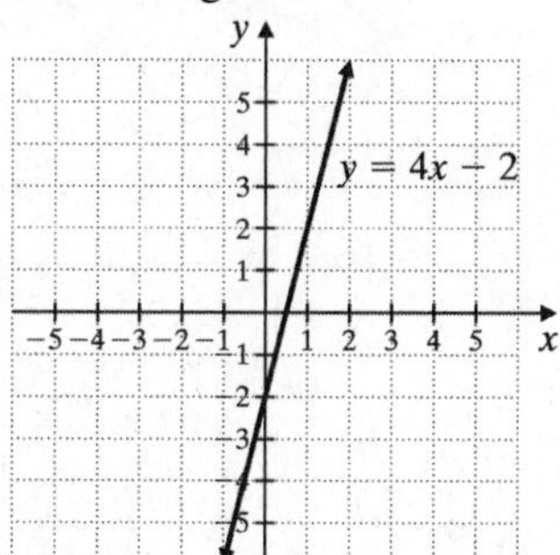

46. $y = 5x$
Find any 3 ordered pair solutions.
Let $x = 0$.
$y = 5x$
$y = 5 \cdot 0$
$y = 0$
(0, 0)
Let $x = 1$.
$y = 5x$
$y = 5(1)$
$y = 5$
(1, 5)
Let $x = -1$.
$y = 5x$
$y = 5(-1)$
$y = -5$
(−1, −5)
Plot (0, 0), (1, 5), and (−1, −5). Then draw a line through them.

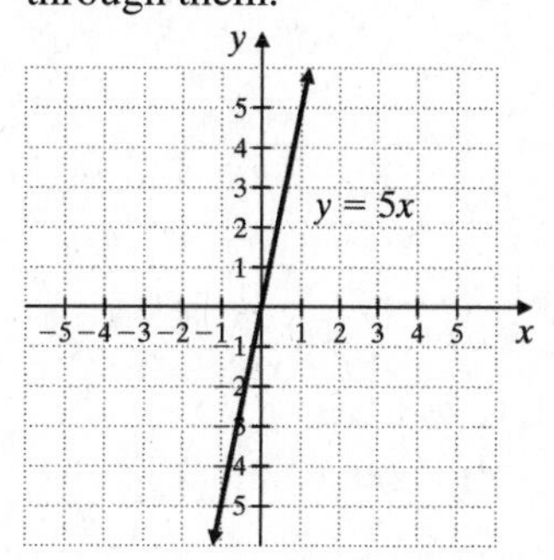

47. $x = -2y$
Find any 3 ordered pair solutions.
Let $y = 0$.
$x = -2y$
$x = -2 \cdot 0$
$x = 0$
(0, 0)
Let $y = -1$.
$x = -2y$
$x = -2(-1)$
$x = 2$
(2, −1)
Let $y = 1$.
$x = -2y$
$x = -2(1)$
$x = -2$
(−2, 1)
Plot (0, 0), (2, −1), and (−2, 1). Then draw a line through them.

Copyright © 2011 Pearson Education, Inc. Publishing as Prentice Hall.

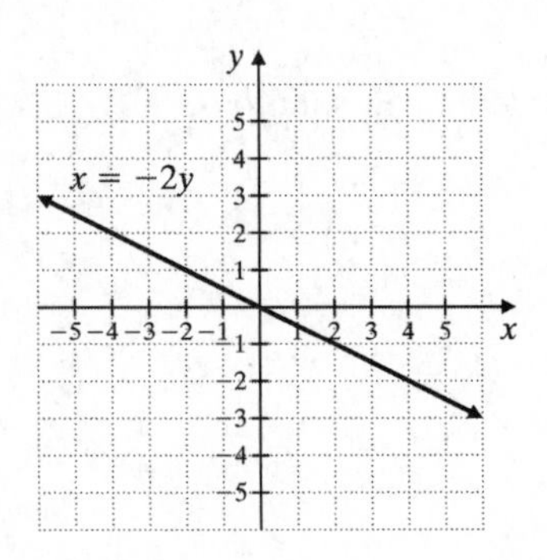

48. $x + y = -1$
Find any 3 ordered pair solutions.
Let $x = 0$.

$$x + y = -1$$
$$0 + y = -1$$
$$y = -1$$

(0, −1)
Let $y = 0$.

$$x + y = -1$$
$$x + 0 = -1$$
$$x = -1$$

(−1, 0)
Let $x = -2$.

$$x + y = -1$$
$$-2 + y = -1$$
$$-2 + 2 + y = -1 + 2$$
$$y = 1$$

(−2, 1)
Plot (0, −1), (−1, 0), and (−2, 1). Then draw a line through them.

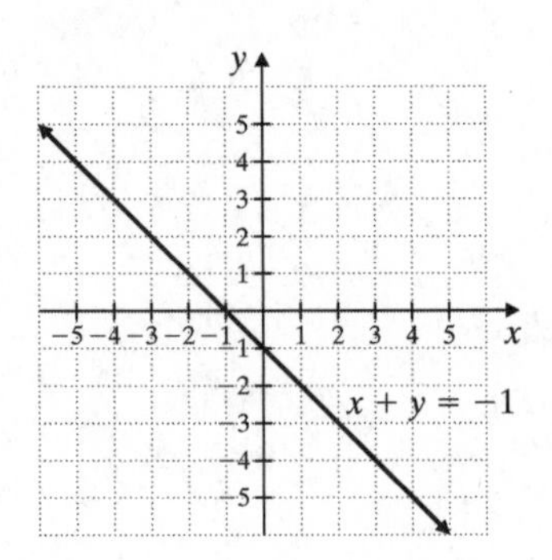

49. $2x - 3y = 12$
Find any 3 ordered pair solutions.
Let $x = 0$.

$$2x - 3y = 12$$
$$2 \cdot 0 - 3y = 12$$
$$0 - 3y = 12$$
$$-3y = 12$$
$$\frac{-3y}{-3} = \frac{12}{-3}$$
$$y = -4$$

(0, −4)
Let $y = 0$.

$$2x - 3y = 12$$
$$2x - 3 \cdot 0 = 12$$
$$2x - 0 = 12$$
$$2x = 12$$
$$\frac{2x}{2} = \frac{12}{2}$$
$$x = 6$$

(6, 0)
Let $x = 3$.

$$2x - 3y = 12$$
$$2 \cdot 3 - 3y = 12$$
$$6 - 3y = 12$$
$$6 - 6 - 3y = 12 - 6$$
$$-3y = 6$$
$$\frac{-3y}{-3} = \frac{6}{-3}$$
$$y = -2$$

(3, −2)
Plot (0, −4), (6, 0), and (3, −2). Then draw a line through them.

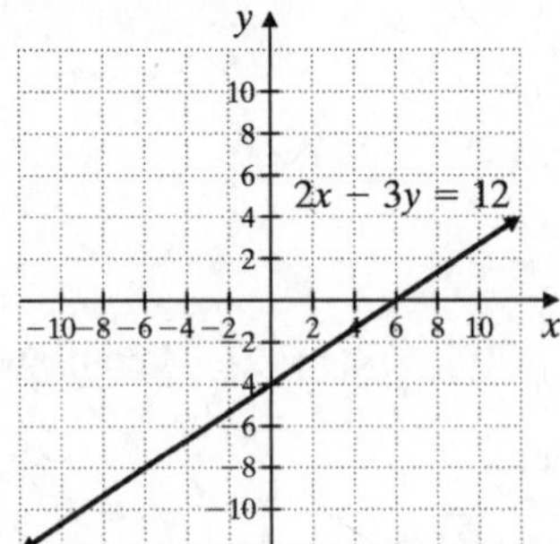

Copyright © 2011 Pearson Education, Inc. Publishing as Prentice Hall.

50. $x = \frac{1}{2}y$

Find any 3 ordered pair solutions.

Let $y = 0$.

$x = \frac{1}{2}y$

$x = \frac{1}{2} \cdot 0$

$x = 0$

(0, 0)

Let $y = 2$.

$x = \frac{1}{2}y$

$x = \frac{1}{2} \cdot 2$

$x = 1$

(1, 2)

Let $y = 4$.

$x = \frac{1}{2}y$

$x = \frac{1}{2} \cdot 4$

$x = 2$

(2, 4)

Plot (0, 0), (1, 2), and (2, 4). Then draw a line through them.

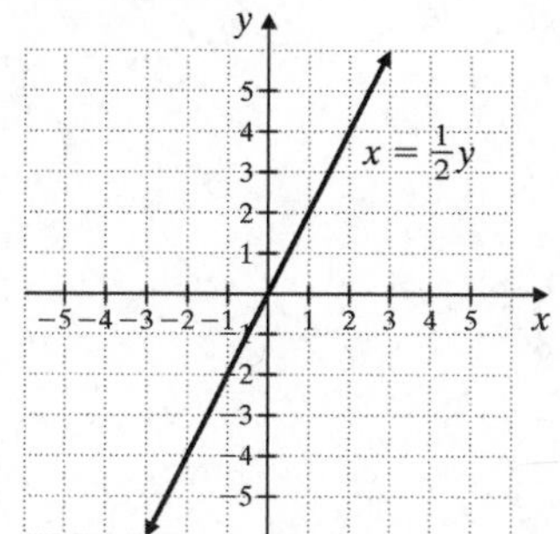

51.

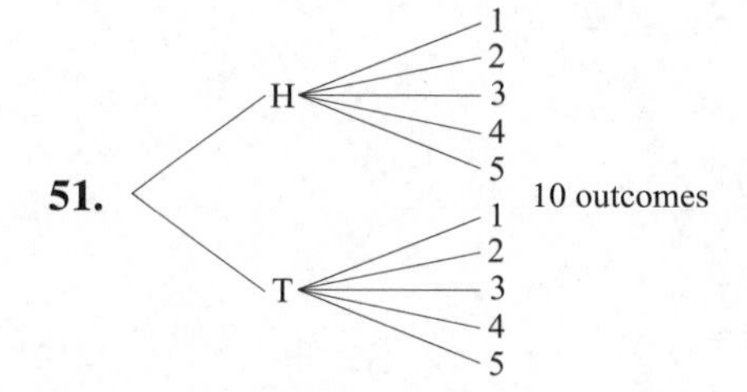

52.

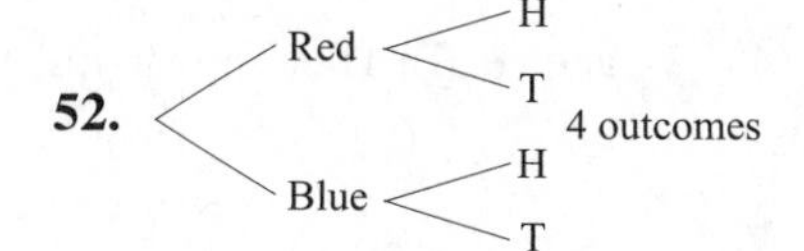

53.

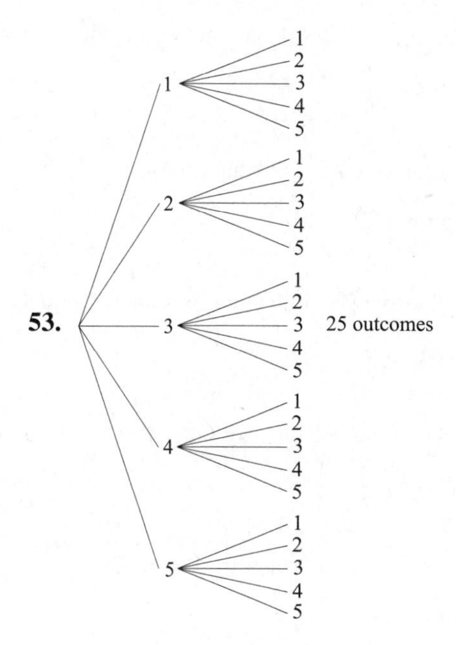

54.

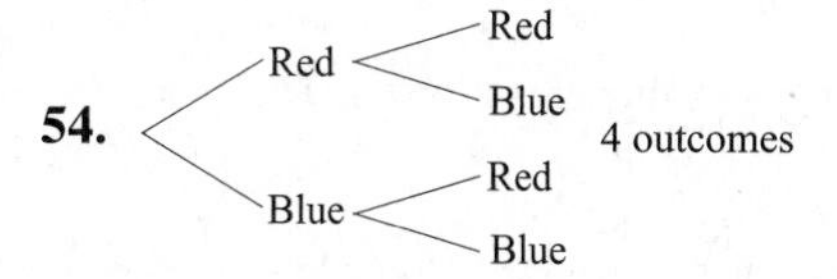

55.

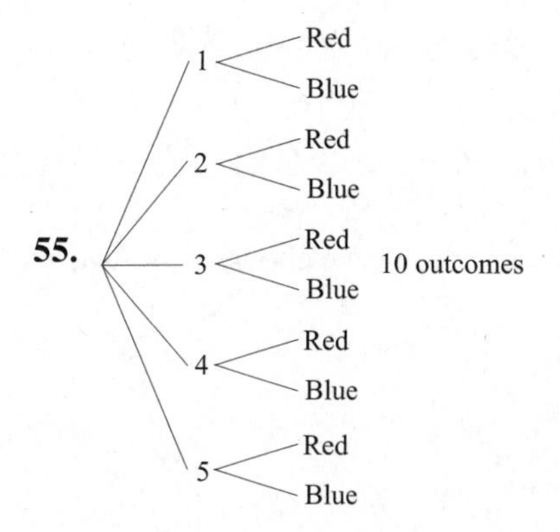

56. One of the six possible outcomes is 4. The probability is $\frac{1}{6}$.

57. One of the six possible outcomes is 3. The probability is $\frac{1}{6}$.

58. One of the five possible outcomes is 4. The probability is $\frac{1}{5}$.

Copyright © 2011 Pearson Education, Inc. Publishing as Prentice Hall.

59. One of the five possible outcomes is 3. The probability is $\frac{1}{5}$.

60. Three of the five possible outcomes are a 1, 3, or 5. The probability is $\frac{3}{5}$.

61. Two of the five possible outcomes are a 2 or a 4. The probability is $\frac{2}{5}$.

62. Two of the eight marbles are blue. The probability is $\frac{2}{8}=\frac{1}{4}$.

63. Three of the eight marbles are yellow. The probability is $\frac{3}{8}$.

64. Two of the eight marbles are red. The probability is $\frac{2}{8}=\frac{1}{4}$.

65. One of the eight marbles is green. The probability is $\frac{1}{8}$.

66. $x=-4$
No matter what y-value we choose, x is always -4.

x	y
−4	−2
−4	0
−4	2

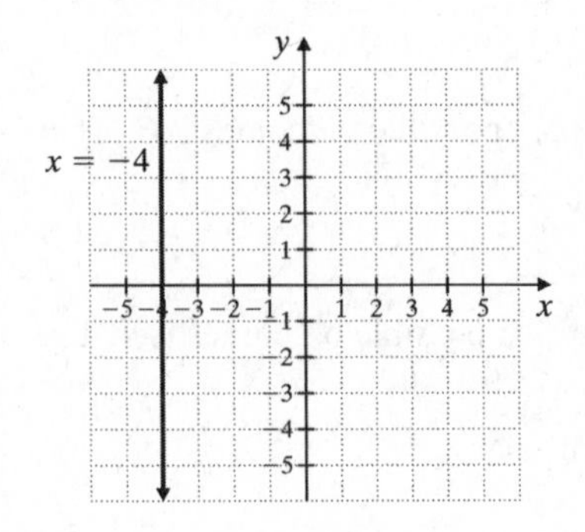

67. $y=3$
No matter what x-value we choose, y is always 3.

x	y
−2	3
0	3
2	3

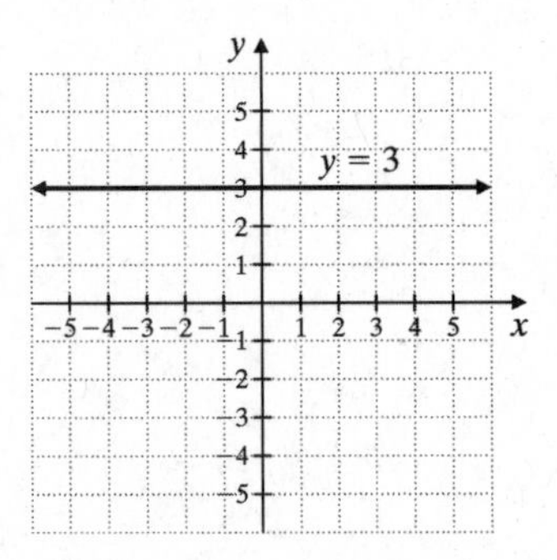

68. $x-2y=8$
Find any 3 ordered pair solutions.
Let $x=0$.

$$x-2y=8$$
$$0-2y=8$$
$$-2y=8$$
$$\frac{-2y}{-2}=\frac{8}{-2}$$
$$y=-4$$

(0, −4)
Let $x=4$.

$$x-2y=8$$
$$4-2y=8$$
$$-4+4-2y=-4+8$$
$$-2y=4$$
$$\frac{-2y}{-2}=\frac{4}{-2}$$
$$y=-2$$

(4, −2)
Let $y=0$.

$$x-2y=8$$
$$x-2\cdot 0=8$$
$$x-0=8$$
$$x=8$$

(8, 0)
Plot (0, −4), (4, −2), and (8, 0). Then draw a line through them.

Copyright © 2011 Pearson Education, Inc. Publishing as Prentice Hall.

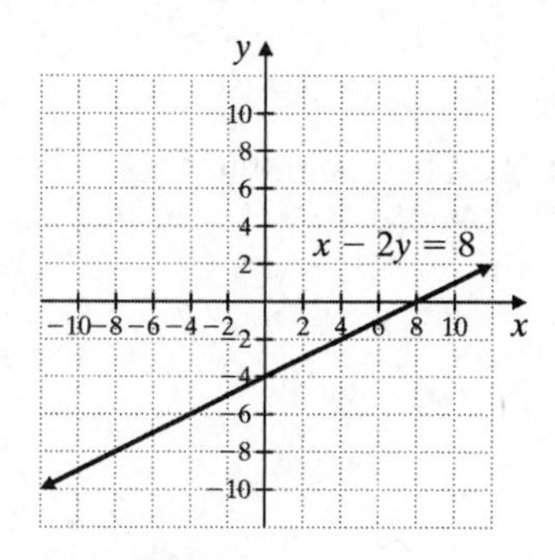

69. $2x + y = 6$
Find any 3 ordered pair solutions.
Let $x = 0$.

$$\begin{aligned} 2x + y &= 6 \\ 2\cdot 0 + y &= 6 \\ 0 + y &= 6 \\ y &= 6 \end{aligned}$$

(0, 6)
Let $y = 0$.

$$\begin{aligned} 2x + y &= 6 \\ 2x + 0 &= 6 \\ 2x &= 6 \\ \frac{2x}{2} &= \frac{6}{2} \\ x &= 3 \end{aligned}$$

(3, 0)
Let $x = -1$.

$$\begin{aligned} 2x + y &= 6 \\ 2(-1) + y &= 6 \\ -2 + y &= 6 \\ 2 - 2 + y &= 2 + 6 \\ y &= 8 \end{aligned}$$

(−1, 8)
Plot (0, 6), (3, 0), and (−1, 8). Then draw a line through them.

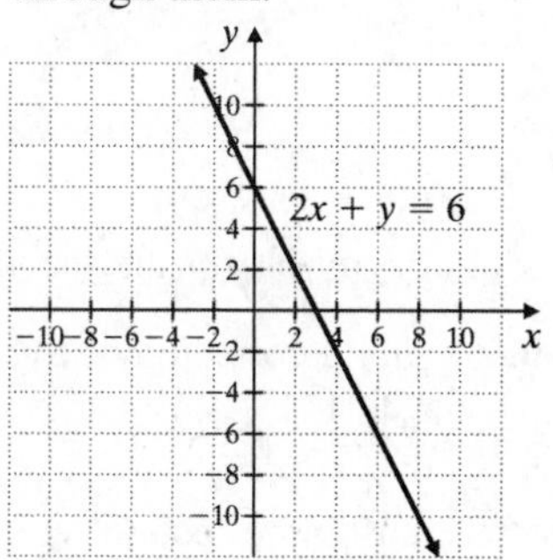

70. $x = y + 3$
Find any 3 ordered pair solutions.
Let $x = 0$.

$$\begin{aligned} x &= y + 3 \\ 0 &= y + 3 \\ 0 - 3 &= y + 3 - 3 \\ -3 &= y \end{aligned}$$

(0, −3)
Let $y = 0$.
$x = y + 3$
$x = 0 + 3$
$x = 3$
(3, 0)
Let $y = -1$.
$x = y + 3$
$x = -1 + 3$
$x = 2$
(2, −1)
Plot (0, −3), (3, 0), and (2, −1). Then draw a line through them.

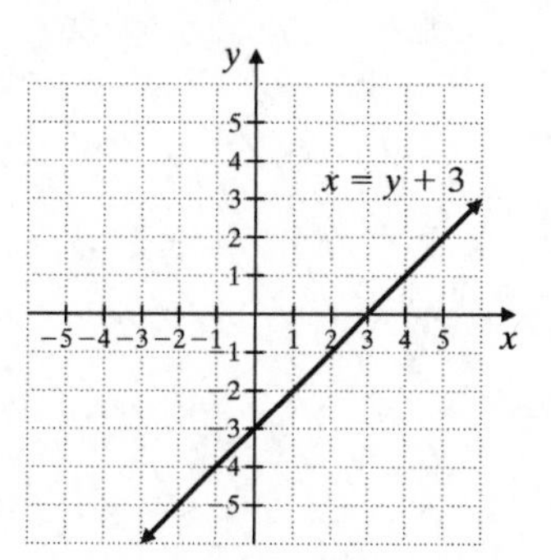

71. $x + y = -4$
Find any 3 ordered pair solutions.
Let $x = 0$.

$$\begin{aligned} x + y &= -4 \\ 0 + y &= -4 \\ y &= -4 \end{aligned}$$

(0, −4)
Let $y = 0$.

$$\begin{aligned} x + y &= -4 \\ x + 0 &= -4 \\ x &= -4 \end{aligned}$$

(−4, 0)
Let $x = -2$.

$$\begin{aligned} x + y &= -4 \\ -2 + y &= -4 \\ 2 - 2 + y &= 2 - 4 \\ y &= -2 \end{aligned}$$

(−2, −2)
Plot (0, −4), (−4, 0), and (−2, −2). Then draw a line through them.

Copyright © 2011 Pearson Education, Inc. Publishing as Prentice Hall.

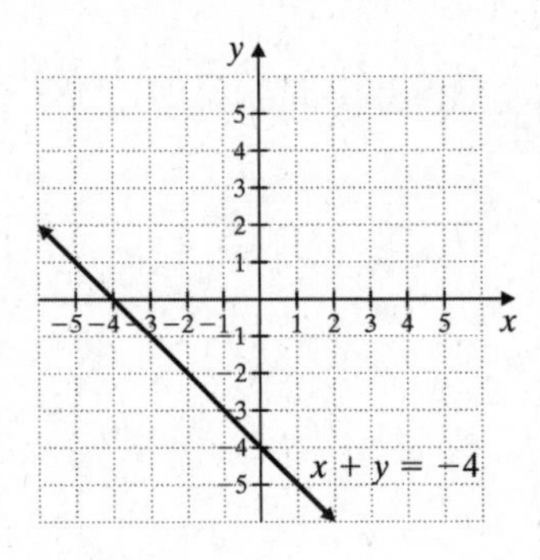

72. $y = \frac{3}{4}x$

Find any 3 ordered pair solutions.

Let $x = 0$.

$y = \frac{3}{4}x$

$y = \frac{3}{4} \cdot 0$

$y = 0$

$(0, 0)$

Let $x = 4$.

$y = \frac{3}{4}x$

$y = \frac{3}{4} \cdot 4$

$y = 3$

$(4, 3)$

Let $x = -4$.

$y = \frac{3}{4}x$

$y = \frac{3}{4}(-4)$

$y = -3$

$(-4, -3)$

Plot $(0, 0)$, $(4, 3)$, and $(-4, -3)$. Then draw a line through them.

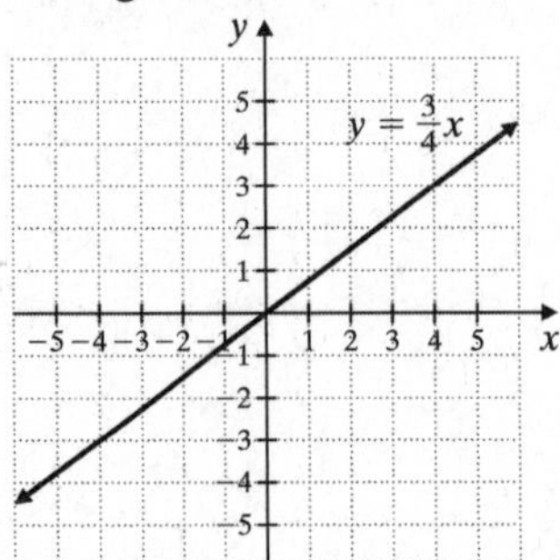

73. $y = -\frac{3}{4}x$

Find any 3 ordered pair solutions.

Let $x = 0$.

$y = -\frac{3}{4}x$

$y = -\frac{3}{4} \cdot 0$

$y = 0$

$(0, 0)$

Let $x = 4$.

$y = -\frac{3}{4}x$

$y = -\frac{3}{4} \cdot 4$

$y = -3$

$(4, -3)$

Let $x = -4$.

$y = -\frac{3}{4}x$

$y = -\frac{3}{4} \cdot (-4)$

$y = 3$

$(-4, 3)$

Plot $(0, 0)$, $(4, -3)$, and $(-4, 3)$. Then draw a line through them.

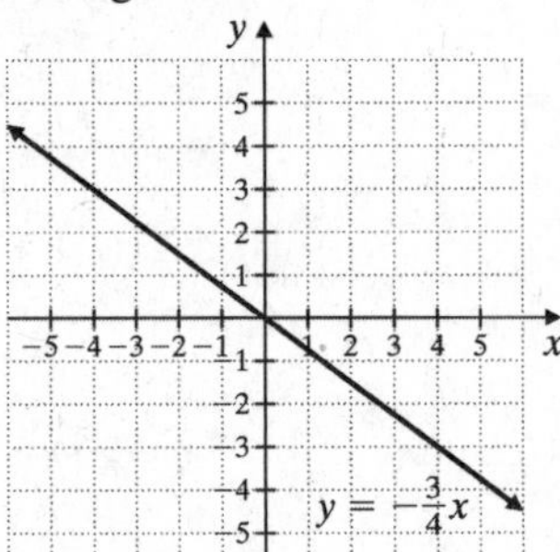

Chapter 8 Test

1. There are $4\frac{1}{2}$ dollar symbols for the second week. Each dollar symbol corresponds to \$50.

   $4\frac{1}{2} \cdot \$50 = \frac{9}{2} \cdot \$50 = \frac{\$450}{2} = \$225$

   \$225 was collected during the second week.

2. Week 3 has the greatest number of dollar symbols. So the most money was collected during the 3rd week. The 3rd week has 7 dollar symbols and each dollar symbol corresponds to \$50, so $7 \cdot \$50 = \$350$ was collected during week 3.

Copyright © 2011 Pearson Education, Inc. Publishing as Prentice Hall.

3. There are a total of 22 dollar symbols and each dollar symbol corresponds to \$50, so a total amount of $22 \cdot \$50 = \$1100$ was collected.

4. Look for the bars whose height is greater than 9. June, August, and September normally have more than 9 centimeters.

5. The shortest bar corresponds to February. The normal monthly rainfall in February in Chicago is 3 centimeters.

6. The bars corresponding to March and November have a height of 7. Thus, during March and November, 7 centimeters of precipitation normally occurs.

7.

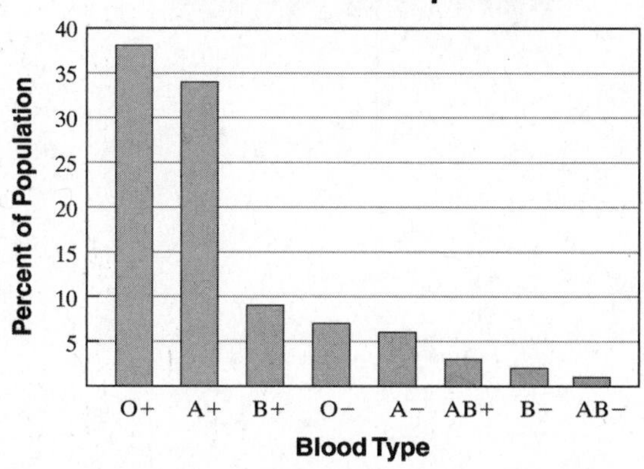

8. The point on the graph corresponding to 2003 is at about 2.25. Thus, the annual inflation rate in 2003 was about 2.25%.

9. The line graph is above the 3 level for 2000, 2005, 2006, and 2008. Thus the inflation rate was greater than 3% in 2000, 2005, 2006, and 2008.

10. Look for the years where the line graph is decreasing. During 1997–1998, 2000–2002, and 2005–2007, the inflation rate was decreasing.

11. $\frac{\text{number who prefer rock music}}{\text{total number}} = \frac{85}{200} = \frac{17}{40}$

The ratio is $\frac{17}{40}$.

12. $\frac{\text{number who prefer country}}{\text{number who prefer jazz}} = \frac{62}{44} = \frac{31}{22}$

The ratio is $\frac{31}{22}$.

13. 14% of 309 million $= 0.14 \cdot 309 \approx 43$ million.
Thus, 43 million people are expected to be in the twenties age group by 2010.

14. The sector corresponding to eighties plus is 4.2%.
4.2% of 309 million $= 0.042 \cdot 309 \approx 13$ million.
Thus, 13 million people are expected to be in the eighties plus group by 2010.

15. The height of the bar for 5'8"–5'11" is 9. Thus, there are 9 students who are 5'8"–5'11" tall.

Copyright © 2011 Pearson Education, Inc. Publishing as Prentice Hall.

16. Add the heights of the bars for 5'0"–5'3" and 5'4"–5'7". There are $5 + 6 = 11$ students who are 5'7" tall or shorter.

17.

Class Interval (Scores)	Tally	Class Frequency (Number of Students)
40–49	|	1
50–59	|||	3
60–69	||||	4
70–79	𝍸	5
80–89	𝍸 |||	8
90–99	||||	4

18.

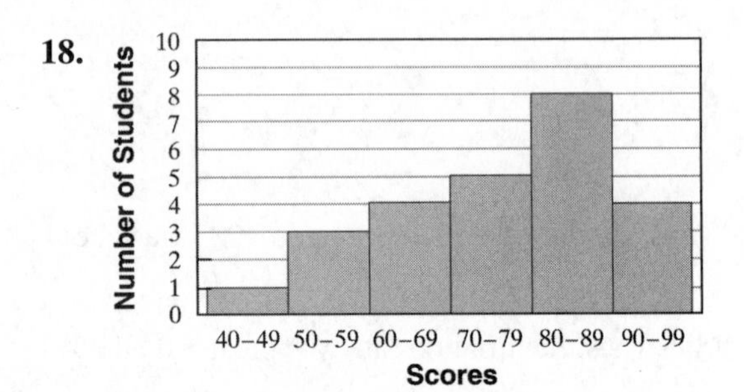

19. Point *A* has coordinates (4, 0).

20. Point *B* has coordinates (0, –3).

21. Point *C* has coordinates (–3, 4).

22. Point *D* has coordinates (–2, –1).

23. $x = -6y$
$0 = -6y$
$\frac{0}{-6} = \frac{-6y}{-6}$
$0 = y$
(0, 0)
$x = -6y$
$x = -6(1)$
$x = -6$
(–6, 1)

$x = -6y$
$12 = -6y$
$\frac{12}{-6} = \frac{-6y}{-6}$
$-2 = y$
(12, –2)

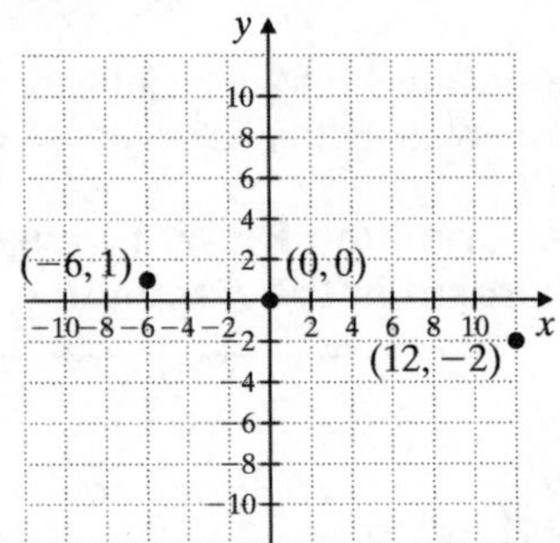

24. $y = 7x - 4$
$y = 7 \cdot 2 - 4$
$y = 14 - 4$
$y = 10$
(2, 10)
$y = 7x - 4$
$y = 7(-1) - 4$
$y = -7 - 4$
$y = -11$
(–1, –11)
$y = 7x - 4$
$y = 7 \cdot 0 - 4$
$y = 0 - 4$
$y = -4$
(0, –4)

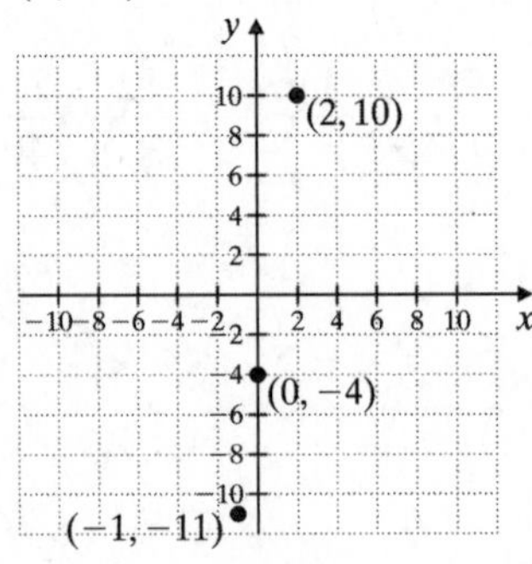

25. $y + x = -4$
Find any 3 ordered pair solutions.
Let $x = 0$.
$y + x = -4$
$y + 0 = -4$
$y = -4$
(0, –4)
Let $y = 0$.

Copyright © 2011 Pearson Education, Inc. Publishing as Prentice Hall.

$y+x=-4$
$0+x=-4$
$x=-4$
$(-4, 0)$
Let $x=-2$.
$y+x=-4$
$y+(-2)=-4$
$y+(-2)+2=-4+2$
$y=-2$

$(-2, -2)$
Plot $(0, -4)$, $(-4, 0)$, and $(-2, -2)$. Then draw a line through them.

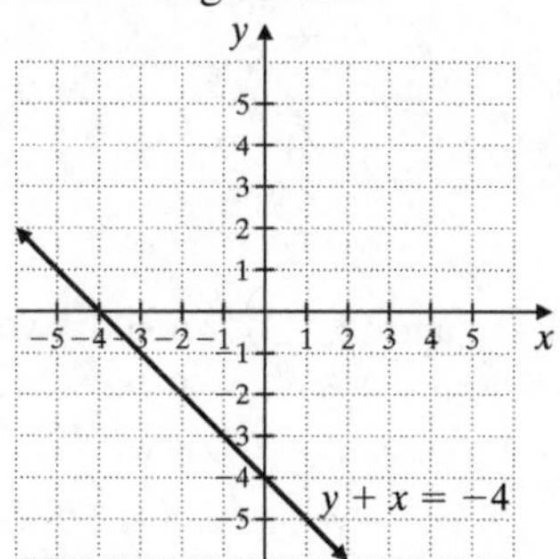

26. $y=-4$
No matter what x-value we choose, y is always -4.

x	y
−2	−4
0	−4
2	−4

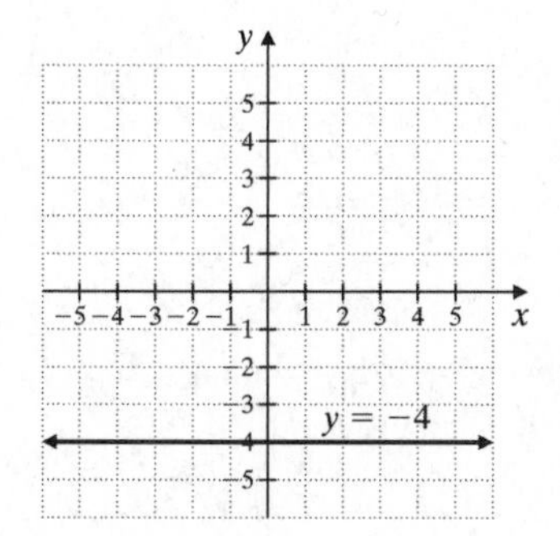

27. $y=3x-5$
Find any 3 ordered pair solutions.
Let $x=0$.
$y=3x-5$
$y=3\cdot 0-5$
$y=0-5$
$y=-5$
$(0, -5)$

Let $x=1$.
$y=3x-5$
$y=3\cdot 1-5$
$y=3-5$
$y=-2$
$(1, -2)$
Let $x=2$.
$y=3x-5$
$y=3\cdot 2-5$
$y=6-5$
$y=1$
$(2, 1)$
Plot $(0, -5)$, $(1, -2)$, and $(2, 1)$. Then draw a line through them.

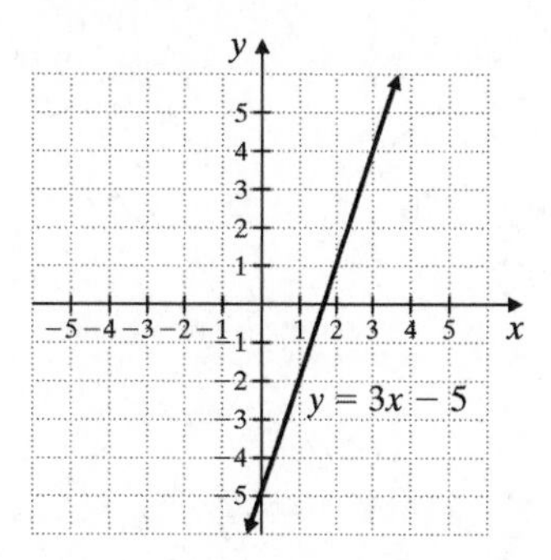

28. $x=5$
No matter what y-value we choose, x is always 5.

x	y
5	−2
5	0
5	2

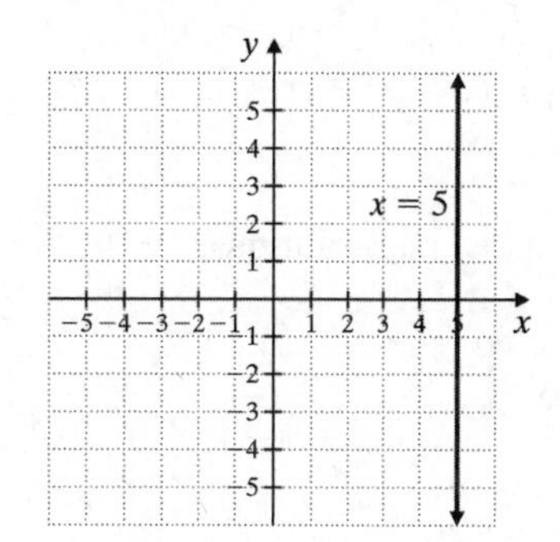

Copyright © 2011 Pearson Education, Inc. Publishing as Prentice Hall.

29. $y = -\frac{1}{2}x$

Find any 3 ordered pair solutions.
Let $x = 0$.

$y = -\frac{1}{2}x$
$y = -\frac{1}{2} \cdot 0$
$y = 0$

(0, 0)
Let $x = -2$.

$y = -\frac{1}{2}x$
$y = -\frac{1}{2} \cdot (-2)$
$y = 1$

(–2, 1)
Let $x = 2$.

$y = -\frac{1}{2}x$
$y = -\frac{1}{2} \cdot 2$
$y = -1$

(2, –1)
Plot (0, 0), (–2, 1), and (2, –1). Then draw a line through them.

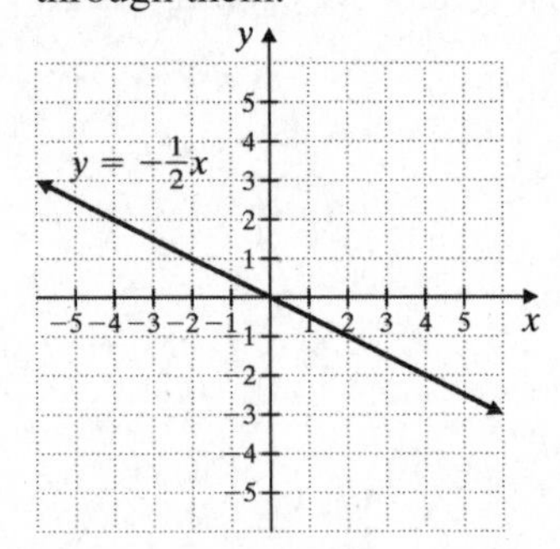

30. $3x - 2y = 12$

Find any 3 ordered pair solutions.
Let $x = 0$.

$3x - 2y = 12$
$3 \cdot 0 - 2y = 12$
$-2y = 12$
$\frac{-2y}{-2} = \frac{12}{-2}$
$y = -6$

(0, –6)
Let $y = 0$.

$3x - 2y = 12$
$3x - 2 \cdot 0 = 12$
$3x = 12$
$\frac{3x}{3} = \frac{12}{3}$
$x = 4$

(4, 0)
Let $x = 2$.

$3x - 2y = 12$
$3 \cdot 2 - 2y = 12$
$6 - 2y = 12$
$6 - 6 - 2y = 12 - 6$
$-2y = 6$
$\frac{-2y}{-2} = \frac{6}{-2}$
$y = -3$

(2, –3)
Plot (0, –6), (4, 0), and (2, –3). Then draw a line through them.

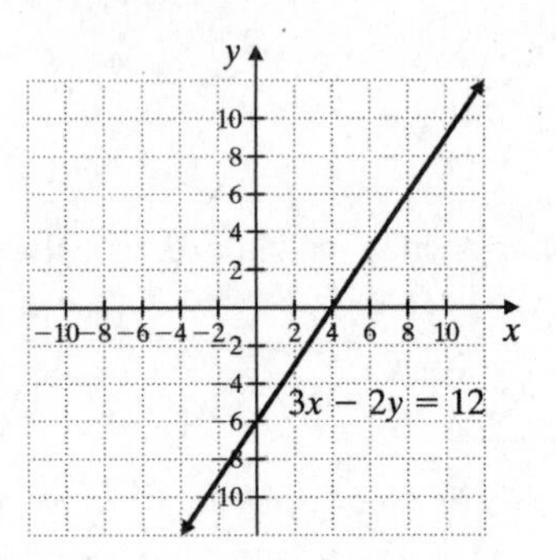

31.

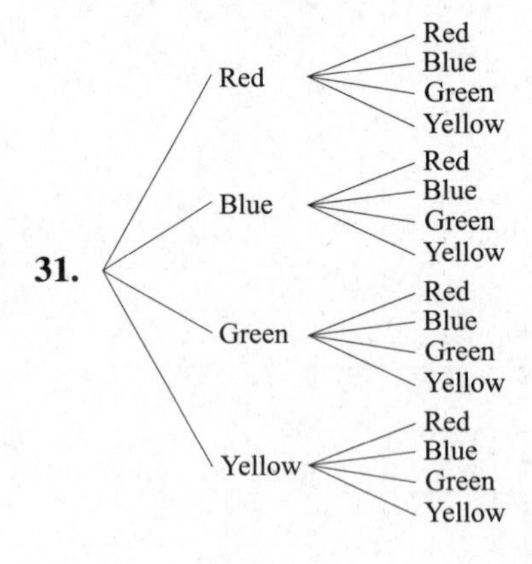

32.

H — H, T
T — H, T

33. One of the ten possible outcomes is a 6. The probability is $\frac{1}{10}$.

Copyright © 2011 Pearson Education, Inc. Publishing as Prentice Hall.

34. Two of the ten possible outcomes are a 3 or a 4.
The probability is $\frac{2}{10}=\frac{1}{5}$.

Cumulative Review Chapters 1–8

1.
$$\begin{aligned}&4^3+[3^2-(10\div 2)]-7\cdot 3\\&=4^3+[3^2-5]-7\cdot 3\\&=4^3+(9-5)-7\cdot 3\\&=4^3+4-7\cdot 3\\&=64+4-7\cdot 3\\&=64+4-21\\&=68-21\\&=47\end{aligned}$$

2.
$$\begin{aligned}&7^2-[5^3+(6\div 3)]+4\cdot 2\\&=7^2-[5^3+2]+4\cdot 2\\&=7^2-(125+2)+4\cdot 2\\&=7^2-127+4\cdot 2\\&=49-127+4\cdot 2\\&=49-127+8\\&=-78+8\\&=-70\end{aligned}$$

3. $x-y=-3-9=-3+(-9)=-12$

4. $x-y=7-(-2)=7+2=9$

5.
$$\begin{aligned}3y-7y&=12\\(3-7)y&=12\\-4y&=12\\\frac{-4y}{-4}&=\frac{12}{-4}\\y&=-3\end{aligned}$$

6.
$$\begin{aligned}2x-6x&=24\\(2-6)x&=24\\-4x&=24\\\frac{-4x}{-4}&=\frac{24}{-4}\\x&=-6\end{aligned}$$

7.
$$\begin{aligned}\frac{x}{6}+1&=\frac{4}{3}\\6\left(\frac{x}{6}+1\right)&=6\left(\frac{4}{3}\right)\\6\cdot\frac{x}{6}+6\cdot 1&=6\cdot\frac{4}{3}\\x+6&=8\\x+6-6&=8-6\\x&=2\end{aligned}$$

8.
$$\begin{aligned}\frac{7}{2}+\frac{a}{4}&=1\\4\left(\frac{7}{2}+\frac{a}{4}\right)&=4(1)\\4\cdot\frac{7}{2}+4\cdot\frac{a}{4}&=4\cdot 1\\14+a&=4\\14-14+a&=4-14\\a&=-10\end{aligned}$$

9.
$$\begin{array}{rr}2\frac{1}{3} & 2\frac{8}{24}\\ +5\frac{3}{8} & +5\frac{9}{24}\\ \hline & 7\frac{17}{24}\end{array}$$

$2\frac{1}{3}+5\frac{3}{8}=7\frac{17}{24}$

10.
$$\begin{array}{rr}3\frac{2}{5} & 3\frac{8}{20}\\ +4\frac{3}{4} & +4\frac{15}{20}\\ \hline & 7\frac{23}{20}=7+1\frac{3}{20}=8\frac{3}{20}\end{array}$$

$3\frac{2}{5}+4\frac{3}{4}=8\frac{3}{20}$

11. $5.9=5\frac{9}{10}$

12. $2.8=2\frac{8}{10}=2\frac{4}{5}$

13.
$$\begin{array}{r}3.500\\-\ 0.068\\\hline 3.432\end{array}$$

Check:
$$\begin{array}{r}3.432\\+\ 0.068\\\hline 3.500\end{array}$$

Copyright © 2011 Pearson Education, Inc. Publishing as Prentice Hall.

14. $\begin{array}{r} 7.400 \\ -\ 0.073 \\ \hline 7.327 \end{array}$ *Check:* $\begin{array}{r} 7.327 \\ +\ 0.073 \\ \hline 7.400 \end{array}$

15. $\begin{array}{r} 0.0531 \\ \times\ 16 \\ \hline 3186 \\ +\ 5310 \\ \hline 0.8496 \end{array}$ 4 decimal places; 4 decimal places

16. $\begin{array}{r} 0.147 \\ \times\ 0.2 \\ \hline 0.0294 \end{array}$ 3 decimal places; 1 decimal place; $3+1=4$ decimal places

17. $\begin{array}{r} 0.052 \\ 115\overline{)\ 5.980} \\ -5\ 75 \\ \hline 230 \\ -230 \\ \hline 0 \end{array}$

$-5.98 \div 115 = -0.052$

18. $\begin{array}{r} 0.136 \\ 205\overline{)\ 27.880} \\ -20\ 5 \\ \hline 7\ 38 \\ -6\ 15 \\ \hline 1\ 230 \\ -1\ 230 \\ \hline 0 \end{array}$

$27.88 \div 205 = 0.136$

19. $(-1.3)^2 + 2.4 = 1.69 + 2.4 = 4.09$

20. $(-2.7)^2 = (-2.7)(-2.7) = 7.29$

21. $\frac{1}{4} = \frac{1 \cdot 25}{4 \cdot 25} = \frac{25}{100} = 0.25$

22. $\frac{3}{8} = \frac{3 \cdot 125}{8 \cdot 125} = \frac{375}{1000} = 0.375$

23.
$$\begin{aligned} 5(x-0.36) &= -x+2.4 \\ 5 \cdot x - 5 \cdot 0.36 &= -x+2.4 \\ 5x-1.8 &= -x+2.4 \\ 5x+x-1.8 &= -x+x+2.4 \\ 6x-1.8 &= 2.4 \\ 6x-1.8+1.8 &= 2.4+1.8 \\ 6x &= 4.2 \\ \frac{6x}{6} &= \frac{4.2}{6} \\ x &= 0.7 \end{aligned}$$

24.
$$\begin{aligned} 4(0.35-x) &= x-7 \\ 4 \cdot 0.35 - 4 \cdot x &= x-7 \\ 1.4-4x &= x-7 \\ 1.4-4x-x &= x-x-7 \\ 1.4-5x &= -7 \\ 1.4-1.4-5x &= -7-1.4 \\ -5x &= -8.4 \\ \frac{-5x}{-5} &= \frac{-8.4}{-5} \\ x &= 1.68 \end{aligned}$$

25. $\sqrt{80} \approx 8.944$

26. $\sqrt{60} \approx 7.746$

27. The ratio of 21 to 29 is $\frac{21}{29}$.

28. The ratio of 7 to 15 is $\frac{7}{15}$.

29. $\begin{array}{r} 22.5 \\ 15\overline{)\ 337.5} \\ -30 \\ \hline 37 \\ -30 \\ \hline 7\ 5 \\ -7\ 5 \\ \hline 0 \end{array}$

$\frac{337.5 \text{ miles}}{15 \text{ gallons}} = \frac{22.5 \text{ miles}}{1 \text{ gallon}}$

The unit rate is 22.5 miles/gallon.

Copyright © 2011 Pearson Education, Inc. Publishing as Prentice Hall.

30. $3\overline{)1.59}$ quotient 0.53

$$\begin{array}{r} 0.53 \\ 3\overline{)\,1.59} \\ \underline{-1\,5} \\ 09 \\ \underline{-9} \\ 0 \end{array}$$

$\frac{\$1.59}{3 \text{ ounces}} = \frac{\$0.53}{1 \text{ ounce}}$

The unit rate is \$0.53 per ounce.

31. $\frac{51}{34} = \frac{-3}{x}$

$51 \cdot x = 34 \cdot (-3)$

$51x = -102$

$\frac{51x}{51} = \frac{-102}{51}$

$x = -2$

The solution is –2.

32. $\frac{8}{5} = \frac{x}{10}$

$8 \cdot 10 = 5 \cdot x$

$80 = 5x$

$\frac{80}{5} = \frac{5x}{5}$

$16 = x$

The solution is 16.

33. $\frac{12 \text{ feet}}{19 \text{ feet}} = \frac{12}{19}$

34. $\frac{4}{9}$

35. $4.6\% = 4.6(0.01) = 0.046$

36. $32\% = 32(0.01) = 0.32$

37. $0.74\% = 0.74(0.01) = 0.0074$

38. $2.7\% = 2.7(0.01) = 0.027$

39. $35\% \text{ of } 60 = 0.35 \cdot 60 = 21$

21 is 35% of 60.

40. $40\% \text{ of } 36 = 0.40 \cdot 36 = 14.4$

14.4 is 40% of 36.

41. $20.8 = 40\% \cdot x$

$20.8 = 0.4x$

$\frac{20.8}{0.4} = \frac{0.4x}{0.4}$

$52 = x$

20.8 is 40% of 52.

42. $9.5 = 25\% \cdot x$

$9.5 = 0.25x$

$\frac{9.5}{0.25} = \frac{0.25x}{0.25}$

$38 = x$

9.5 is 25% of 38.

43. $\$406 \cdot r = \$34.51$

$406r = 34.51$

$\frac{406r}{406} = \frac{34.51}{406}$

$r = 0.085 \text{ or } 8.5\%$

The sales tax rate is 8.5%.

44. $\$2 \cdot r = \$0.13$

$2r = 0.13$

$\frac{2r}{2} = \frac{0.13}{2}$

$r = 0.065 \text{ or } 6.5\%$

The sales tax rate is 6.5%.

45. $A = P\left(1 + \frac{r}{n}\right)^{n \cdot t}$

$= \$4000 \cdot \left(1 + \frac{0.053}{4}\right)^{4 \cdot 10}$

$= \$4000(1.01325)^{40}$

$\approx \$6772.12$

46. $\frac{\text{total}}{12 \text{ months}} = \frac{\$1600 + \$128.60}{12}$

$= \frac{\$1728.60}{12}$

$= \$144.05$

The monthly payment is \$144.05.

47. The numbers are listed in order. Since there are an odd number of values, the median is the middle value, 57.

48. The numbers are listed in order. Since there are an even number of values, the median is $\frac{47 + 50}{2} = \frac{97}{2} = 48.5.$

Copyright © 2011 Pearson Education, Inc. Publishing as Prentice Hall.

49. Two of the six possible outcomes are a 3 or a 4. The probability is $\frac{2}{6} = \frac{1}{3}$.

50. Three of the six possible outcomes are even. The probability is $\frac{3}{6} = \frac{1}{2}$.

Copyright © 2011 Pearson Education, Inc. Publishing as Prentice Hall.

Chapter 9

Section 9.1

Practice Problems

1. Figure (a) is part of a line with one endpoint, so it is a ray. It is ray AB or $\overrightarrow{AB}$.
   Figure (b) has two endpoints, so it is a line segment. It is line segment RS or $\overline{RS}$.
   Figure (c) extends indefinitely in two directions, so it is a line. It is line EF or $\overleftrightarrow{EF}$.
   Figure (d) has two rays with a common endpoint, so it is an angle. It is $\angle HVT$ or $\angle TVH$ or $\angle V$.

2. Two other ways to name $\angle z$ are $\angle RTS$ and $\angle STR$.

3. **a.** $\angle R$ is an obtuse angle. It measures between 90° and 180°.

   b. $\angle N$ is a straight angle. It measures 180°.

   c. $\angle M$ is an acute angle. It measures between 0° and 90°.

   d. $\angle Q$ is a right angle. It measures 90°.

4. The complement of a 29° angle is an angle that measures 90° – 29° = 61°.

5. The supplement of a 67° angle is an angle that measures 180° – 67° = 113°.

6. **a.** $m\angle y = m\angle ADC - m\angle BDC$
   $= 141° - 97°$
   $= 44°$

   b. $m\angle x = 79° - 51° = 28°$

   c. Since the measures of both $\angle x$ and $\angle y$ are between 0° and 90°, they are acute angles.

7. Since $\angle a$ and the angle marked 109° are vertical angles, they have the same measure; so $m\angle a = 109°$.
   Since $\angle a$ and $\angle b$ are adjacent angles, their measures have a sum of 180°. So $m\angle b = 180° - 109° = 71°$.
   Since $\angle b$ and $\angle c$ are vertical angles, they have the same measure; so $m\angle c = 71°$.

8. $\angle w$ and $\angle x$ are vertical angles. $\angle w$ and $\angle y$ are corresponding angles, as are $\angle x$ and $\angle d$. So all of these angles have the same measure: $m\angle x = m\angle y = m\angle d = m\angle w = 45°$.
   $\angle w$ and $\angle a$ are adjacent angles, as are $\angle w$ and $\angle b$, so $m\angle a = m\angle b = 180° - 45° = 135°$.
   $\angle a$ and $\angle c$ are corresponding angles so $m\angle c = m\angle a = 135°$. $\angle c$ and $\angle z$ are vertical angles, so $m\angle z = m\angle c = 135°$.

Vocabulary and Readiness Check

1. A plane is a flat surface that extends indefinitely.

2. A point has no length, no width, and no height.

3. Space extends in all directions indefinitely.

4. A line is a set of points extending indefinitely in two directions.

5. A ray is part of a line with one end point.

6. An angle is made up of two rays that share a common end point. The common end point is called the vertex.

7. A straight angle measures 180°.

8. A right angle measures 90°.

9. An acute angle measures between 0° and 90°.

10. An obtuse angle measures between 90° and 180°.

11. Parallel lines never meet and intersecting lines meet at a point.

12. Two intersecting lines are perpendicular if they form right angles when they intersect.

13. An angle can be measured in degrees.

14. A line that intersects two or more lines at different points is called a transversal.

15. When two lines intersect, four angles are formed, called vertical angles.

16. Two angles that share a common side are called adjacent angles.

Copyright © 2011 Pearson Education, Inc. Publishing as Prentice Hall.

Exercise Set 9.1

1. The figure extends indefinitely in two directions, so it is a line. It is line *CD*, line *l*, or $\overleftrightarrow{CD}$

3. The figure has two end points, so it is a line segment. It is line segment *MN* or $\overline{MN}$.

5. The figure has two rays with a common end point. It is an angle, which can be named $\angle GHI$, $\angle IHG$, or $\angle H$.

7. The figure has one end point and extends indefinitely in one direction, so it is a ray. It is ray *UW* or $\overrightarrow{UW}$.

9. Two other ways to name $\angle x$ are $\angle CPR$ and $\angle RPC$.

11. Two other ways to name $\angle z$ are $\angle TPM$ and $\angle MPT$.

13. $\angle S$ is a straight angle. It measures 180°.

15. $\angle R$ is a right angle. It measures 90°.

17. $\angle Q$ is an obtuse angle. It measures between 90° and 180°.

19. $\angle P$ is an acute angle. It measures between 0° and 90°.

21. The complement of an angle that measures 23° is an angle that measures $90° - 23° = 67°$.

23. The supplement of an angle that measures 17° is an angle that measures $180° - 17° = 163°$.

25. The complement of an angle that measures 58° is an angle that measures $90° - 58° = 32°$.

27. The supplement of an angle that measures 150° is an angle that measures $180° - 150° = 30°$.

29. $52° + 38° = 90°$, so $\angle PNQ$ and $\angle QNR$ are complementary. $60° + 30° = 90°$, so $\angle MNP$ and $\angle RNO$ are complementary.

31. $45° + 135° = 180°$, so there are 4 pairs of supplementary angles: $\angle SPT$ and $\angle RPS$, $\angle SPT$ and $\angle QPT$, $\angle QPR$ and $\angle RPS$, $\angle QPR$ and $\angle QPT$.

33. $m\angle x = 74° - 47° = 27°$

35. $m\angle x = 42° + 90° = 132°$

37. $\angle x$ and the angle marked 150° are supplementary, so $m\angle x = 180° - 150° = 30°$.
$\angle y$ and the angle marked 150° are vertical angles, so $m\angle y = 150°$.
$\angle z$ and $\angle x$ are vertical angles so $m\angle z = m\angle x = 30°$.

39. $\angle x$ and the angle marked 103° are supplementary, so $m\angle x = 180° - 103° = 77°$.
$\angle y$ and the angle marked 103° are vertical angles, so $m\angle y = 103°$.
$\angle x$ and $\angle z$ are vertical angles, so $m\angle z = m\angle x = 77°$.

41. $\angle x$ and the angle marked 80° are supplementary, so $m\angle x = 180° - 80° = 100°$.
$\angle y$ and the angle marked 80° are alternate interior angles, so $m\angle y = 80°$.
$\angle x$ and $\angle z$ are corresponding angles, so $m\angle z = m\angle x = 100°$.

43. $\angle x$ and the angle marked 46° are supplementary, so $m\angle x = 180° - 46° = 134°$.
$\angle y$ and the angle marked 46° are corresponding angles, so $m\angle y = 46°$.
$\angle x$ and $\angle z$ are corresponding angles, so $m\angle z = m\angle x = 134°$.

45. $\angle x$ can also be named $\angle ABC$ or $\angle CBA$.

47. $\angle z$ can also be named $\angle DBE$ or $\angle EBD$.

49. $m\angle ABC = 15°$

51. $m\angle CBD = 50°$

53. $$\begin{aligned} m\angle DBA &= m\angle DBC + m\angle CBA \\ &= 50° + 15° \\ &= 65° \end{aligned}$$

55. $$\begin{aligned} m\angle CBE &= m\angle CBD + m\angle DBE \\ &= 50° + 45° \\ &= 95° \end{aligned}$$

57. $\frac{7}{8} + \frac{1}{4} = \frac{7}{8} + \frac{2}{8} = \frac{9}{8}$ or $1\frac{1}{8}$

Copyright © 2011 Pearson Education, Inc. Publishing as Prentice Hall.

59. $\frac{7}{8}\cdot\frac{1}{4}=\frac{7\cdot 1}{8\cdot 4}=\frac{7}{32}$

61. $3\frac{1}{3}-2\frac{1}{2}=\frac{10}{3}-\frac{5}{2}$
$=\frac{10\cdot 2}{3\cdot 2}-\frac{5\cdot 3}{2\cdot 3}$
$=\frac{20}{6}-\frac{15}{6}$
$=\frac{5}{6}$

63. $3\frac{1}{3}\div 2\frac{1}{2}=\frac{10}{3}\div\frac{5}{2}=\frac{10}{3}\cdot\frac{2}{5}=\frac{5\cdot 2\cdot 2}{3\cdot 5}=\frac{4}{3}$ or $1\frac{1}{3}$

65. The supplement of an angle that measures 125.2° is an angle with measure 180° − 125.2° = 54.8°.

67. False; answers may vary

69. True; answers may vary

71. $\angle a$ and the angle marked 60° are alternate interior angles, so $m\angle a = 60°$. The sum of $\angle a$, $\angle b$, and the angle marked 70° is a straight angle, so $m\angle b = 180° - 60° - 70° = 50°$. $\angle d$ and the angle marked 70° are alternate interior angles, so $m\angle d = 70°$. $\angle c$ and $\angle d$ are supplementary, so $m\angle c = 180° - 70° = 110°$. $\angle e$ and the angle marked 60° are supplementary, so $m\angle e = 180° - 60° = 120°$.

73. no; answers may vary

75. Let x be the measure of each of the angles, in degrees. We are given that $x° + x° = 90°$, so $2x = 90$ and $x = 45$. The angles both measure 45°.

Section 9.2

Practice Problems

1. a. Perimeter $= 10\text{ m} + 10\text{ m} + 18\text{ m} + 18\text{ m}$
$= 56$ meters
The perimeter is 56 meters.

b. Perimeter $= 125\text{ ft} + 125\text{ ft} + 50\text{ ft} + 50\text{ ft}$
$= 350$ ft
The perimeter is 350 feet.

2. $P = 2\cdot l + 2\cdot w$
$= 2\cdot 32\text{ cm} + 2\cdot 15\text{ cm}$
$= 64\text{ cm} + 30\text{ cm}$
$= 94$ centimeters
The perimeter is 94 centimeters.

3. $P = 4\cdot s = 4\cdot 4\text{ ft} = 16$ feet
The perimeter is 16 feet.

4. $P = a + b + c$
$= 6\text{ cm} + 10\text{ cm} + 8\text{ cm}$
$= 24$ centimeters
The perimeter is 24 centimeters.

5. Perimeter $= 6\text{ km} + 4\text{ km} + 9\text{ km} + 4\text{ km}$
$= 23$ kilometers
The perimeter is 23 kilometers.

6. The unmarked horizontal side has length 20 m − 15 m = 5 m. The unmarked vertical side has length 31 m − 6 m = 25 m.
$P = 15\text{ m} + 31\text{ m} + 20\text{ m} + 6\text{ m} + 5\text{ m} + 25\text{ m}$
$= 102$ meters
The perimeter is 102 meters.

7. $P = 2\cdot l + 2\cdot w$
$= 2\cdot 120\text{ feet} + 2\cdot 60\text{ feet}$
$= 240\text{ feet} + 120\text{ feet}$
$= 360$ feet
cost = \$1.90 per foot · 360 feet = \$684
The cost of the fencing is \$684.

8. $C = \pi\cdot d = \pi\cdot 20\text{ yd} = 20\pi\text{ yd} \approx 62.8\text{ yd}$
The exact circumference of the watered region is 20π yards, which is approximately 62.8 yards.

Vocabulary and Readiness Check

1. The <u>perimeter</u> of a polygon is the sum of the lengths of its sides.
2. The distance around a circle is called the <u>circumference</u>.
3. The exact ratio of circumference to diameter is <u>π</u>.
4. The diameter of a circle is double its <u>radius</u>.
5. Both <u>$\frac{22}{7}$ (or 3.14)</u> and <u>3.14 $\left(\text{or } \frac{22}{7}\right)$</u> are approximations for π.
6. The radius of a circle is half its <u>diameter</u>.

Copyright © 2011 Pearson Education, Inc. Publishing as Prentice Hall.

Exercise Set 9.2

1. $P = 2 \cdot l + 2 \cdot w$
$= 2 \cdot 17 \text{ ft} + 2 \cdot 15 \text{ ft}$
$= 34 \text{ ft} + 30 \text{ ft}$
$= 64 \text{ ft}$
The perimeter is 64 feet.

3. $P = 35 \text{ cm} + 25 \text{ cm} + 35 \text{ cm} + 25 \text{ cm}$
$= 120 \text{ cm}$
The perimeter is 120 centimeters.

5. $P = a + b + c$
$= 5 \text{ in.} + 7 \text{ in.} + 9 \text{ in.}$
$= 21 \text{ in.}$
The perimeter is 21 inches.

7. Sum the lengths of the sides.
$P = 10 \text{ ft} + 8 \text{ ft} + 8 \text{ ft} + 15 \text{ ft} + 7 \text{ ft}$
$= 48 \text{ ft}$
The perimeter is 48 feet.

9. All sides of a regular polygon have the same length, so the perimeter is the number of sides multiplied by the length of a side.
$P = 3 \cdot 14 \text{ in.} = 42 \text{ in.}$
The perimeter is 42 inches.

11. All sides of a regular polygon have the same length, so the perimeter is the number of sides multiplied by the length of a side.
$P = 5 \cdot 31 \text{ cm} = 155 \text{ cm}$
The perimeter is 155 centimeters.

13. Sum the lengths of the sides.
$P = 5 \text{ ft} + 3 \text{ ft} + 2 \text{ ft} + 7 \text{ ft} + 4 \text{ ft}$
$= 21 \text{ ft}$
The perimeter is 21 feet.

15. total distance = 2(312) = 624
624 feet of lime powder will be deposited.

17. $P = 2 \cdot l + 2 \cdot w$
$= 2 \cdot 120 \text{ yd} + 2 \cdot 53 \text{ yd}$
$= 240 \text{ yd} + 106 \text{ yd}$
$= 346 \text{ yd}$
The perimeter of the football field is 346 yards.

19. $P = 2 \cdot l + 2 \cdot w$
$= 2 \cdot 8 \text{ ft} + 2 \cdot 3 \text{ ft}$
$= 16 \text{ ft} + 6 \text{ ft}$
$= 22 \text{ ft}$
22 feet of stripping is needed for this project.

21. The amount of stripping needed is 22 feet.
22 feet · \$2.50 per foot = \$55
The total cost of the stripping is \$55.

23. All sides of a regular polygon have the same length, so the perimeter is the number of sides multiplied by the length of a side.
$P = 8 \cdot 9 \text{ in.} = 72 \text{ in.}$
The perimeter is 72 inches.

25. $P = 4 \cdot s = 4 \cdot 7 \text{ in.} = 28 \text{ in.}$
The perimeter is 28 inches.

27. $P = 2 \cdot l + 2 \cdot w$
$= 2 \cdot 11 \text{ ft} + 2 \cdot 10 \text{ ft}$
$= 22 \text{ ft} + 20 \text{ ft}$
$= 42 \text{ ft}$
42 ft · \$0.86 per foot = \$36.12
The cost is \$36.12.

29. The unmarked vertical side has length
28 m − 20 m = 8 m.
The unmarked horizontal side has length
20 m − 17 m = 3m.
$P = 17 \text{ m} + 8 \text{ m} + 3 \text{ m} + 20 \text{ m} + 20 \text{ m} + 28 \text{ m}$
$= 96 \text{ m}$
The perimeter is 96 meters.

31. The unmarked horizontal side has length
(3 + 6 + 4) ft = 13 ft.
$P = (3 + 5 + 6 + 5 + 4 + 15 + 13 + 15) \text{ ft} = 66 \text{ ft}$
The perimeter is 66 feet.

33. The unmarked vertical side has length
5 cm + 14 cm = 19 cm.
The unmarked horizontal side has length
18 cm − 9 cm = 9 cm.
$P = 18 \text{ cm} + 19 \text{ cm} + 9 \text{ cm} + 14 \text{ cm} + 9 \text{ cm} + 5 \text{ cm}$
$= 74 \text{ cm}$
The perimeter is 74 centimeters.

35. $C = \pi \cdot d = \pi \cdot 17 \text{ cm} = 17\pi \text{ cm} \approx 53.38 \text{ cm}$
The circumference is exactly 17π centimeters or approximately 53.38 centimeters.

37. $C = 2 \cdot \pi \cdot r$
$= 2 \cdot \pi \cdot 8 \text{ mi}$
$= 16\pi \text{ mi}$
$\approx 50.24 \text{ mi}$
The circumference is exactly 16π miles, or approximately 50.24 miles.

Copyright © 2011 Pearson Education, Inc. Publishing as Prentice Hall.

39. $C = \pi \cdot d = \pi \cdot 26 \text{ m} = 26\pi \text{ m} \approx 81.64 \text{ m}$
The circumference is exactly 26π meters or approximately 81.64 meters.

41. $\pi \cdot d = \pi \cdot 15 \text{ ft} = 15\pi \text{ ft} \approx 47.1$
He needs 15π feet of netting or 47.1 feet.

43. $C = \pi \cdot d = \pi \cdot 4000 \text{ ft} = 4000\pi \text{ ft} \approx 12{,}560 \text{ ft}$
The distance around is about 12,560 feet.

45. Sum the lengths of the sides.
$$P = 9 \text{ mi} + 6 \text{ mi} + 11 \text{ mi} + 4.7 \text{ mi} = 30.7 \text{ mi}$$
The perimeter is 30.7 miles.

47. $C = \pi \cdot d = \pi \cdot 14 \text{ cm} = 14\pi \text{ cm} \approx 43.96 \text{ cm}$
The circumference is 14π centimeters $\approx$ 43.96 centimeters.

49. $P = 5 \cdot 8 \text{ mm} = 40 \text{ mm}$
The perimeter is 40 millimeters.

51. The unmarked vertical side has length $(22 - 8) \text{ ft} = 14 \text{ ft}$.
The unmarked horizontal side has length $(20 - 7) \text{ ft} = 13 \text{ ft}$.
$P = (7 + 8 + 13 + 14 + 20 + 22) \text{ ft} = 84 \text{ ft}$
The perimeter is 84 feet.

53. $5 + 6 \cdot 3 = 5 + 18 = 23$

55. $(20 - 16) \div 4 = 4 \div 4 = 1$

57. $72 \div (2 \cdot 6) = 72 \div 12 = 6$

59. $(18 + 8) - (12 + 4) = 26 - 16 = 10$

61. a. The first age category that 8-year-old children fit into is "Under 9," thus the minimum width is 30 yards and the minimum length is 40 yards.

b.
$$\begin{aligned} P &= 2 \cdot l + 2 \cdot w \\ &= 2 \cdot 40 \text{ yd} + 2 \cdot 30 \text{ yd} \\ &= 80 \text{ yd} + 60 \text{ yd} \\ &= 140 \text{ yd} \end{aligned}$$
The perimeter of the field is 140 yards.

63. The square's perimeter is $4 \cdot 3 \text{ in.} = 12 \text{ in.}$
The circle's circumference is $\pi \cdot 4 \text{ in.} \approx 12.56 \text{ in.}$
So the circle has the greater distance around; b.

65. a. Smaller circle:
$$\begin{aligned} C &= 2 \cdot \pi \cdot r \\ &= 2 \cdot \pi \cdot 10 \text{ m} \\ &= 20\pi \text{ m} \\ &\approx 62.8 \text{ m} \end{aligned}$$
Larger circle:
$$\begin{aligned} C &= 2 \cdot \pi \cdot r \\ &= 2 \cdot \pi \cdot 20 \text{ m} \\ &= 40\pi \text{ m} \\ &\approx 125.6 \text{ m} \end{aligned}$$

b. Yes, when the radius of a circle is doubled, the circumference is also doubled.

67. answers may vary

69. The length of the curved section at the top is half of the circumference of a circle of diameter 6 meters.
$$\frac{1}{2} \cdot C = \frac{1}{2} \cdot \pi \cdot d = \frac{1}{2} \cdot \pi \cdot 6 \text{ m} = 3\pi \text{ m} \approx 9.4 \text{ meters}$$
The total length of the straight sides is $3 \cdot 6 \text{ m} = 18 \text{ m}$.
The perimeter is the sum of these.
$9.4 \text{ m} + 18 \text{ m} = 27.4 \text{ m}$
The perimeter of the figure is 27.4 meters.

71. The total length of the two straight sections is $2 \cdot 22 \text{ m} = 44 \text{ m}$. The total length of the two curved sections is the circumference of a circle of radius 5 m.
$C = 2 \cdot \pi \cdot r = 2 \cdot \pi \cdot 5 \text{ m} = 10\pi \text{ m}$
The perimeter of the track is the sum of these.
$P = 44 \text{ m} + 10\pi \text{ m} = (44 + 10\pi) \text{ m} \approx 75.4 \text{ m}$

Section 9.3

Practice Problems

1.
$$\begin{aligned} A &= \frac{1}{2}bh \\ &= \frac{1}{2} \cdot 12 \text{ in.} \cdot 8\frac{1}{4} \text{ in.} \\ &= \frac{1}{2} \cdot 12 \text{ in.} \cdot \frac{33}{4} \text{ in.} \\ &= \frac{4 \cdot 3 \cdot 33}{2 \cdot 4} \text{ sq in.} \\ &= 49\frac{1}{2} \text{ sq in.} \end{aligned}$$

The area is $49\frac{1}{2}$ square inches.

Copyright © 2011 Pearson Education, Inc. Publishing as Prentice Hall.

2. $A = \frac{1}{2}(b+B)h$
$= \frac{1}{2}(5 \text{ yd} + 11 \text{ yd})(6.1 \text{ yd})$
$= 48.8 \text{ sq yd}$
The area is 48.8 square yards.

3. Split the rectangle into two pieces, a top rectangle with dimensions 12 m by 24 m, and a bottom rectangle with dimensions 6 m by 18 m. The area of the figure is the sum of the areas of these.
$A = 12 \text{ m} \cdot 24 \text{ m} + 6 \text{ m} \cdot 18 \text{ m}$
$= 288 \text{ sq m} + 108 \text{ sq m}$
$= 396 \text{ sq m}$
The area is 396 square meters.

4. $A = \pi r^2$
$= \pi \cdot (7 \text{ cm})^2$
$= 49\pi \text{ sq cm}$
$\approx 153.86 \text{ sq cm}$
The area is 49π square centimeters, which is approximately 153.86 square centimeters.

5. $V = lwh = 7 \text{ ft} \cdot 3 \text{ ft} \cdot 4 \text{ ft} = 84 \text{ cu ft}$
The volume of the box is 84 cubic feet.
$SA = 2lh + 2wh + 2lw$
$= 2(7 \text{ ft})(4 \text{ ft}) + 2(3 \text{ ft})(4 \text{ ft}) + 2(7 \text{ ft})(3 \text{ ft})$
$= 56 \text{ sq ft} + 24 \text{ sq ft} + 42 \text{ sq ft}$
$= 122 \text{ sq ft}$
The surface area of the box is 122 square feet.

6. $V = \frac{4}{3}\pi r^3 = \frac{4}{3}\pi \cdot \left(\frac{1}{2} \text{ cm}\right)^3$
$= \frac{4}{3}\pi \cdot \frac{1}{8} \text{ cu cm}$
$= \frac{1}{6}\pi \text{ cu cm}$
$\approx \frac{1}{6} \cdot \frac{22}{7} \text{ cu cm}$
$= \frac{11}{21} \text{ cu cm}$
The volume is $\frac{1}{6}\pi$ cubic centimeter, which is approximately $\frac{11}{21}$ cubic centimeter.

$SA = 4\pi r^2 = 4\pi \cdot \left(\frac{1}{2} \text{ cm}\right)^2$
$= 4\pi \cdot \frac{1}{4} \text{ sq cm}$
$= \pi \text{ sq cm}$
$\approx \frac{22}{7} \text{ sq cm or } 3\frac{1}{7} \text{ sq cm}$
The surface area is π square centimeters, which is approximately $3\frac{1}{7}$ square centimeters.

7. $V = \pi r^2 h = \pi \cdot (5 \text{ in.})^2 \cdot 9 \text{ in.}$
$= \pi \cdot 25 \text{ sq in.} \cdot 9 \text{ in.}$
$= 225\pi \text{ cu in.}$
$\approx 706.5 \text{ cu in.}$
The volume is 225π cubic inches, which is approximately 706.5 cubic inches.

8. $V = \frac{1}{3}s^2 h = \frac{1}{3} \cdot (3 \text{ m})^2 \cdot 5.1 \text{ m}$
$= \frac{1}{3} \cdot 9 \text{ sq m} \cdot 5.1 \text{ m}$
$= 15.3 \text{ cu m}$
The volume is 15.3 cubic meters.

Vocabulary and Readiness Check

1. The surface area of a polyhedron is the sum of the areas of its faces.
2. The measure of the amount of space inside a solid is its volume.
3. Area measures the amount of surface enclosed by a region.
4. Volume is measured in cubic units.
5. Area is measured in square units.
6. Surface area is measured in square units.

Exercise Set 9.3

1. $A = l \cdot w = 3.5 \text{ m} \cdot 2 \text{ m} = 7 \text{ sq m}$
The area is 7 square meters.

Copyright © 2011 Pearson Education, Inc. Publishing as Prentice Hall.

3. $A = \frac{1}{2} \cdot b \cdot h$
$= \frac{1}{2} \cdot 6\frac{1}{2} \text{ yd} \cdot 3 \text{ yd}$
$= \frac{1}{2} \cdot \frac{13}{2} \text{ yd} \cdot 3 \text{ yd}$
$= \frac{39}{4} \text{ sq yd}$
$= 9\frac{3}{4} \text{ sq yd}$

The area is $9\frac{3}{4}$ square yards.

5. $A = \frac{1}{2} \cdot b \cdot h = \frac{1}{2} \cdot 6 \text{ yd} \cdot 5 \text{ yd} = 15 \text{ sq yd}$

The area is 15 square yards.

7. $r = d \div 2 = (3 \text{ in.}) \div 2 = 1.5 \text{ in.}$
$A = \pi r^2$
$= \pi(1.5 \text{ in.})^2$
$= 2.25\pi \text{ sq in.}$
$\approx 7.065 \text{ sq in.}$

The area is 2.25π square inches $\approx$ 7.065 square inches.

9. $A = b \cdot h = 7 \text{ ft} \cdot 5.25 \text{ ft} = 36.75 \text{ sq ft}$

The area is 36.75 square feet.

11. $A = \frac{1}{2}(b + B) \cdot h$
$= \frac{1}{2}(5 \text{ m} + 9 \text{ m}) \cdot 4 \text{ m}$
$= \frac{1}{2} \cdot 14 \text{ m} \cdot 4 \text{ m}$
$= 28 \text{ sq m}$

The area is 28 square meters.

13. $A = \frac{1}{2}(b + B) \cdot h$
$= \frac{1}{2}(7 \text{ yd} + 4 \text{ yd}) \cdot 4 \text{ yd}$
$= \frac{1}{2}(11 \text{ yd}) \cdot 4 \text{ yd}$
$= 22 \text{ sq yd}$

The area is 22 square yards.

15. $A = b \cdot h$
$= 7 \text{ ft} \cdot 5\frac{1}{4} \text{ ft}$
$= 7 \text{ ft} \cdot \frac{21}{4} \text{ ft}$
$= \frac{147}{4} \text{ sq ft}$
$= 36\frac{3}{4} \text{ sq ft}$

The area is $36\frac{3}{4}$ square feet.

17. $A = b \cdot h$
$= 5 \text{ in.} \cdot 4\frac{1}{2} \text{ in.}$
$= 5 \text{ in.} \cdot \frac{9}{2} \text{ in.}$
$= \frac{45}{2} \text{ sq in.}$
$= 22\frac{1}{2} \text{ sq in.}$

The area is $22\frac{1}{2}$ square inches.

19. The base of the triangle is
$7 \text{ cm} - 1\frac{1}{2} \text{ cm} - 1\frac{1}{2} \text{ cm} = 4 \text{ cm},$ so its area is
$\frac{1}{2} \cdot 4 \text{ cm} \cdot 2 \text{ cm} = 4 \text{ sq cm}.$
The area of the rectangle is
$7 \text{ cm} \cdot 3 \text{ cm} = 21 \text{ sq cm}.$
The area of the figure is the sum of these.
$A = 4 \text{ sq cm} + 21 \text{ sq cm} = 25 \text{ sq cm}$
The total area is 25 square centimeters.

21. The figure can be divided into two rectangles, one measuring 10 mi by 5 mi and one measuring 12 mi by 3 mi. The area of the figure is the sum of the areas of the two rectangles.
$A = 10 \text{ mi} \cdot 5 \text{ mi} + 12 \text{ mi} \cdot 3 \text{ mi}$
$= 50 \text{ sq mi} + 36 \text{ sq mi}$
$= 86 \text{ sq mi}$
The total area is 86 square miles.

23. The top of the figure is a square with sides of length 3 cm, so its area is
$s^2 = (3 \text{ cm})^2 = 9 \text{ sq cm}.$
The bottom of the figure is a parallelogram with area $b \cdot h = 3 \text{ cm} \cdot 5 \text{ cm} = 15 \text{ sq cm}.$
The area of the figure is the sum of these.
$A = 9 \text{ sq cm} + 15 \text{ sq cm} = 24 \text{ sq cm}$
The total area is 24 square centimeters.

Copyright © 2011 Pearson Education, Inc. Publishing as Prentice Hall.

25. $A = \pi r^2$
$= \pi(6 \text{ in.})^2$
$= 36\pi$ sq in.
$\approx 36 \cdot \frac{22}{7}$ sq in.
$\approx 113\frac{1}{7}$ sq in.

The area is
36π square inches $\approx 113\frac{1}{7}$ square inches.

27. $V = l \cdot w \cdot h = 6 \text{ in.} \cdot 4 \text{ in.} \cdot 3 \text{ in.} = 72$ cu in.
The volume is 72 cubic inches.
$SA = 2lh + 2wh + 2lw$
$= 2 \cdot 6 \text{ in.} \cdot 3 \text{ in.} + 2 \cdot 4 \text{ in.} \cdot 3 \text{ in.} + 2 \cdot 6 \text{ in.} \cdot 4 \text{ in.}$
$= 36 \text{ sq in.} + 24 \text{ sq in.} + 48 \text{ sq in.}$
$= 108$ sq in.
The surface area is 108 square inches.

29. $V = s^3 = (8 \text{ cm})^3 = 512$ cu cm
The volume is 512 cubic centimeters.
$SA = 6s^2 = 6(8 \text{ cm})^2 = 6 \cdot 64 \text{ sq cm} = 384$ sq cm
The surface area is 384 square centimeters.

31. $V = \frac{1}{3} \cdot \pi \cdot r^2 \cdot h$
$= \frac{1}{3} \cdot \pi(2 \text{ yd})^2 (3 \text{ yd})$
$= 4\pi$ cu yd
$\approx 4 \cdot \frac{22}{7}$ cu yd
$= \frac{88}{7}$ cu yd
$= 12\frac{4}{7}$ cu yd

The volume is $12\frac{4}{7}$ cubic yards.

$SA = \pi \cdot r\sqrt{r^2 + h^2} + \pi r^2$
$= \pi \cdot 2 \text{ yd}\sqrt{(2 \text{ yd})^2 + (3 \text{ yd})^2} + \pi(2 \text{ yd})^2$
$= 2\pi\sqrt{13}$ sq yd $+ 4\pi$ sq yd
$= \left(2\sqrt{13}\pi + 4\pi\right)$ sq yd
≈ 35.20 sq yd
The surface area is
$\left(2\sqrt{13}\pi + 4\pi\right)$ square yards ≈ 35.20 square yards.

33. $r = \frac{1}{2} \cdot d = \frac{1}{2} \cdot 10 \text{ in.} = 5$ in.
$V = \frac{4}{3}\pi r^3$
$= \frac{4}{3}\pi(5 \text{ in.})^3$
$= \frac{4}{3}\pi \cdot 125$ cu in.
$= \frac{500}{3}\pi$ cu in.
$\approx \frac{500}{3} \cdot \frac{22}{7}$ cu in. $= 523\frac{17}{21}$ cu in.
The volume is
$\frac{500}{3}\pi$ cubic inches $\approx 523\frac{17}{21}$ cubic inches.
$SA = 4\pi r^2$
$= 4\pi(5 \text{ in.})^2$
$= 4\pi \cdot 25$ sq in.
$= 100\pi$ sq in.
$\approx 100 \cdot \frac{22}{7}$ sq in. $= 314\frac{2}{7}$ sq in.
The surface area is
100π square inches $\approx 314\frac{2}{7}$ square inches.

35. $r = \frac{1}{2}d = \frac{1}{2} \cdot 2 \text{ in.} = 1$ in.
$V = \pi \cdot r^2 \cdot h$
$= \pi(1 \text{ in.})^2 \cdot 9$ in
$= 9\pi$ cu in.
$\approx 9 \cdot \frac{22}{7}$ cu in. $= 28\frac{2}{7}$ cu in.
The volume is
9π cubic inches $\approx 28\frac{2}{7}$ cubic inches.

37. $V = \frac{1}{3}s^2 h$
$= \frac{1}{3}(5 \text{ cm})^2 (9 \text{ cm})$
$= \frac{1}{3} \cdot 25 \text{ sq cm} \cdot 9$ cm
$= 75$ cu cm
The volume is 75 cubic centimeters.

Copyright © 2011 Pearson Education, Inc. Publishing as Prentice Hall.

39. $V = s^3 = \left(1\frac{1}{3}\text{ in.}\right)^3$
$= \left(\frac{4}{3}\text{ in.}\right)^3$
$= \frac{64}{27}\text{ cu in.}$
$= 2\frac{10}{27}\text{ cu in.}$

The volume is $2\frac{10}{27}$ cubic inches.

41. $V = lwh = (2\text{ ft})(1.4\text{ ft})(3\text{ ft}) = 8.4\text{ cu ft}$
The volume is 8.4 cubic feet.
$SA = 2lh + 2wh + 2lw$
$= 2(2\text{ ft})(3\text{ ft}) + 2(1.4\text{ ft})(3\text{ ft}) + 2(2\text{ ft})(1.4\text{ ft})$
$= 12\text{ sq ft} + 8.4\text{ sq ft} + 5.6\text{ sq ft}$
$= 26\text{ sq ft}$
The surface area is 26 square feet.

43. $A = l \cdot w = 505\text{ ft} \cdot 225\text{ ft} = 113{,}625\text{ sq ft}$
The area of the flag is 113,625 square feet.

45. $A = l \cdot w = 7\text{ ft} \cdot 6\text{ ft} = 42\text{ sq ft}$
$4 \cdot 42\text{ sq ft} = 168\text{ sq ft}$
Four panels require 168 square feet of material.

47. $V = \frac{1}{3} \cdot s^2 \cdot h = \frac{1}{3} \cdot (12\text{ cm})^2 \cdot 20\text{ cm} = 960\text{ cu cm}$
The volume is 960 cubic centimeters.

49. The land is in the shape of a trapezoid.
$A = \frac{1}{2}(b + B) \cdot h$
$= \frac{1}{2}(90\text{ ft} + 140\text{ ft}) \cdot 80\text{ ft}$
$= \frac{1}{2} \cdot 230\text{ ft} \cdot 80\text{ ft}$
$= 9200\text{ sq ft}$
There are 9200 square feet of land in the plot.

51. $V = \frac{4}{3} \cdot \pi \cdot r^3$
$= \frac{4}{3}\pi(7\text{ in.})^3$
$= \frac{1372}{3}\pi\text{ cu in. or } 457\frac{1}{3}\pi\text{ cu in.}$

The volume is $\frac{1372}{3}\pi$ cubic inches or

$457\frac{1}{3}\pi$ cubic inches.

$SA = 4 \cdot \pi \cdot r^2 = 4\pi(7\text{ in.})^2 = 196\pi\text{ sq in.}$
The surface area is 196π square inches.

53. a. $A = \frac{1}{2}(b + B) \cdot h$
$= \frac{1}{2}(25\text{ ft} + 36\text{ ft}) \cdot 12\frac{1}{2}\text{ ft}$
$= \frac{1}{2} \cdot 61\text{ ft} \cdot 12\frac{1}{2}\text{ ft}$
$= 381\frac{1}{4}\text{ sq ft}$
To the nearest square foot, the area is 381 square feet.

b. Divide the area of the roof by the area covered by one "square."
$\frac{381}{100} = 3.81$
Since you cannot purchase a part of a square, a total of 4 squares needs to be purchased.

55. $r = \frac{1}{2} \cdot d = \frac{1}{2} \cdot 3\text{ in.} = 1.5\text{ in.}$
$V = \frac{1}{3}\pi \cdot r^2 \cdot h$
$= \frac{1}{3}\pi \cdot (1.5\text{ in.})^2 \cdot 7\text{ in.}$
$= 5.25\pi\text{ cu in.}$
The volume is 5.25π cubic inches.

57. $r = \frac{1}{2} \cdot d = \frac{1}{2} \cdot 4\text{ ft} = 2\text{ ft}$
$A = \pi r^2 = \pi(2\text{ ft})^2 = \pi \cdot 4\text{ sq ft}$
The area of the pizza is 4π square feet, or approximately $4 \cdot 3.14 = 12.56$ square feet.

59. $r = \frac{1}{2} \cdot d = \frac{1}{2} \cdot 3\text{ m} = 1.5\text{ m}$
$V = \frac{4}{3} \cdot \pi \cdot r^3$
$= \frac{4}{3}\pi(1.5\text{ m})^3$
$= 4.5\pi\text{ cu m}$
$\approx 4.5 \cdot 3.14\text{ cu m} = 14.13\text{ cu m}$
The volume is
4.5π cubic meters ≈ 14.13 cubic meters.

Copyright © 2011 Pearson Education, Inc. Publishing as Prentice Hall.

61. $A = l \cdot w = 16 \text{ ft} \cdot 10\frac{1}{2} \text{ ft}$
$= 16 \text{ ft} \cdot \frac{21}{2} \text{ ft}$
$= 168 \text{ sq ft}$
The area of the wall is 168 square feet.

63. $V = \frac{1}{3} \cdot s^2 \cdot h$
$= \frac{1}{3}(5 \text{ in.})^2 \cdot \frac{13}{10} \text{ in.}$
$= \frac{1}{3} \cdot 25 \cdot \frac{13}{10} \text{ cu in.}$
$= 10\frac{5}{6} \text{ cu in.}$
The volume is $10\frac{5}{6}$ cubic inches.

65. $V = lwh = (2 \text{ in.})(2 \text{ in.})(2.2 \text{ in.}) = 8.8 \text{ cu in.}$
The volume is 8.8 cubic inches.

67. $5^2 = 5 \cdot 5 = 25$

69. $3^2 = 3 \cdot 3 = 9$

71. $1^2 + 2^2 = 1 \cdot 1 + 2 \cdot 2 = 1 + 4 = 5$

73. $4^2 + 2^2 = 4 \cdot 4 + 2 \cdot 2 = 16 + 4 = 20$

75. A fence goes around the edge of a yard, thus the situation involves perimeter.

77. Carpet covers the entire floor of a room, so the situation involves area.

79. Paint covers the surface of the wall, thus the situation involves area.

81. A wallpaper border goes around the edge of a room, so the situation involves perimeter.

83. Note that the dimensions given are the diameters of the pizzas.
12-inch pizza:
$r = \frac{1}{2} \cdot d = \frac{1}{2} \cdot 12 \text{ in.} = 6 \text{ in.}$
$A = \pi \cdot r^2 = \pi(6 \text{ in.})^2 = 36\pi \text{ sq in.}$
$\text{Price per square inch} = \frac{\$10}{36\pi \text{ sq in.}} \approx \$0.0884$
8-inch pizzas:
$r = \frac{1}{2} \cdot d = \frac{1}{2} \cdot 8 \text{ in.} = 4 \text{ in.}$
$A = \pi \cdot r^2 = \pi(4 \text{ in.})^2 = 16\pi \text{ sq in.}$
$2 \cdot A = 2 \cdot 16\pi \text{ sq in.} = 32\pi \text{ sq in.}$
Price per square inch: $\frac{\$9}{32\pi \text{ sq in.}} \approx \$0.0895$
Since the price per square inch for the 12-inch pizza is less, the 12-inch pizza is the better deal.

85. $r = \frac{d}{2} = \frac{20 \text{ m}}{2} = 10 \text{ m}$
The volume is half the volume of a sphere.
$V = \frac{1}{2} \cdot \frac{4}{3} \pi r^3$
$= \frac{1}{2} \cdot \frac{4}{3}(3.14)(10 \text{ m})^3$
$\approx 2093.33 \text{ cu m}$
The volume of the hemisphere is about 2093.33 cu m.

87. no; answers may vary

89. The area of the shaded region is the area of the square minus the area of the circle.
Square:
$A = s^2 = (6 \text{ in.})^2 = 36 \text{ sq in.}$
Circle:
$r = \frac{1}{2} \cdot d = \frac{1}{2}(6 \text{ in.}) = 3 \text{ in.}$
$A = \pi \cdot r^2 = \pi(3 \text{ in.})^2 = 9\pi \text{ sq in.} \approx 28.26 \text{ sq in.}$
$36 \text{ sq in.} - 28.26 \text{ sq in.} = 7.74 \text{ sq in.}$
The shaded region has area of approximately 7.74 square inches.

91. The skating area is a rectangle with a half circle on each end.
Rectangle:
$A = l \cdot w = 22 \text{ m} \cdot 10 \text{ m} = 220 \text{ sq m}$
Half circles:
$A = 2 \cdot \frac{1}{2} \cdot \pi \cdot r^2$
$= \pi(5 \text{ m})^2$
$= 25\pi \text{ sq m} \approx 78.5 \text{ sq m}$
$220 \text{ sq m} + 78.5 \text{ sq m} = 298.5 \text{ sq m}$
The skating surface has area of 298.5 square meters.

93. no; answers may vary

Copyright © 2011 Pearson Education, Inc. Publishing as Prentice Hall.

Integrated Review

1. The supplement of a 27° angle measures $180° - 27° = 153°$.
   The complement of a 27° angle measures $90° - 27° = 63°$.

2. $\angle x$ and the angle marked 105° are supplementary angles, so $m\angle x = 180° - 105° = 75°$.
   $\angle y$ and the angle marked 105° are vertical angles, so $m\angle y = 105°$.
   $\angle z$ and the angle marked 105° are supplementary angles, so $m\angle z = 180° - 105° = 75°$.

3. $\angle x$ and the angle marked 52° are supplementary angles, so $m\angle x = 180° - 52° = 128°$.
   $\angle y$ and the angle marked 52° are corresponding angles, so $m\angle y = 52°$.
   $\angle z$ and $\angle y$ are supplementary angles, so $m\angle z = 180° - m\angle y = 180° - 52° = 128°$.

4. The sum of the measures of the angles of a triangle is 180°.
   $m\angle x = 180° - 90° - 38° = 52°$

5. $d = 2 \cdot r = 2 \cdot 2.3 \text{ in.} = 4.6 \text{ in.}$

6. $$\begin{aligned} r &= \frac{1}{2} \cdot d \\ &= \frac{1}{2} \cdot 8\frac{1}{2} \text{ in.} \\ &= \frac{1}{2} \cdot \frac{17}{2} \text{ in.} \\ &= \frac{17}{4} \text{ in.} \\ &= 4\frac{1}{4} \text{ in.} \end{aligned}$$

7. $P = 4 \cdot s = 4 \cdot 5 \text{ m} = 20 \text{ m}$
   The perimeter is 20 meters.
   $A = s^2 = (5 \text{ m})^2 = 25 \text{ sq m}$
   The area is 25 square meters.

8. $P = a + b + c = 4 \text{ ft} + 5 \text{ ft} + 3 \text{ ft} = 12 \text{ ft}$
   The perimeter is 12 feet.
   $A = \frac{1}{2} \cdot b \cdot h = \frac{1}{2} \cdot 3 \text{ ft} \cdot 4 \text{ ft} = 6 \text{ sq ft}$
   The area is 6 square feet.

9. $C = 2 \cdot \pi \cdot r = 2 \cdot \pi \cdot 5 \text{ cm} = 10\pi \text{ cm} \approx 31.4 \text{ cm}$
   The circumference is 10π centimeters ≈ 31.4 centimeters.
   $A = \pi \cdot r^2 = \pi(5 \text{ cm})^2 = 25\pi \text{ sq cm} \approx 78.5 \text{ sq cm}$
   The area is 25π square centimeters ≈ 78.5 square centimeters.

10. $P = 11 \text{ mi} + 5 \text{ mi} + 11 \text{ mi} + 5 \text{ mi} = 32 \text{ mi}$
    The perimeter is 32 miles.
    $A = l \cdot w = 11 \text{ mi} \cdot 4 \text{ mi} = 44 \text{ sq mi}$
    The area is 44 square meters.

11. The unmarked horizontal side has length $17 \text{ cm} - 8 \text{ cm} = 9 \text{ cm}$.
    The unmarked vertical side has length $7 \text{ cm} + 3 \text{ cm} = 10 \text{ cm}$.
    $P = (8 + 3 + 9 + 7 + 17 + 10) \text{ cm} = 54 \text{ cm}$
    The perimeter is 54 centimeters.
    The figure is made up of two rectangles, one with dimensions 10 cm by 8 cm, the other with dimensions 7 cm by 9 cm.
    $A = 10 \text{ cm} \cdot 8 \text{ cm} + 7 \text{ cm} \cdot 9 \text{ cm} = 143 \text{ sq cm}$
    The area is 143 square centimeters.

12. $$\begin{aligned} P &= 2 \cdot l + 2 \cdot w \\ &= 2(17 \text{ ft}) + 2(14 \text{ ft}) \\ &= 34 \text{ ft} + 28 \text{ ft} \\ &= 62 \text{ ft} \end{aligned}$$
    The perimeter is 62 feet.
    $A = l \cdot w = 17 \text{ ft} \cdot 14 \text{ ft} = 238 \text{ sq ft}$
    The area is 238 square feet.

13. $V = s^3 = (4 \text{ in.})^3 = 64 \text{ cu in.}$
    The volume is 64 cubic inches.
    $SA = 6 \cdot s^2 = 6(4 \text{ in.})^2 = 6 \cdot 16 \text{ sq in.} = 96 \text{ sq in.}$
    The surface area is 96 square inches.

14. $V = l \cdot w \cdot h = 3 \text{ ft} \cdot 2 \text{ ft} \cdot 5.1 \text{ ft} = 30.6 \text{ cu ft}$
    The volume is 30.6 cubic feet.
    $$\begin{aligned} SA &= 2lh + 2wh + 2lw \\ &= 2(3 \text{ ft})(5.1 \text{ ft}) + 2(2 \text{ ft})(5.1 \text{ ft}) + 2(3 \text{ ft})(2 \text{ ft}) \\ &= 30.6 \text{ sq ft} + 20.4 \text{ sq ft} + 12 \text{ sq ft} \\ &= 63 \text{ sq ft} \end{aligned}$$
    The surface area is 63 square feet.

15. $V = \frac{1}{3}s^2h = \frac{1}{3}(10 \text{ cm})^2(12 \text{ cm}) = 400 \text{ cu cm}$
    The volume is 400 cubic centimeters.

Copyright © 2011 Pearson Education, Inc. Publishing as Prentice Hall.

16. $r = \frac{1}{2} \cdot d = \frac{1}{2} \cdot 3 \text{ mi} = \frac{3}{2} \text{ mi}$

$$V = \frac{4}{3}\pi r^3 = \frac{4}{3}\pi\left(\frac{3}{2} \text{ mi}\right)^3 = \frac{9}{2}\pi \text{ cu mi} = 4\frac{1}{2}\pi \text{ cu mi} \approx \frac{9}{2} \cdot \frac{22}{7} \text{ cu mi} = 14\frac{1}{7} \text{ cu mi}$$

The volume is $4\frac{1}{2}\pi$ cubic miles $\approx 14\frac{1}{7}$ cubic miles.

Section 9.4

Practice Problems

1. $6 \text{ ft} = \frac{6 \text{ ft}}{1} \cdot 1 = \frac{6 \text{ ft}}{1} \cdot \frac{12 \text{ in.}}{1 \text{ ft}} = 6 \cdot 12 \text{ in.} = 72 \text{ in.}$

2. $8 \text{ yd} = \frac{8 \text{ yd}}{1} \cdot 1 = \frac{8 \text{ yd}}{1} \cdot \frac{3 \text{ ft}}{1 \text{ yd}} = 8 \cdot 3 \text{ ft} = 24 \text{ ft}$

3. $18 \text{ in.} = \frac{18 \text{ in.}}{1} \cdot \frac{1 \text{ ft}}{12 \text{ in.}} = \frac{18}{12} \text{ ft} = 1.5 \text{ ft}$

4. $68 \text{ in.} = \frac{68 \text{ in.}}{1} \cdot \frac{1 \text{ ft}}{12 \text{ in.}} = \frac{68}{12} \text{ ft}$

$$\begin{array}{r} 5 \\ 12\overline{)\,68} \\ -60 \\ \hline 8 \end{array}$$

Thus, 68 in. = 5 ft 8 in.

5. $5 \text{ yd} = \frac{5 \text{ yd}}{1} \cdot \frac{3 \text{ ft}}{1 \text{ yd}} = 15 \text{ ft}$

5 yd 2 ft = 15 ft + 2 ft = 17 ft

6.
$$\begin{array}{r} 4 \text{ ft } \ 8 \text{ in.} \\ +\ 8 \text{ ft } 11 \text{ in.} \\ \hline 12 \text{ ft } 19 \text{ in.} \end{array}$$

Since 19 inches is the same as 1 ft 7 in., we have 12 ft 19 in. = 12 ft + 1 ft 7 in. = 13 ft 7 in.

7.
$$\begin{array}{r} 4 \text{ ft } \ 7 \text{ in.} \\ \times \qquad 4 \\ \hline 16 \text{ ft } 28 \text{ in.} \end{array}$$

Since 28 in. is the same as 2 ft 4 in., we simplify as 16 ft 28 in. = 16 ft + 2 ft 4 in. = 18 ft 4 in.

8.
$$\begin{array}{r} 5 \text{ ft } 8 \text{ in.} \\ -\ 1 \text{ ft } 9 \text{ in.} \\ \hline \end{array} \rightarrow \begin{array}{r} 4 \text{ ft } 20 \text{ in.} \\ -\ 1 \text{ ft } \ 9 \text{ in.} \\ \hline 3 \text{ ft } 11 \text{ in.} \end{array}$$

The remaining board length is 3 ft 11 in.

9. $2.5 \text{ m} = \frac{2.5 \text{ m}}{1} \cdot \frac{1000 \text{ mm}}{1 \text{ m}} = 2500 \text{ mm}$

10. 3500 m = 3.500 km or 3.5 km

11. 640 m = 0.64 km

$$\begin{array}{r} 2.10 \text{ km} \\ -\ 0.64 \text{ km} \\ \hline 1.46 \text{ km} \end{array}$$

2.1 km = 2100 m

$$\begin{array}{r} 2100 \text{ m} \\ -\ 640 \text{ m} \\ \hline 1460 \text{ m} \end{array}$$

12.
$$\begin{array}{r} 18.3 \text{ hm} \\ \times\ 5 \quad \\ \hline 91.5 \text{ hm} \end{array}$$

13. 0.8 m = 80 cm
80 cm + 45 cm = 125 cm
The scarf will be 125 cm or 1.25 m.

Vocabulary and Readiness Check

1. The basic unit of length in the metric system is the meter.
2. The expression $\frac{1 \text{ foot}}{12 \text{ inches}}$ is an example of a unit fraction.
3. A meter is slightly longer than a yard.
4. One foot equals 12 inches.
5. One yard equals 3 feet.
6. One yard equals 36 inches.
7. One mile equals 5280 feet.

Copyright © 2011 Pearson Education, Inc. Publishing as Prentice Hall.

Exercise Set 9.4

1. $60 \text{ in.} = \frac{60 \text{ in.}}{1} \cdot \frac{1 \text{ ft}}{12 \text{ in.}} = \frac{60}{12} \text{ ft} = 5 \text{ ft}$

3. $12 \text{ yd} = \frac{12 \text{ yd}}{1} \cdot \frac{3 \text{ ft}}{1 \text{ yd}} = 12 \cdot 3 \text{ ft} = 36 \text{ ft}$

5. $$\begin{aligned} 42,240 \text{ ft} &= \frac{42,240 \text{ ft}}{1} \cdot \frac{1 \text{ mi}}{5280 \text{ ft}} \\ &= \frac{42,240}{5280} \text{ mi} \\ &= 8 \text{ mi} \end{aligned}$$

7. $8\frac{1}{2} \text{ ft} = \frac{8.5 \text{ ft}}{1} \cdot \frac{12 \text{ in.}}{1 \text{ ft}} = 8.5 \cdot 12 \text{ in.} = 102 \text{ in.}$

9. $10 \text{ ft} = \frac{10 \text{ ft}}{1} \cdot \frac{1 \text{ yd}}{3 \text{ ft}} = \frac{10}{3} \text{ yd} = 3\frac{1}{3} \text{ yd}$

11. $$\begin{aligned} 6.4 \text{ mi} &= \frac{6.4 \text{ mi}}{1} \cdot \frac{5280 \text{ ft}}{1 \text{ mi}} \\ &= 6.4 \cdot 5280 \text{ ft} \\ &= 33,792 \text{ ft} \end{aligned}$$

13. $$\begin{aligned} 162 \text{ in.} &= \frac{162 \text{ in.}}{1} \cdot \frac{1 \text{ ft}}{12 \text{ in.}} \cdot \frac{1 \text{ yd}}{3 \text{ ft}} \\ &= \frac{162}{36} \text{ yd} \\ &= 4.5 \text{ yd} \end{aligned}$$

15. $3 \text{ in.} = \frac{3 \text{ in.}}{1} \cdot \frac{1 \text{ ft}}{12 \text{ in.}} = \frac{3}{12} \text{ ft} = 0.25 \text{ ft}$

17. $40 \text{ ft} = \frac{40 \text{ ft}}{1} \cdot \frac{1 \text{ yd}}{3 \text{ ft}} = \frac{40}{3} \text{ yd}$

$$\begin{array}{r} 13 \text{ yd } 1 \text{ ft} \\ 3\overline{)\ 40} \\ \underline{-3} \\ 10 \\ \underline{-9} \\ 1 \end{array}$$

19. $85 \text{ in.} = \frac{85 \text{ in.}}{1} \cdot \frac{1 \text{ ft}}{12 \text{ in.}} = \frac{85}{12} \text{ ft}$

$$\begin{array}{r} 7 \text{ ft } 1 \text{ in.} \\ 12\overline{)\ 85} \\ \underline{-84} \\ 1 \end{array}$$

21. $10,000 \text{ ft} = \frac{10,000 \text{ ft}}{1} \cdot \frac{1 \text{ mi}}{5280 \text{ ft}} = \frac{10,000}{5280} \text{ mi}$

$$\begin{array}{r} 1 \text{ mi } 4720 \text{ ft} \\ 5280\overline{)10,000} \\ \underline{-5280} \\ 4720 \end{array}$$

23. $$\begin{aligned} 5 \text{ ft } 2 \text{ in.} &= \frac{5 \text{ ft}}{1} \cdot \frac{12 \text{ in.}}{1 \text{ ft}} + 2 \text{ in.} \\ &= 60 \text{ in.} + 2 \text{ in.} \\ &= 62 \text{ in.} \end{aligned}$$

25. $$\begin{aligned} 8 \text{ yd } 2 \text{ ft} &= \frac{8 \text{ yd}}{1} \cdot \frac{3 \text{ ft}}{1 \text{ yd}} + 2 \text{ ft} \\ &= 24 \text{ ft} + 2 \text{ ft} \\ &= 26 \text{ ft} \end{aligned}$$

27. $2 \text{ yd } 1 \text{ ft} = \frac{2 \text{ yd}}{1} \cdot \frac{3 \text{ ft}}{1 \text{ yd}} + 1 \text{ ft} = 6 \text{ ft} + 1 \text{ ft} = 7 \text{ ft}$

$7 \text{ ft} = \frac{7 \text{ ft}}{1} \cdot \frac{12 \text{ in.}}{1 \text{ ft}} = 7 \cdot 12 \text{ in.} = 84 \text{ in.}$

29. $$\begin{aligned} 3 \text{ ft } 10 \text{ in.} + 7 \text{ ft } 4 \text{ in.} &= 10 \text{ ft } 14 \text{ in.} \\ &= 10 \text{ ft} + 1 \text{ ft } 2 \text{ in.} \\ &= 11 \text{ ft } 2 \text{ in.} \end{aligned}$$

31. $$\begin{aligned} 12 \text{ yd } 2 \text{ ft} + 9 \text{ yd } 2 \text{ ft} &= 21 \text{ yd } 4 \text{ ft} \\ &= 21 \text{ yd} + 1 \text{ yd } 1 \text{ ft} \\ &= 22 \text{ yd } 1 \text{ ft} \end{aligned}$$

33. $$\begin{array}{r} 22 \text{ ft } 8 \text{ in.} \\ \underline{-\ 16 \text{ ft } 3 \text{ in.}} \\ 6 \text{ ft } 5 \text{ in.} \end{array}$$

35. $$\begin{array}{r} 18 \text{ ft } 3 \text{ in.} \\ \underline{-\ 10 \text{ ft } 9 \text{ in.}} \end{array} \rightarrow \begin{array}{r} 17 \text{ ft } 15 \text{ in.} \\ \underline{-\ 10 \text{ ft } \ \ 9 \text{ in.}} \\ 7 \text{ ft } \ \ 6 \text{ in.} \end{array}$$

37. $28 \text{ ft } 8 \text{ in.} \div 2 = 14 \text{ ft } 4 \text{ in.}$

39. $$\begin{array}{l} \quad 16 \text{ yd } \ 2 \text{ ft} \\ \underline{\times \qquad\qquad\ 5} \\ 80 \text{ yd } 10 \text{ ft} = 80 \text{ yd} + 3 \text{ yd } 1 \text{ ft} = 83 \text{ yd } 1 \text{ ft} \end{array}$$

41. $60 \text{ m} = \frac{60 \text{ m}}{1} \cdot \frac{100 \text{ cm}}{1 \text{ m}} = 6000 \text{ cm}$

43. $40 \text{ mm} = \frac{40 \text{ mm}}{1} \cdot \frac{1 \text{ cm}}{10 \text{ mm}} = 4 \text{ cm}$

Copyright © 2011 Pearson Education, Inc. Publishing as Prentice Hall.

45. $500\text{ m} = \frac{500\text{ m}}{1} \cdot \frac{1\text{ km}}{1000\text{ m}} = \frac{500}{1000}\text{ km} = 0.5\text{ km}$

47. $1700\text{ mm} = \frac{1700\text{ mm}}{1} \cdot \frac{1\text{ m}}{1000\text{ mm}} = 1.7\text{ m}$

49. $1500\text{ cm} = \frac{1500\text{ cm}}{1} \cdot \frac{1\text{ m}}{100\text{ cm}} = \frac{1500}{100}\text{ m} = 15\text{ m}$

51. $0.42\text{ km} = \frac{0.42\text{ km}}{1} \cdot \frac{100{,}000\text{ cm}}{1\text{ km}} = 42{,}000\text{ cm}$

53. $7\text{ km} = \frac{7\text{ km}}{1} \cdot \frac{1000\text{ m}}{1\text{ km}} = 7000\text{ m}$

55. $8.3\text{ cm} = \frac{8.3\text{ cm}}{1} \cdot \frac{10\text{ mm}}{1\text{ cm}} = 83\text{ mm}$

57. $$\begin{aligned} 20.1\text{ mm} &= \frac{20.1\text{ mm}}{1} \cdot \frac{1\text{ dm}}{100\text{ mm}} \\ &= \frac{20.1}{100}\text{ dm} \\ &= 0.201\text{ dm} \end{aligned}$$

59. $0.04\text{ m} = \frac{0.04\text{ m}}{1} \cdot \frac{1000\text{ mm}}{1\text{ m}} = 40\text{ mm}$

61. $$\begin{array}{r} 8.60\text{ m} \\ +\ 0.34\text{ m} \\ \hline 8.94\text{ m} \end{array}$$

63. $$\begin{array}{r} 2.9\text{ m} \\ +\ 40.0\text{ mm} \\ \hline \end{array} \quad \begin{array}{r} 2.90\text{ m} \\ +\ 0.04\text{ m} \\ \hline 2.94\text{ m} \end{array} \text{ or } \begin{array}{r} 2900\text{ mm} \\ +\ 40\text{ mm} \\ \hline 2940\text{ mm} \end{array}$$

65. $$\begin{array}{r} 24.8\text{ mm} \\ -\ 1.19\text{ cm} \\ \hline \end{array} \quad \begin{array}{r} 24.8\text{ mm} \\ -\ 11.9\text{ mm} \\ \hline 12.9\text{ mm} \end{array} \text{ or } \begin{array}{r} 2.48\text{ cm} \\ -\ 1.19\text{ cm} \\ \hline 1.29\text{ cm} \end{array}$$

67. $$\begin{array}{r} 15\text{ km} \\ -\ 2360\text{ m} \\ \hline \end{array} \quad \begin{array}{r} 15.00\text{ km} \\ -\ 2.36\text{ km} \\ \hline 12.64\text{ km} \end{array} \text{ or } \begin{array}{r} 15{,}000\text{ m} \\ -\ 2360\text{ m} \\ \hline 12{,}640\text{ m} \end{array}$$

69. $18.3\text{ m} \times 3 = 54.9\text{ m}$

71. $6.2\text{ km} \div 4 = 1.55\text{ km}$

Copyright © 2011 Pearson Education, Inc. Publishing as Prentice Hall.

		Yards	Feet	Inches
73.	Chrysler Building in New York City	$348\frac{2}{3}$	1046	12,552
75.	Python length	$11\frac{2}{3}$	35	420

		Meters	Millimeters	Kilometers	Centimeters
77.	Length of elephant	5	5000	0.005	500
79.	Tennis ball diameter	0.065	65	0.000065	6.5
81.	Distance from London to Paris	342,000	342,000,000	342	34,200,000

83.
$$\begin{array}{r} 6 \text{ ft } 10 \text{ in.} \\ +\ 3 \text{ ft } \ 8 \text{ in.} \\ \hline 9 \text{ ft } 18 \text{ in.} \end{array} = 9 \text{ ft} + 1 \text{ ft } 6 \text{ in.} = 10 \text{ ft } 6 \text{ in.}$$
The bamboo is 10 ft 6 in. now.

85.
$$\begin{array}{r} 6000 \text{ ft} \\ -\ 900 \text{ ft} \\ \hline 5100 \text{ ft} \end{array}$$
The Grand Canyon of the Colorado River is 5100 feet deeper than the Grand Canyon of the Yellowstone River.

87. 8 ft 11 in. ÷ 22.5 in. = 107 in. ÷ 22.5 in. ≈ 4.8
Robert is 4.8 times as tall as Gul.

89.
$$\begin{array}{r} 80 \text{ mm} \\ -\ 5.33 \text{ cm} \\ \hline \end{array} \quad \begin{array}{r} 80.0 \text{ mm} \\ -\ 53.3 \text{ mm} \\ \hline 26.7 \text{ mm} \end{array}$$
The ice must be 26.7 mm thicker before skating is allowed.

91.
$$\begin{aligned} 1 \text{ ft } 9 \text{ in.} \times 9 &= 9 \text{ ft } 81 \text{ in.} \\ &= 9 \text{ ft} + 6 \text{ ft } 9 \text{ in.} \\ &= 15 \text{ ft } 9 \text{ in.} \end{aligned}$$
The stacks extend 15 ft 9 in. from the wall.

93.
$$\begin{array}{r} 3.35 \\ 20\overline{)\,67.00} \\ \underline{-60} \\ 7\,0 \\ \underline{-6\,0} \\ 1\,00 \\ \underline{-1\,00} \\ 0 \end{array}$$
Each piece will be 3.35 meters long.

Copyright © 2011 Pearson Education, Inc. Publishing as Prentice Hall.

95. $182 \text{ ft} \times 2 = 364 \text{ ft}$
Two trucks are 364 feet long.
$$364 \text{ ft} = \frac{364 \text{ ft}}{1} \cdot \frac{1 \text{ yd}}{3 \text{ ft}} = 121\frac{1}{3} \text{ yd}$$
Two trucks are $121\frac{1}{3}$ yards long.

97. $0.21 = \frac{21}{100}$

99. $\frac{13}{100} = 0.13$

101. $\frac{1}{4} = \frac{1}{4} \cdot \frac{25}{25} = \frac{25}{100} = 0.25$

103. No, the width of a twin-size bed being 20 meters is not reasonable.

105. Yes, glass for a drinking glass being 2 millimeters thick is reasonable.

107. No, the distance across the Colorado River being 50 kilometers is not reasonable.

109. 5 yd 2 in. is close to 5 yd. 7 yd 30 in. is close to 7 yd 36 in. = 8 yd.
Estimate: 5 yd + 8 yd = 13 yd.

111. answers may vary; for example,
$$4 \text{ ft} = \frac{4 \text{ ft}}{1} \cdot \frac{1 \text{ yd}}{3 \text{ ft}} = \frac{4}{3} \text{ yd} = 1\frac{1}{3} \text{ yd}$$
$$4 \text{ ft} = \frac{4 \text{ ft}}{1} \cdot \frac{12 \text{ in.}}{1 \text{ ft}} = 48 \text{ in.}$$

113. answers may vary

115. $18.3 \text{ m} \times 18.3 \text{ m} = 334.89 \text{ sq m}$
The area of the sign is 334.89 square meters.

Section 9.5

Practice Problems

1.
$$\begin{aligned} 6500 \text{ lb} &= \frac{6500 \text{ lb}}{1} \cdot \frac{1 \text{ ton}}{2000 \text{ lb}} \\ &= \frac{6500}{2000} \text{ tons} \\ &= \frac{13}{4} \text{ tons or } 3\frac{1}{4} \text{ tons} \end{aligned}$$

2. $72 \text{ oz} = \frac{72 \text{ oz}}{1} \cdot \frac{1 \text{ lb}}{16 \text{ oz}} = \frac{72}{16} \text{ lb} = \frac{9}{2} \text{ lb or } 4\frac{1}{2} \text{ lb}$

3. $47 \text{ oz} = 47 \text{ oz} \cdot \frac{1 \text{ lb}}{16 \text{ oz}} = \frac{47}{16} \text{ lb}$
$$\begin{array}{r} 2 \text{ lb } 15 \text{ oz} \\ 16\overline{)\;47} \\ \underline{-32} \\ 15 \end{array}$$
Thus, 47 oz = 2 lb 15 oz

4.
$$\begin{array}{r} 8 \text{ tons} \quad 100 \text{ lb} \\ \underline{-\ 5 \text{ tons} \quad 1200 \text{ lb}} \end{array} \rightarrow \begin{array}{r} 7 \text{ tons} \quad 2100 \text{ lb} \\ \underline{-\ 5 \text{ tons} \quad 1200 \text{ lb}} \\ 2 \text{ tons} \quad 900 \text{ lb} \end{array}$$

5.
$$\begin{array}{rr} 1 \text{ lb} & 6 \text{ oz} \\ 4\overline{)\;5 \text{ lb}} & \overline{8 \text{ oz}} \\ \underline{-\ 4 \text{ lb}} & \\ 1 \text{ lb} = & \underline{16 \text{ oz}} \\ & 24 \text{ oz} \end{array}$$

6.
$$\begin{array}{r} \text{batch weight} \\ \underline{+ \text{ container weight}} \\ \text{total weight} \end{array} \rightarrow \begin{array}{r} 5 \text{ lb } 14 \text{ oz} \\ \underline{+ \qquad 6 \text{ oz}} \\ 5 \text{ lb } 20 \text{ oz} \end{array}$$
5 lb 20 oz = 5 lb + 1 lb 4 oz = 6 lb 4 oz
The total weight is 6 lb 4 oz.

7. $3.41 \text{ g} = \frac{3.41 \text{ g}}{1} \cdot \frac{1000 \text{ mg}}{1 \text{ g}} = 3410 \text{ mg}$

8. $56.2 \text{ cg} = 56.2 \text{ cg} = 0.562 \text{ g}$

9. $3.1 \text{ dg} = 0.31 \text{ g}$ or $2.5 \text{ g} = 25 \text{ dg}$
$$\begin{array}{r} 2.50 \text{ g} \\ \underline{-\ 0.31 \text{ g}} \\ 2.19 \text{ g} \end{array} \qquad \begin{array}{r} 25.0 \text{ dg} \\ \underline{-\ 3.1 \text{ dg}} \\ 21.9 \text{ dg} \end{array}$$

10.
$$\begin{array}{r} 22.9 \quad \approx 23 \\ 24\overline{)\;550.0 \text{ kg}} \\ \underline{-48}\phantom{0.0 \text{ kg}} \\ 70\phantom{.0 \text{ kg}} \\ \underline{-48}\phantom{.0 \text{ kg}} \\ 22\;0\phantom{\text{ kg}} \\ \underline{-21\;6}\phantom{\text{ kg}} \\ 4\phantom{\text{ kg}} \end{array}$$
Each bag weighs about 23 kg.

Vocabulary and Readiness Check

1. Mass is a measure of the amount of substance in an object. This measure does not change.

Copyright © 2011 Pearson Education, Inc. Publishing as Prentice Hall.

2. Weight is the measure of the pull of gravity.

3. The basic unit of mass in the metric system is the gram.

4. One pound equals 16 ounces.

5. One ton equals 2000 pounds.

Exercise Set 9.5

1. $2 \text{ lb} = \frac{2 \text{ lb}}{1} \cdot \frac{16 \text{ oz}}{1 \text{ lb}} = 2 \cdot 16 \text{ oz} = 32 \text{ oz}$

3. $$\begin{aligned} 5 \text{ tons} &= \frac{5 \text{ tons}}{1} \cdot \frac{2000 \text{ lb}}{1 \text{ ton}} \\ &= 5 \cdot 2000 \text{ lb} \\ &= 10,000 \text{ lb} \end{aligned}$$

5. $$\begin{aligned} 18,000 \text{ lb} &= \frac{18,000 \text{ lb}}{1} \cdot \frac{1 \text{ ton}}{2000 \text{ lb}} \\ &= \frac{18,000}{2000} \text{ tons} \\ &= 9 \text{ tons} \end{aligned}$$

7. $60 \text{ oz} = \frac{60 \text{ oz}}{1} \cdot \frac{1 \text{ lb}}{16 \text{ oz}} = \frac{60}{16} \text{ lb} = \frac{15}{4} \text{ lb} = 3\frac{3}{4} \text{ lb}$

9. $$\begin{aligned} 3500 \text{ lb} &= \frac{3500 \text{ lb}}{1} \cdot \frac{1 \text{ ton}}{2000 \text{ lb}} \\ &= \frac{3500}{2000} \text{ tons} \\ &= \frac{7}{4} \text{ tons} \\ &= 1\frac{3}{4} \text{ tons} \end{aligned}$$

11. $$\begin{aligned} 12.75 \text{ lb} &= \frac{12.75 \text{ lb}}{1} \cdot \frac{16 \text{ oz}}{1 \text{ lb}} \\ &= 12.75 \cdot 16 \text{ oz} \\ &= 204 \text{ oz} \end{aligned}$$

13. $$\begin{aligned} 4.9 \text{ tons} &= \frac{4.9 \text{ tons}}{1} \cdot \frac{2000 \text{ lb}}{1 \text{ ton}} \\ &= 4.9 \cdot 2000 \text{ lb} \\ &= 9800 \text{ lb} \end{aligned}$$

15. $$\begin{aligned} 4\frac{3}{4} \text{ lb} &= \frac{19}{4} \text{ lb} \\ &= \frac{\frac{19}{4} \text{ lb}}{1} \cdot \frac{16 \text{ oz}}{1 \text{ lb}} \\ &= \frac{19}{4} \cdot 16 \text{ oz} \\ &= 76 \text{ oz} \end{aligned}$$

17. $$\begin{aligned} 2950 \text{ lb} &= \frac{2950 \text{ lb}}{1} \cdot \frac{1 \text{ ton}}{2000 \text{ lb}} \\ &= \frac{2950}{2000} \text{ tons} \\ &= \frac{59}{40} \text{ tons} \\ &= 1.475 \text{ tons} \\ &\approx 1.5 \text{ tons} \end{aligned}$$

19. $\frac{4}{5} \text{ oz} = \frac{\frac{4}{5}}{1} \cdot \frac{1 \text{ lb}}{16 \text{ oz}} = \frac{4}{5} \cdot \frac{1}{16} \text{ lb} = \frac{1}{20} \text{ lb}$

21. $$\begin{aligned} 5\frac{3}{4} \text{ lb} &= \frac{23}{4} \text{ lb} \\ &= \frac{\frac{23}{4} \text{ lb}}{1} \cdot \frac{16 \text{ oz}}{1 \text{ lb}} \\ &= \frac{23}{4} \cdot 16 \text{ oz} \\ &= 92 \text{ oz} \end{aligned}$$

23. $$\begin{aligned} 10 \text{ lb } 1 \text{ oz} &= 10 \cdot 16 \text{ oz} + 1 \text{ oz} \\ &= 160 \text{ oz} + 1 \text{ oz} \\ &= 161 \text{ oz} \end{aligned}$$

25. $89 \text{ oz} = \frac{89 \text{ oz}}{1} \cdot \frac{1 \text{ lb}}{16 \text{ oz}} = \frac{89}{16} \text{ lb}$

$$\begin{array}{r} 5 \text{ lb } 9 \text{ oz} \\ 16 \overline{)\ 89} \\ \underline{-80} \\ 9 \end{array}$$

89 oz = 5 lb 9 oz

27. $$\begin{aligned} 34 \text{ lb } 12 \text{ oz} + 18 \text{ lb } 14 \text{ oz} &= 52 \text{ lb } 26 \text{ oz} \\ &= 52 \text{ lb} + 1 \text{ lb } 10 \text{ oz} \\ &= 53 \text{ lb } 10 \text{ oz} \end{aligned}$$

29. 3 tons 1820 lb + 4 tons 930 lb
= 7 tons 2750 lb
= 7 tons + 1 ton 750 lb
= 8 tons 750 lb

Copyright © 2011 Pearson Education, Inc. Publishing as Prentice Hall.

31.
$$\begin{array}{r} 5 \text{ tons } 1050 \text{ lb} \\ -\ 2 \text{ tons } \ 875 \text{ lb} \\ \hline 3 \text{ tons } \ 175 \text{ lb} \end{array}$$

33.
$$\begin{array}{r} 12 \text{ lb } 4 \text{ oz} \\ -\ 3 \text{ lb } 9 \text{ oz} \\ \hline \end{array} \quad \begin{array}{r} 11 \text{ lb } 20 \text{ oz} \\ -\ 3 \text{ lb } \ 9 \text{ oz} \\ \hline 8 \text{ lb } 11 \text{ oz} \end{array}$$

35.
$$\begin{aligned} 5 \text{ lb } 3 \text{ oz} \times 6 &= 30 \text{ lb } 18 \text{ oz} \\ &= 30 \text{ lb} + 1 \text{ lb } 2 \text{ oz} \\ &= 31 \text{ lb } 2 \text{ oz} \end{aligned}$$

37.
$$\begin{aligned} 6 \text{ tons } 1500 \text{ lb } \div 5 &= \frac{6}{5} \text{ tons } 300 \text{ lb} \\ &= 1\frac{1}{5} \text{ tons } 300 \text{ lb} \\ &= 1 \text{ ton} + \frac{2000 \text{ lb}}{5} + 300 \text{ lb} \\ &= 1 \text{ ton} + 400 \text{ lb} + 300 \text{ lb} \\ &= 1 \text{ ton } 700 \text{ lb} \end{aligned}$$

39. $500 \text{ g} = \frac{500 \text{ g}}{1} \cdot \frac{1 \text{ kg}}{1000 \text{ g}} = \frac{500}{1000} \text{ kg} = 0.5 \text{ kg}$

41. $4 \text{ g} = \frac{4 \text{ g}}{1} \cdot \frac{1000 \text{ mg}}{1 \text{ g}} = 4 \cdot 1000 \text{ mg} = 4000 \text{ mg}$

43. $25 \text{ kg} = \frac{25 \text{ kg}}{1} \cdot \frac{1000 \text{ g}}{1 \text{ kg}} = 25 \cdot 1000 \text{ g} = 25{,}000 \text{ g}$

45. $48 \text{ mg} = \frac{48 \text{ mg}}{1} \cdot \frac{1 \text{ g}}{1000 \text{ mg}} = \frac{48}{1000} \text{ g} = 0.048 \text{ g}$

47. $6.3 \text{ g} = \frac{6.3 \text{ g}}{1} \cdot \frac{1 \text{ kg}}{1000 \text{ g}} = \frac{6.3}{1000} \text{ kg} = 0.0063 \text{ kg}$

49.
$$\begin{aligned} 15.14 \text{ g} &= \frac{15.14 \text{ g}}{1} \cdot \frac{1000 \text{ mg}}{1 \text{ g}} \\ &= 15.14 \cdot 1000 \text{ mg} \\ &= 15{,}140 \text{ mg} \end{aligned}$$

51.
$$\begin{aligned} 6.25 \text{ kg} &= \frac{6.25 \text{ kg}}{1} \cdot \frac{1000 \text{ g}}{1 \text{ kg}} \\ &= 6.25 \cdot 1000 \text{ g} \\ &= 6250 \text{ g} \end{aligned}$$

53.
$$\begin{aligned} 35 \text{ hg} &= \frac{35 \text{ hg}}{1} \cdot \frac{10{,}000 \text{ cg}}{1 \text{ hg}} \\ &= 35 \cdot 10{,}000 \text{ cg} \\ &= 350{,}000 \text{ cg} \end{aligned}$$

55.
$$\begin{array}{r} 3.8 \text{ mg} \\ +\ 9.7 \text{ mg} \\ \hline 13.5 \text{ mg} \end{array}$$

57. 205 mg + 5.61 g = 0.205 g + 5.61 g = 5.815 g
or
$$\begin{aligned} 205 \text{ mg} + 5.61 \text{ g} &= 205 \text{ mg} + 5610 \text{ mg} \\ &= 5815 \text{ mg} \end{aligned}$$

59. 9 g − 7150 mg
$$\begin{array}{r} 9000 \text{ mg} \\ -\ 7150 \text{ mg} \\ \hline 1850 \text{ mg} \end{array} \quad \text{or} \quad \begin{array}{r} 9.000 \text{ g} \\ -\ 7.150 \text{ g} \\ \hline 1.850 \text{ g} \end{array} \text{ or } 1.85 \text{ g}$$

61. 1.61 kg − 250 g = 1.61 kg − 0.250 kg = 1.36 kg
or
1.61 kg − 250 g = 1610 g − 250 g = 1360 g

63.
$$\begin{array}{r} 5.2 \text{ kg} \\ \times\ 2.6 \\ \hline 13.52 \text{ kg} \end{array}$$

65. $17 \text{ kg} \div 8 = \frac{17}{8} \text{ kg}$

$$\begin{array}{r} 2.125 \\ 8 \overline{)\ 17.000} \\ \underline{-16} \\ 1\,0 \\ \underline{-8} \\ 20 \\ \underline{-16} \\ 40 \\ \underline{-40} \\ 0 \end{array}$$

17 kg ÷ 8 = 2.125 kg

Copyright © 2011 Pearson Education, Inc. Publishing as Prentice Hall.

	Object	Tons	Pounds	Ounces
67.	Statue of Liberty—weight of copper sheeting	100	200,000	3,200,000
69.	A 12-inch cube of osmium (heaviest metal)	$\frac{269}{400}$ or 0.6725	1345	21,520

	Object	Grams	Kilograms	Milligrams	Centigrams
71.	Capsule of Amoxicillin (antibiotic)	0.5	0.0005	500	50
73.	A six-year-old boy	21,000	21	21,000,000	2,100,000

75. $\begin{array}{r} 336 \\ \times\ 24 \\ \hline 1344 \\ 6720 \\ \hline 8064 \end{array}$

$8064\text{ g} = \frac{8064\text{ g}}{1} \cdot \frac{1\text{ kg}}{1000\text{ g}} = \frac{8064}{1000}\text{ kg} = 8.064\text{ kg}$

24 cans weigh 8.064 kg.

77. $\begin{aligned} 0.09\text{ g} &= \frac{0.09\text{ g}}{1} \cdot \frac{1000\text{ mg}}{1\text{ g}} \\ &= 0.09 \cdot 1000\text{ mg} \\ &= 90\text{ mg} \end{aligned}$

90 mg − 60 mg = 30 mg

The extra-strength tablet contains 30 mg more medication.

79. $\begin{array}{r} 1\text{ lb } 10\text{ oz} \\ +\ 3\text{ lb } 14\text{ oz} \\ \hline 4\text{ lb } 24\text{ oz} \end{array} = 4\text{ lb} + 1\text{ lb } 8\text{ oz} = 5\text{ lb } 8\text{ oz}$

The total amount of rice is 5 lb 8 oz.

81. $\begin{array}{r} 64\text{ lb } \ 8\text{ oz} \\ -\ 28\text{ lb } 10\text{ oz} \\ \hline \end{array} \qquad \begin{array}{r} 63\text{ lb } 24\text{ oz} \\ -\ 28\text{ lb } 10\text{ oz} \\ \hline 35\text{ lb } 14\text{ oz} \end{array}$

Carla's zucchini was 35 lb 14 oz lighter than the record weight.

83. $\begin{array}{r} 7\text{ lb } \ \ 8\text{ oz} \\ -\qquad 8.6\text{ oz} \\ \hline \end{array} \qquad \begin{array}{r} 6\text{ lb } \ 24\text{ oz} \\ -\qquad 8.6\text{ oz} \\ \hline 6\text{ lb } 15.4\text{ oz} \end{array}$

This is 6 lb 15.4 oz lighter.

85. $3 \times 16 = 48$

3 cartons contain 48 boxes of fruit.

$3\text{ mg} \times 48 = 144\text{ mg}$

3 cartons contain 144 mg of preservatives.

87. $26\text{ g} \times 12 = 312\text{ g}$

$\frac{312\text{ g}}{1} \cdot \frac{1\text{ kg}}{1000\text{ g}} = 0.312\text{ kg}$ of packaging in a carton

$6.432\text{ kg} - 0.312\text{ kg} = 6.12\text{ kg}$

The actual weight of the oatmeal is 6.12 kg.

Copyright © 2011 Pearson Education, Inc. Publishing as Prentice Hall.

89. $3 \text{ lb } 4 \text{ oz} \times 10 = 30 \text{ lb } 40 \text{ oz}$
$= 30 \text{ lb} + 2 \text{ lb } 8 \text{ oz}$
$= 32 \text{ lb } 8 \text{ oz}$
Each box weighs 32 lb 8 oz.
$32 \text{ lb } 8 \text{ oz} \times 4 = 128 \text{ lb } 32 \text{ oz}$
$= 128 \text{ lb} + 2 \text{ lb}$
$= 130 \text{ lb}$
4 boxes of meat weigh 130 lb.

91. $\begin{array}{r} 55 \text{ lb } 4 \text{ oz} \\ -\ 2 \text{ lb } 8 \text{ oz} \\ \hline \end{array}$ $\quad \begin{array}{r} 54 \text{ lb } 20 \text{ oz} \\ -\ 2 \text{ lb } \ 8 \text{ oz} \\ \hline 52 \text{ lb } 12 \text{ oz} \end{array}$

$\begin{array}{r} 52 \text{ lb } 12 \text{ oz} \\ \times \qquad 4 \\ \hline 208 \text{ lb } 48 \text{ oz} \end{array} = 208 \text{ lb} + 3 \text{ lb} = 211 \text{ lb}$
4 cartons contain 211 lb of pineapple.

93. $\frac{4}{25} = \frac{4}{25} \cdot \frac{4}{4} = \frac{16}{100} = 0.16$

95. $\frac{7}{8} = \frac{7}{8} \cdot \frac{125}{125} = \frac{875}{1000} = 0.875$

97. No, a pill containing 2 kg of medication is not reasonable.

99. Yes, a bag of flour weighing 4.5 kg is reasonable.

101. No, a professor weighing less than 150 g is not reasonable.

103. answers may vary; for example 250 mg or 0.25 g

105. True, a kilogram is 1000 grams.

107. answers may vary

Section 9.6

Practice Problems

1. $43 \text{ pt} = \frac{43 \text{ pt}}{1} \cdot \frac{1 \text{ qt}}{2 \text{ pt}} = \frac{43}{2} \text{ qt} = 21\frac{1}{2} \text{ qt}$

2. $26 \text{ qt} = \frac{26 \text{ qt}}{1} \cdot \frac{4 \text{ c}}{1 \text{ qt}} = 26 \cdot 4 \text{ c} = 104 \text{ c}$

3. $\begin{array}{r} 1 \text{ gal } 1 \text{ qt} \\ -\qquad 2 \text{ qt} \\ \hline \end{array} \rightarrow \begin{array}{r} 5 \text{ qt} \\ -\ 2 \text{ qt} \\ \hline 3 \text{ qt} \end{array}$

4. $\begin{array}{r} 15 \text{ gal } 3 \text{ qt} \\ +\ 4 \text{ gal } 3 \text{ qt} \\ \hline 19 \text{ gal } 6 \text{ qt} \end{array} = 19 \text{ gal} + 1 \text{ gal } 2 \text{ qt} = 20 \text{ gal } 2 \text{ qt}$
The total amount of oil will be 20 gal 2 qt.

5. $2100 \text{ ml} = \frac{2100 \text{ ml}}{1} \cdot \frac{1 \text{ L}}{1000 \text{ ml}} = \frac{2100}{1000} \text{ L} = 2.1 \text{ L}$

6. $2.13 \text{ dal} = \frac{2.13 \text{ dal}}{1} \cdot \frac{10 \text{ L}}{1 \text{ dal}} = 2.13 \cdot 10 \text{ L} = 21.3 \text{ L}$

7. $1250 \text{ ml} = 1.250 \text{ L}$ $\quad 2.9 \text{ L} = 2900 \text{ ml}$
$\begin{array}{r} 1.25 \text{ L} \\ +\ 2.9\ \ \text{L} \\ \hline 4.15 \text{ L} \end{array} \qquad \begin{array}{r} 1250 \text{ ml} \\ +\ 2900 \text{ ml} \\ \hline 4150 \text{ ml} \end{array}$
The total is 4.15 L or 4150 ml.

8. $\begin{array}{r} 28.6 \text{ L} \\ \times\ \ 85 \\ \hline 143\ 0 \\ 2288\ 0 \\ \hline 2431.0 \text{ L} \end{array}$
Thus, 2431 L can be pumped in 85 minutes.

Vocabulary and Readiness Check

1. Units of capacity are generally used to measure liquids.
2. The basic unit of capacity in the metric system is the liter.
3. One cup equals 8 fluid ounces.
4. One quart equals 2 pints.
5. One pint equals 2 cups.
6. One quart equals 4 cups.
7. One gallon equals 4 quarts.

Exercise Set 9.6

1. $32 \text{ fl oz} = \frac{32 \text{ fl oz}}{1} \cdot \frac{1 \text{ c}}{8 \text{ fl oz}} = \frac{32}{8} \text{ c} = 4 \text{ c}$

3. $8 \text{ qt} = \frac{8 \text{ qt}}{1} \cdot \frac{2 \text{ pt}}{1 \text{ qt}} = 8 \cdot 2 \text{ pt} = 16 \text{ pt}$

5. $14 \text{ qt} = \frac{14 \text{ qt}}{1} \cdot \frac{1 \text{ gal}}{4 \text{ qt}} = \frac{14}{4} \text{ gal} = 3\frac{1}{2} \text{ gal}$

Copyright © 2011 Pearson Education, Inc. Publishing as Prentice Hall.

7. $80 \text{ fl oz} = \frac{80 \text{ fl oz}}{1} \cdot \frac{1 \text{ c}}{8 \text{ fl oz}} = \frac{80}{8} \text{ c} = 10 \text{ c}$

$10 \text{ c} = \frac{10 \text{ c}}{1} \cdot \frac{1 \text{ pt}}{2 \text{ c}} = \frac{10}{2} \text{ pt} = 5 \text{ pt}$

9. $2 \text{ qt} = \frac{2 \text{ qt}}{1} \cdot \frac{2 \text{ pt}}{1 \text{ qt}} \cdot \frac{2 \text{ c}}{1 \text{ pt}} = 2 \cdot 2 \cdot 2 \text{ c} = 8 \text{ c}$

11. $120 \text{ fl oz} = \frac{120 \text{ fl oz}}{1} \cdot \frac{1 \text{ c}}{8 \text{ fl oz}} \cdot \frac{1 \text{ pt}}{2 \text{ c}} \cdot \frac{1 \text{ qt}}{2 \text{ pt}}$

$= \frac{120}{8 \cdot 2 \cdot 2} \text{ qt}$

$= \frac{15}{4} \text{ qt}$

$= 3\frac{3}{4} \text{ qt}$

13. $42 \text{ c} = \frac{42 \text{ c}}{1} \cdot \frac{1 \text{ qt}}{4 \text{ c}} = \frac{42}{4} \text{ qt} = 10\frac{1}{2} \text{ qt}$

15. $4\frac{1}{2} \text{ pt} = \frac{9}{2} \text{ pt} = \frac{\frac{9}{2} \text{ pt}}{1} \cdot \frac{2 \text{ c}}{1 \text{ pt}} = \frac{9}{2} \cdot 2 \text{ c} = 9 \text{ c}$

17. $5 \text{ gal } 3 \text{ qt} = \frac{5 \text{ gal}}{1} \cdot \frac{4 \text{ qt}}{1 \text{ gal}} + 3 \text{ qt}$

$= 5 \cdot 4 \text{ qt} + 3 \text{ qt}$

$= 20 \text{ qt} + 3 \text{ qt}$

$= 23 \text{ qt}$

19. $\frac{1}{2} \text{ c} = \frac{\frac{1}{2} \text{ c}}{1} \cdot \frac{1 \text{ pt}}{2 \text{ c}} = \frac{1}{2} \cdot \frac{1}{2} \text{ pt} = \frac{1}{4} \text{ pt}$

21. $58 \text{ qt} = 56 \text{ qt} + 2 \text{ qt}$

$= \frac{56 \text{ qt}}{1} \cdot \frac{1 \text{ gal}}{4 \text{ qt}} + 2 \text{ qt}$

$= \frac{56}{4} \text{ gal} + 2 \text{ qt}$

$= 14 \text{ gal } 2 \text{ qt}$

23. $39 \text{ pt} = 38 \text{ pt} + 1 \text{ pt}$

$= \frac{38 \text{ pt}}{1} \cdot \frac{1 \text{ qt}}{2 \text{ pt}} + 1 \text{ pt}$

$= 19 \text{ qt} + 1 \text{ pt}$

$= 16 \text{ qt} + 3 \text{ qt} + 1 \text{ pt}$

$= \frac{16 \text{ qt}}{1} \cdot \frac{1 \text{ gal}}{4 \text{ qt}} + 3 \text{ qt} + 1 \text{ pt}$

$= 4 \text{ gal} + 3 \text{ qt} + 1 \text{ pt}$

$= 4 \text{ gal } 3 \text{ qt } 1 \text{ pt}$

25. $2\frac{3}{4} \text{ gal} = \frac{11}{4} \text{ gal}$

$= \frac{\frac{11}{4} \text{ gal}}{1} \cdot \frac{4 \text{ qt}}{1 \text{ gal}} \cdot \frac{2 \text{ pt}}{1 \text{ qt}}$

$= \frac{11}{4} \cdot 4 \cdot 2 \text{ pt}$

$= 22 \text{ pt}$

27.

$$\begin{array}{r} 5 \text{ gal } 3 \text{ qt} \\ +\ 7 \text{ gal } 3 \text{ qt} \\ \hline 12 \text{ gal } 6 \text{ qt} \end{array} = 12 \text{ gal} + 1 \text{ gal } 2 \text{ qt} = 13 \text{ gal } 2 \text{ qt}$$

29. $1 \text{ c } 5 \text{ fl oz} + 2 \text{ c } 7 \text{ fl oz} = 3 \text{ c } 12 \text{ fl oz}$

$= 3 \text{ c} + 1 \text{ c } 4 \text{ fl oz}$

$= 4 \text{ c } 4 \text{ fl oz}$

31.

$$\begin{array}{r} 3 \text{ gal} \\ -\ 1 \text{ gal } 3 \text{ qt} \\ \hline \end{array} \qquad \begin{array}{r} 2 \text{ gal } 4 \text{ qt} \\ -\ 1 \text{ gal } 3 \text{ qt} \\ \hline 1 \text{ gal } 1 \text{ qt} \end{array}$$

33.

$$\begin{array}{r} 3 \text{ gal } 1 \text{ qt} \\ -\ 1 \text{ qt } 1 \text{ pt} \\ \hline \end{array} \qquad \begin{array}{r} 2 \text{ gal } 5 \text{ qt} \\ -\ 1 \text{ qt } 1 \text{ pt} \\ \hline \end{array} \qquad \begin{array}{r} 2 \text{ gal } 4 \text{ qt } 2 \text{ pt} \\ -\ 1 \text{ qt } 1 \text{ pt} \\ \hline 2 \text{ gal } 3 \text{ qt } 1 \text{ pt} \end{array}$$

35. $8 \text{ gal } 2 \text{ qt} \times 2 = 16 \text{ gal } 4 \text{ qt}$

$= 16 \text{ gal} + 1 \text{ gal}$

$= 17 \text{ gal}$

37. $9 \text{ gal } 2 \text{ qt} \div 2 = (8 \text{ gal } 4 \text{ qt} + 2 \text{ qt}) \div 2$

$= 8 \text{ gal } 6 \text{ qt} \div 2$

$= 4 \text{ gal } 3 \text{ qt}$

39. $5\text{L} = \frac{5\text{L}}{1} \cdot \frac{1000 \text{ ml}}{1 \text{ L}} = 5000 \text{ ml}$

41. $0.16 \text{ L} = \frac{0.16 \text{ L}}{1} \cdot \frac{1 \text{ kl}}{1000 \text{ L}}$

$= \frac{0.16}{1000} \text{ kl}$

$= 0.00016 \text{ kl}$

43. $5600 \text{ ml} = \frac{5600 \text{ ml}}{1} \cdot \frac{1 \text{ L}}{1000 \text{ ml}} = \frac{5600}{1000} \text{ L} = 5.6 \text{ L}$

45. $3.2 \text{ L} = \frac{3.2 \text{ L}}{1} \cdot \frac{100 \text{ cl}}{1 \text{ L}} = 3.2 \cdot 100 \text{ cl} = 320 \text{ cl}$

47. $410 \text{ L} = \frac{410 \text{ L}}{1} \cdot \frac{1 \text{ kl}}{1000 \text{ L}} = \frac{410}{1000} \text{ kl} = 0.41 \text{ kl}$

Copyright © 2011 Pearson Education, Inc. Publishing as Prentice Hall.

49. $64 \text{ ml} = \frac{64 \text{ ml}}{1} \cdot \frac{1 \text{ L}}{1000 \text{ ml}} = \frac{64}{1000} \text{ L} = 0.064 \text{ L}$

51. $$\begin{aligned} 0.16 \text{ kl} &= \frac{0.16 \text{ kl}}{1} \cdot \frac{1000 \text{ L}}{1 \text{ kl}} \\ &= 0.16 \cdot 1000 \text{ L} \\ &= 160 \text{ L} \end{aligned}$$

53. $$\begin{aligned} 3.6 \text{ L} &= \frac{3.6 \text{ L}}{1} \cdot \frac{1000 \text{ ml}}{1 \text{ L}} \\ &= 3.6 \cdot 1000 \text{ ml} \\ &= 3600 \text{ ml} \end{aligned}$$

55. 3.4 L + 15.9 L = 19.3 L

57. 2700 ml + 1.8 L = 2.7 L + 1.8 L = 4.5 L
or
2700 ml + 1.8 L = 2700 ml + 1800 ml = 4500 ml

59. $$\begin{array}{r} 8.6 \text{ L} \\ -\ 190 \text{ ml} \\ \hline \end{array} \quad \begin{array}{r} 8600 \text{ ml} \\ -\ 190 \text{ ml} \\ \hline 8410 \text{ ml} \end{array} \quad \text{or} \quad \begin{array}{r} 8.60 \text{ L} \\ -\ 0.19 \text{ L} \\ \hline 8.41 \text{ L} \end{array}$$

61. $$\begin{aligned} 17{,}500 \text{ ml} - 0.9 \text{ L} &= 17{,}500 \text{ ml} - 900 \text{ ml} \\ &= 16{,}600 \text{ ml} \end{aligned}$$
or
17,500 ml − 0.9 L = 17.5 L − 0.9 L = 16.6 L

63. 480 ml × 8 = 3840 ml

65. $$\begin{aligned} 81.2 \text{ L} \div 0.5 &= 81.2 \text{ L} \div \frac{1}{2} \\ &= 81.2 \text{ L} \cdot 2 \\ &= 162.4 \text{ L} \end{aligned}$$

	Capacity	Cups	Gallons	Quarts	Pints
67.	An average-size bath of water	336	21	84	168
69.	Your kidneys filter about this amount of blood every minute	4	$\frac{1}{4}$	1	2

71. $$\begin{array}{r} 2 \text{ L} \\ -\ 410 \text{ ml} \\ \hline \end{array} \quad \begin{array}{r} 2.000 \text{ L} \\ -\ 0.410 \text{ L} \\ \hline 1.590 \text{ L} \end{array}$$
There was 1.59 L left in the bottle.

73. 354 ml + 18.6 L = 0.354 L + 18.6 L = 18.954 L
There were 18.954 liters of gasoline in her tank.

Copyright © 2011 Pearson Education, Inc. Publishing as Prentice Hall.

75. $\frac{1}{30}\text{ gal} = \frac{\frac{1}{30}\text{ gal}}{1} \cdot \frac{4\text{ qt}}{1\text{ gal}} \cdot \frac{2\text{ pt}}{1\text{ qt}} \cdot \frac{2\text{ c}}{1\text{ pt}} \cdot \frac{8\text{ fl oz}}{1\text{ c}}$

$= \frac{1}{30} \cdot 128\text{ fl oz}$

$\approx 4.3\text{ fl oz}$

$\frac{1}{30}$ gal is about 4.3 fluid ounces.

77. $5\text{ pt }1\text{ c} + 2\text{ pt }1\text{ c} = 7\text{ pt }2\text{ c}$

$= 7\text{ pt} + 1\text{ pt}$

$= 8\text{ pt}$

$= \frac{8\text{ pt}}{1} \cdot \frac{1\text{ qt}}{2\text{ pt}}$

$= \frac{8}{2}\text{ qt}$

$= \frac{4\text{ qt}}{1} \cdot \frac{1\text{ gal}}{4\text{ qt}}$

$= \frac{4}{4}\text{ gal}$

$= 1\text{ gal}$

Yes, the liquid can be poured into the container without causing it to overflow.

79. $44.3\overline{)14.0}$ becomes $443\overline{)140.0000}$, quotient $0.3160 \approx 0.316$

```
         0.3160 ≈ 0.316
443 ) 140.0000
     -132 9
        7 10
       -4 43
        2 670
       -2 658
           120
            -0
           120
```

$\frac{\$14}{44.3\text{ L}} \approx \frac{\$0.316}{1\text{ L}}$

The price was \$0.316 per liter.

81. $\frac{20}{25} = \frac{4 \cdot 5}{5 \cdot 5} = \frac{4}{5}$

83. $\frac{27}{45} = \frac{3 \cdot 9}{5 \cdot 9} = \frac{3}{5}$

85. $\frac{72}{80} = \frac{8 \cdot 9}{8 \cdot 10} = \frac{9}{10}$

87. No, a 2 L dose of cough medicine is not reasonable.

89. No, a tub filled with 3000 ml of hot water is not reasonable.

91. less than; answers may vary

93. answers may vary

95. $1\text{ gal} = \frac{1\text{ gal}}{1} \cdot \frac{4\text{ qt}}{1\text{ gal}} \cdot \frac{2\text{ pt}}{1\text{ qt}} \cdot \frac{2\text{ c}}{1\text{ pt}} \cdot \frac{8\text{ fl oz}}{1\text{ c}}$

$= 1 \cdot 4 \cdot 2 \cdot 2 \cdot 8\text{ fl oz}$

$= 128\text{ fl oz}$

There are 128 fl oz in 1 gallon.

97. B indicates 1.5 cc.

99. D indicates 2.7 cc.

101. B indicates 54 u or 0.54 cc.

103. D indicates 86 u or 0.86 cc.

Section 9.7

Practice Problems

1. $1.5\text{ cm} = \frac{1.5\text{ cm}}{1} \cdot \frac{1\text{ in.}}{2.54\text{ cm}} = \frac{1.5}{2.54}\text{ in.} \approx 0.59\text{ in.}$

2. $8\text{ oz} \approx \frac{8\text{ oz}}{1} \cdot \frac{28.35\text{ g}}{1\text{ oz}} = 8 \cdot 28.35\text{ g} = 226.8\text{ g}$

3. $237\text{ ml} \approx \frac{237\text{ ml}}{1} \cdot \frac{1\text{ fl oz}}{29.57\text{ ml}}$

   $= \frac{237}{29.57}\text{ fl oz}$

   $\approx 8\text{ fl oz}$

4. $F = \frac{9}{5} \cdot C + 32 = \frac{9}{5} \cdot 60 + 32 = 108 + 32 = 140$

   Thus, 60°C is equivalent to 140°F.

5. $F = 1.8 \cdot C + 32 = 1.8 \cdot 32 + 32 = 57.6 + 32 = 89.6$

   Therefore, 32°C is the same as 89.6°F.

6. $C = \frac{5}{9} \cdot (F - 32) = \frac{5}{9} \cdot (68 - 32) = \frac{5}{9} \cdot (36) = 20$

   Therefore, 68°F is the same temperature as 20°C.

7. $C = \frac{5}{9} \cdot (F - 32) = \frac{5}{9} \cdot (113 - 32) = \frac{5}{9} \cdot (81) = 45$

   Therefore, 113°F is 45°C.

Copyright © 2011 Pearson Education, Inc. Publishing as Prentice Hall.

8. $C = \frac{5}{9} \cdot (F - 32)$
$= \frac{5}{9} \cdot (102.8 - 32)$
$= \frac{5}{9} \cdot (70.8)$
$= 39.3$
Albert's temperature is 39.3°C.

Exercise Set 9.7

1. $756 \text{ ml} \approx \frac{756 \text{ ml}}{1} \cdot \frac{1 \text{ fl oz}}{29.57 \text{ ml}}$
$= \frac{756}{29.57} \text{ fl oz}$
$\approx 25.57 \text{ fl oz}$

3. $86 \text{ in.} = \frac{86 \text{ in.}}{1} \cdot \frac{2.54 \text{ cm}}{1 \text{ in.}}$
$= 86 \cdot 2.54 \text{ cm}$
$= 218.44 \text{ cm}$

5. $1000 \text{ g} \approx \frac{1000 \text{ g}}{1} \cdot \frac{0.04 \text{ oz}}{1 \text{ g}}$
$= 1000 \cdot 0.04 \text{ oz}$
$= 40 \text{ oz}$

7. $93 \text{ km} \approx \frac{93 \text{ km}}{1} \cdot \frac{0.62 \text{ mi}}{1 \text{ km}}$
$= 93 \cdot 0.62 \text{ mi}$
$= 57.66 \text{ mi}$

9. $14.5 \text{ L} \approx \frac{14.5 \text{ L}}{1} \cdot \frac{0.26 \text{ gal}}{1 \text{ L}} \approx 3.77 \text{ gal}$

11. $30 \text{ lb} \approx \frac{30 \text{ lb}}{1} \cdot \frac{0.45 \text{ kg}}{1 \text{ lb}} = 30 \cdot 0.45 \text{ kg} = 13.5 \text{ kg}$

		Meters	Yards	Centimeters	Feet	Inches
13.	The Height of a Woman	1.5	$1\frac{2}{3}$	150	5	60
15.	Leaning Tower of Pisa	55	60	5500	180	2160

17. $10 \text{ cm} = \frac{10 \text{ cm}}{1} \cdot \frac{1 \text{ in.}}{2.54 \text{ cm}} \approx 3.94 \text{ in.}$
The balance beam is approximately 3.94 inches wide.

Copyright © 2011 Pearson Education, Inc. Publishing as Prentice Hall.

19. $50 \text{ mph} \approx \frac{50 \text{ mph}}{1} \cdot \frac{1.61 \text{ km}}{1 \text{ mi}} = 80.5 \text{ kph}$
The speed limit is approximately 80.5 kilometers per hour.

21. $200 \text{ mg} = 0.2 \text{ g} \approx \frac{0.2 \text{ g}}{1} \cdot \frac{0.04 \text{ oz}}{1 \text{ g}} = 0.008 \text{ oz}$

23. $100 \text{ kg} \approx \frac{100 \text{ kg}}{1} \cdot \frac{2.2 \text{ lb}}{1 \text{ kg}} = 100 \cdot 2.2 \text{ lb} = 220 \text{ lb}$

15 stone 10 lb

$$\begin{aligned} &= \frac{15 \text{ stone}}{1} \cdot \frac{14 \text{ lb}}{1 \text{ stone}} + 10 \text{ lb} \\ &= 15 \cdot 14 \text{ lb} + 10 \text{ lb} \\ &= 210 \text{ lb} + 10 \text{ lb} \\ &= 220 \text{ lb} \end{aligned}$$

Yes; the stamp is approximately correct.

25. $4500 \text{ km} = \frac{4500 \text{ km}}{1} \cdot \frac{0.62 \text{ mi}}{1 \text{ km}} = 2790 \text{ mi}$
The trip is about 2790 miles.

27. $3\frac{1}{2} \text{ in.} = \frac{3\frac{1}{2} \text{ in.}}{1} \cdot \frac{2.54 \text{ cm}}{1 \text{ in.}} = 8.89 \text{ cm}$

$8.89 \text{ cm} = \frac{8.89 \text{ cm}}{1} \cdot \frac{10 \text{ mm}}{1 \text{ cm}} = 88.9 \text{ mm} \approx 90 \text{ mm}$
The width is approximately 90 mm.

29. $1.5 \text{ lb} - 1.25 \text{ lb} = 0.25 \text{ lb}$

$0.25 \text{ lb} \approx \frac{0.25 \text{ lb}}{1} \cdot \frac{0.45 \text{ kg}}{1 \text{ lb}} \cdot \frac{1000 \text{ g}}{1 \text{ kg}} \approx 112.5 \text{ g}$
The difference is approximately 112.5 g.

31. $167 \text{ kmh} \approx \frac{167 \text{ kmh}}{1} \cdot \frac{0.62 \text{ mi}}{1 \text{ km}} \approx 104 \text{ mph}$
The sneeze is approximately 104 miles per hour.

33. $8 \text{ m} \approx \frac{8 \text{ m}}{1} \cdot \frac{3.28 \text{ ft}}{1 \text{ m}} \approx 26.24 \text{ ft}$
The base diameter is approximately 26.24 ft.

35. $4.5 \text{ km} \approx \frac{4.5 \text{ km}}{1} \cdot \frac{0.62 \text{ mi}}{1 \text{ km}} \approx 3 \text{ mi}$
The track is approximately 3 mi.

37. One dose every 4 hours results in $\frac{24}{4} = 6$ doses per day and $6 \times 7 = 42$ doses per week.
$5 \text{ ml} \times 42 = 210 \text{ ml}$

$210 \text{ ml} \approx \frac{210 \text{ ml}}{1} \cdot \frac{1 \text{ fl oz}}{29.57 \text{ ml}} \approx 7.1 \text{ fl oz}$
8 fluid ounces of medicine should be purchased.

39. This math book has a height of about 28 cm; b.

41. A liter has greater capacity than a quart; b.

43. A kilogram weighs greater than a pound; c

45. An $8\frac{1}{2}$-ounce glass of water has a capacity of about 250 ml $\left(\frac{1}{4} \text{ L}\right)$; d.

47. The weight of an average man is about 70 kg ($70 \text{ kg} \approx 2.2 \cdot 70 \text{ lb} = 154 \text{ lb}$); d

49. $C = \frac{5}{9}(F - 32) = \frac{5}{9}(77 - 32) = \frac{5}{9}(45) = 25$
77°F is 25°C.

51. $C = \frac{5}{9}(F - 32) = \frac{5}{9}(104 - 32) = \frac{5}{9}(72) = 40$
104°F is 40°C.

53. $F = \frac{9}{5}C + 32 = \frac{9}{5}(50) + 32 = 90 + 32 = 122$
50°C is 122°F.

55. $F = \frac{9}{5}C + 32 = \frac{9}{5}(115) + 32 = 207 + 32 = 239$
115°C is 239°F.

57. $C = \frac{5}{9}(F - 32) = \frac{5}{9}(20 - 32) = \frac{5}{9}(-12) \approx -6.7$
20°F is −6.7°C.

59.
$$\begin{aligned} C &= \frac{5}{9}(F - 32) \\ &= \frac{5}{9}(142.1 - 32) \\ &= \frac{5}{9}(110.1) \\ &\approx 61.2 \end{aligned}$$
142.1°F is 61.2°C.

Copyright © 2011 Pearson Education, Inc. Publishing as Prentice Hall.

61. $F = 1.8C + 32$
$= 1.8(92) + 32$
$= 165.6 + 32$
$= 197.6$
92°C is 197.6°F.

63. $F = 1.8C + 32$
$= 1.8(12.4) + 32$
$= 22.32 + 32$
≈ 54.3
12.4°C is 54.3°F.

65. $C = \frac{5}{9}(F - 32)$
$= \frac{5}{9}(134 - 32)$
$= \frac{5}{9}(102)$
≈ 56.7
134°F is 56.7°C.

67. $F = 1.8C + 32 = 1.8(27) + 32 = 48.6 + 32 = 80.6$
27°C is 80.6°F.

69. $C = \frac{5}{9}(F - 32) = \frac{5}{9}(70 - 32) = \frac{5}{9}(38) \approx 21.1$
70°F is 21.1°C.

71. $F = 1.8C + 32$
$= 1.8(118) + 32$
$= 212.4 + 32$
$= 244.4$
118°C is 244.4°F.

73. $F = 1.8C + 32$
$= 1.8(4000) + 32$
$= 7200 + 32$
$= 7232$
4000°C is 7232°F.

75. $6 \cdot 4 + 5 \div 1 = 24 + 5 \div 1 = 24 + 5 = 29$

77. $3[(1 + 5) \cdot (8 - 6)] = 3(6 \cdot 2) = 3(12) = 36$

79. Yes, a 72°F room feels comfortable.

81. No, a fever of 40°F is not reasonable.

83. No, an overcoat is not needed when the temperature is 30°C.

85. Yes, a fever of 40°C is reasonable.

87. $\text{BSA} = \sqrt{\frac{90 \times 182}{3600}} \approx 2.13$
The BSA is approximately 2.13 sq m.

89. $40 \text{ in.} = \frac{40 \text{ in.}}{1} \cdot \frac{2.54 \text{ cm}}{1 \text{ in.}} = 101.6 \text{ cm}$
$\text{BSA} = \sqrt{\frac{50 \times 101.6}{3600}} \approx 1.19$
The BSA is approximately 1.19 sq m.

91. $60 \text{ in.} = \frac{60 \text{ in.}}{1} \cdot \frac{2.54 \text{ cm}}{1 \text{ in.}} = 152.4 \text{ cm}$
$150 \text{ lb} \approx \frac{150 \text{ lb}}{1} \cdot \frac{0.45 \text{ kg}}{1 \text{ lb}} = 67.5 \text{ kg}$
$\text{BSA} \approx \sqrt{\frac{67.5 \times 152.4}{3600}} \approx 1.69$
The BSA is approximately 1.69 sq m.

93. $C = \frac{5}{9}(F - 32)$
$= \frac{5}{9}(918,000,000 - 32)$
$= \frac{5}{9}(917,999,968)$
$\approx 510,000,000$
918,000,000°F is approximately 510,000,000°C.

95. answers may vary

Chapter 9 Vocabulary Check

1. Weight is a measure of the pull of gravity.
2. Mass is a measure of the amount of substance in an object. This measure does not change.
3. The basic unit of length in the metric system is the meter.
4. To convert from one unit of length to another, unit fractions may be used.
5. The gram is the basic unit of mass in the metric system.
6. The liter is the basic unit of capacity in the metric system.
7. A line segment is a piece of a line with two endpoints.

Copyright © 2011 Pearson Education, Inc. Publishing as Prentice Hall.

8. Two angles that have a sum of 90° are called complementary angles.

9. A line is a set of points extending indefinitely in two directions.

10. The perimeter of a polygon is the distance around the polygon.

11. An angle is made up of two rays that share the same end point. The common end point is called the vertex.

12. Area measures the amount of surface of a region.

13. A ray is a part of a line with one end point. A ray extends indefinitely in one direction.

14. A line that intersects two or more lines at different points is called a transversal.

15. An angle that measures 180° is called a straight angle.

16. The measure of the space of a solid is called its volume.

17. When two lines intersect, four angles are formed. Two of these angles that are opposite each other are called vertical angles.

18. Two of the angles from #17 that share a common side are called adjacent angles.

19. An angle whose measure is between 90° and 180° is called an obtuse angle.

20. An angle that measures 90° is called a right angle.

21. An angle whose measure is between 0° and 90° is called an acute angle.

22. Two angles that have a sum of 180° are called supplementary angles.

23. The surface area of a polyhedron is the sum of the areas of the faces of the polyhedron.

Chapter 9 Review

1. $\angle A$ is a right angle. It measures 90°.

2. $\angle B$ is a straight angle. It measures 180°.

3. $\angle C$ is an acute angle. It measures between 0° and 90°.

4. $\angle D$ is an obtuse angle. It measures between 90° and 180°.

5. The complement of a 25° angle has measure $90° - 25° = 65°$.

6. The supplement of a 105° angle has measure $180° - 105° = 75°$.

7. $m\angle x = 90° - 32° = 58°$

8. $m\angle x = 180° - 82° = 98°$

9. $m\angle x = 105° - 15° = 90°$

10. $m\angle x = 45° - 20° = 25°$

11. $47° + 133° = 180°$, so $\angle a$ and $\angle b$ are supplementary. So are $\angle b$ and $\angle c$, $\angle c$ and $\angle d$, and $\angle d$ and $\angle a$.

12. $47° + 43° = 90°$, so $\angle x$ and $\angle w$ are complementary. Also, $58° + 32° = 90°$, so $\angle y$ and $\angle z$ are complementary.

13. $\angle x$ and the angle marked 100° are vertical angles, so $m\angle x = 100°$.
$\angle x$ and $\angle y$ are adjacent angles, so $m\angle y = 180° - 100° = 80°$.
$\angle y$ and $\angle z$ are vertical angles, so $m\angle z = m\angle y = 80°$.

14. $\angle x$ and the angle marked 25° are adjacent angles, so $m\angle x = 180° - 25° = 155°$.
$\angle x$ and $\angle y$ are vertical angles, so $m\angle y = m\angle x = 155°$.
$\angle z$ and the angle marked 25° are vertical angles, so $m\angle z = 25°$.

15. $\angle x$ and the angle marked 53° are vertical angles, so $m\angle x = 53°$.
$\angle x$ and $\angle y$ are alternate interior angles, so $m\angle y = m\angle x = 53°$.
$\angle y$ and $\angle z$ are adjacent angles, so $m\angle z = 180° - m\angle y = 180° - 53° = 127°$.

16. $\angle x$ and the angle marked 42° are vertical angles, so $m\angle x = 42°$.
$\angle x$ and $\angle y$ are alternate interior angles, so $m\angle y = m\angle x = 42°$.
$\angle y$ and $\angle z$ are adjacent angles, so $m\angle z = 180° - m\angle y = 180° - 42° = 138°$.

Copyright © 2011 Pearson Education, Inc. Publishing as Prentice Hall.

17. $P = 23 \text{ m} + 11\frac{1}{2} \text{ m} + 23 \text{ m} + 11\frac{1}{2} \text{ m} = 69 \text{ m}$
The perimeter is 69 meters.

18. $P = 11 \text{ cm} + 7.6 \text{ cm} + 12 \text{ cm} = 30.6 \text{ cm}$
The perimeter is 30.6 centimeters.

19. The unmarked vertical side has length $8 \text{ m} - 5 \text{ m} = 3 \text{ m}$. The unmarked horizontal side has length $10 \text{ m} - 7 \text{ m} = 3 \text{ m}$.
$P = (7 + 3 + 3 + 5 + 10 + 8) \text{ m} = 36 \text{ m}$
The perimeter is 36 meters.

20. The unmarked vertical side has length $5 \text{ ft} + 4 \text{ ft} + 11 \text{ ft} = 20 \text{ ft}$.
$P = (22 + 20 + 22 + 11 + 3 + 4 + 3 + 5) \text{ ft} = 90 \text{ ft}$
The perimeter is 90 feet.

21. $P = 2 \cdot l + 2 \cdot w = 2 \cdot 10 \text{ ft} + 2 \cdot 6 \text{ ft} = 32 \text{ ft}$
The perimeter is 32 feet.

22. $P = 4 \cdot s = 4 \cdot 110 \text{ ft} = 440 \text{ ft}$
The perimeter is 440 feet.

23. $C = \pi \cdot d = \pi \cdot 1.7 \text{ in.} \approx 3.14 \cdot 1.7 \text{ in.} = 5.338 \text{ in.}$
The circumference is 5.338 inches.

24.
$$\begin{aligned} C &= 2 \cdot \pi \cdot r \\ &= 2 \cdot \pi \cdot 5 \text{ yd} \\ &= \pi \cdot 10 \text{ yd} \\ &\approx 3.14 \cdot 10 \text{ yd} \\ &= 31.4 \text{ yd} \end{aligned}$$
The circumference is 31.4 yards.

25.
$$\begin{aligned} A &= \frac{1}{2} \cdot (b + B) \cdot h \\ &= \frac{1}{2} \cdot (12 \text{ ft} + 36 \text{ ft}) \cdot 10 \text{ ft} \\ &= \frac{1}{2} \cdot 48 \text{ ft} \cdot 10 \text{ ft} \\ &= 240 \text{ sq ft} \end{aligned}$$
The area is 240 square feet.

26. $A = b \cdot h = 21 \text{ yd} \cdot 9 \text{ yd} = 189 \text{ sq yd}$
The area is 189 square yards.

27. $A = l \cdot w = 40 \text{ cm} \cdot 15 \text{ cm} = 600 \text{ sq cm}$
The area is 600 square centimeters.

28. $A = s^2 = (9.1 \text{ m})^2 = 82.81 \text{ sq m}$
The area is 82.81 square meters.

29. $A = \pi \cdot r^2 = \pi \cdot (7 \text{ ft})^2 = 49\pi \text{ sq ft} \approx 153.86 \text{ sq ft}$
The area is 49π square feet $\approx$ 153.86 square feet.

30. $A = \pi \cdot r^2 = \pi(3 \text{ in.})^2 = 9\pi \text{ sq in.} \approx 28.26 \text{ sq in.}$
The area is
9π square inches $\approx$ 28.26 square inches.

31. $A = \frac{1}{2} \cdot b \cdot h = \frac{1}{2} \cdot 34 \text{ in.} \cdot 7 \text{ in.} = 119 \text{ sq in.}$
The area is 119 square inches.

32. $A = \frac{1}{2} \cdot b \cdot h = \frac{1}{2} \cdot 20 \text{ m} \cdot 14 \text{ m} = 140 \text{ sq m}$
The area is 140 square meters.

33. The unmarked horizontal side has length $13 \text{ m} - 3 \text{ m} = 10 \text{ m}$. The unmarked vertical side has length $12 \text{ m} - 4 \text{ m} = 8 \text{ m}$. The area is the sum of the areas of the two rectangles.
$$\begin{aligned} A &= 12 \text{ m} \cdot 10 \text{ m} + 8 \text{ m} \cdot 3 \text{ m} \\ &= 120 \text{ sq m} + 24 \text{ sq m} \\ &= 144 \text{ sq m} \end{aligned}$$
The area is 144 square meters.

34. The unmarked vertical side has length $30 \text{ cm} - 5 \text{ cm} = 25 \text{ cm}$.
The unmarked horizontal side has length $60 \text{ cm} - 35 \text{ cm} = 25 \text{ cm}$.
The area is the sum of the areas of the two rectangles.
$A = 30 \text{ cm} \cdot 25 \text{ cm} + 35 \text{ cm} \cdot 25 \text{ cm} = 1625 \text{ sq cm}$
The area is 1625 square centimeters.

35. $A = l \cdot w = 36 \text{ ft} \cdot 12 \text{ ft} = 432 \text{ sq ft}$
The area of the driveway is 432 square feet.

36. $A = 10 \text{ ft} \cdot 13 \text{ ft} = 130 \text{ sq ft}$
130 square feet of carpet are needed.

37.
$$\begin{aligned} V &= s^3 \\ &= \left(2\frac{1}{2} \text{ in.}\right)^3 \\ &= \left(\frac{5}{2} \text{ in.}\right)^3 \\ &= \frac{125}{8} \text{ cu in.} \\ &= 15\frac{5}{8} \text{ cu in.} \end{aligned}$$
The volume is $15\frac{5}{8}$ cubic inches.

Copyright © 2011 Pearson Education, Inc. Publishing as Prentice Hall.

$$\begin{aligned}SA &= 6s^2\\ &= 6\left(2\frac{1}{2}\text{ in.}\right)^2\\ &= 6\left(\frac{5}{2}\text{ in.}\right)^2\\ &= 6\left(\frac{25}{4}\right)\text{ sq in.}\\ &= \frac{75}{2}\text{ sq in.}\\ &= 37\frac{1}{2}\text{ sq in.}\end{aligned}$$

The surface area is $37\frac{1}{2}$ square inches.

38. $V = l \cdot w \cdot h = 2\text{ ft} \cdot 7\text{ ft} \cdot 6\text{ ft} = 84\text{ cu ft}$
The volume is 84 cubic feet.

$$\begin{aligned}SA &= 2lh + 2wh + 2lw\\ &= 2\cdot 7\text{ ft}\cdot 6\text{ ft} + 2\cdot 2\text{ ft}\cdot 6\text{ ft} + 2\cdot 7\text{ ft}\cdot 2\text{ ft}\\ &= 84\text{ sq ft} + 24\text{ sq ft} + 28\text{ sq ft}\\ &= 136\text{ sq ft}\end{aligned}$$

The surface area is 136 square feet.

39.
$$\begin{aligned}V &= \pi\cdot r^2\cdot h\\ &= \pi\cdot(20\text{ cm})^2\cdot 50\text{ cm}\\ &= 20{,}000\pi\text{ cu cm}\\ &\approx 62{,}800\text{ cu cm}\end{aligned}$$

The volume is $20{,}000\pi$ cubic centimeters $\approx 62{,}800$ cubic centimeters.

40.
$$\begin{aligned}V &= \frac{4}{3}\cdot\pi\cdot r^3\\ &= \frac{4}{3}\cdot\pi\cdot\left(\frac{1}{2}\text{ km}\right)^3\\ &= \frac{1}{6}\pi\text{ cu km}\\ &\approx \frac{11}{21}\text{ cu km}\end{aligned}$$

The volume is $\frac{1}{6}\pi$ cubic kilometers $\approx \frac{11}{21}$ cubic kilometers.

41.
$$\begin{aligned}V &= \frac{1}{3}\cdot s^2\cdot h\\ &= \frac{1}{3}\cdot(2\text{ ft})^2\cdot 2\text{ ft}\\ &= \frac{8}{3}\text{ cu ft}\\ &= 2\frac{2}{3}\text{ cu ft}\end{aligned}$$

The volume of the pyramid is $2\frac{2}{3}$ cubic feet.

42.
$$\begin{aligned}V &= \pi\cdot r^2\cdot h\\ &= \pi\cdot(3.5\text{ in.})^2\cdot 8\text{ in.}\\ &= 98\pi\text{ cu in.}\\ &\approx 307.72\text{ cu in.}\end{aligned}$$

The volume of the can is about 307.72 cubic inches.

43. Find the volume of each drawer.

$$\begin{aligned}V &= l\cdot w\cdot h\\ &= \left(2\frac{1}{2}\text{ ft}\right)\cdot\left(1\frac{1}{2}\text{ ft}\right)\cdot\left(\frac{2}{3}\text{ ft}\right)\\ &= \frac{5}{2}\cdot\frac{3}{2}\cdot\frac{2}{3}\text{ cu ft}\\ &= \frac{5}{2}\text{ cu ft}\end{aligned}$$

The three drawers have volume $3\cdot\frac{5}{2} = \frac{15}{2} = 7\frac{1}{2}$ cubic feet.

44. $r = d \div 2 = 1\text{ ft} \div 2 = 0.5\text{ ft}$
$V = \pi\cdot r^2\cdot h = \pi\cdot(0.5\text{ ft})^2\cdot 2\text{ ft} = 0.5\pi\text{ cu ft}$
The volume of the canister is 0.5π cubic feet or $\frac{1}{2}\pi$ cubic feet.

45. $108\text{ in.} = \frac{108\text{ in.}}{1}\cdot\frac{1\text{ ft}}{12\text{ in.}} = \frac{108}{12}\text{ ft} = 9\text{ ft}$

46. $72\text{ ft} = \frac{72\text{ ft}}{1}\cdot\frac{1\text{ yd}}{3\text{ ft}} = \frac{72}{3}\text{ yd} = 24\text{ yd}$

47. $1.5\text{ mi} = \frac{1.5\text{ mi}}{1}\cdot\frac{5280\text{ ft}}{1\text{ mi}} = 1.5\cdot 5280\text{ ft} = 7920\text{ ft}$

48. $\frac{1}{2}\text{ yd} = \frac{\frac{1}{2}\text{ yd}}{1}\cdot\frac{3\text{ ft}}{1\text{ yd}}\cdot\frac{12\text{ in.}}{1\text{ ft}} = \frac{1}{2}\cdot 3\cdot 12\text{ in.} = 18\text{ in.}$

Copyright © 2011 Pearson Education, Inc. Publishing as Prentice Hall.

49. $52 \text{ ft} = 51 \text{ ft} + 1 \text{ ft}$
$= \frac{51 \text{ ft}}{1} \cdot \frac{1 \text{ yd}}{3 \text{ ft}} + 1 \text{ ft}$
$= \frac{51}{3} \text{ yd} + 1 \text{ ft}$
$= 17 \text{ yd } 1 \text{ ft}$

50. $46 \text{ in.} = 36 \text{ in.} + 10 \text{ in.}$
$= \frac{36 \text{ in.}}{1} \cdot \frac{1 \text{ ft}}{12 \text{ in.}} + 10 \text{ in.}$
$= \frac{36}{12} \text{ ft} + 10 \text{ in.}$
$= 3 \text{ ft } 10 \text{ in.}$

51. $42 \text{ m} = \frac{42 \text{ m}}{1} \cdot \frac{100 \text{ cm}}{1 \text{ m}} = 42 \cdot 100 \text{ cm} = 4200 \text{ cm}$

52. $82 \text{ cm} = \frac{82 \text{ cm}}{1} \cdot \frac{10 \text{ mm}}{1 \text{ cm}} = 82 \cdot 10 \text{ mm} = 820 \text{ mm}$

53. $12.18 \text{ mm} = \frac{12.18 \text{ mm}}{1} \cdot \frac{1 \text{ m}}{1000 \text{ mm}}$
$= \frac{12.18}{1000} \text{ m}$
$= 0.01218 \text{ m}$

54. $2.31 \text{ m} = \frac{2.31 \text{ m}}{1} \cdot \frac{1 \text{ km}}{1000 \text{ m}}$
$= \frac{2.31}{1000} \text{ km}$
$= 0.00231 \text{ km}$

55.
$$\begin{array}{r} 4 \text{ yd } 2 \text{ ft} \\ +\,16 \text{ yd } 2 \text{ ft} \\ \hline 20 \text{ yd } 4 \text{ ft} \end{array} = 20 \text{ yd} + 1 \text{ yd } 1 \text{ ft} = 21 \text{ yd } 1 \text{ ft}$$

56. $7 \text{ ft } 4 \text{ in.} \div 2 = (6 \text{ ft} + 1 \text{ ft } 4 \text{ in.}) \div 2$
$= (6 \text{ ft} + 16 \text{ in.}) \div 2$
$= \frac{6}{2} \text{ ft} + \frac{16}{2} \text{ in.}$
$= 3 \text{ ft } 8 \text{ in.}$

57. 8 cm = 80 mm or 15 mm = 1.5 cm
$$\begin{array}{r} 80 \text{ mm} \\ +\,15 \text{ mm} \\ \hline 95 \text{ mm} \end{array} \quad \text{or} \quad \begin{array}{r} 8.0 \text{ cm} \\ +\,1.5 \text{ cm} \\ \hline 9.5 \text{ cm} \end{array}$$

58. 4 m = 400 cm or 126 cm = 1.26 m
$$\begin{array}{r} 400 \text{ cm} \\ -\,126 \text{ cm} \\ \hline 274 \text{ cm} \end{array} \quad \text{or} \quad \begin{array}{r} 4.00 \text{ m} \\ -\,1.26 \text{ m} \\ \hline 2.74 \text{ m} \end{array}$$

59.
$$\begin{array}{r} 333 \text{ yd } 1 \text{ ft} \\ -\,163 \text{ yd } 2 \text{ ft} \\ \hline \end{array} \qquad \begin{array}{r} 332 \text{ yd } 4 \text{ ft} \\ -\,163 \text{ yd } 2 \text{ ft} \\ \hline 169 \text{ yd } 2 \text{ ft} \end{array}$$
The amount of material that remains is 169 yd 2 ft.

60.
$$\begin{array}{r} 5 \text{ ft} \quad 2 \text{ in.} \\ \times \qquad 50 \\ \hline 250 \text{ ft } 100 \text{ in.} \end{array}$$
$= 250 \text{ ft} + 96 \text{ in.} + 4 \text{ in.}$
$= 250 \text{ ft} + 8 \text{ ft } 4 \text{ in.}$
$= 258 \text{ ft } 4 \text{ in.}$
The sashes require 258 ft 4 in. of material.

61.
$$\begin{array}{r} 217 \text{ km} \\ \times\ 2 \\ \hline 434 \text{ km} \end{array}$$
$434 \text{ km} \div 4 = \frac{434}{4} \text{ km} = 108.5 \text{ km}$
Each must drive 108.5 km.

62.
$$\begin{array}{r} 0.8 \text{ m} \\ \times\,30 \text{ cm} \\ \hline \end{array} \qquad \begin{array}{r} 0.8 \text{ m} \\ \times\quad 0.3 \text{ m} \\ \hline 0.24 \text{ sq m} \end{array}$$
The area is 0.24 sq m.

63. $66 \text{ oz} = \frac{66 \text{ oz}}{1} \cdot \frac{1 \text{ lb}}{16 \text{ oz}} = \frac{66}{16} \text{ lb} = \frac{33}{8} \text{ lb} = 4\frac{1}{8} \text{ lb}$

64. $2.3 \text{ tons} = \frac{2.3 \text{ tons}}{1} \cdot \frac{2000 \text{ lb}}{1 \text{ ton}}$
$= 2.3 \cdot 2000 \text{ lb}$
$= 4600 \text{ lb}$

65. $52 \text{ oz} = 48 \text{ oz} + 4 \text{ oz}$
$= \frac{48 \text{ oz}}{1} \cdot \frac{1 \text{ lb}}{16 \text{ oz}} + 4 \text{ oz}$
$= \frac{48}{16} \text{ lb} + 4 \text{ oz}$
$= 3 \text{ lb } 4 \text{ oz}$

66. $10,300 \text{ lb} = 10,000 \text{ lb} + 300 \text{ lb}$
$= \frac{10,000 \text{ lb}}{1} \cdot \frac{1 \text{ ton}}{2000 \text{ lb}} + 300 \text{ lb}$
$= \frac{10,000}{2000} \text{ tons} + 300 \text{ lb}$
$= 5 \text{ tons } 300 \text{ lb}$

67. $27 \text{ mg} = \frac{27 \text{ mg}}{1} \cdot \frac{1 \text{ g}}{1000 \text{ mg}} = \frac{27}{1000} \text{ g} = 0.027 \text{ g}$

Copyright © 2011 Pearson Education, Inc. Publishing as Prentice Hall.

68. $40\text{ kg} = \frac{40\text{ kg}}{1} \cdot \frac{1000\text{ g}}{1\text{ kg}} = 40 \cdot 1000\text{ g} = 40{,}000\text{ g}$

69. $2.1\text{ hg} = \frac{2.1\text{ hg}}{1} \cdot \frac{10\text{ dag}}{1\text{ hg}} = 2.1 \cdot 10\text{ dag} = 21\text{ dag}$

70.
$$\begin{aligned}0.03\text{ mg} &= \frac{0.03\text{ mg}}{1} \cdot \frac{1\text{ dg}}{100\text{ mg}}\\ &= \frac{0.03}{100}\text{ dg}\\ &= 0.0003\text{ dg}\end{aligned}$$

71.
$$\begin{array}{r}6\text{ lb}\ \ 5\text{ oz}\\ -\ 2\text{ lb}\ 12\text{ oz}\\ \hline\end{array} \qquad \begin{array}{r}5\text{ lb}\ 21\text{ oz}\\ -\ 2\text{ lb}\ 12\text{ oz}\\ \hline 3\text{ lb}\ \ 9\text{ oz}\end{array}$$

72.
$$\begin{array}{r}8\text{ lb}\ 6\text{ oz}\\ \times\qquad 4\\ \hline 32\text{ lb}\ 24\text{ oz}\end{array} = 32\text{ lb} + 1\text{ lb}\ 8\text{ oz} = 33\text{ lb}\ 8\text{ oz}$$

73.
$$\begin{array}{r}4.3\text{ mg}\\ \times\ 5\quad\\ \hline 21.5\text{ mg}\end{array}$$

74. $4.8\text{ kg} = 4800\text{ g}$ $4200\text{ g} = 4.2\text{ kg}$
$$\begin{array}{r}4800\text{ g}\\ -\ 4200\text{ g}\\ \hline 600\text{ g}\end{array} \quad \text{or} \quad \begin{array}{r}4.8\text{ kg}\\ -\ 4.2\text{ kg}\\ \hline 0.6\text{ kg}\end{array}$$

75.
$$\begin{array}{r}1\text{ lb}\ 12\text{ oz}\\ +\ 2\text{ lb}\ \ 8\text{ oz}\\ \hline 3\text{ lb}\ 20\text{ oz}\end{array} = 3\text{ lb} + 1\text{ lb}\ 4\text{ oz} = 4\text{ lb}\ 4\text{ oz}$$
The total weight was 4 lb 4 oz.

76.
$$\begin{aligned}38\text{ tons } 300\text{ lb} \div 4 &= \frac{38}{4}\text{ tons } \frac{300}{4}\text{ lb}\\ &= 9\frac{1}{2}\text{ tons } 75\text{ lb}\\ &= 9\text{ tons} + \frac{1}{2}\text{ ton} + 75\text{ lb}\\ &= 9\text{ tons} + 1000\text{ lb} + 75\text{ lb}\\ &= 9\text{ tons } 1075\text{ lb}\end{aligned}$$
They each receive 9 tons 1075 lb.

77. $28\text{ pt} = \frac{28\text{ pt}}{1} \cdot \frac{1\text{ qt}}{2\text{ pt}} = \frac{28}{2}\text{ qt} = 14\text{ qt}$

78. $40\text{ fl oz} = \frac{40\text{ fl oz}}{1} \cdot \frac{1\text{ c}}{8\text{ fl oz}} = \frac{40}{8}\text{ c} = 5\text{ c}$

79.
$$\begin{aligned}3\text{ qt } 1\text{ pt} &= \frac{3\text{ qt}}{1} \cdot \frac{2\text{ pt}}{1\text{ qt}} + 1\text{ pt}\\ &= 3 \cdot 2\text{ pt} + 1\text{ pt}\\ &= 6\text{ pt} + 1\text{ pt}\\ &= 7\text{ pt}\end{aligned}$$

80. $18\text{ qt} = \frac{18\text{ qt}}{1} \cdot \frac{2\text{ pt}}{1\text{ qt}} \cdot \frac{2\text{ c}}{1\text{ pt}} = 18 \cdot 2 \cdot 2\text{ c} = 72\text{ c}$

81.
$$\begin{aligned}9\text{ pt} &= 8\text{ pt} + 1\text{ pt}\\ &= \frac{8\text{ pt}}{1} \cdot \frac{1\text{ qt}}{2\text{ pt}} + 1\text{ pt}\\ &= \frac{8}{2}\text{ qt} + 1\text{ pt}\\ &= 4\text{ qt } 1\text{ pt}\end{aligned}$$

82.
$$\begin{aligned}15\text{ qt} &= 12\text{ qt} + 3\text{ qt}\\ &= \frac{12\text{ qt}}{1} \cdot \frac{1\text{ gal}}{4\text{ qt}} + 3\text{ qt}\\ &= \frac{12}{4}\text{ gal} + 3\text{ qt}\\ &= 3\text{ gal } 3\text{ qt}\end{aligned}$$

83.
$$\begin{aligned}3.8\text{ L} &= \frac{3.8\text{ L}}{1} \cdot \frac{1000\text{ ml}}{1\text{ L}}\\ &= 3.8 \cdot 1000\text{ ml}\\ &= 3800\text{ ml}\end{aligned}$$

84. $14\text{ hl} = \frac{14\text{ hl}}{1} \cdot \frac{1\text{ kl}}{10\text{ hl}} = \frac{14}{10}\text{ kl} = 1.4\text{ kl}$

85. $30.6\text{ L} = \frac{30.6\text{ L}}{1} \cdot \frac{100\text{ cl}}{1\text{ L}} = 30.6 \cdot 100\text{ cl} = 3060\text{ cl}$

86. $2.45\text{ ml} = \frac{2.45\text{ ml}}{1} \cdot \frac{1\text{ L}}{1000\text{ ml}} = 0.00245\text{ L}$

87.
$$\begin{array}{r}1\text{ qt}\ 1\text{ pt}\\ +\ 3\text{ qt}\ 1\text{ pt}\\ \hline 4\text{ qt}\ 2\text{ pt}\end{array} = 4\text{ qt} + 1\text{ qt} = 1\text{ gal}\ 1\text{ qt}$$

88.
$$\begin{array}{r}3\text{ gal}\ 2\text{ qt}\\ \times\qquad 2\\ \hline 6\text{ gal}\ 4\text{ qt}\end{array} = 6\text{ gal} + 1\text{ gal} = 7\text{ gal}$$

89. $0.946\text{ L} = 946\text{ ml}$ $210\text{ ml} = 0.21\text{ L}$
$$\begin{array}{r}946\text{ ml}\\ -\ 210\text{ ml}\\ \hline 736\text{ ml}\end{array} \quad \text{or} \quad \begin{array}{r}0.946\text{ L}\\ -\ 0.210\text{ L}\\ \hline 0.736\text{ L}\end{array}$$

Copyright © 2011 Pearson Education, Inc. Publishing as Prentice Hall.

90. 6.1 L = 6100 ml   9400 ml = 9.4 L

$$\begin{array}{r} 6100 \text{ ml} \\ +\,9400 \text{ ml} \\ \hline 15{,}500 \text{ ml} \end{array} \quad \text{or} \quad \begin{array}{r} 6.1 \text{ L} \\ +\,9.4 \text{ L} \\ \hline 15.5 \text{ L} \end{array}$$

91.
$$\begin{array}{r} 4 \text{ gal } 2 \text{ qt} \\ -\,1 \text{ gal } 3 \text{ qt} \\ \hline \end{array} \qquad \begin{array}{r} 3 \text{ gal } 6 \text{ qt} \\ -\,1 \text{ gal } 3 \text{ qt} \\ \hline 2 \text{ gal } 3 \text{ qt} \end{array}$$

There are 2 gal 3 qt of tea remaining.

92.
$$\begin{aligned} 1 \text{ c } 4 \text{ fl oz} \div 2 &= (8 \text{ fl oz} + 4 \text{ fl oz}) \div 2 \\ &= 12 \text{ fl oz} \div 2 \\ &= 6 \text{ fl oz} \end{aligned}$$

Use 6 fl oz of stock for half of a recipe.

93. $85 \text{ ml} \times 8 \times 16 = 10{,}880 \text{ ml}$

$$\frac{10{,}880 \text{ ml}}{1} \cdot \frac{1 \text{ L}}{1000 \text{ ml}} = \frac{10{,}880}{1000} \text{ L} = 10.88 \text{ L}$$

There are 10.88 L of polish in 8 boxes.

94.
$$\begin{aligned} 6 \text{ L} + 1300 \text{ ml} + 2.6 \text{ L} &= 6 \text{ L} + 1.3 \text{ L} + 2.6 \text{ L} \\ &= 9.9 \text{ L} \end{aligned}$$

Since 9.9 L is less than 10 L, yes it will fit.

95. $7 \text{ m} \approx \frac{7 \text{ m}}{1} \cdot \frac{3.28 \text{ ft}}{1 \text{ m}} = 22.96 \text{ ft}$

96. $11.5 \text{ yd} \approx \frac{11.5 \text{ yd}}{1} \cdot \frac{1 \text{ m}}{1.09 \text{ yd}} \approx 10.55 \text{ m}$

97. $17.5 \text{ L} \approx \frac{17.5 \text{ L}}{1} \cdot \frac{0.26 \text{ gal}}{1 \text{ L}} = 4.55 \text{ gal}$

98. $7.8 \text{ L} \approx \frac{7.8 \text{ L}}{1} \cdot \frac{1.06 \text{ qt}}{1 \text{ L}} = 8.268 \text{ qt}$

99. $15 \text{ oz} \approx \frac{15 \text{ oz}}{1} \cdot \frac{28.35 \text{ g}}{1 \text{ oz}} = 425.25 \text{ g}$

100. $23 \text{ lb} \approx \frac{23 \text{ lb}}{1} \cdot \frac{0.45 \text{ kg}}{1 \text{ lb}} = 10.35 \text{ kg}$

101. $1.2 \text{ mm} \times 50 = 60 \text{ mm}$

$$60 \text{ mm} = \frac{60 \text{ mm}}{1} \cdot \frac{1 \text{ cm}}{10 \text{ mm}} = 6 \text{ cm}$$

$$6 \text{ cm} = \frac{6 \text{ cm}}{1} \cdot \frac{1 \text{ in.}}{2.54 \text{ cm}} \approx 2.36 \text{ in.}$$

The height of the stack is approximately 2.36 in.

102. $82 \text{ kg} \approx \frac{82 \text{ kg}}{1} \cdot \frac{2.20 \text{ lb}}{1 \text{ kg}} = 180.4 \text{ lb}$

The person weighs approximately 180.4 lb.

103.
$$\begin{aligned} F &= 1.8C + 32 \\ &= 1.8(42) + 32 \\ &= 75.6 + 32 \\ &= 107.6 \end{aligned}$$

42°C is 107.6°F.

104. $F = 1.8C + 32 = 1.8(160) + 32 = 288 + 32 = 320$

160°C is 320°F.

105. $C = \frac{5}{9}(F - 32) = \frac{5}{9}(41.3 - 32) = \frac{5}{9}(9.3) \approx 5.2$

41.3°F is 5.2°C.

106. $C = \frac{5}{9}(F - 32) = \frac{5}{9}(80 - 32) = \frac{5}{9}(48) \approx 26.7$

80°F is 26.7°C.

107. $C = \frac{5}{9}(F - 32) = \frac{5}{9}(35 - 32) = \frac{5}{9}(3) \approx 1.7$

35°F is 1.7°C.

108. $F = 1.8C + 32 = 1.8(165) + 32 = 297 + 32 = 329$

165°C is 329°F.

109. The supplement of a 72° angle is an angle that measures 180° – 72° = 108°.

110. The complement of a 1° angle is an angle that measures 90° – 1° = 89°.

111. $\angle x$ and the angle marked 85° are adjacent angles, so $m\angle x = 180° - 85° = 95°$.

112. Let $\angle y$ be the angle corresponding to $\angle x$ at the bottom intersection. Then $\angle y$ and the angle marked 123° are adjacent angles, so $m\angle x = m\angle y = 180° - 123° = 57°$.

113. P = 7 in. + 11.2 in. + 9.1 in. = 27.3 in.

The perimeter is 27.3 inches.

114. The unmarked horizontal side has length 40 ft – 22 ft – 11 ft = 7 ft.

$$\begin{aligned} P &= (22 + 15 + 7 + 15 + 11 + 42 + 40 + 42) \text{ ft} \\ &= 194 \text{ ft} \end{aligned}$$

The perimeter is 194 feet.

Copyright © 2011 Pearson Education, Inc. Publishing as Prentice Hall.

115. The unmarked horizontal side has length 43 m − 13 m = 30 m. The unmarked vertical side has length 42 m − 14 m = 28 m. The area is the sum of the areas of the two rectangles.

$A = 28\text{ m}\cdot 13\text{ m} + 42\text{ m}\cdot 30\text{ m}$
$= 364\text{ sq m} + 1260\text{ sq m}$
$= 1624\text{ sq m}$

The area is 1624 square meters.

116. $A = \pi\cdot r^2 = \pi\cdot(3\text{ m})^2 = 9\pi\text{ sq m} \approx 28.26\text{ sq m}$

The area is 9π square meters ≈ 28.26 square meters.

117.
$$V = \frac{1}{3}\cdot\pi\cdot r^2\cdot h$$
$$= \frac{1}{3}\cdot\pi\cdot\left(5\frac{1}{4}\text{ in.}\right)^2\cdot 12\text{ in.}$$
$$= \frac{1}{3}\cdot\pi\cdot\left(\frac{21}{4}\text{ in.}\right)^2\cdot 12\text{ cu in.}$$
$$= \frac{441}{4}\pi\text{ cu in.}$$
$$\approx \frac{441}{4}\cdot\frac{22}{7}\text{ cu in.} = 346\frac{1}{2}\text{ cu in.}$$

The volume is $346\frac{1}{2}$ cubic inches.

118. $V = l\cdot w\cdot h = 5\text{ in.}\cdot 4\text{ in.}\cdot 7\text{ in.} = 140\text{ cu in.}$

The volume is 140 cubic inches.

$SA = 2lh + 2wh + 2lw$
$= 2\cdot 7\text{ in.}\cdot 5\text{ in.} + 2\cdot 4\text{ in.}\cdot 5\text{ in.} + 2\cdot 7\text{ in.}\cdot 4\text{ in.}$
$= 70\text{ sq in.} + 40\text{ sq in.} + 56\text{ sq in.}$
$= 166\text{ sq in.}$

The surface area is 166 square inches.

119. $6.25\text{ ft} = \frac{6.25\text{ ft}}{1}\cdot\frac{12\text{ in.}}{1\text{ ft}} = 75\text{ in.}$

120. $8200\text{ lb} = 8000\text{ lb} + 200\text{ lb}$
$= \frac{8000\text{ lb}}{1}\cdot\frac{1\text{ ton}}{2000\text{ lb}} + 200\text{ lb}$
$= 4\text{ tons }200\text{ lb}$

121. $5\text{ m} = \frac{5\text{ m}}{1}\cdot\frac{100\text{ cm}}{1\text{ m}} = 500\text{ cm}$

122. $286\text{ mm} = \frac{286\text{ mm}}{1}\cdot\frac{1\text{ km}}{1{,}000{,}000\text{ mm}}$
$= 0.000286\text{ km}$

123. $1400\text{ mg} = \frac{1400\text{ mg}}{1}\cdot\frac{1\text{ g}}{1000\text{ mg}} = 1.4\text{ g}$

124. $6.75\text{ gal} = \frac{6.75\text{ gal}}{1}\cdot\frac{4\text{ qt}}{1\text{ gal}} = 27\text{ qt}$

125. $F = 1.8C + 32 = 1.8(86) + 32 = 154.8 + 32 = 186.8$

86°C is 186.8°F.

126. $C = \frac{5}{9}(F - 32) = \frac{5}{9}(51.8 - 32) = \frac{5}{9}(19.8) = 11$

51.8°F is 11°C.

127. 9.3 km = 9300 m

$$\begin{array}{r} 9300\text{ m} \\ -\ 183\text{ m} \\ \hline 9117\text{ m} \end{array}$$

or 183 m = 0.183 km

$$\begin{array}{r} 9.300\text{ km} \\ -\ 0.183\text{ km} \\ \hline 9.117\text{ km} \end{array}$$

128. 35 L = 35,000 ml

$$\begin{array}{r} 35{,}000\text{ ml} \\ +\ \ 700\text{ ml} \\ \hline 35{,}700\text{ ml} \end{array}$$

700 ml = 0.7 L

$$\begin{array}{r} 35.0\text{ L} \\ +\ 0.7\text{ L} \\ \hline 35.7\text{ L} \end{array}$$

129.
$$\begin{array}{r} 3\text{ gal }3\text{ qt} \\ +\ 4\text{ gal }2\text{ qt} \\ \hline 7\text{ gal }5\text{ qt} \end{array} = 7\text{ gal} + 1\text{ gal }1\text{ qt} = 8\text{ gal }1\text{ qt}$$

130.
$$\begin{array}{r} 3.2\text{ kg} \\ \times\ \ 4 \\ \hline 12.8\text{ kg} \end{array}$$

Chapter 9 Test

1. The complement of an angle that measures 78° is an angle that measures 90° − 78° = 12°.

2. The supplement of a 124° angle is an angle that measures 180° − 124° = 56°.

3. $m\angle x = 90° - 40° = 50°$

4. $\angle x$ and the angle marked 62° are adjacent angles, so $m\angle x = 180° - 62° = 118°$.
   $\angle y$ and the angle marked 62° are vertical angles, so $m\angle y = 62°$.
   $\angle x$ and $\angle z$ are vertical angles, so $m\angle z = m\angle x = 118°$.

Copyright © 2011 Pearson Education, Inc. Publishing as Prentice Hall.

5. $\angle x$ and the angle marked 73° are vertical angles, so $m\angle x = 73°$.
$\angle x$ and $\angle y$ are alternate interior angles, so $m\angle y = m\angle x = 73°$.
$\angle x$ and $\angle z$ are corresponding angles, so $m\angle z = m\angle x = 73°$.

6. $d = 2 \cdot r = 2 \cdot 3.1 \text{ m} = 6.2 \text{ m}$

7. $r = d \div 2 = 20 \text{ in.} \div 2 = 10 \text{ in.}$

8. Circumference:
$$\begin{aligned} C &= 2 \cdot \pi \cdot r \\ &= 2 \cdot \pi \cdot 9 \text{ in.} \\ &= 18\pi \text{ in.} \\ &\approx 56.52 \text{ in.} \end{aligned}$$
The circumference is 18π inches ≈ 56.52 inches.
Area:
$$\begin{aligned} A &= \pi r^2 \\ &= \pi(9 \text{ in.})^2 \\ &= 81\pi \text{ sq in.} \\ &\approx 254.34 \text{ sq in.} \end{aligned}$$
The area is 81π square inches ≈ 254.34 square inches.

9.
$$\begin{aligned} P &= 2 \cdot l + 2 \cdot w \\ &= 2(7 \text{ yd}) + 2(5.3 \text{ yd}) \\ &= 14 \text{ yd} + 10.6 \text{ yd} \\ &= 24.6 \text{ yd} \end{aligned}$$
The perimeter is 24.6 yards.
$A = l \cdot w = 7 \text{ yd} \cdot 5.3 \text{ yd} = 37.1 \text{ sq yd}$
The area is 37.1 square yards.

10. The unmarked vertical side has length 11 in. − 7 in. = 4 in. The unmarked horizontal side has length 23 in. − 6 in. = 17 in.
$P = (6 + 4 + 17 + 7 + 23 + 11) \text{ in.} = 68 \text{ in.}$
The perimeter is 68 inches.
Extending the unmarked vertical side downward divides the region into two rectangles. The region's area is the sum of the areas of these:
$$\begin{aligned} A &= 11 \text{ in.} \cdot 6 \text{ in.} + 7 \text{ in.} \cdot 17 \text{ in.} \\ &= 66 \text{ sq in.} + 119 \text{ sq in.} \\ &= 185 \text{ sq in.} \end{aligned}$$
The area is 185 square inches.

11.
$$\begin{aligned} V &= \pi \cdot r^2 \cdot h \\ &= \pi \cdot (2 \text{ in.})^2 \cdot 5 \text{ in.} \\ &= 20\pi \text{ cu in.} \\ &\approx 20 \cdot \frac{22}{7} \text{ cu in.} = 62\frac{6}{7} \text{ cu in.} \end{aligned}$$
The volume is $62\frac{6}{7}$ cubic inches.

12. $V = l \cdot w \cdot h = 5 \text{ ft} \cdot 3 \text{ ft} \cdot 2 \text{ ft} = 30 \text{ cu ft}$
The volume is 30 cubic feet.

13. $P = 4 \cdot s = 4 \cdot 4 \text{ in.} = 16 \text{ in.}$
The perimeter of the frame is 16 inches.

14. $V = l \cdot w \cdot h = 3 \text{ ft} \cdot 3 \text{ ft} \cdot 2 \text{ ft} = 18 \text{ cu ft}$
18 cubic feet of soil are needed.

15.
$$\begin{aligned} P &= 2 \cdot l + 2 \cdot w \\ &= 2 \cdot 18 \text{ ft} + 2 \cdot 13 \text{ ft} \\ &= 36 \text{ ft} + 26 \text{ ft} \\ &= 62 \text{ ft} \end{aligned}$$
$$\begin{aligned} \text{cost} &= P \cdot \$1.87 \text{ per ft} \\ &= 62 \text{ ft} \cdot \$1.87 \text{ per ft} \\ &= \$115.94 \end{aligned}$$
62 feet of baseboard are needed, at a total cost of $115.94.

16.
$$\begin{array}{r} 23 \\ 12\overline{)\,280} \\ \underline{-24} \\ 40 \\ \underline{-36} \\ 4 \end{array}$$
280 inches = 23 ft 4 in.

17. $2\frac{1}{2} \text{ gal} = \frac{2\frac{1}{2} \text{ gal}}{1} \cdot \frac{4 \text{ qt}}{1 \text{ gal}} = 2\frac{1}{2} \cdot 4 \text{ qt} = 10 \text{ qt}$

18. $30 \text{ oz} = \frac{30 \text{ oz}}{1} \cdot \frac{1 \text{ lb}}{16 \text{ oz}} = \frac{30}{16} \text{ lb} = \frac{15}{8} \text{ lb} = 1\frac{7}{8} \text{ lb}$

19.
$$\begin{aligned} 2.8 \text{ tons} &= \frac{2.8 \text{ tons}}{1} \cdot \frac{2000 \text{ lb}}{1 \text{ ton}} \\ &= 2.8 \cdot 2000 \text{ lb} \\ &= 5600 \text{ lb} \end{aligned}$$

Copyright © 2011 Pearson Education, Inc. Publishing as Prentice Hall.

20. $38 \text{ pt} = \frac{38 \text{ pt}}{1} \cdot \frac{1 \text{ qt}}{2 \text{ pt}} \cdot \frac{1 \text{ gal}}{4 \text{ qt}}$
$= \frac{38}{8} \text{ gal}$
$= \frac{19}{4} \text{ gal}$
$= 4\frac{3}{4} \text{ gal}$

21. $40 \text{ mg} = \frac{40 \text{ mg}}{1} \cdot \frac{1 \text{ g}}{1000 \text{ mg}} = \frac{40}{1000} \text{ g} = 0.04 \text{ g}$

22. $2.4 \text{ kg} = \frac{2.4 \text{ kg}}{1} \cdot \frac{1000 \text{ g}}{1 \text{ kg}} = 2.4 \cdot 1000 \text{ g} = 2400 \text{ g}$

23. $3.6 \text{ cm} = \frac{3.6 \text{ cm}}{1} \cdot \frac{10 \text{ mm}}{1 \text{ cm}} = 3.6 \cdot 10 \text{ mm} = 36 \text{ mm}$

24. $4.3 \text{ dg} = \frac{4.3 \text{ dg}}{1} \cdot \frac{1 \text{ g}}{10 \text{ dg}} = \frac{4.3}{10} \text{ g} = 0.43 \text{ g}$

25. $0.83 \text{ L} = \frac{0.83 \text{ L}}{1} \cdot \frac{1000 \text{ ml}}{1 \text{ L}} = 0.83 \cdot 1000 = 830 \text{ ml}$

26.
3 qt 1 pt
+ 2 qt 1 pt
5 qt 2 pt = 4 qt + 1 qt + 2 pt
= 1 gal + 1 qt + 1 qt
= 1 gal 2 qt

27.
8 lb 6 oz → 7 lb 22 oz
− 4 lb 9 oz → − 4 lb 9 oz
3 lb 13 oz

28. 2 ft 9 in. × 3 = 6 ft 27 in.
= 6 ft + 2 ft 3 in.
= 8 ft 3 in.

29. 5 gal 2 qt ÷ 2 = 4 gal 6 qt ÷ 2
$= \frac{4}{2} \text{ gal } \frac{6}{2} \text{ qt}$
= 2 gal 3 qt

30. 8 cm = 80 mm; 14 mm = 1.4 cm
80 mm − 14 mm = 66 mm or 8.0 cm − 1.4 cm = 6.6 cm

31. 1.8 km = 1800 m; 456 m = 0.456 km
1800 m + 456 m = 2256 m or 1.800 km + 0.456 km = 2.256 km

32. $C = \frac{5}{9}(F-32)$
$= \frac{5}{9}(84-32)$
$= \frac{5}{9}(52)$
≈ 28.9
84°F is 28.9°C.

33. $F = 1.8C + 32$
$= 1.8(12.6) + 32$
$= 22.68 + 32$
≈ 54.7
12.6°C is 54.7°F

34. $8.4 \text{ m} \cdot \frac{2}{3} = \frac{8.4}{1} \cdot \frac{2}{3} \text{ m} = 5.6 \text{ m}$
The trees will be 5.6 m tall.

35.
20 gal − 15 gal 1 qt → 19 gal 4 qt − 15 gal 1 qt = 4 gal 3 qt
Thus, 4 gal 3 qt remains in the container.

36. 88 m + 340 cm = 88 m + 3.40 m = 91.4 m
The span is 91.4 meters

37.
2 ft 9 in.
× 6
12 ft 54 in. = 12 ft + 4 ft 6 in. = 16 ft 6 in.
Thus, 16 ft 6 in. of material is needed.

Cumulative Review Chapters 1–9

1. $3a - 6 = a + 4$
$3a - a - 6 = a - a + 4$
$2a - 6 = 4$
$2a - 6 + 6 = 4 + 6$
$2a = 10$
$\frac{2a}{2} = \frac{10}{2}$
$a = 5$

Copyright © 2011 Pearson Education, Inc. Publishing as Prentice Hall.

2.
$$\begin{aligned} 2x+1 &= 3x-5 \\ 2x-2x+1 &= 3x-2x-5 \\ 1 &= x-5 \\ 1+5 &= x-5+5 \\ 6 &= x \end{aligned}$$

3. a. $\left(\frac{2}{5}\right)^4 = \frac{2}{5}\cdot\frac{2}{5}\cdot\frac{2}{5}\cdot\frac{2}{5} = \frac{2^4}{5^4} = \frac{16}{625}$

b. $\left(-\frac{1}{4}\right)^2 = \left(-\frac{1}{4}\right)\left(-\frac{1}{4}\right) = \frac{1}{16}$

4. a. $\left(-\frac{1}{3}\right)^3 = \left(-\frac{1}{3}\right)\left(-\frac{1}{3}\right)\left(-\frac{1}{3}\right) = -\frac{1}{27}$

b. $\left(\frac{3}{7}\right)^2 = \frac{3}{7}\cdot\frac{3}{7} = \frac{9}{49}$

5.
$$\begin{array}{rr} 2\frac{4}{5} & 2\frac{8}{10} \\ 5 & 5 \\ +1\frac{1}{2} & +1\frac{5}{10} \\ \hline & 8\frac{13}{10} = 8+1\frac{3}{10} = 9\frac{3}{10} \end{array}$$

6.
$$\begin{array}{rr} 2\frac{1}{3} & 2\frac{5}{15} \\ 4\frac{2}{5} & 4\frac{6}{15} \\ +3 & +3 \\ \hline & 9\frac{11}{15} \end{array}$$

7. $$\begin{aligned} 11.1x-6.3+8.9x-4.6 &= 11.1x+8.9x-6.3-4.6 \\ &= 20x-10.9 \end{aligned}$$

8. $$\begin{aligned} 2.5y+3.7-1.3y-1.9 &= 2.5y-1.3y+3.7-1.9 \\ &= 1.2y+1.8 \end{aligned}$$

9. $$\begin{aligned} \frac{5.68+(0.9)^2 \div 100}{0.2} &= \frac{5.68+0.81\div 100}{0.2} \\ &= \frac{5.68+0.0081}{0.2} \\ &= \frac{5.6881}{0.2} \\ &= 28.4405 \end{aligned}$$

10. $\frac{0.12+0.96}{0.5} = \frac{1.08}{0.5} = 2.16$

11.
$$\begin{array}{r} 0.77... \\ 9\overline{)\,7.00} \\ \underline{-63} \\ 70 \\ \underline{-63} \\ 7 \end{array}$$

Thus $0.\overline{7} = \frac{7}{9}$.

12.
$$\begin{array}{r} 0.4 \\ 5\overline{)\,2.0} \\ \underline{-2\,0} \\ 0 \end{array}$$

Thus $0.43 > \frac{2}{5}$.

13.
$$\begin{aligned} 0.5y+2.3 &= 1.65 \\ 0.5y+2.3-2.3 &= 1.65-2.3 \\ 0.5y &= -0.65 \\ \frac{0.5y}{0.5} &= \frac{-0.65}{0.5} \\ y &= -1.3 \end{aligned}$$

14.
$$\begin{aligned} 0.4x-9.3 &= 2.7 \\ 0.4x-9.3+9.3 &= 2.7+9.3 \\ 0.4x &= 12 \\ \frac{0.4x}{0.4} &= \frac{12}{0.4} \\ x &= 30 \end{aligned}$$

15. Use $a^2+b^2=c^2$ where $a=b=300$.
$$\begin{aligned} 300^2+300^2 &= c^2 \\ 90{,}000+90{,}000 &= c^2 \\ 180{,}000 &= c^2 \\ \sqrt{180{,}000} &= c \\ 424 &\approx c \end{aligned}$$

The length of the diagonal is approximately 424 feet.

Copyright © 2011 Pearson Education, Inc. Publishing as Prentice Hall.

16. Use $a^2 + b^2 = c^2$ where $a = 200$ and $b = 125$.

$$\begin{aligned} 200^2 + 125^2 &= c^2 \\ 40,000 + 15,625 &= c^2 \\ 55,625 &= c^2 \\ \sqrt{55,625} &= c \\ 236 &\approx c \end{aligned}$$

The length of the diagonal is approximately 236 feet.

17. a. $\frac{\text{width}}{\text{length}} = \frac{5 \text{ feet}}{7 \text{ feet}} = \frac{5}{7}$

b. $$\begin{aligned} P &= 2 \cdot l + 2 \cdot w \\ &= 2(7 \text{ feet}) + 2(5 \text{ feet}) \\ &= 14 \text{ feet} + 10 \text{ feet} \\ &= 24 \text{ feet} \end{aligned}$$

$\frac{\text{length}}{\text{perimeter}} = \frac{7 \text{ feet}}{24 \text{ feet}} = \frac{7}{24}$

18. a. $P = 4s = 4(9 \text{ in.}) = 36 \text{ in.}$

$\frac{\text{side}}{\text{perimeter}} = \frac{9 \text{ inches}}{36 \text{ inches}} = \frac{9}{36} = \frac{1}{4}$

b. $A = s^2 = (9 \text{ in.})^2 = 81 \text{ sq in.}$

$\frac{\text{perimeter}}{\text{area}} = \frac{36 \text{ inches}}{81 \text{ sq inches}} = \frac{36}{81} = \frac{4}{9}$

19. $\frac{2160 \text{ dollars}}{12 \text{ weeks}} = \frac{180 \text{ dollars}}{1 \text{ week}}$

20. $\frac{8 \text{ chaperones}}{40 \text{ students}} = \frac{1 \text{ chaperone}}{5 \text{ students}}$

21. $$\begin{aligned} \frac{1.6}{1.1} &= \frac{x}{0.3} \\ 1.6 \cdot 0.3 &= 1.1 \cdot x \\ 0.48 &= 1.1x \\ \frac{0.48}{1.1} &= \frac{1.1x}{1.1} \\ 0.44 &\approx x \end{aligned}$$

22. $$\begin{aligned} \frac{2.4}{3.5} &= \frac{0.7}{x} \\ 2.4 \cdot x &= 3.5 \cdot 0.7 \\ 2.4x &= 2.45 \\ \frac{2.4x}{2.4} &= \frac{2.45}{2.4} \\ x &\approx 1.02 \end{aligned}$$

23. Let x be the dose for a 140-lb woman.

$$\begin{aligned} \frac{4 \text{ cc}}{25 \text{ lb}} &= \frac{x \text{ cc}}{140 \text{ lb}} \\ \frac{4}{25} &= \frac{x}{140} \\ 4 \cdot 140 &= 25 \cdot x \\ 560 &= 25x \\ \frac{560}{25} &= \frac{25x}{25} \\ 22.4 &= x \end{aligned}$$

The dose is 22.4 cc.

24. Let x be the amount for 5 pie crusts.

$$\begin{aligned} \frac{3 \text{ c}}{2 \text{ crusts}} &= \frac{x \text{ c}}{5 \text{ crusts}} \\ \frac{3}{2} &= \frac{x}{5} \\ 3 \cdot 5 &= 2 \cdot x \\ 15 &= 2x \\ \frac{15}{2} &= \frac{2x}{2} \\ 7.5 &= x \end{aligned}$$

5 pie crusts require 7.5 cups of flour.

25. $\frac{17}{100} = 17\%$

17% of the people surveyed drive blue cars.

26. $\frac{38}{100} = 38\%$

38% of the shoppers used only cash.

27. $$\begin{aligned} 13 &= 6\frac{1}{2}\% \cdot x \\ 13 &= 0.065x \\ \frac{13}{0.065} &= \frac{0.065x}{0.065} \\ 200 &= x \end{aligned}$$

13 is $6\frac{1}{2}\%$ of 200.

28. $$\begin{aligned} 54 &= 4\frac{1}{2}\% \cdot x \\ 54 &= 0.045x \\ \frac{54}{0.045} &= \frac{0.045x}{0.045} \\ 1200 &= x \end{aligned}$$

54 is $4\frac{1}{2}\%$ of 1200.

Copyright © 2011 Pearson Education, Inc. Publishing as Prentice Hall.

29. $x = 30\% \cdot 9$
$x = 0.3 \cdot 9$
$x = 2.7$
2.7 is 30% of 9.

30. $x = 42\% \cdot 30$
$x = 0.42 \cdot 30$
$x = 12.6$
12.6 is 42% of 30.

31. $$\text{percent increase} = \frac{\text{amount of increase}}{\text{original amount}} = \frac{45-34}{34} = \frac{11}{34} \approx 0.32$$
The scholarship applications increased by 32%.

32. $$\text{percent increase} = \frac{\text{amount of increase}}{\text{original amount}} = \frac{19-15}{15} = \frac{4}{15} \approx 0.27$$
The price of the paint increased by 27%.

33. sales tax = $85.50 \cdot 0.075 = 6.4125$
The sales tax is $6.41.
$85.50 + 6.41 = 91.91$
The total price is $91.91.

34. sales tax = $375 \cdot 0.08 = 30$
The sales tax is $30.
$375 + 30 = 405$
The total price is $405.

35. Point A is 4 units left of the y-axis and 2 units above the x-axis. The coordinates are $(-4, 2)$.
Point B is 1 unit right of the y-axis and 2 units above the x-axis. The coordinates are $(1, 2)$.
Point C is on the y-axis and 1 unit above the x-axis. The coordinates are $(0, 1)$.
Point D is 3 units left of the y-axis and on the x-axis. The coordinates are $(-3, 0)$.
Point E is 5 units right of the y-axis and 4 units below the x-axis. The coordinates are $(5, -4)$.

36. Point A is 2 units right of the y-axis and 3 units below the x-axis. The coordinates are $(2, -3)$.
Point B is 5 units left of the y-axis and on the x-axis. The coordinates are $(-5, 0)$.
Point C is on the y-axis and 4 units above the x-axis. The coordinates are $(0, 4)$.
Point D is 3 units left of the y-axis and 2 units below the x-axis. The coordinates are $(-3, -2)$.

37. No matter what x-value we choose, y is always 4.

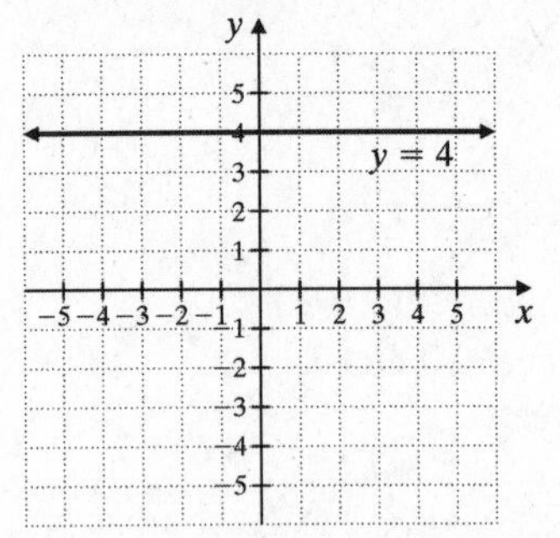

38. No matter what x-value we choose, y is always -2.

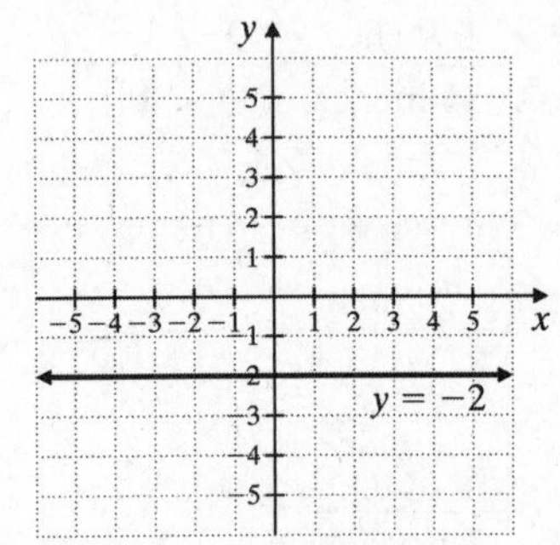

39. The seven numbers are listed in order. The median is the middle number, 57.

40. The five numbers in order are:
60, 72, 83, 89, 95.
The median is the middle number, 83.

41. There is 1 red marble and $1 + 1 + 2 = 4$ total marbles. The probability is $\frac{1}{4}$.

42. There are 2 nickels and $2 + 2 + 3 = 7$ total coins. The probability is $\frac{2}{7}$.

43. The complement of a 48° angle is an angle that has measure $90° - 48° = 42°$.

44. The supplement of a 137° angle is an angle that has measure $180° - 137° = 43°$.

Copyright © 2011 Pearson Education, Inc. Publishing as Prentice Hall.

45. $8 \text{ ft} = \frac{8 \text{ ft}}{1} \cdot \frac{12 \text{ in.}}{1 \text{ ft}} = 8 \cdot 12 \text{ in.} = 96 \text{ in.}$

46. $7 \text{ yd} = \frac{7 \text{ yd}}{1} \cdot \frac{3 \text{ ft}}{1 \text{ yd}} = 7 \cdot 3 \text{ ft} = 21 \text{ ft}$

47. $A = \pi r^2 = \pi \cdot 3^2 = 9\pi \approx 9 \cdot 3.14 = 28.26$
The area is exactly 9π square feet or approximately 28.26 square feet.

48. $A = \pi r^2 = \pi \cdot 2^2 = 4\pi \approx 4 \cdot 3.14 = 12.56$
The area is exactly 4π square miles or approximately 12.56 square miles.

49. $\text{C} = \frac{5}{9}(\text{F} - 32) = \frac{5}{9}(59 - 32) = \frac{5}{9}(27) = 15$
59°F is 15°C.

50. $\text{C} = \frac{5}{9}(\text{F} - 32) = \frac{5}{9}(86 - 32) = \frac{5}{9}(54) = 30$
86°F is 30°C.

Copyright © 2011 Pearson Education, Inc. Publishing as Prentice Hall.

Chapter 10

Section 10.1

Practice Problems

1. $(3y+7)+(-9y-14)=(3y-9y)+(7-14)$
$=(-6y)+(-7)$
$=-6y-7$

2. $(x^2-4x-3)+(5x^2-6x)$
$=x^2+5x^2-4x-6x-3$
$=6x^2-10x-3$

3. $(-z^2-4.2z+11)+(9z^2-1.9z+6.3)$
$=-z^2+9z^2-4.2z-1.9z+11+6.3$
$=8z^2-6.1z+17.3$

4. $$\begin{array}{r} x^2 \quad -x+1.1 \\ +\ -8x^2 \quad -x-6.7 \\ \hline -7x^2-2x-5.6 \end{array}$$

5. $-(3y^2+y-2)=-1(3y^2+y-2)$
$=-1(3y^2)+(-1)(y)+(-1)(-2)$
$=-3y^2-y+2$

6. $(9b+8)-(11b-20)=(9b+8)+(-11b+20)$
$=9b-11b+8+20$
$=-2b+28$

7. $(11x^2+7x+2)-(15x^2+4x)$
$=(11x^2+7x+2)+(-15x^2-4x)$
$=11x^2-15x^2+7x-4x+2$
$=-4x^2+3x+2$

8. $(-3y^2+5y)-(-7y^2+y-4)$
$=(-3y^2+5y)+(7y^2-y+4)$
$=-3y^2+7y^2+5y-y+4$
$=4y^2+4y+4$

9. $$\begin{array}{r} -4x^2+20x+17 \\ -\ (3x^2-12x) \\ \hline \end{array} \qquad \begin{array}{r} -4x^2+20x+17 \\ -3x^2+12x \\ \hline -7x^2+32x+17 \end{array}$$

10. $2y^3+y^2-6=2(3)^3+(3)^2-6$
$=2(27)+9-6$
$=54+9-6$
$=57$
The value of $2y^3+y^2-6$ when $y=3$ is 57.

11. $-16t^2+530=-16(1)^2+530=-16+530=514$
The height of the object at 1 second is 514 feet.
$-16t^2+530=-16(4)^2+530$
$=-16(16)+530$
$=-256+530$
$=274$
The height of the object at 4 seconds is 274 feet.

Vocabulary and Readiness Check

1. The addends of an algebraic expression are the terms of the expression.

2. A polynomial with exactly one term is called a monomial.

3. A polynomial with exactly two terms is called a binomial.

4. A polynomial with exactly three terms is called a trinomial.

5. To add polynomials, combine like terms.

6. To subtract polynomials, change the sign of the terms of the polynomial being subtracted; then add.

Exercise Set 10.1

1. $(2x+3)+(-7x-27)=2x-7x+3-27$
$=-5x-24$

3. $(-3z^2+5z-5)+(-8z^2-8z+4)$
$=-3z^2-8z^2+5z-8z-5+4$
$=-11z^2-3z-1$

5. $(12y-20)+(9y^2+13y-20)$
$=9y^2+12y+13y-20-20$
$=9y^2+25y-40$

Copyright © 2011 Pearson Education, Inc. Publishing as Prentice Hall.

7. $(4.3a^4+5)+(-8.6a^4-2a^2+4)$
$=4.3a^4-8.6a^4-2a^2+5+4$
$=-4.3a^4-2a^2+9$

9. $-(9x-16)=-1(9x-16)$
$=-1(9x)+(-1)(-16)$
$=-9x+16$

11. $-(-3z^2+z-7)=-1(-3z^2+z-7)$
$=(-1)(-3z^2)+(-1)(z)+(-1)(-7)$
$=3z^2-z+7$

13. $(8a-5)-(3a+8)=(8a-5)+(-3a-8)$
$=8a-3a-5-8$
$=5a-13$

15. $(3x^2-2x+1)-(5x^2-6x)$
$=(3x^2-2x+1)+(-5x^2+6x)$
$=3x^2-5x^2-2x+6x+1$
$=-2x^2+4x+1$

17. $(10y^2-7)-(20y^3-2y^2-3)$
$=(10y^2-7)+(-20y^3+2y^2+3)$
$=-20y^3+10y^2+2y^2-7+3$
$=-20y^3+12y^2-4$

19.
$$\begin{array}{r} 2x+12 \\ \underline{-\ (9x^2+3x\ -4)} \end{array} \qquad \begin{array}{r} 2x+12 \\ \underline{+\ -9x^2\ -3x\ +4} \\ -9x^2\ \ -x+16 \end{array}$$

21.
$$\begin{array}{r} 13y^2-6y-14 \\ \underline{-(5y^2+4y-6)} \end{array} \qquad \begin{array}{r} 13y^2\ -6y-14 \\ \underline{+\ -5y^2\ -4y\ +6} \\ 8y^2-10y\ \ -8 \end{array}$$

23. $(25x-5)+(-20x-7)=25x-20x-5-7$
$=5x-12$

25. $(4y+4)-(3y+8)=(4y+4)+(-3y-8)$
$=4y-3y+4-8$
$=y-4$

27. $(3x^2+3x-4)+(-8x^2+9)$
$=3x^2-8x^2+3x-4+9$
$=-5x^2+3x+5$

29. $(5x+4.5)+(-x-8.6)=5x-x+4.5-8.6$
$=4x-4.1$

31. $(a-5)-(-3a+2)=(a-5)+(3a-2)$
$=a+3a-5-2$
$=4a-7$

33. $(21y-4.6)-(36y-8.2)$
$=(21y-4.6)+(-36y+8.2)$
$=21y-36y-4.6+8.2$
$=-15y+3.6$

35. $(18t^2-4t+2)-(-t^2+7t-1)$
$=(18t^2-4t+2)+(t^2-7t+1)$
$=18t^2+t^2-4t-7t+2+1$
$=19t^2-11t+3$

37. $(2b^3+5b^2-5b-8)+(8b^2+9b+6)$
$=2b^3+5b^2+8b^2-5b+9b-8+6$
$=2b^3+13b^2+4b-2$

39.
$$\begin{array}{r} 6x^2\qquad\quad -7 \\ \underline{+\ -11x^2-11x+20} \\ -5x^2-11x+13 \end{array}$$

41.
$$\begin{array}{r} 3z+\frac{6}{7} \\ \underline{-\left(3z-\frac{3}{7}\right)} \end{array} \qquad \begin{array}{r} 3z+\frac{6}{7} \\ \underline{+\ -3z+\frac{3}{7}} \\ \frac{9}{7} \end{array}$$

43. $-2x+9=-2(2)+9=-4+9=5$

45. $x^2-6x+3=(2)^2-6(2)+3=4-12+3=-5$

47. $\frac{3x^2}{2}-14=\frac{3(2)^2}{2}-14=\frac{12}{2}-14=6-14=-8$

49. $2x+10=2(5)+10=10+10=20$

51. $x^2=5^2=25$

53. $2x^2+4x-20=2(5)^2+4(5)-20$
$=2(25)+20-20$
$=50+20-20$
$=50$

Copyright © 2011 Pearson Education, Inc. Publishing as Prentice Hall.

55. $16t^2 = 16(6)^2 = 16(36) = 576$
In 6 seconds, the object travels 576 feet.

57. $\begin{aligned} 3000 + 20x &= 3000 + 20(10) \\ &= 3000 + 200 \\ &= 3200 \end{aligned}$
The cost for 10 file cabinets is \$3200.

59. $\begin{aligned} 867 - 16t^2 &= 867 - 16(4)^2 \\ &= 867 - 16(16) \\ &= 867 - 256 \\ &= 611 \end{aligned}$
After 4 seconds, the chalk was 611 feet above the ground.

61. $\begin{aligned} 1053 - 16t^2 &= 1053 - 16(3)^2 \\ &= 1053 - 16(9) \\ &= 1053 - 144 \\ &= 909 \end{aligned}$
After 3 seconds, the height of the object is 909 feet.

63. 2010 corresponds to $x = 10$.
$\begin{aligned} 0.5x^2 + 17x + 97 &= 0.5(10)^2 + 17(10) + 97 \\ &= 0.5(100) + 170 + 97 \\ &= 50 + 170 + 97 \\ &= 317 \end{aligned}$
If the rate of growth continues as is, we can expect 317 million cellular subscribers in 2010.

65. $3^4 = 3 \cdot 3 \cdot 3 \cdot 3 = 81$

67. $(-5)^2 = (-5)(-5) = 25$

69. $x \cdot x \cdot x = x^3$

71. $2 \cdot 2 \cdot a \cdot a \cdot a \cdot a = 2^2 a^4$

73. $\begin{aligned} P &= (5x - 10) + (2x + 1) + (x + 11) \\ &= 5x + 2x + x - 10 + 1 + 11 \\ &= 8x + 2 \end{aligned}$
The perimeter is $(8x + 2)$ inches.

75. $\begin{aligned} (7x - 10) - (3x + 5) &= (7x - 10) + (-3x - 5) \\ &= 7x - 3x - 10 - 5 \\ &= 4x - 15 \end{aligned}$
The missing length is $(4x - 15)$ units.

77.
$$\begin{array}{r} 3x^2 + \underline{\quad}x - \underline{\quad} \\ + \underline{\quad}x^2 - \ 6x + \ 2 \\ \hline 5x^2 + 14x - \ 4 \end{array}$$
Since $3x^2 + 2x^2 = 5x^2$, $20x - 6x = 14x$ and $-6 + 2 = -4$, the missing numbers are 20, 6, and 2.
$(3x^2 + \underline{20}x - \underline{6}) + (\underline{2}x^2 - 6x + 2) = 5x^2 + 14x - 4$

79. $\begin{aligned} 7a^4 - 6a^2 + 2a - 1 &= 7(1.2)^4 - 6(1.2)^2 + 2(1.2) - 1 \\ &= 7.2752 \end{aligned}$

81. $1053 - 16t^2 = 1053 - 16(8)^2 = 1053 - 1024 = 29$
$1053 - 16t^2$ when $t = 8$ is 29 feet.
$\begin{aligned} 1053 - 16t^2 &= 1053 - 16(9)^2 \\ &= 1053 - 1296 \\ &= -243 \end{aligned}$
$1053 - 16t^2$ when $t = 9$ is -243 feet.
answers may vary

Section 10.2

Practice Problems

1. $z^5 \cdot z^6 = z^{5+6} = z^{11}$

2. $8y^5 \cdot 4y^9 = (8 \cdot 4)(y^5 \cdot y^9) = 32y^{5+9} = 32y^{14}$

3. $\begin{aligned} (-4r^6s^2)(-3r^2s^5) &= (-4 \cdot -3)(r^6 \cdot r^2)(s^2 \cdot s^5) \\ &= 12r^{6+2}s^{2+5} \\ &= 12r^8s^7 \end{aligned}$

4. $11y^5 \cdot 3y^2 \cdot y = (11 \cdot 3)(y^5 \cdot y^2 \cdot y^1) = 33y^8$

5. $(z^3)^6 = z^{3 \cdot 6} = z^{18}$

6. $(z^4)^5 \cdot (z^3)^7 = (z^{20})(z^{21}) = z^{20+21} = z^{41}$

7. $(3b)^4 = 3^4 \cdot b^4 = 81b^4$

8. $(4x^2y^6)^3 = 4^3(x^2)^3(y^6)^3 = 64x^6y^{18}$

Copyright © 2011 Pearson Education, Inc. Publishing as Prentice Hall.

9. $(2x^2y^4)^4(3x^6y^9)^2$
$= 2^4(x^2)^4(y^4)^4 \cdot 3^2(x^6)^2(y^9)^2$
$= 16x^8y^{16} \cdot 9x^{12}y^{18}$
$= (16 \cdot 9)(x^8 \cdot x^{12})(y^{16} \cdot y^{18})$
$= 144x^{20}y^{34}$

Vocabulary and Readiness Check

1. In $7x^2$, the 2 is called the exponent.

2. To simplify $x^4 \cdot x^3$, we add the exponents.

3. To simplify $(x^4)^3$, we multiply the exponents.

4. The expression $(6x)^2$ simplifies to $36x^2$.

Exercise Set 10.2

1. $x^5 \cdot x^9 = x^{5+9} = x^{14}$

3. $a^3 \cdot a = a^{3+1} = a^4$

5. $3z^3 \cdot 5z^2 = (3 \cdot 5)(z^3 \cdot z^2) = 15z^5$

7. $-4x \cdot 10x = (-4 \cdot 10)(x \cdot x) = -40x^2$

9. $2x \cdot 3x \cdot 7x = (2 \cdot 3 \cdot 7)(x \cdot x \cdot x) = 42x^3$

11. $a \cdot 4a^{11} \cdot 3a^5 = (4 \cdot 3)(a^1 \cdot a^{11} \cdot a^5) = 12a^{17}$

13. $(-5x^2y^3)(-5x^4y) = (-5 \cdot -5)(x^2 \cdot x^4)(y^3 \cdot y^1)$
$= 25x^6y^4$

15. $(7ab)(4a^4b^5) = (7 \cdot 4)(a^1 \cdot a^4)(b^1 \cdot b^5) = 28a^5b^6$

17. $(x^5)^3 = x^{5 \cdot 3} = x^{15}$

19. $(z^3)^{10} = z^{3 \cdot 10} = z^{30}$

21. $(b^7)^6 \cdot (b^2)^{10} = b^{7 \cdot 6} \cdot b^{2 \cdot 10}$
$= b^{42} \cdot b^{20}$
$= b^{42+20}$
$= b^{62}$

23. $(3a)^4 = 3^4 \cdot a^4 = 81a^4$

25. $(a^{11}b^8)^3 = (a^{11})^3(b^8)^3 = a^{11 \cdot 3}b^{8 \cdot 3} = a^{33}b^{24}$

27. $(10x^5y^3)^3 = 10^3(x^5)^3(y^3)^3$
$= 1000x^{5 \cdot 3}y^{3 \cdot 3}$
$= 1000x^{15}y^9$

29. $(-3y)(2y^7)^3 = (-3y) \cdot 2^3(y^7)^3$
$= (-3y) \cdot 8y^{21}$
$= (-3)(8)(y^1 \cdot y^{21})$
$= -24y^{22}$

31. $(4xy)^3(2x^3y^5)^2 = (4^3x^3y^3)[2^2(x^3)^2(y^5)^2]$
$= (64x^3y^3)(4x^6y^{10})$
$= (64 \cdot 4)(x^3 \cdot x^6)(y^3 \cdot y^{10})$
$= 256x^9y^{13}$

33. $7(x - 3) = 7 \cdot x - 7 \cdot 3 = 7x - 21$

35. $-2(3a + 2b) = -2 \cdot 3a + (-2)(2b) = -6a - 4b$

37. $9(x+2y-3) = 9 \cdot x + 9 \cdot 2y - 9 \cdot 3$
$= 9x + 18y - 27$

39. area $= s^2$
$= (4x^6)^2$
$= 4^2(x^6)^2$
$= 16x^{12}$

The area is $16x^{12}$ square inches.

41. area $= \frac{1}{2}bh$
$= \frac{1}{2} \cdot (6a^3b^4) \cdot (4ab)$
$= \left(\frac{1}{2} \cdot 6 \cdot 4\right)(a^3 \cdot a^1)(b^4 \cdot b^1)$
$= 12a^4b^5$

The area is $12a^4b^5$ square meters.

43. $(14a^7b^6)^3(9a^6b^3)^4$
$= 14^3(a^7)^3(b^6)^3 \cdot 9^4(a^6)^4(b^3)^4$
$= 2744a^{21}b^{18} \cdot 6561a^{24}b^{12}$
$= (2744 \cdot 6561)(a^{21} \cdot a^{24})(b^{18} \cdot b^{12})$
$= 18{,}003{,}384a^{45}b^{30}$

45. $(8.1x^{10})^5 = 8.1^5(x^{10})^5 = 34{,}867.84401x^{50}$

Copyright © 2011 Pearson Education, Inc. Publishing as Prentice Hall.

47. $(x^{90}y^{72})^3 = x^{90\cdot 3}y^{72\cdot 3} = x^{270}y^{216}$

49. answers may vary

Integrated Review

1. $(3x+5)+(-x-8) = 3x-x+5-8 = 2x-3$

2. $(15y-7)+(5y-4) = 15y+5y-7-4 = 20y-11$

3. $(7x+1)-(-3x-2) = (7x+1)+(3x+2)$
$= 7x+3x+1+2$
$= 10x+3$

4. $(14y-6)-(19y-2) = (14y-6)+(-19y+2)$
$= 14y-19y-6+2$
$= -5y-4$

5. $(a^4+5a)-(3a^4-3a^2-4a)$
$= a^4+5a-3a^4+3a^2+4a$
$= a^4-3a^4+3a^2+5a+4a$
$= -2a^4+3a^2+9a$

6. $(2a^3-6a^2+11)-(6a^3+6a^2+11)$
$= (2a^3-6a^2+11)+(-6a^3-6a^2-11)$
$= 2a^3-6a^3-6a^2-6a^2+11-11$
$= -4a^3-12a^2$

7. $(4.5x^2+8.1x)+(2.8x^2-12.3x-5.3)$
$= 4.5x^2+2.8x^2+8.1x-12.3x-5.3$
$= 7.3x^2-4.2x-5.3$

8. $(1.2y^2-3.6y)+(0.6y^2+1.2y-5.6)$
$= 1.2y^2+0.6y^2-3.6y+1.2y-5.6$
$= 1.8y^2-2.4y-5.6$

9.
$$\begin{array}{r} 8x+1 \\ -\ (2x-6) \\ \hline \end{array} \qquad \begin{array}{r} 8x+1 \\ +\ -2x+6 \\ \hline 6x+7 \end{array}$$

10.
$$\begin{array}{r} 5x^2+2x-10 \\ -\ (3x^2\ \ -x\ \ +2) \\ \hline \end{array} \qquad \begin{array}{r} 5x^2+2x-10 \\ +\ -3x^2\ \ +x\ \ -2 \\ \hline 2x^2+3x-12 \end{array}$$

11. $2x-7 = 2(3)-7 = 6-7 = -1$

12. $x^2+5x+2 = 3^2+5(3)+2 = 9+15+2 = 26$

13. $x^9 \cdot x^{11} = x^{9+11} = x^{20}$

14. $x^5 \cdot x^5 = x^{5+5} = x^{10}$

15. $y^3 \cdot y = y^{3+1} = y^4$

16. $a \cdot a^{10} = a^{1+10} = a^{11}$

17. $(x^7)^{11} = x^{7\cdot 11} = x^{77}$

18. $(x^6)^6 = x^{6\cdot 6} = x^{36}$

19. $(x^3)^4 \cdot (x^5)^6 = x^{3\cdot 4} \cdot x^{5\cdot 6} = x^{12} \cdot x^{30} = x^{42}$

20. $(y^2)^9 \cdot (y^3)^3 = y^{2\cdot 9} \cdot y^{3\cdot 3} = y^{18} \cdot y^9 = y^{27}$

21. $(5x)^3 = 5^3x^3 = 125x^3$

22. $(2y)^5 = 2^5y^5 = 32y^5$

23. $(-6xy^2)(2xy^5) = (-6\cdot 2)(x\cdot x)(y^2\cdot y^5)$
$= -12x^2y^7$

24. $(-4a^2b^3)(-3ab) = (-4\cdot -3)(a^2\cdot a^1)(b^3\cdot b^1)$
$= 12a^3b^4$

25. $(y^{11}z^{13})^3 = (y^{11})^3(z^{13})^3 = y^{11\cdot 3}z^{13\cdot 3} = y^{33}z^{39}$

26. $(a^5b^{12})^4 = (a^5)^4(b^{12})^4 = a^{5\cdot 4}b^{12\cdot 4} = a^{20}b^{48}$

27. $(10x^2y)^2(3y) = 10^2(x^2)^2y^2\cdot 3y$
$= 100x^4y^2\cdot 3y$
$= (100\cdot 3)x^4(y^2\cdot y^1)$
$= 300x^4y^3$

28. $(8y^3z)^2(2z^5) = 8^2(y^3)^2z^2\cdot 2z^5$
$= 64y^6z^2\cdot 2z^5$
$= (64\cdot 2)y^6(z^2\cdot z^5)$
$= 128y^6z^7$

29. $(2a^5b)^4(3a^9b^4)^2 = 2^4(a^5)^4b^4\cdot 3^2(a^9)^2(b^4)^2$
$= 16a^{20}b^4\cdot 9a^{18}b^8$
$= (16\cdot 9)(a^{20}\cdot a^{18})(b^4\cdot b^8)$
$= 144a^{38}b^{12}$

Copyright © 2011 Pearson Education, Inc. Publishing as Prentice Hall.

30. $(5x^4y^6)^3(x^2y^2)^5 = 5^3(x^4)^3(y^6)^3 \cdot (x^2)^5(y^2)^5$
$= 125x^{12}y^{18} \cdot x^{10}y^{10}$
$= 125(x^{12} \cdot x^{10})(y^{18} \cdot y^{10})$
$= 125x^{22}y^{28}$

Section 10.3

Practice Problems

1. $4y(8y^2+5) = 4y \cdot 8y^2 + 4y \cdot 5 = 32y^3 + 20y$

2. $3r(8r^2 - r + 11) = 3r \cdot 8r^2 - 3r \cdot r + 3r \cdot 11$
$= 24r^3 - 3r^2 + 33r$

3. $(b+3)(b+5) = b(b+5) + 3(b+5)$
$= b \cdot b + b \cdot 5 + 3 \cdot b + 3 \cdot 5$
$= b^2 + 5b + 3b + 15$
$= b^2 + 8b + 15$

4. $(7x-1)(5x+4) = 7x(5x+4) - 1(5x+4)$
$= 7x \cdot 5x + 7x \cdot 4 - 1 \cdot 5x - 1 \cdot 4$
$= 35x^2 + 28x - 5x - 4$
$= 35x^2 + 23x - 4$

5. $(6y-1)^2 = (6y-1)(6y-1)$
$= 6y(6y-1) - 1(6y-1)$
$= 6y \cdot 6y + 6y(-1) - 1 \cdot 6y - 1(-1)$
$= 36y^2 - 6y - 6y + 1$
$= 36y^2 - 12y + 1$

6. $(10x-7)(2x+3)$
$= 10x \cdot 2x + 10x \cdot 3 + (-7)(2x) + (-7)(3)$
$= 20x^2 + 30x - 14x - 21$
$= 20x^2 + 16x - 21$

7. $(3x+2)^2 = (3x+2)(3x+2)$
$= 3x \cdot 3x + 3x \cdot 2 + 2 \cdot 3x + 2 \cdot 2$
$= 9x^2 + 6x + 6x + 4$
$= 9x^2 + 12x + 4$

8. $(2x+5)(x^2+4x-1)$
$= 2x(x^2+4x-1) + 5(x^2+4x-1)$
$= 2x \cdot x^2 + 2x \cdot 4x + 2x(-1) + 5 \cdot x^2 + 5 \cdot 4x + 5(-1)$
$= 2x^3 + 8x^2 - 2x + 5x^2 + 20x - 5$
$= 2x^3 + 13x^2 + 18x - 5$

9.
$$\begin{array}{r} x^2 + 3x - 2 \\ \times \qquad 3x + 4 \\ \hline 4x^2 + 12x - 8 \\ 3x^3 + 9x^2 \quad\; - 6x \quad \\ \hline 3x^3 + 13x^2 + 6x - 8 \end{array}$$

Exercise Set 10.3

1. $3x(9x^2-3) = 3x \cdot 9x^2 - 3x \cdot 3 = 27x^3 - 9x$

3. $-3a(2a^2 - 3a - 5)$
$= -3a \cdot 2a^2 - (-3a)(3a) - (-3a)(5)$
$= -6a^3 + 9a^2 + 15a$

5. $7x^2(6x^2 - 5x + 7)$
$= (7x^2)(6x^2) - (7x^2)(5x) + (7x^2)(7)$
$= 42x^4 - 35x^3 + 49x^2$

7. $(x+3)(x+10) = x(x+10) + 3(x+10)$
$= x \cdot x + x \cdot 10 + 3 \cdot x + 3 \cdot 10$
$= x^2 + 10x + 3x + 30$
$= x^2 + 13x + 30$

9. $(2x-6)(x+4) = 2x(x+4) - 6(x+4)$
$= 2x \cdot x + 2x \cdot 4 - 6 \cdot x - 6 \cdot 4$
$= 2x^2 + 8x - 6x - 24$
$= 2x^2 + 2x - 24$

11. $(6a+4)^2 = (6a+4)(6a+4)$
$= 6a(6a+4) + 4(6a+4)$
$= 6a \cdot 6a + 6a \cdot 4 + 4 \cdot 6a + 4 \cdot 4$
$= 36a^2 + 24a + 24a + 16$
$= 36a^2 + 48a + 16$

13. $(a+6)(a^2 - 6a + 3)$
$= a(a^2 - 6a + 3) + 6(a^2 - 6a + 3)$
$= a \cdot a^2 + a(-6a) + a \cdot 3 + 6 \cdot a^2 + 6(-6a) + 6 \cdot 3$
$= a^3 - 6a^2 + 3a + 6a^2 - 36a + 18$
$= a^3 - 33a + 18$

Copyright © 2011 Pearson Education, Inc. Publishing as Prentice Hall.

15. $(4x-5)(2x^2+3x-10) = 4x(2x^2+3x-10)-5(2x^2+3x-10)$
$= 4x \cdot 2x^2 + 4x \cdot 3x + 4x(-10) + (-5)(2x^2) + (-5)(3x) - (-5)(10)$
$= 8x^3 + 12x^2 - 40x - 10x^2 - 15x + 50$
$= 8x^3 + 2x^2 - 55x + 50$

17. $(x^3+2x+x^2)(3x+1+x^2) = x^3(3x+1+x^2) + 2x(3x+1+x^2) + x^2(3x+1+x^2)$
$= x^3 \cdot 3x + x^3 \cdot 1 + x^3 \cdot x^2 + 2x \cdot 3x + 2x \cdot 1 + 2x \cdot x^2 + x^2 \cdot 3x + x^2 \cdot 1 + x^2 \cdot x^2$
$= 3x^4 + x^3 + x^5 + 6x^2 + 2x + 2x^3 + 3x^3 + x^2 + x^4$
$= x^5 + 4x^4 + 6x^3 + 7x^2 + 2x$

19. $10r(-3r+2) = 10r \cdot (-3r) + 10r \cdot 2$
$= -30r^2 + 20r$

21. $-2y^2(3y+y^2-6) = -2y^2 \cdot 3y + (-2y^2) \cdot y^2 - (-2y^2)(6)$
$= -6y^3 - 2y^4 + 12y^2$

23. $(x+2)(x+12) = x(x+12) + 2(x+12)$
$= x \cdot x + x \cdot 12 + 2 \cdot x + 2 \cdot 12$
$= x^2 + 12x + 2x + 24$
$= x^2 + 14x + 24$

25. $(2a+3)(2a-3) = 2a(2a-3) + 3(2a-3)$
$= 2a \cdot 2a + 2a(-3) + 3 \cdot 2a + 3(-3)$
$= 4a^2 - 6a + 6a - 9$
$= 4a^2 - 9$

27. $(x+5)^2 = (x+5)(x+5)$
$= x(x+5) + 5(x+5)$
$= x \cdot x + x \cdot 5 + 5 \cdot x + 5 \cdot 5$
$= x^2 + 5x + 5x + 25$
$= x^2 + 10x + 25$

29. $\left(b+\frac{3}{5}\right)\left(b+\frac{4}{5}\right) = b\left(b+\frac{4}{5}\right) + \frac{3}{5}\left(b+\frac{4}{5}\right)$
$= b^2 + \frac{4}{5}b + \frac{3}{5}b + \frac{3}{5} \cdot \frac{4}{5}$
$= b^2 + \frac{7}{5}b + \frac{12}{25}$

31. $(6x+1)(x^2+4x+1) = 6x(x^2+4x+1) + 1(x^2+4x+1)$
$= 6x^3 + 24x^2 + 6x + x^2 + 4x + 1$
$= 6x^3 + 25x^2 + 10x + 1$

Copyright © 2011 Pearson Education, Inc. Publishing as Prentice Hall.

33. $(7x+5)^2 = (7x+5)(7x+5)$
$= 7x(7x+5)+5(7x+5)$
$= 49x^2+35x+35x+25$
$= 49x^2+70x+25$

35. $(2x-1)^2 = (2x-1)(2x-1)$
$= 2x(2x-1)+(-1)(2x-1)$
$= 4x^2-2x-2x+1$
$= 4x^2-4x+1$

37. $(2x^2-3)(4x^3+2x-3)$
$= 2x^2(4x^3+2x-3)-3(4x^3+2x-3)$
$= 8x^5+4x^3-6x^2-12x^3-6x+9$
$= 8x^5-8x^3-6x^2-6x+9$

39. $(x^3+x^2+x)(x^2+x+1)$
$= x^3(x^2+x+1)+x^2(x^2+x+1)+x(x^2+x+1)$
$= x^5+x^4+x^3+x^4+x^3+x^2+x^3+x^2+x$
$= x^5+2x^4+3x^3+2x^2+x$

41.
$$\begin{array}{rrrr} & & 2z^2 & -z+1 \\ & \times & 5z^2 & +z-2 \\ \hline & & -4z^2 & +2z-2 \\ & 2z^3 & -z^2 & +z \\ 10z^4 & -5z^3 & +5z^2 & \\ \hline 10z^4 & -3z^3 & & +3z-2 \end{array}$$

43. $50 = 2\cdot 5\cdot 5 = 2\cdot 5^2$

45. $72 = 2\cdot 2\cdot 2\cdot 3\cdot 3 = 2^3\cdot 3^2$

47. $200 = 2\cdot 2\cdot 2\cdot 5\cdot 5 = 2^3\cdot 5^2$

49. $(y-6)(y^2+3y+2)$
$= y(y^2+3y+2)-6(y^2+3y+2)$
$= y^3+3y^2+2y-6y^2-18y-12$
$= y^3-3y^2-16y-12$
The area is $(y^3-3y^2-16y-12)$ square feet.

51. $(x^2-1)(x^2-1)-(x\cdot x)$
$= x^2(x^2-1)+(-1)(x^2-1)-x^2$
$= x^4-x^2-x^2+1-x^2$
$= x^4-3x^2+1$
The area of the shaded figure is (x^4-3x^2+1) square meters.

53. answers may vary

Section 10.4

Practice Problems

1. $42 = 2\cdot 3\cdot 7$
$28 = 2\cdot 2\cdot 7$
↓ ↓
$2 \cdot 7$
The GCF is $2\cdot 7 = 14$.

2. The GCF of z^7, z^8, and $z = z^1$ is $z^1 = z$ since 1 is the smallest exponent to which z is raised.

3. The GCF of 6, 3, and 15 is 3.
The GCF of a^4, a^5, and a^2 is a^2.
The GCF of $6a^4$, $3a^5$, and $15a^2$ is $3a^2$.

4. The GCF of $10y^7$ and $5y^9$ is $5y^7$.
$10y^7+5y^9 = 5y^7\cdot 2+5y^7\cdot y^2 = 5y^7(2+y^2)$

5. The GCF of the terms is 2.
$4z^2-12z+2 = 2\cdot 2z^2-2\cdot 6z+2\cdot 1$
$= 2(2z^2-6z+1)$

6. $-3y^2-9y+15x^2$
$= 3\cdot -y^2+3\cdot -3y+3\cdot 5x^2$
$= 3(-y^2-3y+5x^2)$ or $-3(y^2+3y-5x^2)$

Vocabulary and Readiness Check

1. In $-3\cdot x^4 = -3x^4$, the -3 and the x^4 are each called a factor and $-3x^4$ is called a product.

2. The greatest common factor (GCF) of a list of integers is the largest integer that is a factor of all integers in the list.

Copyright © 2011 Pearson Education, Inc. Publishing as Prentice Hall.

3. The GCF of a list of variables raised to powers in the variable raised to the smallest exponent in the list.

4. Factoring is the process of writing an expression as a product.

Exercise Set 10.4

1. $48 = 2 \cdot 2 \cdot 2 \cdot 2 \cdot 3$
$15 = 3 \cdot 5$
GCF = 3

3. $60 = 2 \cdot 2 \cdot 3 \cdot 5$
$72 = 2 \cdot 2 \cdot 2 \cdot 3 \cdot 3$
$\text{GCF} = 2 \cdot 2 \cdot 3 = 12$

5. $12 = 2 \cdot 2 \cdot 3$
$20 = 2 \cdot 2 \cdot 5$
$36 = 2 \cdot 2 \cdot 3 \cdot 3$
$\text{GCF} = 2 \cdot 2 = 4$

7. $8 = 2 \cdot 2 \cdot 2$
$32 = 2 \cdot 2 \cdot 2 \cdot 2 \cdot 2$
$100 = 2 \cdot 2 \cdot 5 \cdot 5$
$\text{GCF} = 2 \cdot 2 = 4$

9. $y^7 = y^2 \cdot y^5$
$y^2 = y^2$
$y^{10} = y^2 \cdot y^8$
$\text{GCF} = y^2$

11. $a^5 = a^5$
$a^5 = a^5$
$a^5 = a^5$
$\text{GCF} = a^5$

13. $x^3y^2 = x \cdot x^2 \cdot y^2$
$xy^2 = x \cdot y^2$
$x^4y^2 = x \cdot x^3 \cdot y^2$
$\text{GCF} = x \cdot y^2 = xy^2$

15. $3x^4 = 3 \cdot x \cdot x^3$
$5x^7 = 5 \cdot x \cdot x^6$
$10x = 2 \cdot 5 \cdot x$
$\text{GCF} = x$

17. $2z^3 = 2 \cdot z^3$
$14z^5 = 2 \cdot 7 \cdot z^3 \cdot z^2$
$18z^3 = 2 \cdot 3 \cdot 3 \cdot z^3$
$\text{GCF} = 2z^3$

19. $3y^2 = 3 \cdot y \cdot y$
$18y = 2 \cdot 3 \cdot 3 \cdot y$
$\text{GCF} = 3y$
$3y^2 + 18y = 3y \cdot y + 3y \cdot 6 = 3y(y+6)$

21. $10a^6 = 5a^6 \cdot 2$
$5a^8 = 5a^6 \cdot a^2$
$\text{GCF} = 5a^6$
$10a^6 - 5a^8 = 5a^6 \cdot 2 - 5a^6 \cdot a^2$
$= 5a^6(2 - a^2)$

23. $4x^3 = 4x \cdot x^2$
$12x^2 = 4x \cdot 3x$
$20x = 4x \cdot 5$
$\text{GCF} = 4x$
$4x^3 + 12x^2 + 20x = 4x \cdot x^2 + 4x \cdot 3x + 4x \cdot 5$
$= 4x(x^2 + 3x + 5)$

25. $z^7 = z^5 \cdot z^2$
$6z^5 = z^5 \cdot 6$
$\text{GCF} = z^5$
$z^7 - 6z^5 = z^5 \cdot z^2 - z^5 \cdot 6 = z^5(z^2 - 6)$

27. $-35 = -7 \cdot 5$ or $7 \cdot -5$
$14y = -7 \cdot -2 \cdot y$ or $7 \cdot 2 \cdot y$
$-7y^2 = -7 \cdot y^2$ or $7 \cdot -y^2$
GCF = −7 or 7
$-35 + 14y - 7y^2$
$= -7 \cdot 5 + (-7)(-2y) + (-7)(y^2)$
$= -7(5 - 2y + y^2)$ or $7(-5 + 2y - y^2)$

29. $12a^5 = 12a^5 \cdot 1$
$36a^6 = 12a^5 \cdot 3a$
$\text{GCF} = 12a^5$
$12a^5 - 36a^6 = 12a^5 \cdot 1 - 12a^5 \cdot 3a$
$= 12a^5(1 - 3a)$

Copyright © 2011 Pearson Education, Inc. Publishing as Prentice Hall.

31. $30\% \cdot 120 = x$
$0.30 \cdot 120 = x$
$36 = x$
30% of 120 is 36.

33. $80\% = \frac{80}{100} = \frac{4 \cdot 20}{5 \cdot 20} = \frac{4}{5}$

35. $\frac{3}{8} = \frac{3}{8} \cdot \frac{100\%}{1} = \frac{300}{8}\% = 37.5\%$

37. area on the left: $x \cdot x = x^2$
area on the right: $2 \cdot x = 2x$
total area: $x^2 + 2x$
Notice that $x(x+2) = x^2 + 2x$.

39. answers may vary

41. Let $x = 2$ and $z = 7$.

$$\begin{aligned}(xy+z)^x &= (2y+7)^2 \\ &= (2y+7)(2y+7) \\ &= 2y(2y+7)+7(2y+7) \\ &= 4y^2+14y+14y+49 \\ &= 4y^2+28y+49\end{aligned}$$

Chapter 10 Vocabulary Check

1. Factoring is the process of writing an expression as a product.
2. The greatest common factor of a list of terms is the product of all common factors.
3. The FOIL method may be used when multiplying two binomials.
4. A polynomial with exactly 3 terms is called a trinomial.
5. A polynomial with exactly 2 terms is called a binomial.
6. A polynomial with exactly 1 term is called a monomial.
7. Monomials, binomials, and trinomials are all examples of polynomials.
8. In $5x^3$, the 3 is called an exponent.

Chapter 10 Review

1. $(2b+7)+(8b-10) = 2b+8b+7-10$
$= 10b-3$

2. $(7s-6)+(14s-9) = 7s+14s-6-9$
$= 21s-15$

3. $(3x+0.2)-(4x-2.6) = (3x+0.2)+(-4x+2.6)$
$= 3x-4x+0.2+2.6$
$= -x+2.8$

4. $(10y-6)-(11y+6) = (10y-6)+(-11y-6)$
$= 10y-11y-6-6$
$= -y-12$

5. $(4z^2+6z-1)+(5z-5) = 4z^2+6z+5z-1-5$
$= 4z^2+11z-6$

6. $(17a^3+11a^2+a)+(14a^2-a)$
$= 17a^3+11a^2+14a^2+a-a$
$= 17a^3+25a^2$

7. $\left(9y^2-y+\frac{1}{2}\right)-\left(20y^2-\frac{1}{4}\right)$
$= \left(9y^2-y+\frac{1}{2}\right)+\left(-20y^2+\frac{1}{4}\right)$
$= 9y^2-20y^2-y+\frac{1}{2}+\frac{1}{4}$
$= -11y^2-y+\frac{3}{4}$

8.

$$\begin{array}{r} x^2-6x+1 \\ -\quad (x-2) \\ \hline \end{array} \qquad \begin{array}{r} x^2-6x+1 \\ +\quad -x+2 \\ \hline x^2-7x+3 \end{array}$$

9. $5x^2 = 5(3)^2 = 5 \cdot 9 = 45$

10. $2 - 7x = 2 - 7(3) = 2 - 21 = -19$

11. $(3x + 16) + (10x - 2) + (3x + 16) + (10x - 2)$
$= 3x + 10x + 3x + 10x + 16 - 2 + 16 - 2$
$= 26x + 28$
The perimeter is $(26x + 28)$ feet.

Copyright © 2011 Pearson Education, Inc. Publishing as Prentice Hall.

12. Perimeter
$= (4x^2+1)+(4x^2+1)+(4x^2+1)+(4x^2+1)$
$= 4x^2+4x^2+4x^2+4x^2+1+1+1+1$
$= 16x^2+4$
The perimeter is $(16x^2+4)$ meters.

13. $x^{10}\cdot x^{14} = x^{10+14} = x^{24}$

14. $y\cdot y^6 = y^{1+6} = y^7$

15. $4z^2\cdot 6z^5 = 4\cdot 6z^{2+5} = 24z^7$

16. $(-3x^2y)(5xy^4) = -3\cdot 5x^{2+1}y^{1+4} = -15x^3y^5$

17. $(a^5)^7 = a^{5\cdot 7} = a^{35}$

18. $(x^2)^4\cdot(x^{10})^2 = x^{2\cdot 4}\cdot x^{10\cdot 2}$
$= x^8\cdot x^{20}$
$= x^{8+20}$
$= x^{28}$

19. $(9b)^2 = 9^2\cdot b^2 = 81b^2$

20. $(a^4b^2c)^5 = (a^4)^5\cdot(b^2)^5\cdot(c)^5$
$= a^{4\cdot 5}\cdot b^{2\cdot 5}\cdot c^5$
$= a^{20}b^{10}c^5$

21. $(7x)(2x^5)^3 = 7x\cdot 2^3(x^5)^3$
$= 7x\cdot 8x^{5\cdot 3}$
$= 7x\cdot 8x^{15}$
$= 56x^{16}$

22. $(3x^6y^5)^3(2x^6y^5)^2$
$= 3^3(x^6)^3(5^5)^3\cdot 2^2(x^6)^2(y^5)^2$
$= 27x^{18}y^{15}\cdot 4x^{12}y^{10}$
$= 108x^{30}y^{25}$

23. $A = s^2 = (9a^7)(9a^7) = 9\cdot 9a^{7+7} = 81a^{14}$
The area is $81a^{14}$ square miles.

24. $A = lw = 3x^4\cdot 9x = 3\cdot 9x^{4+1} = 27x^5$
The area is $27x^5$ square inches.

25. $2a(5a^2-6) = 2a\cdot 5a^2 - 2a\cdot 6 = 10a^3 - 12a$

26. $-3y^2(y^2-2y+1)$
$= -3y^2\cdot y^2 - (-3y^2)(2y) + (-3y^2)(1)$
$= -3y^4 + 6y^3 - 3y^2$

27. $(x+2)(x+6) = x\cdot x + x\cdot 6 + 2\cdot x + 2\cdot 6$
$= x^2+6x+2x+12$
$= x^2+8x+12$

28. $(3x-1)(5x-9) = 3x\cdot 5x + 3x(-9) - 1\cdot 5x - 1(-9)$
$= 15x^2 - 27x - 5x + 9$
$= 15x^2 - 32x + 9$

29. $(y-5)^2 = (y-5)(y-5)$
$= y\cdot y + y(-5) - 5\cdot y - 5(-5)$
$= y^2 - 5y - 5y + 25$
$= y^2 - 10y + 25$

30. $(7a+1)^2 = (7a+1)(7a+1)$
$= 7a\cdot 7a + 7a\cdot 1 + 1\cdot 7a + 1\cdot 1$
$= 49a^2 + 7a + 7a + 1$
$= 49a^2 + 14a + 1$

31. $(x+1)(x^2-2x+3)$
$= x(x^2-2x+3) + 1(x^2-2x+3)$
$= x^3 - 2x^2 + 3x + x^2 - 2x + 3$
$= x^3 - x^2 + x + 3$

32. $(4y^2-3)(2y^2+y+1)$
$= 4y^2(2y^2+y+1) - 3(2y^2+y+1)$
$= 8y^4 + 4y^3 + 4y^2 - 6y^2 - 3y - 3$
$= 8y^4 + 4y^3 - 2y^2 - 3y - 3$

33. $(3z^2+2z+1)(z^2+z+1)$
$= 3z^2(z^2+z+1) + 2z(z^2+z+1) + 1(z^2+z+1)$
$= 3z^4 + 3z^3 + 3z^2 + 2z^3 + 2z^2 + 2z + z^2 + z + 1$
$= 3z^4 + 5z^3 + 6z^2 + 3z + 1$

34. $(a+6)(a^2-a+1) = a(a^2-a+1) + 6(a^2-a+1)$
$= a^3 - a^2 + a + 6a^2 - 6a + 6$
$= a^3 + 5a^2 - 5a + 6$

The area is (a^3+5a^2-5a+6) square centimeters.

Copyright © 2011 Pearson Education, Inc. Publishing as Prentice Hall.

35. $20 = 2 \cdot 2 \cdot 5$
$35 = 5 \cdot 7$
$\text{GCF} = 5$

36. $12 = 2 \cdot 2 \cdot 3$
$32 = 2 \cdot 2 \cdot 2 \cdot 2 \cdot 2$
$\text{GCF} = 2 \cdot 2 = 4$

37. $24 = 2 \cdot 2 \cdot 2 \cdot 3$
$30 = 2 \cdot 3 \cdot 5$
$60 = 2 \cdot 2 \cdot 3 \cdot 5$
$\text{GCF} = 2 \cdot 3 = 6$

38. $10 = 2 \cdot 5$
$20 = 2 \cdot 2 \cdot 5$
$25 = 5 \cdot 5$
$\text{GCF} = 5$

39. $x^3 = x^2 \cdot x$
$x^2 = x^2$
$x^{10} = x^2 \cdot x^8$
$\text{GCF} = x^2$

40. $y^{10} = y^7 \cdot y^3$
$y^7 = y^7$
$y^7 = y^7$
$\text{GCF} = y^7$

41. $xy^2 = x \cdot y \cdot y$
$xy = x \cdot y$
$x^3y^3 = x \cdot x^2 \cdot y \cdot y^2$
$\text{GCF} = xy$

42. $a^5b^4 = a^5 \cdot b^2 \cdot b^2$
$a^6b^3 = a^5 \cdot a \cdot b^2 \cdot b$
$a^7b^2 = a^5 \cdot a^2 \cdot b^2$
$\text{GCF} = a^5b^2$

43. $5a^3 = 5 \cdot a \cdot a^2$
$10a = 2 \cdot 5 \cdot a$
$20a^4 = 2 \cdot 2 \cdot 5 \cdot a \cdot a^3$
$\text{GCF} = 5a$

44. $12y^2z = 4y^2z \cdot 3$
$20y^2z = 4y^2z \cdot 5$
$24y^5z = 4y^2z \cdot 6y^3$
$\text{GCF} = 4y^2z$

45. $2x^2 = 2x \cdot x$
$12x = 2x \cdot 6$
$\text{GCF} = 2x$
$2x^2 + 12x = 2x \cdot x + 2x \cdot 6 = 2x(x+6)$

46. $6a^2 = 6a \cdot a$
$12a = 6a \cdot 2$
$\text{GCF} = 6a$
$6a^2 - 12a = 6a \cdot a - 6a \cdot 2 = 6a(a-2)$

47. $6y^4 = y^4 \cdot 6$
$y^6 = y^4 \cdot y^2$
$\text{GCF} = y^4$
$6y^4 - y^6 = y^4 \cdot 6 - y^4 \cdot y^2 = y^4(6-y^2)$

48. $7x^2 = 7 \cdot x^2$
$14x = 7 \cdot 2x$
$7 = 7 \cdot 1$
$\text{GCF} = 7$
$7x^2 - 14x + 7 = 7 \cdot x^2 - 7 \cdot 2x + 7 \cdot 1$
$= 7(x^2 - 2x + 1)$

49. $5a^7 = a^3 \cdot 5a^4$
$a^4 = a^3 \cdot a^1$
$a^3 = a^3 \cdot 1$
$\text{GCF} = a^3$
$5a^7 - a^4 + a^3 = a^3 \cdot 5a^4 - a^3 \cdot a + a^3 \cdot 1$
$= a^3(5a^4 - a + 1)$

50. $10y^6 = 10y \cdot y^5$
$10y = 10y \cdot 1$
$\text{GCF} = 10y$
$10y^6 - 10y = 10y \cdot y^5 - 10y \cdot 1 = 10y(y^5 - 1)$

51.
$$\begin{array}{r} z^2 - 5z + 8 \\ + \quad 6z - 4 \\ \hline z^2 + z + 4 \end{array}$$

52.
$$\begin{array}{r} 8y - 5 \\ -\ (12y - 3) \\ \hline \end{array} \qquad \begin{array}{r} 8y - 5 \\ + -12y + 3 \\ \hline -4y - 2 \end{array}$$

53. $x^5 \cdot x^{16} = x^{5+16} = x^{21}$

54. $y^8 \cdot y = y^{8+1} = y^9$

Copyright © 2011 Pearson Education, Inc. Publishing as Prentice Hall.

55. $(a^3b^5c)^6 = (a^3)^6(b^5)^6c^6 = a^{18}b^{30}c^6$

56. $(9x^2)\cdot(3x^2)^2 = (9x^2)(9x^4) = 81x^6$

57. $3a(4a^3-5) = 3a\cdot 4a^3 + 3a(-5) = 12a^4 - 15a$

58. $(x+4)(x+5) = x^2+5x+4x+20 = x^2+9x+20$

59. $(3x+4)^2 = (3x+4)(3x+4)$
$= 9x^2+12x+12x+16$
$= 9x^2+24x+16$

60. $(6z+5)(z-2) = 6z^2-12z+5z-10$
$= 6z^2-7z-10$

61. $28 = 2\cdot 2\cdot 7$
$32 = 2\cdot 2\cdot 2\cdot 2\cdot 2$
$40 = 2\cdot 2\cdot 2\cdot 5$
$\text{GCF} = 2\cdot 2 = 4$

62. $5z^5 = 5\cdot z^4\cdot z$
$12z^8 = 2\cdot 2\cdot 3\cdot z^4\cdot z^4$
$3z^4 = 3\cdot z^4$
$\text{GCF} = z^4$

63. $z^9 = z^7\cdot z^2$
$4z^7 = z^7\cdot 4$
$\text{GCF} = z^7$
$z^9 - 4z = z^7\cdot z^2 - z^7\cdot 4 = z^7(z^2-4)$

64. $x^{12} = x^5\cdot x^7$
$6x^5 = x^5\cdot 6$
$\text{GCF} = x^5$
$x^{12}+6x^5 = x^5\cdot x^7 + x^5\cdot 6 = x^5(x^7+6)$

65. $15a^4 = 15a^4\cdot 1$
$45a^5 = 15a^4\cdot 3a$
$\text{GCF} = 15a^4$
$15a^4 + 45a^5 = 15a^4\cdot 1 + 15a^4\cdot 3 = 15a^4(1+3a)$

66. $16z^5 = 8z^5\cdot 2$
$24z^8 = 8z^5\cdot 3z^3$
$\text{GCF} = 8z^5$
$16z^5 - 24z^8 = 8z^5\cdot 2 - 8z^5\cdot 3z^3 = 8z^5(2-3z^3)$

Chapter 10 Test

1. $(11x-3)+(4x-1) = 11x+4x-3-1 = 15x-4$

2. $(11x-3)-(4x-1) = (11x-3)+(-4x+1)$
$= 11x-4x-3+1$
$= 7x-2$

3. $(1.3y^2+5y)+(2.1y^2-3y-3)$
$= 1.3y^2+2.1y^2+5y-3y-3$
$= 3.4y^2+2y-3$

4.
$$\begin{array}{r} 6a^2+2a+1 \\ \underline{-(8a^2\ \ +a)} \end{array} \qquad \begin{array}{r} 6a^2+2a+1 \\ \underline{+\ -8a^2\ \ -a} \\ -2a^2\ \ +a+1 \end{array}$$

5. $x^2-6x+1 = (8)^2-6(8)+1$
$= 64-48+1$
$= 17$

6. $y^3\cdot y^{11} = y^{3+11} = y^{14}$

7. $(y^3)^{11} = y^{3\cdot 11} = y^{33}$

8. $(2x^2)^4 = 2^4\cdot(x^2)^4 = 16\cdot x^{2\cdot 4} = 16x^8$

9. $(6a^3)(-2a^7) = (6)(-2)(a^3\cdot a^7) = -12a^{10}$

10. $(p^6)^7(p^2)^6 = p^{6\cdot 7}\cdot p^{2\cdot 6}$
$= p^{42}\cdot p^{12}$
$= p^{42+12}$
$= p^{54}$

11. $(3a^4b)^2(2ba^4)^3 = (3^2a^{4\cdot 2}b^2)(2^3b^3a^{4\cdot 3})$
$= 9a^8b^2\cdot 8b^3a^{12}$
$= 9\cdot 8a^{8+12}b^{2+3}$
$= 72a^{20}b^5$

12. $5x(2x^2+1.3) = 5x\cdot 2x^2 + 5x\cdot 1.3 = 10x^3+6.5x$

13. $-2y(y^3+6y^2-4)$
$= -2y\cdot y^3 - 2y\cdot 6y^2 - 2y\cdot(-4)$
$= -2y^4 - 12y^3 + 8y$

Copyright © 2011 Pearson Education, Inc. Publishing as Prentice Hall.

14. $(x-3)(x+2) = x(x+2)-3(x+2)$
$= x\cdot x + x\cdot 2 - 3\cdot x - 3\cdot 2$
$= x^2 + 2x - 3x - 6$
$= x^2 - x - 6$

15. $(5x+2)^2 = (5x+2)(5x+2)$
$= 5x(5x+2)+2(5x+2)$
$= 5x\cdot 5x + 5x\cdot 2 + 2\cdot 5x + 2\cdot 2$
$= 25x^2 + 10x + 10x + 4$
$= 25x^2 + 20x + 4$

16. $(a+2)(a^2-2a+4)$
$= a(a^2-2a+4)+2(a^2-2a+4)$
$= a\cdot a^2 + a(-2a) + a\cdot 4 + 2\cdot a^2 + 2(-2a) + 2\cdot 4$
$= a^3 - 2a^2 + 4a + 2a^2 - 4a + 8$
$= a^3 + 8$

17. Area:
$(x+7)(5x-2) = x(5x-2)+7(5x-2)$
$= 5x^2 - 2x + 35x - 14$
$= 5x^2 + 33x - 14$

The area is $(5x^2+33x-14)$ square inches.
Perimeter:
$2(2x) + 2(5x - 2) = 4x + 10x - 4 = 14x - 4$
The perimeter is $(14x - 4)$ inches.

18. $45 = 3\cdot 3\cdot 5$
$60 = 2\cdot 2\cdot 3\cdot 5$
$\text{GCF} = 3\cdot 5 = 15$

19. $6y^3 = 2\cdot 3\cdot y^3$
$9y^5 = 3\cdot 3\cdot y^3\cdot y^2$
$18y^4 = 2\cdot 3\cdot 3\cdot y^3\cdot y$
$\text{GCF} = 3y^3$

20. $3y^2 = 3y\cdot y$
$15y = 3y\cdot 5$
$\text{GCF} = 3y$
$3y^2 - 15y = 3y\cdot y - 3y\cdot 5 = 3y(y-5)$

21. $10a^2 = 2a\cdot 5a$
$12a = 2a\cdot 6$
$\text{GCF} = 2a$
$10a^2 + 12a = 2a\cdot 5a + 2a\cdot 6 = 2a(5a+6)$

22. $6x^2 = 6\cdot x^2$
$12x = 6\cdot 2x$
$30 = 6\cdot 5$
$\text{GCF} = 6$
$6x^2 - 12x - 30 = 6\cdot x^2 - 6\cdot 2x - 6\cdot 5$
$= 6(x^2 - 2x - 5)$

23. $7x^6 = x^3\cdot 7x^3$
$6x^4 = x^3\cdot 6x$
$x^3 = x^3\cdot 1$
$\text{GCF} = x^3$
$7x^6 - 6x^4 + x^3 = x^3\cdot 7x^3 - x^3\cdot 6x + x^3\cdot 1$
$= x^3(7x^3 - 6x + 1)$

Cumulative Review Chapters 1–10

1. Area = 380 · 280 = 106,400
The area of Colorado is 106,400 square miles.

2. $21\times 7 = 147$
There are 147 pecan trees.

3. $1+(-10)+(-8)+9 = -9+(-8)+9$
$= -17+9$
$= -8$

4. $-2+(-7)+3+(-4) = -9+3+(-4)$
$= -6+(-4)$
$= -10$

5. $8 - 15 = 8 + (-15) = -7$

6. $4 - 7 = 4 + (-7) = -3$

7. $-4 - (-5) = -4 + 5 = 1$

8. $3 - (-2) = 3 + 2 = 5$

9. $7x = 6x + 4$
$7x - 6x = 6x + 4 - 6x$
$x = 4$

10. $4x = -2 + 3x$
$4x - 3x = -2 + 3x - 3x$
$x = -2$

Copyright © 2011 Pearson Education, Inc. Publishing as Prentice Hall.

11.
$$\begin{aligned} 17-7x+3 &= -3x+21-3x \\ 20-7x &= -6x+21 \\ 20-7x+7x &= -6x+21+7x \\ 20 &= x+21 \\ 20-21 &= x+21-21 \\ -1 &= x \end{aligned}$$

12.
$$\begin{aligned} 20-6x+4 &= -2x+18+2x \\ 24-6x &= 18 \\ 24-6x-24 &= 18-24 \\ -6x &= -6 \\ \frac{-6x}{-6} &= \frac{-6}{-6} \\ x &= 1 \end{aligned}$$

13. $\frac{2x}{15}+\frac{3x}{10}=\frac{2x}{15}\cdot\frac{2}{2}+\frac{3x}{10}\cdot\frac{3}{3}=\frac{4x}{30}+\frac{9x}{30}=\frac{13x}{30}$

14. $\frac{5}{7y}-\frac{9}{14y}=\frac{5}{7y}\cdot\frac{2}{2}-\frac{9}{14y}=\frac{10}{14y}-\frac{9}{14y}=\frac{1}{14y}$

15. To round 736.2359 to the nearest tenth, notice that the digit in the hundredths place is 3. Since this digit is less than 5, we do not add 1 to the digit in the tenths place. 736.2359 rounded to the nearest tenth is 736.2.

16. To round 328.174 to the nearest tenth, notice that the digit in the hundredths place is 7. Since this digit is at least 5, we add 1 to the digit in the tenths place. 328.174 rounded to the nearest tenth is 328.2.

17.
$$\begin{array}{r} 23.850 \\ +\ 1.604 \\ \hline 25.454 \end{array}$$

18.
$$\begin{array}{r} 12.762 \\ +\ 4.290 \\ \hline 17.052 \end{array}$$

19.
$$\begin{aligned} 3.7y &= -3.33 \\ 3.7(-9) &\stackrel{?}{=} -3.33 \\ -33.3 &= -3.33 \quad \text{False} \end{aligned}$$
No, −9 is not a solution.

20.
$$\begin{aligned} 2.8x &= 16.8 \\ 2.8(6) &\stackrel{?}{=} 16.8 \\ 16.8 &= 16.8 \quad \text{True} \end{aligned}$$
Yes, 6 is a solution.

21. $\frac{786.1}{1000}=0.7861$

22. $\frac{818}{1000}=0.818$

23. $\frac{0.12}{10}=0.012$

24. $\frac{5.03}{100}=0.0503$

25. $-2x+5=-2(3.8)+5=-7.6+5=-2.6$

26. $6x-1=6(-2.1)-1=-12.6-1=-13.6$

27.
$$\begin{array}{r} 3.142 \approx 3.14 \\ 7\overline{)\,22.000} \\ \underline{-21} \\ 1\,0 \\ \underline{-\ 7} \\ 30 \\ \underline{-\ 28} \\ 20 \\ \underline{-\ 14} \\ 6 \end{array}$$

$\frac{22}{7}\approx 3.14$

28.
$$\begin{array}{r} 1.9473 \approx 1.947 \\ 19\overline{)\,37.0000} \\ \underline{-19} \\ 18\,0 \\ \underline{-17\,1} \\ 90 \\ \underline{-76} \\ 140 \\ \underline{-133} \\ 70 \\ \underline{-57} \\ 13 \end{array}$$

$\frac{37}{19}\approx 1.947$

29. $\sqrt{\frac{1}{36}}=\frac{1}{6}$ since $\left(\frac{1}{6}\right)^2=\frac{1}{6}\cdot\frac{1}{6}=\frac{1}{36}.$

Copyright © 2011 Pearson Education, Inc. Publishing as Prentice Hall.

30. $\sqrt{\frac{4}{25}} = \frac{2}{5}$ since $\left(\frac{2}{5}\right)^2 = \frac{2}{5} \cdot \frac{2}{5} = \frac{4}{25}$.

31. Let x be the height of the tree.

$$\frac{6}{9} = \frac{x}{69}$$
$$6 \cdot 69 = 9x$$
$$414 = 9x$$
$$\frac{414}{9} = \frac{9x}{9}$$
$$46 = x$$

The height of the tree is 46 feet.

32. Let x be the height of the hydrant.

$$\frac{1}{2} = \frac{x}{6}$$
$$1 \cdot 6 = 2 \cdot x$$
$$6 = 2x$$
$$\frac{6}{2} = \frac{2x}{2}$$
$$3 = x$$

The height of the fire hydrant is 3 feet.

33. $1.2 = 30\% \cdot x$

34. $9 = 45\% \cdot x$

35.
$$x \cdot 50 = 8$$
$$50x = 8$$
$$\frac{50x}{50} = \frac{8}{50}$$
$$x = 0.16$$
$$x = 16\%$$

16% of 50 is 8.

36.
$$x \cdot 16 = 4$$
$$16x = 4$$
$$\frac{16x}{16} = \frac{4}{16}$$
$$x = 0.25$$
$$x = 25\%$$

25% of 16 is 4.

37.
$$31 = 4\% \cdot x$$
$$31 = 0.04x$$
$$\frac{31}{0.04} = \frac{0.04x}{0.04}$$
$$775 = x$$

There are 775 freshman.

38.
$$2\% \cdot x = 29$$
$$0.02x = 29$$
$$\frac{0.02x}{0.02} = \frac{29}{0.02}$$
$$x = 1450$$

There are 1450 apples in the shipment.

39.
$$\text{simple interest} = P \cdot R \cdot T$$
$$= \$2400 \cdot 10\% \cdot \frac{8}{12}$$
$$= \$2400 \cdot 0.10 \cdot \frac{2}{3}$$
$$= \$160$$

The interest is $160.

40.
$$\text{simple interest} = P \cdot R \cdot T$$
$$= \$1000 \cdot 3\% \cdot \frac{10}{12}$$
$$= \$1000 \cdot 0.03 \cdot \frac{5}{6}$$
$$= \$25$$

The interest is $25.

41. $25\% + 32\% = 57\%$

57% of visitors came from Mexico and Canada.

42. $20\% + 11\% = 31\%$

31% of visitors came from Europe and Asia.

43. $P = 2l + 2w = 2(11) + 2(3) = 22 + 6 = 28$

The perimeter is 28 inches.

44. $P = 6 + 8 + 11 = 25$

The perimeter is 25 feet.

45. $A = bh = 1.5 \cdot 3.4 = 5.1$

The area is 5.1 square miles.

46. $A = \frac{1}{2}bh = \frac{1}{2}(17)(8) = 68$

The area of the triangle is 68 square inches.

47.

$$\begin{array}{r} 8 \text{ tons } 1000 \text{ lb} \\ -\ 3 \text{ tons } 1350 \text{ lb} \\ \hline \end{array} \qquad \begin{array}{r} 7 \text{ tons } 3000 \text{ lb} \\ -\ 3 \text{ tons } 1350 \text{ lb} \\ \hline 4 \text{ tons } 1650 \text{ lb} \end{array}$$

Copyright © 2011 Pearson Education, Inc. Publishing as Prentice Hall.

48.
$$\begin{array}{l} \quad 5 \text{ tons } 700 \text{ lb} \\ \underline{\times \qquad\qquad\quad 3} \\ 15 \text{ tons } 2100 \text{ lb} = 15 \text{ tons} + 1 \text{ ton } 100 \text{ lb} \\ \qquad\qquad\qquad\quad = 16 \text{ tons } 100 \text{ lb} \end{array}$$

49. $3210 \text{ ml} = \frac{3210 \text{ ml}}{1} \cdot \frac{1 \text{ L}}{1000 \text{ ml}} = \frac{3210}{1000} \text{ L} = 3.21 \text{ L}$

50. $4321 \text{ cl} = \frac{4321 \text{ cl}}{1} \cdot \frac{1 \text{ L}}{100 \text{ cl}} = \frac{4321}{100} \text{ L} = 43.21 \text{ L}$

51. $(3x - 1) + (-6x + 2) = 3x - 6x - 1 + 2 = -3x + 1$

52.
$$\begin{aligned} (7a+4)-(3a-8) &= (7a+4)+(-3a+8) \\ &= 7a-3a+4+8 \\ &= 4a+12 \end{aligned}$$

53.
$$\begin{aligned} (x+2)(x+3) &= x(x+3)+2(x+3) \\ &= x \cdot x + x \cdot 3 + 2 \cdot x + 2 \cdot 3 \\ &= x^2 + 3x + 2x + 6 \\ &= x^2 + 5x + 6 \end{aligned}$$

54.
$$\begin{aligned} (2x+5)(x+7) &= 2x(x+7)+5(x+7) \\ &= 2x \cdot x + 2x \cdot 7 + 5 \cdot x + 5 \cdot 7 \\ &= 2x^2 + 14x + 5x + 35 \\ &= 2x^2 + 19x + 35 \end{aligned}$$

Copyright © 2011 Pearson Education, Inc. Publishing as Prentice Hall.

Appendices

Appendix B

Practice Problems

1. $\frac{y^{10}}{y^6} = y^{10-6} = y^4$

2. $\frac{5^{11}}{5^8} = 5^{11-8} = 5^3 = 125$

3. $\frac{12a^4b^{11}}{ab} = 12 \cdot \frac{a^4}{a^1} \cdot \frac{b^{11}}{b^1}$
$= 12 \cdot (a^{4-1}) \cdot (b^{11-1})$
$= 12a^3b^{10}$

4. $6^0 = 1$

5. $(-8)^0 = 1$

6. $-8^0 = -1 \cdot 8^0 = -1 \cdot 1 = -1$

7. $7y^0 = 7 \cdot y^0 = 7 \cdot 1 = 7$

8. $5^{-2} = \frac{1}{5^2} = \frac{1}{25}$

9. $5x^{-2} = 5 \cdot \frac{1}{x^2} = \frac{5}{x^2}$

10. $4^{-1} + 3^{-1} = \frac{1}{4} + \frac{1}{3} = \frac{3}{12} + \frac{4}{12} = \frac{7}{12}$

11. $\left(\frac{6}{7}\right)^{-2} = \frac{6^{-2}}{7^{-2}} = \frac{6^{-2}}{1} \cdot \frac{1}{7^{-2}} = \frac{1}{6^2} \cdot \frac{7^2}{1} = \frac{7^2}{6^2} = \frac{49}{36}$

12. $\frac{x}{x^{-4}} = \frac{x^1}{x^{-4}} = x^{1-(-4)} = x^5$

13. $\frac{y^{-4}}{y^6} = y^{-4-6} = y^{-10} = \frac{1}{y^{10}}$

14. $y^{-6} \cdot y^3 \cdot y^{-4} = y^{-6+3} \cdot y^{-4}$
$= y^{-3} \cdot y^{-4}$
$= y^{-3+(-4)}$
$= y^{-7}$
$= \frac{1}{y^7}$

15. $(a^6b^{-4})(a^{-3}b^8) = a^{6+(-3)} \cdot b^{-4+8} = a^3b^4$

16. $(3y^9z^{10})(2y^3z^{-12}) = 3 \cdot 2 \cdot y^{9+3} \cdot z^{10+(-12)}$
$= 6 \cdot y^{12} \cdot z^{-2}$
$= \frac{6y^{12}}{z^2}$

Appendix B Exercise Set

1. $\frac{x^3}{x} = \frac{x^3}{x^1} = x^{3-1} = x^2$

3. $\frac{9^8}{9^6} = 9^{8-6} = 9^2 = 81$

5. $\frac{p^7q^{20}}{pq^{15}} = \frac{p^7}{p^1} \cdot \frac{q^{20}}{q^{15}} = p^{7-1} \cdot q^{20-15} = p^6q^5$

7. $\frac{7x^3y^6}{14x^2y^3} = \frac{7}{14} \cdot \frac{x^3}{x^2} \cdot \frac{y^6}{y^3}$
$= \frac{1}{2} \cdot x^{3-2} \cdot y^{6-3}$
$= \frac{1}{2} \cdot x^1 \cdot y^3$
$= \frac{xy^3}{2}$

9. $7^0 = 1$

11. $2x^0 = 2 \cdot x^0 = 2 \cdot 1 = 2$

13. $-7^0 = -1 \cdot 7^0 = -1 \cdot 1 = -1$

15. $(-7)^0 = 1$

Copyright © 2011 Pearson Education, Inc. Publishing as Prentice Hall.

17. $4^{-3} = \frac{1}{4^3} = \frac{1}{64}$

19. $7x^{-3} = 7 \cdot \frac{1}{x^3} = \frac{7}{x^3}$

21. $3^{-1} + 2^{-1} = \frac{1}{3^1} + \frac{1}{2^1} = \frac{1}{3} + \frac{1}{2} = \frac{2}{6} + \frac{3}{6} = \frac{5}{6}$

23. $\frac{1}{p^{-3}} = p^3$

25. $\frac{x^{-2}}{x} = \frac{x^{-2}}{x^1} = x^{-2-1} = x^{-3} = \frac{1}{x^3}$

27. $\frac{z^{-4}}{z^{-7}} = z^{-4-(-7)} = z^3$

29. $3^{-2} + 3^{-1} = \frac{1}{3^2} + \frac{1}{3^1} = \frac{1}{9} + \frac{1}{3} = \frac{1}{9} + \frac{3}{9} = \frac{4}{9}$

31. $\left(\frac{5}{y}\right)^{-2} = \frac{5^{-2}}{y^{-2}} = \frac{5^{-2}}{1} \cdot \frac{1}{y^{-2}} = \frac{1}{5^2} \cdot \frac{y^2}{1} = \frac{y^2}{25}$

33. $\frac{1}{p^{-4}} = p^4$

35.
$$\begin{aligned} a^2 \cdot a^{-9} \cdot a^{13} &= a^{2+(-9)} \cdot a^{13} \\ &= a^{-7} \cdot a^{13} \\ &= a^{-7+13} \\ &= a^6 \end{aligned}$$

37.
$$\begin{aligned} (x^8 y^{-6})(x^{-2} y^{12}) &= x^{8+(-2)} \cdot y^{-6+12} \\ &= x^6 y^6 \end{aligned}$$

39.
$$\begin{aligned} x^{-7} \cdot x^{-8} \cdot x^4 &= x^{-7+(-8)} \cdot x^4 \\ &= x^{-15} \cdot x^4 \\ &= x^{-15+4} \\ &= x^{-11} \\ &= \frac{1}{x^{11}} \end{aligned}$$

41.
$$\begin{aligned} (5x^{-7})(3x^4) &= 5 \cdot 3 \cdot x^{-7+4} \\ &= 15 \cdot x^{-3} \\ &= 15 \cdot \frac{1}{x^3} \\ &= \frac{15}{x^3} \end{aligned}$$

43.
$$\begin{aligned} y^5 \cdot y^{-7} \cdot y^{-10} &= y^{5+(-7)} \cdot y^{-10} \\ &= y^{-2} \cdot y^{-10} \\ &= y^{-2+(-10)} \\ &= y^{-12} \\ &= \frac{1}{y^{12}} \end{aligned}$$

45.
$$\begin{aligned} (8m^5 n^{-1})(7m^2 n^{-4}) &= 8 \cdot 7 \cdot m^{5+2} n^{-1+(-4)} \\ &= 56 \cdot m^7 \cdot n^{-5} \\ &= 56 \cdot m^7 \cdot \frac{1}{n^5} \\ &= \frac{56m^7}{n^5} \end{aligned}$$

47. $\frac{x^{15}}{x^8} = x^{15-8} = x^7$

49. $\frac{a^9 b^{14}}{ab} = \frac{a^9}{a^1} \cdot \frac{b^{14}}{b^1} = a^{9-1} \cdot b^{14-1} = a^8 b^{13}$

51. $\frac{x^3}{x^9} = x^{3-9} = x^{-6} = \frac{1}{x^6}$

53. $3z^0 = 3 \cdot z^0 = 3 \cdot 1 = 3$

55. $5^{-3} = \frac{1}{5^3} = \frac{1}{125}$

57. $8x^{-9} = 8 \cdot \frac{1}{x^9} = \frac{8}{x^9}$

59. $5^{-1} + 10^{-1} = \frac{1}{5^1} + \frac{1}{10^1} = \frac{1}{5} + \frac{1}{10} = \frac{2}{10} + \frac{1}{10} = \frac{3}{10}$

61. $\frac{z^{-8}}{z^{-1}} = z^{-8-(-1)} = z^{-7} = \frac{1}{z^7}$

Copyright © 2011 Pearson Education, Inc. Publishing as Prentice Hall.

63. $x^{-7} \cdot x^5 \cdot x^{-7} = x^{-7+5} \cdot x^{-7}$
$= x^{-2} \cdot x^{-7}$
$= x^{-2+(-7)}$
$= x^{-9}$
$= \frac{1}{x^9}$

65. $(a^{-2}b^3)(a^{10}b^{-11}) = a^{-2+10} \cdot b^{3+(-11)}$
$= a^8 \cdot b^{-8}$
$= a^8 \cdot \frac{1}{b^8}$
$= \frac{a^8}{b^8}$

67. $(3x^{20}y^{-1})(10x^{-11}y^{-5}) = 3 \cdot 10 \cdot x^{20+(-11)}y^{-1+(-5)}$
$= 30 \cdot x^9 \cdot y^{-6}$
$= 30x^9 \cdot \frac{1}{y^6}$
$= \frac{30x^9}{y^6}$

Appendix C

Practice Problems

1. a. $760,000 = 7.6 \times 10^5$

b. $0.00035 = 3.5 \times 10^{-4}$

2. a. $9.062 \times 10^{-4} = 0.0009062$

b. $8.002 \times 10^6 = 8,002,000$

3. a. $(8 \times 10^7)(3 \times 10^{-9}) = 8 \cdot 3 \cdot 10^7 \cdot 10^{-9}$
$= 24 \times 10^{-2}$
$= 0.24$

b. $\frac{8 \times 10^4}{2 \times 10^{-3}} = \frac{8}{2} \times 10^{4-(-3)}$
$= 4 \times 10^7$
$= 40,000,000$

Appendix C Exercise Set

1. $78,000 = 7.8 \times 10^4$

3. $0.00000167 = 1.67 \times 10^{-6}$

5. $0.00635 = 6.35 \times 10^{-3}$

7. $1,160,000 = 1.16 \times 10^6$

9. $13,600 = 1.36 \times 10^4$

11. $8.673 \times 10^{-10} = 0.0000000008673$

13. $3.3 \times 10^{-2} = 0.033$

15. $2.032 \times 10^4 = 20,320$

17. $7.0 \times 10^8 = 700,000,000$

19. $940,000,000 = 9.4 \times 10^8$

21. $1.23 \times 10^{12} = 1,230,000,000,000$

23. $23,000,000,000 = 2.3 \times 10^{10}$

25. $(1.2 \times 10^{-3})(3 \times 10^{-2}) = 1.2 \cdot 3 \times 10^{-3+(-2)}$
$= 3.6 \times 10^{-5}$
$= 0.000036$

27. $(4 \times 10^{-10})(7 \times 10^{-9}) = 4 \cdot 7 \cdot 10^{-10+(-9)}$
$= 28 \times 10^{-19}$
$= 2.8 \times 10^1 \times 10^{-19}$
$= 2.8 \times 10^{-18}$
$= 0.0000000000000000028$

29. $\frac{8 \times 10^{-1}}{16 \times 10^5} = \frac{8}{16} \times 10^{-1-5}$
$= 0.5 \times 10^{-6}$
$= 5 \times 10^{-1} \times 10^{-6}$
$= 5 \times 10^{-7}$
$= 0.0000005$

31. $\frac{1.4 \times 10^{-2}}{7 \times 10^{-8}} = \frac{1.4}{7} \times 10^{-2-(-8)}$
$= 0.2 \times 10^6$
$= 200,000$

Copyright © 2011 Pearson Education, Inc. Publishing as Prentice Hall.

33. $(7.5\times10^5)(3600) = (7.5\times10^5)(3.6\times10^3)$
$= 7.5\cdot3.6\times10^{5+3}$
$= 27\times10^8$
$= 2.7\times10^1\times10^8$
$= 2.7\times10^9$

On average, 2.7×10^9 gallons of water flow over Niagara Falls each hour.

Practice Final Exam

1. $2^3\cdot5^2 = 2\cdot2\cdot2\cdot5\cdot5 = 200$

2. $16+9\div3\cdot4-7 = 16+3\cdot4-7$
$= 16+12-7$
$= 28-7$
$= 21$

3. $18 - 24 = 18 + (-24) = -6$

4. $5\cdot(-20) = -100$

5. $\sqrt{49} = 7$ because $7^2 = 49$.

6. $(-5)^3 - 24\div(-3) = -125-24\div(-3)$
$= -125-(-8)$
$= -125+8$
$= -117$

7. $0\div49 = 0$

8. $62\div0$ is undefined.

9. $-\frac{8}{15y}-\frac{2}{15y} = \frac{-8-2}{15y} = \frac{-10}{15y} = -\frac{2\cdot5}{3\cdot5\cdot y} = -\frac{2}{3y}$

10. $\frac{11}{12}-\frac{3}{8}+\frac{5}{24} = \frac{11\cdot2}{12\cdot2}-\frac{3\cdot3}{8\cdot3}+\frac{5}{24}$
$= \frac{22}{24}-\frac{9}{24}+\frac{5}{24}$
$= \frac{22-9+5}{24}$
$= \frac{18}{24}$
$= \frac{3\cdot6}{4\cdot6}$
$= \frac{3}{4}$

11. $\frac{3a}{8}\cdot\frac{16}{6a^3} = \frac{3a\cdot16}{8\cdot6a^3} = \frac{3\cdot a\cdot8\cdot2}{8\cdot2\cdot3\cdot a\cdot a\cdot a} = \frac{1}{a\cdot a} = \frac{1}{a^2}$

12. $-\frac{16}{3}\div-\frac{3}{12} = -\frac{16}{3}\cdot-\frac{12}{3}$
$= \frac{16\cdot12}{3\cdot3}$
$= \frac{16\cdot3\cdot4}{3\cdot3}$
$= \frac{64}{3}$ or $21\frac{1}{3}$

13.
$$\begin{array}{r} 19 \\ -\,2\frac{3}{11} \\ \hline \end{array} \qquad \begin{array}{r} 18\frac{11}{11} \\ -\,2\frac{3}{11} \\ \hline 16\frac{8}{11} \end{array}$$

14. $\frac{0.23+1.63}{-0.3} = \frac{1.86}{-0.3} = -6.2$

15.
$$\begin{array}{rl} 10.2 & \text{1 decimal place} \\ \times\ 4.01 & \text{2 decimal places} \\ \hline 102 & \\ 40\ 800 & \\ \hline 40.902 & 1+2=3 \text{ decimal places} \end{array}$$

16. $0.6\% = 0.6(0.01) = 0.006$

17. $6.1 = 6.1(100\%) = 610\%$

18. $\frac{3}{8} = \frac{3}{8}\cdot\frac{100}{1}\% = \frac{300}{8}\% = 37.5\%$

19. $0.345 = \frac{345}{1000} = \frac{5\cdot69}{5\cdot200} = \frac{69}{200}$

20. $-\frac{13}{26} = -\frac{1\cdot13}{2\cdot13} = -\frac{1}{2} = -\frac{1\cdot5}{2\cdot5} = -\frac{5}{10} = -0.5$

21. 34.8923 rounded to the nearest tenth is 34.9.

22. Let x be 2.
$5(x^3-2) = 5(2^3-2) = 5(8-2) = 5(6) = 30$

23. $10-y^2 = 10-(-3)^2 = 10-9 = 1$

Copyright © 2011 Pearson Education, Inc. Publishing as Prentice Hall.

24. $x \div y = \frac{1}{2} \div 3\frac{7}{8}$
$= \frac{1}{2} \div \frac{31}{8}$
$= \frac{1}{2} \cdot \frac{8}{31}$
$= \frac{1 \cdot 8}{2 \cdot 31}$
$= \frac{1 \cdot 2 \cdot 4}{2 \cdot 31}$
$= \frac{4}{31}$

25. $-(3z+2)-5z-18 = -1(3z+2)-5z-18$
$= -1 \cdot 3z + (-1) \cdot 2 - 5z - 18$
$= -3z - 2 - 5z - 18$
$= -3z - 5z - 2 - 18$
$= -8z - 20$

26. perimeter = $3(5x + 5) = 3 \cdot 5x + 3 \cdot 5 = 15x + 15$
The perimeter is $(15x + 15)$ inches.

27. $\frac{n}{-7} = 4$
$-7 \cdot \frac{n}{-7} = -7 \cdot 4$
$\frac{-7}{-7} \cdot n = -7 \cdot 4$
$n = -28$
The solution is −28.

28. $-4x + 7 = 15$
$-4x + 7 - 7 = 15 - 7$
$-4x = 8$
$\frac{-4x}{-4} = \frac{8}{-4}$
$x = -2$

29. $-4(x-11) - 34 = 10 - 12$
$-4x + 44 - 34 = 10 - 12$
$-4x + 10 = -2$
$-4x + 10 - 10 = -2 - 10$
$-4x = -12$
$\frac{-4x}{-4} = \frac{-12}{-4}$
$x = 3$

30. $\frac{x}{5} + x = -\frac{24}{5}$
$5\left(\frac{x}{5} + x\right) = 5\left(-\frac{24}{5}\right)$
$5 \cdot \frac{x}{5} + 5 \cdot x = -24$
$x + 5x = -24$
$6x = -24$
$\frac{6x}{6} = \frac{-24}{6}$
$x = -4$

31. $2(x + 5.7) = 6x - 3.4$
$2x + 11.4 = 6x - 3.4$
$2x + 11.4 - 11.4 = 6x - 3.4 - 11.4$
$2x = 6x - 14.8$
$2x - 6x = 6x - 6x - 14.8$
$-4x = -14.8$
$\frac{-4x}{-4} = \frac{-14.8}{-4}$
$x = 3.7$

32. $\frac{8}{x} = \frac{11}{6}$
$8 \cdot 6 = x \cdot 11$
$48 = 11x$
$\frac{48}{11} = \frac{11x}{11}$
$\frac{48}{11} = x$
$4\frac{4}{11} = x$

33. Perimeter = $(20 + 10 + 20 + 10)$ yards = 60 yards
Area = (length)(width)
= (20 yards)(10 yards)
= 200 square yards

34. average $= \frac{-12 + (-13) + 0 + 9}{4} = \frac{-16}{4} = -4$

35. The difference of three times a number and five times the same number is 4 translates to
$3x - 5x = 4$
$-2x = 4$
$\frac{-2x}{-2} = \frac{4}{-2}$
$x = -2$
The number is −2.

Copyright © 2011 Pearson Education, Inc. Publishing as Prentice Hall.

36. $258 \div 10\frac{3}{4} = \frac{258}{1} \div \frac{43}{4} = \frac{258}{1} \cdot \frac{4}{43} = \frac{43 \cdot 6 \cdot 4}{1 \cdot 43} = 24$

Expect to travel 24 miles on 1 gallon of gas.

37. Let x be the number of women runners entered in the race. Since the number of men entered in the race is 112 more than the number of women, the number of men is $x + 112$. Since the total number of runners in the race is 600, the sum of x and $x + 112$ is 600.

$$x + x + 112 = 600$$
$$2x + 112 = 600$$
$$2x + 112 - 112 = 600 - 112$$
$$2x = 488$$
$$\frac{2x}{2} = \frac{488}{2}$$
$$x = 244$$

244 women entered the race.

38. Let x be the number of grams.

$$\begin{matrix}\text{grams} \rightarrow \\ \text{pounds} \rightarrow\end{matrix} \frac{10}{15} = \frac{x}{80} \begin{matrix}\leftarrow \text{grams} \\ \leftarrow \text{pounds}\end{matrix}$$
$$10 \cdot 80 = 15 \cdot x$$
$$800 = 15x$$
$$\frac{800}{15} = \frac{15x}{15}$$
$$53\frac{1}{3} = x$$

The standard dose for an 80-pound dog is $53\frac{1}{3}$ grams.

39. Amount of discount $= 15\% \cdot \$120$
$= 0.15 \cdot \$120$
$= \$18$

Sale price = \$120 – \$18 = \$102

The amount of the discount is \$18; the sale price is \$102.

40. $y + x = -4$

Find any 3 ordered-pair solutions.

Let $x = 0$.

$$y + x = -4$$
$$y + 0 = -4$$
$$y = -4$$

$(0, -4)$

Let $y = 0$.

$$y + x = -4$$
$$0 + x = -4$$
$$x = -4$$

$(-4, 0)$

Let $x = -2$.

$$y + x = -4$$
$$y + (-2) = -4$$
$$y + (-2) + 2 = -4 + 2$$
$$y = -2$$

$(-2, -2)$

Plot $(0, -4)$, $(-4, 0)$, and $(-2, -2)$. Then draw the line through them.

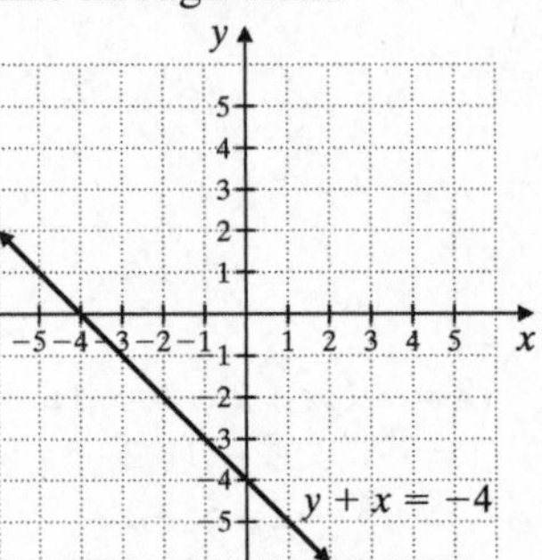

41. $y = 3x - 5$

Find any 3 ordered-pair solutions.

Let $x = 0$.

$y = 3x - 5$
$y = 3 \cdot 0 - 5$
$y = 0 - 5$
$y = -5$
$(0, -5)$

Let $x = 1$.

$y = 3x - 5$
$y = 3 \cdot 1 - 5$
$y = 3 - 5$
$y = -2$
$(1, -2)$

Let $x = 2$.

$y = 3x - 5$
$y = 3 \cdot 2 - 5$
$y = 6 - 5$
$y = 1$
$(2, 1)$

Plot $(0, -5)$, $(1, -2)$, and $(2, 1)$. Then draw the line through them.

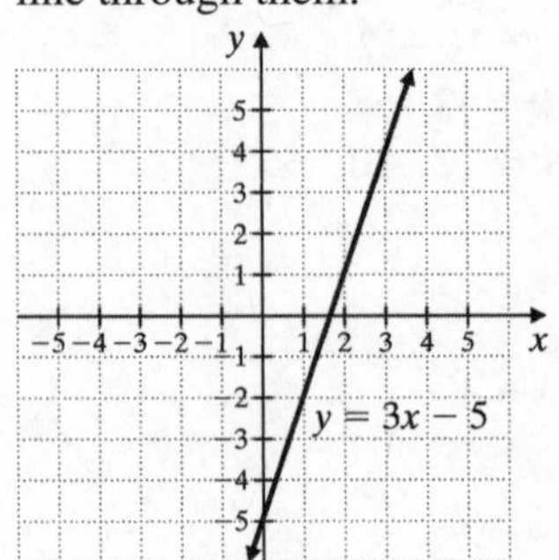

Copyright © 2011 Pearson Education, Inc. Publishing as Prentice Hall.

42. $y = -4$
No matter what x-value we choose, y is always -4.

x	y
−2	−4
0	−4
2	−4

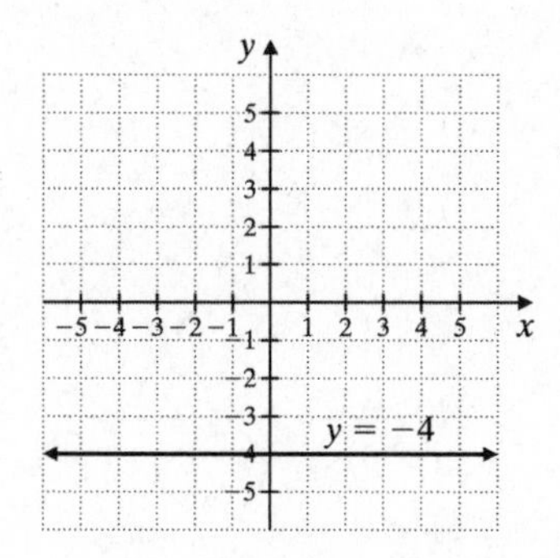

43. $(11x-3)+(4x-1)=11x+4x-3-1=15x-4$

44.
$$\begin{array}{r} 6a^2+2a+1 \\ -(8a^2 \quad +a) \\ \hline \end{array} \qquad \begin{array}{r} 6a^2+2a+1 \\ +\,-8a^2 \quad -a \\ \hline -2a^2 \quad +a+1 \end{array}$$

45. $(6a^3)(-2a^7)=(6)(-2)(a^3\cdot a^7)=-12a^{10}$

46.
$$\begin{aligned}(3a^4b)^2(2ba^4)^3 &= (3^2a^{4\cdot 2}b^2)(2^3b^3a^{4\cdot 3}) \\ &= 9a^8b^2\cdot 8b^3a^{12} \\ &= 9\cdot 8a^{8+12}b^{2+3} \\ &= 72a^{20}b^5\end{aligned}$$

47.
$$\begin{aligned}(x-3)(x+2) &= x(x+2)-3(x+2) \\ &= x\cdot x+x\cdot 2-3\cdot x-3\cdot 2 \\ &= x^2+2x-3x-6 \\ &= x^2-x-6\end{aligned}$$

48. $3y^2 = 3y\cdot y$
$15y = 3y\cdot 5$
$\text{GCF} = 3y$
$3y^2-15y = 3y\cdot y-3y\cdot 5 = 3y(y-5)$

49. The complement of an angle that measures 78° is an angle that measures 90° − 78° = 12°.

50. $\angle x$ and the angle marked 73° are vertical angles, so $m\angle x = 73°$.
$\angle x$ and $\angle y$ are alternate interior angles, so $m\angle y = m\angle x = 73°$.
$\angle x$ and $\angle z$ are corresponding angles, so $m\angle z = m\angle x = 73°$.

51. The unmarked vertical side has length 11 in. − 7 in. = 4 in. The unmarked horizontal side has length 23 in. − 6 in. = 17 in.
$P = (6+4+17+7+23+11)$ in. = 68 in.
Extending the unmarked vertical side downward divides the region into two rectangles. The region's area is the sum of the areas of these:
$$\begin{aligned}A &= 11\text{ in.}\cdot 6\text{ in.}+7\text{ in.}\cdot 17\text{ in.} \\ &= 66\text{ sq in.}+119\text{ sq in.} \\ &= 185\text{ sq in.}\end{aligned}$$

52. Circumference:
$$\begin{aligned}C &= 2\cdot\pi\cdot r \\ &= 2\cdot\pi\cdot 9\text{ in.} \\ &= 18\pi\text{ in.} \\ &\approx 56.52\text{ in.}\end{aligned}$$
Area:
$$\begin{aligned}A &= \pi r^2 \\ &= \pi(9\text{ in.})^2 \\ &= 81\pi\text{ sq in.} \\ &\approx 254.34\text{ sq in.}\end{aligned}$$

53. $2\frac{1}{2}\text{ gal} = \frac{2\frac{1}{2}\text{ gal}}{1}\cdot\frac{4\text{ qt}}{1\text{ gal}} = 2\frac{1}{2}\cdot 4\text{ qt} = 10\text{ qt}$

54. $2.4\text{ kg} = \frac{2.4\text{ kg}}{1}\cdot\frac{1000\text{ g}}{1\text{ kg}} = 2.4\cdot 1000\text{ g} = 2400\text{ g}$

Copyright © 2011 Pearson Education, Inc. Publishing as Prentice Hall.